Environmental Biotechnology: For Sustainable Future

Ranbir Chander Sobti • Naveen Kumar Arora
Richa Kothari
Editors

Environmental Biotechnology: For Sustainable Future

Springer

Editors
Ranbir Chander Sobti
Babasaheb Bhimrao Ambedkar University
Lucknow, Uttar Pradesh, India

Naveen Kumar Arora
Department of Environmental Science
Babasaheb Bhimrao Ambedkar University
Lucknow, Uttar Pradesh, India

Richa Kothari
Department of Environmental Sciences
Central University of Jammu
Jammu and Kashmir, India

Babasaheb Bhimrao Ambedkar University
Lucknow, Uttar Pradesh, India

ISBN 978-981-10-7283-3 ISBN 978-981-10-7284-0 (eBook)
https://doi.org/10.1007/978-981-10-7284-0

Library of Congress Control Number: 2018957141

This Springer imprint is published by the registered company Springer Nature Singapore Pte Ltd.
The registered company address is: 152 Beach Road, #21-01/04 Gateway East, Singapore 189721, Singapore

Contents

About the Editors

Prof. (Dr.) Ranbir Chander Sobti is former Vice Chancellor of Babasaheb Bhimrao Ambedkar University, Lucknow (UP), India, and Panjab University, Chandigarh, India. He has been proactively involved in cancer biology, with a focus on finding out novel tumour markers for cancer detection, and is currently engaged in the development of whole organ by tissue/organ culture through custom designed decellularization protocol. His intense scientific contributions have resulted in around 300 high-impact research publications, more than 40 books and 23 sponsored research projects. He is fellow of leading national and international academies of science and is associated with many other academic associations and institutions in the domain of higher education and research in India and abroad and was General President of the Indian Science Congress Association in 2014. He has been bestowed with a large number of prestigious awards and medals including the INSA Young Scientist Medal (1978), UGC Career Research Award, Punjab Rattan Award and JC Bose Oration Award. The Government of India bestowed upon him the honour of "Padma Shri" in 2009 by way of rightfully acknowledging his great contributions to higher education in India. On 2 July 2016, he was also honoured with "Bharat Gaurav" Lifetime Achievement Award at the House of Commons, British Parliament, London, UK.

Prof. (Dr.) Naveen Kumar Arora, PhD in Microbiology Professor in the Department of Environmental Science, Babasaheb Bhimrao Ambedkar University (a Central University), Lucknow, Uttar Pradesh, India, is a renowned researcher in the field of Environmental Microbiology and Biotechnology. His specific area of research is rhizosphere biology and plant growth-promoting rhizobacteria (PGPR). He has more than 60 research papers published in premium international journals and several articles published in magazines and dailies. He is editor of 11 books, published by Springer, and member of several national and international societies, in editorial board of 4 journals, and reviewer of several international journals. He is also the Editor-in-Chief of the journal *Environmental Sustainability* published by Springer Nature. He has delivered lectures in conferences and seminars around the globe. He has been advisor to 118 postgraduate and 9 doctoral students. He has also received awards for excellence in research by the Honourable Governor of Uttar Pradesh and Asian PGPR Society. Although an academician and researcher by profession, he has a huge obsession for the wildlife and its conservation and has authored a book, *Splendid Wilds*. He is President of Society for Conservation of Wildlife and has a dedicated website www.naveenarora.co.in for the cause of wildlife and environment conservation. He is also General Secretary of the Society for Environmental Sustainability.

Dr. Richa Kothari, M.Phil. and Ph.D. in Energy and Environment, is working as an Associate Professor at the Department of Environmental Sciences, Central University of Jammu, Samba, Jammu (J&K), India. She is awarded by one of the prestigious Indo-US Science and Technology Forum (IUSSTF) for Water Advanced Research and Innovation Fellowship Program in the year 2016 at the Robert B. Daugherty Water for Food Institute, University of Nebraska-Lincoln, Lincoln, NE, USA. She has guided more than 30 MSc and MTech students and 2 doctoral students. She has more than 60 articles in reputed peer-reviewed international and national journals with high-impact factors and chapters in edited books. She is also a reviewer of well-reputed Scopus journals. She is part of editorial teams of many national and international journals. Her research interests are bioenergy production approaches and low-cost wastewater treatment technologies.

Part I
Biodegradation and Bioremediation

Chapter 1
Biochar for Effective Cleaning of Contaminated Dumpsite Soil: A Sustainable and Cost-Effective Remediation Technique for Developing Nations

Paromita Chakraborty, Moitraiyee Mukhopadhyay, R. Shruthi, Debayan Mazumdar, Daniel Snow, and Jim Jian Wang

Contents

P. Chakraborty (✉)
Department Of Civil Engineering, SRM Institute of Science and Technology, Kancheepuram district, Tamil Nadu, India

SRM Research Institute, SRM Institute of Science and Technology, Kancheepuram district, Tamil Nadu, India

Water Science Laboratory, University of Nebraska, Lincoln, NE, USA
e-mail: paromita.c@res.srmuniv.ac.in

M. Mukhopadhyay · R. Shruthi
Department Of Civil Engineering, SRM Institute of Science and Technology, Kancheepuram district, Tamil Nadu, India

D. Mazumdar
Department of Chemical Engineering, SRM Institute of Science and Technology, Kancheepuram district, Tamil Nadu, India

D. Snow
Water Science Laboratory, University of Nebraska, Lincoln, NE, USA

J. J. Wang
School of Plant, Environmental and Soil Sciences, Louisiana State University Agricultural Center, Baton Rouge, LA, USA

R. C. Sobti et al. (eds.), *Environmental Biotechnology: For Sustainable Future*,
https://doi.org/10.1007/978-981-10-7284-0_1

Abstract Several studies have reported that open municipal dumpsites in developing countries are acting as a major source for a wide variety of pollutants. In developing nations, many dumpsites are located in the urban centers or even within the residential boundaries. Contaminants released during incomplete combustion of municipal solid waste have profound adverse impact on human health and the environment. Hence there is an urgent need to identify a low-cost technique to decontaminate such heavily polluted sites. In this chapter, we have reviewed several papers and discussed how different types of engineered biochars can be effectively used to adsorb contaminants from dumpsite soil. Biochars are basically carbon-rich solids treated by high-temperature pyrolysis. Biochars are obtained by heating biomass in presence of less oxygen or in anaerobic conditon. Properly pyrolysed mixtures of organic and cellulosic wastes are capable of adsorbing a wide variety of organic contaminants from wastewater, sludge and soil prior to the release or disposal in engineered landfills. Biochar produced from waste organic material such as coconut shells, sugarcane bagasse and straw has been reported with high adsorption capacity. Because locally produced waste organic material can be utilized for production of these low-cost adsorbents, they are especially attractive for remediation and treatment systems in developing countries. Pyrolytic temperature is believed to be the most important factor affecting the sorption capacity of biochar, followed by grinding to increase the surface area. Holding and adsorption capacity of the biochar for treating contaminants in soil could be a limiting factor of these materials. Some studies have shown that less than 5–7% (m/m) mixing of biochar and soil resulted in higher water retention capacity leading to increased potential for biodegradation. We therefore suggest that improved low-cost processing methods should be investigated so that biochar can be exploited as an adsorptive medium for remediating and treating contaminated soils in these regions.

Keywords Open municipal dumpsites · Biochar · Pyrolytic temperature · Cost-effective · Adsorption capacity · Incomplete combustion · Organic content

Introduction

Urbanization and economic development have increased municipal solid waste (MSW) generation across the globe. In the twenty-first century, the treatment of MSW has become a serious environmental issue, and MSW management continues to be an important environmental challenge. As shown in Fig. 1.1, East Asia and Pacific regions lead in the generation of waste followed by Latin America and Caribbean region, Eastern and Central Asian countries, South Asia, Middle East and

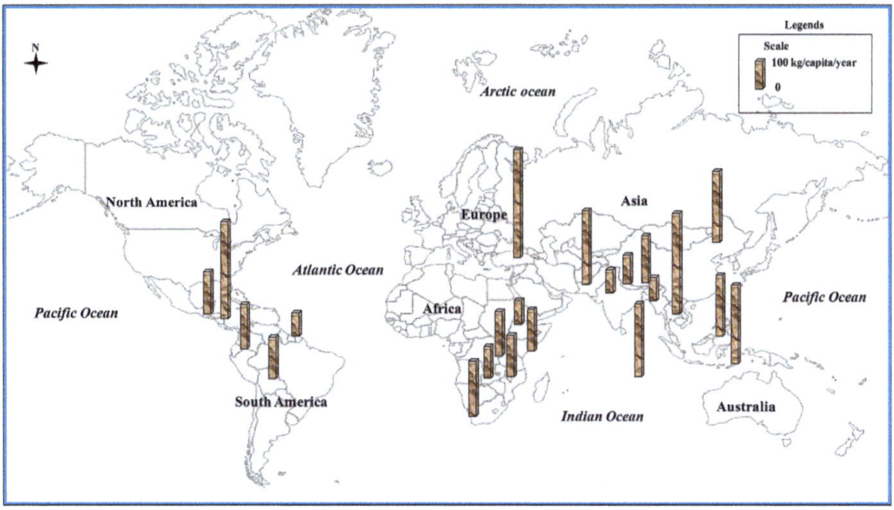

Fig. 1.1 Global map showing the quantity of annual solid waste generation (in kg/capita/year)

North African nations and finally the sub-Saharan African nations. It is projected that the global MSW generation levels will increase from 1.3 billion tons per year to 2.2 billion tons per year (World Bank 2012). The global waste generation rate is estimated to increase by about one million tons/day due to the current trend in economic growth (Inanc et al. 2004). The current situation is very serious in developing economies, as wastes have been poorly managed for many years. Even if waste is properly collected, often it does not reach legal disposal sites and is instead discarded in scattered, unregulated dumps. In addition to the existing large number of unregulated dumpsites, industrial, municipal and hazardous wastes are often mixed and disposed together, creating dangerous, toxic conditions because of the mixing of many different types of wastes. Improper location of disposal sites and the scarcity of modern engineering designs (liners and collection systems for leachate) threaten groundwater supplies and are a serious issue for those regions depending almost solely on groundwater sources. Simple waste management practices, such as covering wastes, weighing garbage and fences around dumps, are often not practiced in the developing countries.

A crucial, but often missing aspect, for MSW handling in developing countries is the establishment of sustainable approaches for treating these wastes in place. Waste tariffs only cover 30–40% of operating costs, leaving no funds available for capital investment. The shortfall is covered with money from the central or local budgets. Unless measures are taken, problems with waste disposal will only get worse. This problem is particularly serious in cities with limited capacity for dumpsite expansion. According to the United Nations (UN) Centre for Human Settlements (UNCHS), management of solid waste in developing and under developed nations is one of the most poorly treated services. Waste management systems are non-technical, obsolescent and inefficient resulting in indiscriminate open dumping of solid wastes (UNCHS; Habitat Refuse Collection Vehicles for Developing Countries; Nairobi 1991). Increasing imports of electronic waste (e-waste) for recycling by many Asian and African countries are contributing to improper handling of waste and haphazard dumping of refuse into low-lying areas of open land. To reduce the quantity of wastes in dumpsites, open burning is widely practised in landfills and dump yards particularly in the developing nations leading to the release of toxic pollutants such as dioxins, furans and heavy metals to the environment, thereby affecting human health (Chakraborty et al. 2018; Frazzoli et al. 2010; Robinson 2009; Thanh and Matsui 2011; Shih et al. 2016). For example, high concentrations of dioxin-like polychlorinated biphenyls (dl-PCBs) have been reported in human milk from mothers residing in and around the dumpsite of Kolkata due to the impact of fish diet (Someya et al. 2010).

Various technologies may be employed to remediate soil contaminated with mobile pollutants, including vitrification, mechanical separation, pyrometallurgic separation, phytoremediation, chemical treatment, electrokinetics, biochemical processes, soil flushing and soil washing (Mulligan et al. 2001). However, all these techniques have several drawbacks and point towards finding alternative techniques with lesser demerits for dumpsite soil remediation. Emission of toxic gases during vitrification process, pretreatment requirements in pyrometallurgic separation, release of toxic byproducts in biochemical processes and incapability of uptaking heavy contamination in phytoremediation are few drawbacks of the above-mentioned techniques (Schnoor et al. 1995; Mulligan et al. 2001). Various researchers have pointed out that remediation of pollutants by adsorption on biochar derived from biomasses could be an economical approach towards management of contaminated soils (Beesley et al. 2011; Zhang et al. 2013). Biochar has also proven to play an important role in sequestration of atmospheric carbon dioxide along with rehabilitating degraded land (Barrow 2012). Being highly recalcitrant in soil, the residence time for wood biochar is reported to be 10–1000 times more than the residence times of most soil organic matter (SOM), suggesting that biochar addition could provide a possible sink for carbon (Duku et al. 2011).

Biochar produced from dairy wastes was found to adsorb 100% and 77% of lead (Pb) and atrazine, respectively, from aqueous solutions (Cao et al. 2009). Freely dissolved concentrations of polycyclic aromatic hydrocarbons (PAHs) in sewage sludge reduced up to 57% by addition of different amount of biochar (Oleszczuk et al. 2012). The addition of woodchip biochar also reduced the off-site transport of antibiotics in soils amended by animal wastes (Jeong et al. 2012). Biochar addition has a long history in improving the quality and fertility of soil, but recent studies have supported the use of this material in reducing the bioavailability of organic and inorganic pollutants. Heavy metals were found to get stabilized in soil due to the increase in pH caused due to the addition of biochar (Zhang et al. 2013). Thus with the help of biochar, a sustainable approach for disposal of various organic refuse such as agricultural wastes, manure, industrial wastes, etc. could be attained which will not only lead to a reduction in greenhouse gas production but also will result in reduction of the prevalence of groundwater and surface water contamination (Barrow 2012). The aim of this chapter is to present an overview of (1) types of wastes that may contribute toxicants in open dumpsites particularly in developing nations, (2) effectiveness of biochar as an adsorbent for these toxicants, (3) remediation approaches for contaminated soil using biochar, (4) retention capacity of biochar and (5) future prospects for using biochar in remediation and managing contaminated dumpsite soil in developing countries.

Sources of Organic and Inorganic Wastes in Dumpsite Soil

A cocktail of wastes from multifarious sources end up in the dump yards of the developing nations primarily due to the absence of segregation of waste. In reports by the World Bank, the global waste composition consists of organics (46%), paper (17%), plastic (10%), glass (5%), metals (4%) and others (18%) (Urban Development Series Knowledge; World Bank 2012). Other categories include textile, leather, e-waste, appliances and inert materials. Waste composition can vary considerably depending on the level of economic development, geography, cultural norms and energy sources in each region. Accumulation of these waste results in the release of a wide range of organic and inorganic pollutants due to various activities practised in dumpsites. Contaminated soil in open dumpsites act as a secondary source of persistent organic pollutants (POPs) by re-emission of organic pollutants, like organochlorine pesticides (OCPs) (Chakraborty et al. 2015) or polychlorinated biphenyls (PCBs) (Chakraborty et al. 2013, 2016). The binding capacity of heavy metals and most of the organic pollutants in the soil depends on the pH, total organic carbon (TOC) and quantity of organic matter present in soil (Korthals et al. 1996; Tao et al. 2005; Jiang et al. 2012a). The phenomenon of long-range atmospheric transport (LRAT) of POPs released from a point source results in their atmospheric transportation and re-deposition leading to detection of POPs even in remote arctic areas (Yang et al. 2005).

Sources of Organic Contaminants in Dumpsite Soil

Organic pollutants entering MSW dumpsites can be from industrial, domestic or agricultural sources. Waste from textile industries consists of dye materials and other undesirable chemicals. Major colour effluents come from dyeing and printing process (Tan et al. 2000). Even low concentrations of such dyes in wastewater treatment plant (WWTP) effluents are undesirable due to resistance to degradation and difficulty in decolourization (Willmott et al. 1998; Nigam et al. 2000). Pharmaceutical and personal care products (PPCPs) are a class of emerging pollutants that has recently gained attention due to their ubiquitous presence and biological activities, including antibiotic resistance and effect to the endocrine system. This class of contaminants mainly enters waste streams from biomedical, domestic and industrial rejects from pharmaceutical and cosmetic industry (Jiang et al. 2012b). Bisphenol

A (BPA) and phthalic acid esters (PAEs) are compounds that come under the category of endocrine disruptors (EDCs) and found in variety of general use items like plastic bottles, audio and video technologies, pipe materials, sports equipment, etc. Sewage sludge from sewage treatment plants are also disposed by landfilling or spreading in dumpsites usually in developing countries. Sewage sludge is also a potential source for the release of PPCPs in the dumpsite. A major part of the effluent concentration of most PPCPs in WWTPs is sorbed to sludge and suspended solids. Changes in pH or redox conditions in dumpsites may result in release of contaminants after disposal.

Among developing countries, China is the largest exporter and importer of e-waste and receives about 70% of the total e-waste exported from developed nations. Other countries like India, Pakistan, Vietnam, the Philippines and Malaysia also import a considerable amount of hazardous e-wastes from developed nations (Robinson 2009; Frazzoli et al. 2010). The incomplete combustion of dumpsite waste releases PAHs in soil and air. Open burning of municipal waste has been found to be an important source for dioxin-like compounds in Asian countries (Minh et al. 2003) and PCBs in Indian cities (Chakraborty et al. 2013; Chakraborty et al. 2016). Discarded waste from the informal e-waste recycling workshops, the residues and leftover components are usually dumped in landfills and open dumpsites (Chakraborty et al. 2018). In Bangladesh, for example, around 20–35% of the e-waste were laid in landfills or simply dumped in rivers, open dumps, ponds, drains etc. (Islam et al. 2016).

Sources of Inorganic Contaminants in Dumpsite Soil

Heavy metals from various industries such as plating, plastics, glass, rubber, leather, etc are inorganic waste (Thitame et al. 2010). Environmental problems such as soil pollution, water pollution and groundwater contamination can occur by the release of inorganic pollutants due to inefficient disposal of these inorganic wastes. The methods employed by developing countries such as open burning of electronic equipment are carried out in dumpsites thus making such sites acts as a major source for release of heavy metals to the ambient environment (Olafisoye et al. 2013). Livestock and municipal sludge are also sources of inorganic contaminants. Significant levels of Cd, Cu, Pb, Cr, Ni and Zn are reported in livestock manures such as poultry, pig and cattle slurries (Luo et al. 2009). Buzier et al. (2006) observed that the global yield of WWTPs for metals like Cr, Cd and Pb to be often greater than 75% as sludge is particularly rich in metals. Sludge removed from many of these WWTPs are spread in local dumpsites for final disposal, resulting in the direct release of heavy metals into the environment. Waste from inorganic fertilizers and fungicides are another source of heavy metals that can reach the dumpsites. Except in some of the poorest developing countries of the world where less fertilizers are used, all other countries apply macronutrient fertilizers on agricultural soil. Heavy metals like As, Cd, U, Th, Hg, Ba and Zn are present in phosphatic fertilizers, while nitrogen fertilizers, lime fertilizers and manures are also known to contain metals like Cr, Hg, Co, Cu and As (Kabata-Pendias and Pendias 2001; Eckel et al. 2008). In fields and agroecosystems, various inorganic heavy metals and organometallic compounds are being used as fungicides. Pb and Cu arsenates, Bordeaux mix, Cu oxychloride and phenyl mercuric chloride are some among the fungicides that have been reported from dumpsites. Clearly, soil in dumpsites can be contaminated by heavy metals leaching from a variety of wastes.

Biochar as an Adsorbing Material

What Is Biochar?

Organic material when decomposed thermally in controlled supply of oxygen produces carbon dioxide, some combustible gases (mainly H_2, CO, CH_4) and a solid, black coloured substance, which is rich in carbon referred to as "char". Being slightly different from char, biochar so far lacks a proper definition in the scientific community. According to the international biochar initiative, biochar can be defined as "a solid material produced from biomass (any substance having good organic content) when heated at pyrolytic temperatures". Biochar has been available for centuries for use as horticulture and soil amendments but only recently came to light for contaminant remediation. Biochar is composed of three types of carbon (C): recalcitrant C, labile or leachable C and ash. Recalcitrant C is that organic carbon that is tolerant to degradation dominated by charcoal and is unavailable for microbes, whereas labile C are those fractions that are readily leached and mineralizable. Their oxidization drives the flux of CO_2 between soil and atmosphere. The third component, ash, contains macro- and micronutrients for biological uptake. The presence of fused aromatic C structures is the unique natural design that distinguishes biochar from any other organic matter. The occurrence of these structures contributes to the stable nature of biochar (Lehmann et al. 2011). Decarboxylation and demethoxylation along with dehydroxylation are major mechanisms towards the formation of more recalcitrant C fraction during pyrolysis process (Jung et al. 2016).

Different Methods for Production of Biochar

The low oxic or anoxic conditions during pyrolysis result in the thermal decomposition of biomass which eventually lead to the formation of biochar (Kloss et al. 2012). Biochar pyrolysis is a carbon negative activity as more carbon dioxide is sequestered in soil than what is released in the atmosphere (Barrow 2012). There are two types of pyrolysis, namely, conventional/slow pyrolysis and fast pyrolysis. In slow pyrolysis, the biomass is heated slowly to about 500 °C in the absence of air; this would not allow the vapours to escape rapidly. The risk of contamination of the biochar with the production of dioxins and harmful PAHs is reduced to a great extent using slow pyrolysis (Barrow 2012). Fast pyrolysis uses dry feedstock and provides rapid heat transfer which results in rapid escape of vapours. A higher char aromaticity is obtained in biochars prepared by slow pyrolysis (Mohan et al. 2014). There are a wide variety of carbon sources and methods to produce biochar. Some of the methods employed to produce biochar include slow pyrolysis reactors, fluidized bedfast pyrolysis reactors, screw pyrolyzers, hydrothermal carbonization, etc. (Lehmann et al. 2003; Sun et al. 2014). Slow pyrolysis reactors typically produce 15–25% of biochar depending on the feedstock and operating conditions. The biochar produced from fluidized bedfast pyrolysis reactors has distinct properties from those produced using slow pyrolysis due to the relatively high flow rate of gas and low residence time of biochar in the reactor bed, but the process is usually difficult to handle, and there is an interference of sand particles into the biochar during production (Lehmann and Joseph 2015). Screw pyrolyzers are used in biochar production at small scales. The flash carbonizer that uses ignition of a flash fire at elevated pressure in a packed bed of biomass results in significant improvement in yields (Lehmann et al. 2003; Lehmann and Joseph 2015). Biochar produced by the method of hydrothermal carbonization showed relatively high production rate compared to slow pyrolysis inside a furnace in a nitrogen rich environment (Sun et al. 2014).

Properties of feedstock and the pyrolysis conditions have high influence on the physical and chemical properties of biochar (Downie et al. 2009). Feedstocks differ from each other in their elemental compositions, which is attributed to the presence of soil and dust particles, lignin, cellulose and hemicellulose and moisture content. This elemental composition eventually determines the properties of biochar formed (Ubbelohde and Lewis 1960; Boehm 1994; Yip et al. 2007; Alexis et al. 2007). Pyrolysis causes volatilization which results in mass loss, volume reduction and shrinking of the biomass without altering its original structure. Also, pyrolysis alters the C/N, O/C and H/C ratios, porosity, surface area, cation exchange capacity, crystallinity and functional groups in the biomass (Kloss et al. 2012). Increasing the pyrolytic temperature is linked with the increase in specific surface area which in turn leads to an increase in adsorption (Zhang et al. 2013). Also the biochars produced at higher temperatures contain mainly micropores, while the biochars produced at lower temperature are not microporous which indicates that adsorption capacity could be more for biochars prepared at higher temperatures (Zhang et al. 2013). Table 1.1 gives the specific surface area attained by biochar produced from different feedstock at different temperatures. It has been observed that generally, as the pyrolytic temperature increases, the specific area of biochar increases for the same biochar feedstock considered. Thus, as the pyrolytic temperature increases, the specific surface area of biochar increases and ultimately increases the adsorption capacity of biochar. Novak et al. (2009) worked on biochar produced from peanut hulls, pecan shells, poultry litter and switch grass at different pyrolytic temperatures and observed that the biochar produced at higher temperatures attained higher specific surface areas. Chen et al. (2008) studied the characteristics of pine needle biochar by increasing the temperatures from 100 °C to 700 °C and found that the surface area increased from 0.65 m^2/g to 490.8 m^2/g. Thus, it can be concluded that the temperature range of 500–700 °C can increase the adsorption capacity of biochar due to increase in pore volume attained (Table 1.1). According to the data provided in Table 1.2, Ma et al. (2007) observed that application of bamboo-derived biochar on soil helped in the removal of extractable Cd by 79.6% within 12 days of application, while hardwood-derived biochar produced at 450 °C reduced Pb in soil pore water by tenfolds and Zn concentrations by 300 fold in column leaching tests (Beesley et al. 2011). Beesley et al. (2011) reported that hardwood biochar enhances the soil mobility of As and Cu, while wood biochar reduced the Zn and Cd leaching loss by greater than 90%. Cotton stalk- and hardwood-derived biochars reduced the bioavailability of Cd and As, respectively (Zhou et al. 2008; Hartley et al. 2009). Similarly, biochar derived from eucalyptus, orchard prune residue, chicken manure, green waste, rice straw, quail litter, oakwood, etc. reduced the bioavailability of various metals like Cd, Cu, Pb and Cr (Table 1.2). Among these source materials, hardwood biochar was more commonly used and has been found to be more effective especially at higher pyrolytic temperatures (500–700 °C) due to increased surface area for adsorption. Biochar produced at these temperatures also showed maximum cation exchange capacity (Jung et al. 2016). Yu et al. (2009) and Spokas and Reicosky (2009) concluded that high pyrolytic temperature and higher application rates of biochar increases the sorption of organic compounds like carbofurans, acetochlor, diuron, chlorpyrifos and atrazine. Dairy manure, pinewood, green waste, pine needle, eucalyptus, wheat straw and swine manure were used in treating compounds like atrazine, terbuthylazine, phenanthrene, and carbaryl. Also, the bioavailability of organic compounds like pentachlorophenol, PAHs, chlorobenzene, chlorpyrifos, fipronil and diuron were reduced by applying biochar derived from eucalyptus, hardwood, cotton straw, wheat straw and bamboo produced at high temperatures. It can be concluded that biochar produced at 600 °C can be efficiently used in remediating many organic pollutants and this can be attributed to the high surface area attained during its production (Table 1.2).

Developing nations are facing a serious problem with the huge amount of waste generated and improper disposal techniques. Biochar has proven to have equivalent or even greater sorption efficiency for both organic and inorganic pollutants from waste resources as shown in Table 1.2. In developing nations, biomass resources are available, such as forestry residues, wood waste, MSW, industrial wastewater and manure which contribute to country's primary energy supply. Biomass waste generally has an elemental composition of $CH_{1.4}O_{0.6}$ and mainly composed of cellulose, hemicelluloses, lignin and small amount of extractives (Duku et al. 2011). Biochar production from MSW is a safe and beneficial disposal option than the conventional methods such as incineration, landfilling, aerobic/anaerobic digestion, open air burning and composting (Serio et al. 2000). Open burning that is widely practised in developing nations lead to the emission of dangerous contaminants such as POPs, thereby degrading the environment. Biochar typically has more hydrogen and oxygen in its structure making it less carbonized than activated carbon. Biochar could potentially replace coal, coconut shell and wood-based activated carbons as a low-cost sorbent for contaminants and pathogens (Mohan et al. 2014). Theoretically

Table 1.1 Influence of pyrolytic temperature on the specific surface area of biochar produced from different feedstock material

Biochar Feedstock	Temperature (°C)	Surface Area (m^2/g)	References
Cow manure	500	21.9	Zhao et al. (2013)
Shrimp hull	500	13.3	
Bone dregs	500	113	
Wastewater sludge	500	71.6	
Waste paper	500	133	
Saw dust	500	203	
Grass	500	3.33	
Peanut shell	500	43.5	
Chlorella	500	2.78	
Water weeds	500	3.78	
Pig manure	200	3.59	
	350	4.26	
	500	47.4	
	650	42.4	
Wheat straw	200	2.53	
	350	3.48	
	500	33.2	
	650	182	
Pine needle	100	0.65	Chen et al. (2008)
	200	6.22	
	250	9.52	
	300	19.92	
	400	112.4	
	500	236.4	
	600	206.7	
	700	490.8	
Peanut hull	400	0.52	Novak et al. (2009a)
	500	1.33	
Pean shell	350	1.01	
	700	222	
Poultry litter	350	1.1	
	700	9	
Switchgrass	250	0.4	
	500	62.2	
Soya bean stover	300	5.61	Ahmad et al. (2012)
	700	420.3	
Peanut shell	300	3.14	
	700	448.2	

(continued)

Table 1.1 (continued)

Biochar Feedstock	Temperature (°C)	Surface Area (m^2/g)	References
Cotton seed hulls	350	4.7 ± 8	Tang et al. (2013)
	500	0	
	650	34 ± 3	
	800	322 ± 1	
Oak wood	350	450	
	600	642	
Corn Stover	350	293	
	600	527	
Broiler litter manure	350	59.5 ± 19.7	
	700	94.2 ± 5.1	
Soya bean stalk	300	144.14	
	400	138.76	
	500	152.98	
	600	179.03	
	700	250.23	
Broiler litter	350	60	Uchimiya et al. (2010)
	700	94	
Feed lot	350	1.3	Cantrell et al. (2012)
	700	145.2	
Fescue straw	100	1.8	
	200	3.3	
	300	4.5	
	400	8.7	
	500	50	
	600	75	
	700	139	
Oak bark	450	1.9	Mohan et al. (2011)
Oakwood	400–450	2.7	
Orange peel	150	22.8	Chen and Chen (2009)
	200	7.8	
	250	33.3	
	300	32.3	
	350	51	
	400	34	
	500	42.4	
	600	7.8	
	700	201	

the primary cost for biochar production is the cost incurred in feedstock collection, processing and pyrolysis operations in which the cost for transportation of the feedstock and the produced biochar is negligible in the total cost (Inyang and Dickenson 2015).The estimated break-even price for biochar is US $246/t, which is approximately one sixth of commercially available activated carbon (US $1500/t) (McCarl

Table 1.2 Effect of biochar on organic and inorganic pollutants in soil

Feedstock	Production temperature	Contaminant	Effect	References
Effect of biochar application on the mobility of heavy metals in soils				
Bamboo	Not available	Cd	Combined effect of electrokinetics, removal of extractable Cd by 79.6% within 12 days	Ma et al. (2007)
Hardwood	450 °C	As, Cd, Cu, Zn	Reduction in Cd in soil pore water by tenfold; Zn concentrations reduced 300- and 45-fold, respectively, in column leaching tests	Beesley and Dickinson (2011) and Beesley and Marmiroli (2011)
Hardwood	450 °C	As, Cd, Cu, Pb, Zn	Biochar surface mulching enhanced As and Cu mobility in the soil profile; little effect on Cd and Pb	Beesley and Dickinson (2011)
Wood	200 °C and 400 °C	Cd, Zn	Reduction in Zn and Cd leaching loss by >90%	Debela et al. (2012)
Effect of biochar application on the bioavailability of heavy metals in soils				
Cotton stalks	450 °C	Cd	Reduction of the bioavailability of Cd in soil by adsorption or co-precipitation	Zhou et al. (2008)
Hardwood-derived biochar	400 °C	As	Significant reduction of As in the foliage of *Miscanthus*	Hartley et al. (2009)
Eucalyptus	550 °C	As, Cd, Cu, Pb, Zn	Decrease in As, Cd, Cu and Pb in maize shoots	Namgay et al. (2010)
Orchard prune residue	500 °C	Cd, Cr, Cu, Ni, Pb, Zn	Significant reduction of the bioavailable Cd, Pb and Zn, with Cd showing the greatest reduction; an increase in the pH, CEC and water-holding capacity	Fellet et al. (2011)
Chicken manure and green waste	550 °C	Cd, Cu, Pb	Significant reduction of Cd, Cu and Pb accumulation by Indian mustard	Park et al. (2011)
Chicken manure	550 °C	Cr	Enhanced soil Cr(VI) reduction to Cr(III)	Choppala et al. (2012)
Sewage sludge	500 °C	Cu, Ni, Zn, Cd, Pb	Significant reduction in plant availability of the metals studied	Méndez et al. (2012)

(continued)

Table 1.2 (continued)

Feedstock	Production temperature	Contaminant	Effect	References
Rice straw	Not clear	Cu, Pb, Cd	Significant reduction in concentrations of free Cu, Pb and Cd in contaminated soils; identification of functional groups on biochar with high adsorption affinity to Cu	Jiang et al. (2012a)
Quail litter	500 °C	Cd	Reduction of the concentration of Cd in physic nut; greater reduction with the higher application rates	Suppadit et al. (2012)
Oakwood	400 °C	Pb	Bioavailability reduction by 75.8%; bioaccessibility reduction by 12.5%	Ahmad et al. (2012)
Effect of biochar application on sorption of organic pollutants in soils				
Eucalyptus wood	450 °C and 850 °C	Diuron, chlorpyrifos and carbofuran	Higher pyrolysis temperature and higher rates of biochar applied to soils result in stronger adsorption and weaker desorption of pesticides	Yu et al. (2006)
Woodchip	500 °C	Atrazine and acetochlor	Acetochlor adsorption increased 1.5 times; atrazine adsorption also increased	Spokas and Reicosky (2009)
Dairy manure	200 °C and 350 °C	Atrazine	At 200 °C, partitioning of atrazine is positively related to biochar carbon content	Cao et al. (2009)
Pinewood	350 °C and 700 °C	Terbuthylazine	Soil sorption increased 2.7- and 63-folds in the BC350 and BC700 treatments, respectively	Wang et al. (2010)
Green wastes	450 °C	Atrazine	Biochar-enhanced adsorption of pesticide	Zheng et al. (2010)
Pinewood	350 °C and 700 °C	Phenanthrene	Biochar produced at 700 °C showed a greater ability at enhancing a soil's sorption ability than that prepared at 350 °C	Zhang et al. (2010)

(continued)

Table 1.2 (continued)

Feedstock	Production temperature	Contaminant	Effect	References
Pine needles	100 °C, 300 °C, 400 °C, and 700 °C	PAHs	Sorption capacity increased with pyrolysis temperature	Chen and Yuan (2011)
Eucalyptus woodchips	850 °C	Diuron	Pesticide absorption increases with the biochar contact time with soil and application rate	Yu et al. (2011)
Poultry litter, wheat straw and swine manure	250 °C and 400 °C	Herbicides	Biochars showed high sorption ability for two herbicides, fluridone and norflurazon	Sun et al. (2012)
Swine manure	350 °C and 700 °C	Carbaryl	At low carbaryl concentrations, the sorption capacity BC700 > BC350; similar sorption capacity at high carbaryl concentrations	Zhang et al. (2013)
MgO-impregnated magnetic biochar, sugarcane harvest residue biochar, magnetic biochar	550 ° C	Phosphate	Adsorption was higher in MgO-impregnated magnetic biochar	Li et al. (2016)
Effect of biochar application on bioavailability of organic pollutants in soils				
Eucalyptus	450 °C and 850 °C	Diuron, chlorpyrifos and carbofuran	Reductions of chlorpyrifos and carbofuran in total plant residues, respectively	Yu et al. (2009)
Hardwood	450 °C	PAHs	Pore water concentrations of PAHs were reduced by biochar, with greater than 50% decrease of the heavier, more toxicologically relevant PAHs	Beesley et al. (2010)
Cotton straw	450 °C and 850 °C	Chlorpyrifos and fipronil	Chinese chive uptake of fipronil and chlorpyrifos reduced by 52% and 81%, respectively, with 1% of 850 °C biochar addition	Yang et al. (2010)
Bamboo	600 °C	Pentachlorophenol	Biochar reduced PCP bioavailability in soil	Xu et al. (2012)

(continued)

Table 1.2 (continued)

Feedstock	Production temperature	Contaminant	Effect	References
Hardwood	600 °C	PAHs	Biochar application reduced concentration and biological activity of PAHs in soil	Gomez-Eyles et al. (2011)
Wheat straw	500 °C	Chlorobenzenes (CBs)	Biochar amendment significantly reduced the bioavailability of CBs	Song et al. (2012b)
Wheat straw	250 °C, 300 °C, and 500 °C	Hexachlorobenzene (HCB)	Biochar amendment of soil resulted in a rapid reduction in the bioavailability of HCB, even at 0.1% biochar application rate	Song et al. (2012a)

et al. 2009; Ahmad et al. 2012, 2014). Furthermore due to the ubiquity and cost-effectiveness of fresh biochar, exhausted biochar can be replaced easily and can be recycled by burning it to produce ash for use as liming agent in acidic soil (Feng et al. 2013).

Why Biochar Acts as a Potential Adsorbent?

Biochars, generally used for soil conditioning and carbon sequestration, are now finding use in contaminant remediation (Mohan et al. 2014). Biochar can be considered as a sustainable material as it requires less investment and, unlike activated carbon, the hydrogen and carbon remain in its structure along with the ash that originates from its biomass. One of the most important properties of biochar is its adsorption capacity. Biochars with high specific area are usually used as sorbents (Zhao et al. 2013). Several studies have demonstrated the effectivity of biochar in stabilizing some inorganic pollutants like heavy metals. They can stabilize the heavy metals in polluted soils, thus improving the soil quality and reducing heavy metals uptake by crops. Also, studies have demonstrated that the high surface area, high aromatic nature, micropore volume and abundance of polar functional groups present in biochar have aided in effective sorption of organic contaminants like POPs viz., PCBs, PAHs and emerging pollutants such as steroid hormones and pharmaceuticals (Zhang et al. 2013; Zhao et al. 2013; Arun et al. 2017). Sludge-derived biochar was found to be an excellent adsorbent for amoxicillin in wastewater mainly due to large Brunauer–Emmett–Teller (BET) specific surface area (Arun et al. 2017). As shown in Fig. 1.2, scanning electronic microscope (SEM) images showed the distinct pattern and pores available before exposure of the biochar to water containing amoxicillin. Reduction of the particle size indicates the adsorption of amoxicillin to the pores of the biochar, and this property may be important for

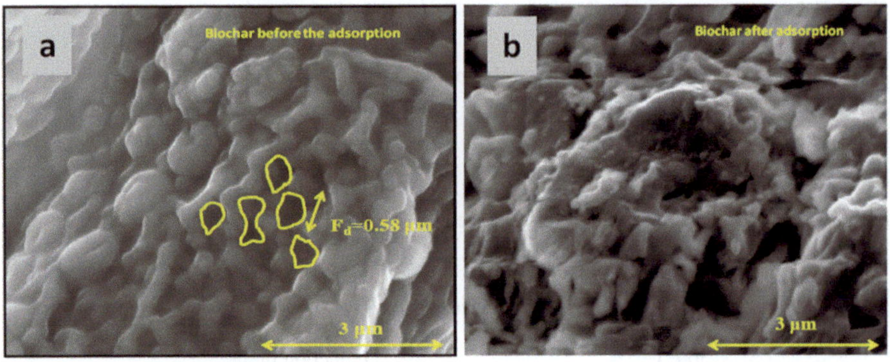

Fig. 1.2 SEM images given by Arun et al. 2017, for sludge-derived biochar (**a**) before amoxicillin exposure and (**b**) after amoxicillin exposure

sequestration of a variety of contaminants. Li et al. (2016) reported that increasing the Mg content in MgO-impregnated magnetic biochar (MMSB) increased the adsorption of phosphate compared to sugarcane harvest residue biochar and magnetic biochar without Mg. Also studies on Mg/Al-layered double hydroxide-modified biochar showed that the Mg/Al ratio and pH of solution affected the adsorption of phosphates from aqueous solution (Li et al. 2016). One of the main aspects to be considered is the effect of biochar properties on the bioavailability of these contaminants. The risk of these organic and inorganic pollutants from entering the food chain, surface runoff, and leaching to groundwater is highly reduced due to adsorption. The effect of biochar on the bioavailability of metals depends on the feedstock materials used for preparing biochar and the type of heavy metal considered (Zhang et al. 2013). The amendment of soil contaminated by Cd and Zn using hardwood-derived biochar reduced the concentration of both the metals in pore water (Beesley et al. 2010; Karami et al. 2011).

Soil Treatment Using Biochar

Remediation of Organic Pollutants

POPs are those categories of pollutants that resist photolytic, biological or chemical degradation to a varying degree. Characterized by low water solubility and high lipid solubility, many are found to deposit in fatty tissues of organisms leading to their bioaccumulation in the environment due to their persistent nature. Some classes of organic pollutants well known for their persistence and toxicity are PCBs, OCPs and PAHs. Due to their everlasting effects on human health and environment, the need for cost-effective sustainable methods to remediate soils, especially dumpsite soils where they are detected frequently, is a necessity. Several studies

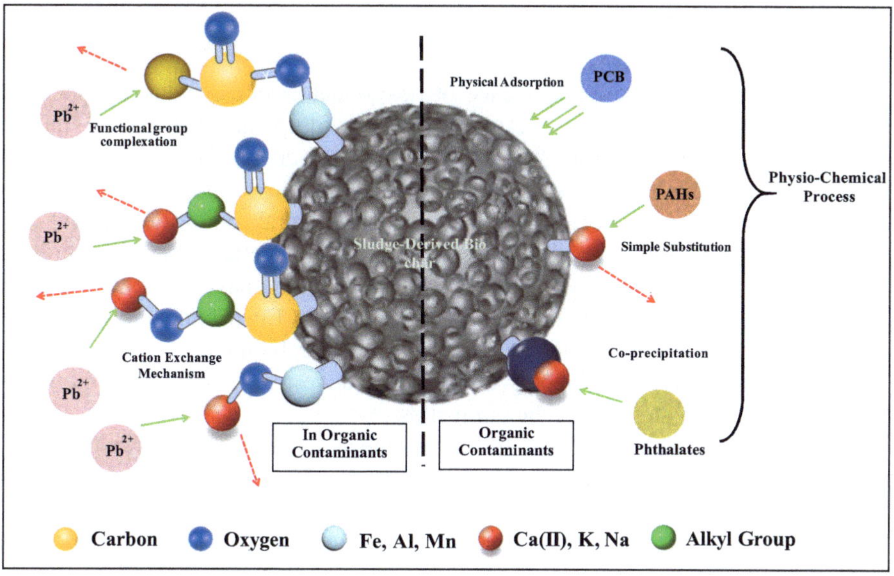

Fig. 1.3 Mechanism on the use of biochar for the removal of inorganic and organic pollutants as given by Zhang et al. 2013

have proved the effectivity of biochar in uptake of such POPs (Tong et al. 2011; Ahmad et al. 2014). The high aromaticity, high surface area, micropore volume and abundance of polar functional groups in biochar are the major factors that suffice the adsorption of organic pollutants by biochar. As shown in Fig. 1.3, the main phenomena governing are physical adsorption, co-precipitation and simple substitution. Arun et al. (2017) reported sludge-derived biochar produced at pyrolytic temperature of 300 °C to be effective in removal of amoxicillin by adsorption from wastewater, which was evident from the distinct pattern observed before and after adsorption in UV spectroscopy. Beesley et al. (2011) have observed that the PAH concentration in soil pore water reduced to 50% after treating with biochar. The reduction in bioavailability of various organic pollutants due to adsorption in biochar derived from different sources have been given in Table 1.2. When soil polluted with chlorpyrifos and fipronil was treated with biochar produced from corn straw, it was observed that uptake rate of chlorpyrifos and fipronil by plant reduced to 52% and 81%, respectively (Yang et al. 2010). A 95% removal of malachite green (MG) was attained within 40 min by adsorption into rice straw biochar applied on 25 mg/L of MG solution (Hameed and El-Khaiary 2008). Acid-treated straw biochar used to remove reactive brilliant blue and rhodamine B resulted in high adsorption by biochar than the activated charcoal (AC) accounting to the high surface area and less carbonization of AC compared to charcoal. Studies across a variety of locations globally suggest that biochar can be a better option in amending soil polluted with organic pollutants (Chun et al. 2004; Qiu et al. 2009).

Remediation of Toxic Metals

The indiscriminate use and occurrence of heavy metals in the industrial, medical and technological sector has resulted in its widespread distribution in the environment, raising concerns over their possible effects on the environment and human health. Some of the sources of heavy metals release into the environment include geogenic, industrial, pharmaceutical, agricultural, domestic effluent and atmospheric sources (He et al. 2005). Among heavy metals arsenic, cadmium, mercury, lead and chromium are the top priority as they have been identified as possible human carcinogens by the United States Environmental Protection Agency (USEPA) and International Agency for Research on Cancer. Thus, there is a necessity to develop eco-friendly and cost-effective methods to deal with these pollutants that ultimately reach the dumpsites.

As stated by Zhang et al. 2013, the various mechanisms involved in stabilization of heavy metals using biochar are (a) ion exchange between the cations associated with biochar and concerned heavy metals along with co-precipitation and inner sphere complexation with humic matter and mineral oxides of biochar and (b) surface complexation – metals undergo surface complexation with several functional groups and inner sphere complexation with free hydroxyl present in the mineral oxides (Fig. 1.3). Physical adsorption is a process involving simple adsorption and surface precipitation that lead to the stabilization of the heavy metal content (Lu et al. 2012). Heavy metal stabilization can further depend on the type of soil considered and the cations present in both biochar and soil (Zhang et al. 2013). Ahmad et al. (2014) observed that the bioavailability of Pb in military shooting range soil reduced to 75.8% after the treatment. Chen et al. (2016) found that biochar prepared from corn straw could remove around 95% of Cu and 90% of Zn. On the other hand, pinewood biochar showed adsorption capacities of 4.13 mg/g, 1.2 mg/g and 2.62 mg/g, while oakwood char showed 0.37 mg/g, 5.85 mg/g and 3 mg/g for Pb, Cd and As, respectively (Mohan et al. 2014). The adsorption capacity of Cu by biochar prepared from peanut straw, soybean straw and canola straw was 0.09 mg/g, 0.05 mg/g and 0.04 mg/g, respectively (Tong et al. 2011). Recently, research has focused on enhancing biochar's affinity for contaminant oxyanions such as AsO_4^{3-}, AsO_3^{3-} and CrO_4^{2-} due to generally negative charge of pristine biochar. Development of metal oxide-impregnated biochar composite has shown promising adsorption of these metals present as oxyanions (Agrafioti et al. 2014; Li et al. 2016; Wang et al. 2016).

Retention Capacity of Biochar

The retention capacity of biochar depends on the pyrolytic temperature, specific surface area, total pore volume, mechanism employed in adsorption, etc. Wood-derived activated charcoal and dairy manure biochar showed that despite having lower surface area than activated charcoal, biochar could retain Pb six times more than activated carbon (Cao et al. 2009). The charge and surface area properties are the factors that usually help in reducing nutrient losses from biochar when it is used in soil (Glaser et al. 2002; Lehmann et al. 2003). The presence of functional groups in biochar affects the sorption capacity depending on the surface charge on it, thus helping in the adsorption and retention of both transitional and nontransitional metals on the biochar particles (Amonette and Joseph 2009). Interaction of biochar with natural organic molecules and clay minerals present in soil can suppress the sorption of the organic pollutants from soil (Pignatello et al. 2006). Wang et al. (2010) observed that the retention capacity of herbicide terbuthylazine decreased due to high organic content in the soil because dissolved organic carbon competed for sorption sites thereby reducing the retention capacity.

Future Prospect for Remediation of Dumpsite Soils

Municipal waste is a growing health and ecological problem particularly in the developing countries because of the toxic contaminants that enter poorly maintained dumpsites. Several developing nations have used biochar to decontaminate the surface soil. Dumpsite soil quality can be easily improved through the targeted use of low-cost biochar soil amendments that sequester and treat these contaminants. Ageing is a phenomenon where the presence of natural organic matter and clay minerals in the soil clogs the micropores of biochar and decreases the sorption of organic pollutants by biochar. Such high concentrations of organic matter in soil compete with the organic pollutants for sorption sites eventually decreasing the adsorption efficiency of biochar (Zhang et al. 2013). One important aspect yet to be investigated is the impact of biochar on the soil fauna as there are reports showing adverse effect of biochar on earthworms and soil microbes (Beesley et al. 2011). A high rate of biochar application can decrease the soil fertility by adsorbing the required soil nutrients (Frazzoli et al. 2010). Furthermore there are evidences for release of toxic pollutants like PAHs from the biochar itself due to high rate of application on soil (Thies and Rillig 2009). Based on literature, it seems that biochar can be more widely used in remediation of the organic and inorganic pollutants from soil by addition in layers on the polluted dumpsite soil. Incorporated biochar would act as an adsorptive medium for sequestering pollutants. Thus, biochar can be produced at high pyrolytic temperatures which would improve its adsorption characteristics by increasing the specific surface area and total pore volume available for adsorption of pollutants. Biochar has to be incorporated and maintained in the soil for the optimum time period. Bioaugmentation can be used for bioremediation of the biochar which has adsorbed different types of pollutants from the dumpsite soil. Bioaugmentation employs soil microorganisms that would pre-concentrate the contaminants on the biochar. Later on the immobilized microbial organisms would degrade the pollutants (Zhao et al. 2013).

Biochar amendments prepared using wooden pellets for amending landfill soil covers facilitated the growth of methanotrophic bacteria that are effective in reducing methane emissions. The highest oxidation rates were observed in the upper layers of amended soil (up to 30 cm depth) with more oxygen availability (Reddy et al. 2014). Sea mango-based biochar utilized for removal of organic and inorganic pollutants from landfill leachates gave the highest adsorptive removal for colour (95.1%), chemical oxygen demand (COD) (84.94%) and NH_3-N (95.77%) (Shehzad et al. 2016). Asia alone generates 4.4 billion tones and 790 MT of solid waste and MSW, respectively, per year, of which 6% is attributed to the Indian subcontinent. The land allotment for 48MT of waste from India is only 20.2 km^2 which should be increased to 169.6 km^2 by the end of 2047 with a projected amount of 300 MT of waste generated. Urban waste is predominantly rich in organic matter (46%) followed by paper (6%), glass (0.7%), rags (3.2%) and plastic (1%) and the rest is moisture. About 600 MT of waste is generated in India alone from agricultural waste of which sugar industries contribute 90 MT. With such rich amount of organic waste composition in the MSW, it can very obviously be used for the production of high-quality biochar for the removal of organic and inorganic contaminants as discussed in the previous sections. Furthermore, 730 Tg of biomass are burnt in Asia of which 250 Tg come from agricultural burning that leads to the emission of SOx, NOx, CO, CO_2, PAHs and PCDD/Fs (Gadde et al. 2009). Biochar production can thereby reduce the load of open burning of crop residues and therefore, significantly contribute in arresting air pollution and concurrently help in conserving carbon. The versatility of biochar technologies also offers the potential for equitable technology transfer and use in developing countries (Pratt and Moran 2010). Cost-effectiveness in developed and developing nations using marginal abatement cost curve (MACC) was done by Pratt and Moran (2010). Several considerations were taken in to account, including price of electricity, carbon, biochar application rates and potential yield gain. Electricity generated from the biochar plants as byproduct such as syngas and bio-oil were used to operate the plants in developed nations. On the other hand, taking into consideration the developing nations where biochar kiln and stoves are employed, there was far better abatement of carbon emissions in comparison to fossil fuels. In developed nations with the advantage of adequate infrastructure and abundance of biomass for feedstock, it may appear that due to the high carbon market prices, biochar-producing plants will be highly cost-effective. However, MACC output clearly suggests otherwise and even without substantial infrastructure, waste management in developing nations was found to be more profitable by offering more abatement potential at lower costs than carbon capture and storage (CCS) (Pratt and Moran 2010).

Conclusion

With consideration of relatively simple low-cost methods for producing biochar and its remarkable adsorption capacity, it seems likely that biochar would provide a cost-effective solution for cleaning and managing contaminated soil in addition to its soil amendment benefits for dumpsite soils in developing and underdeveloped countries. The major influential factors governing the characteristics of biochar produced are the pyrolytic temperature employed and the feedstock utilized for its preparation. More in-depth research is needed to help facilitate production of biochar in these countries to increase the retention capacity of the pollutants. Furthermore, local municipalities would be expected to create a demand for improved biochar as an effective remediation technique and dumpsite management alternative. To meet the increasing demand, there can be a new market venture for such inexpensive remediation technique in the developing world.

Acknowledgements The authors would like to thank MoEFCC (Ministry of Environment, Forest and Climate Change) for the project given to Dr. Paromita Chakraborty No. Q-14011/43/2013-CPW (EHC). PC would also like to acknowledge the Water Advanced Research and Innovation (WARI) Fellowship Program supported by the Department of Science and Technology, Govt. of India, the University of Nebraska-Lincoln (UNL), the Daugherty Water for Food Institute (DWFI) and the Indo-US Science and Technology Forum (IUSSTF).

References

Agrafioti, E., Kalderis, D., & Diamadopoulos, E. (2014). Ca and Fe modified biochars as adsorbents of arsenic and chromium in aqueous solutions. *Journal of Environmental Management, 146*, 444–450.

Ahmad, M., Lee, S. S., Dou, X., Mohan, D., Sung, J.-K., Yang, J. E., & Ok, Y. S. (2012). Effects of pyrolysis temperature on soybean stover-and peanut shell-derived biochar properties and TCE adsorption in water. *Bioresource Technology, 118*, 536–544.

Ahmad, M., Rajapaksha, A. U., Lim, J. E., Zhang, M., Bolan, N., Mohan, D., Vithanage, M., Lee, S. S., & Ok, Y. S. (2014). Biochar as a sorbent for contaminant management in soil and water: A review. *Chemosphere, 99*, 19–33.

Alexis, M., Rasse, D. P., Rumpel, C., Bardoux, G., Péchot, N., Schmalzer, P., Drake, B., & Mariotti, A. (2007). Fire impact on C and N losses and charcoal production in a scrub oak ecosystem. *Biogeochemistry, 82*, 201–216.

Amonette, J. E., & Joseph, S. (2009). Characteristics of biochar: Microchemical properties. In *Biochar for environmental management. Science and Technology* (p. 33). London: Earthscan.

Arun, S., Kothari, K., Mazumdar, D., Mukhopadhyay, M., & Chakraborty, P. (2017). Biochar production from domestic sludge: A cost-effective, recycled product for removal of amoxicillin in wastewater. *IOP Conference Series: Materials Science and Engineering, 225*, 012164.

Barrow, C. (2012). Biochar: Potential for countering land degradation and for improving agriculture. *Applied Geography, 34*, 21–28.

Beesley, L., Moreno-Jiménez, E., & Gomez-Eyles, J. L. (2010). Effects of biochar and greenwaste compost amendments on mobility, bioavailability and toxicity of inorganic and organic contaminants in a multi-element polluted soil. *Environmental Pollution, 158*, 2282–2287.

Beesley, L., & Dickinson, N. (2011). Carbon and trace element fluxes in the pore water of an urban soil following greenwaste compost, woody and biochar amendments, inoculated with the earthworm Lumbricus terrestris. *Soil Biology & Biochemistry, 43*, 188–196.

Beesley, L., & Marmiroli, M. (2011). The immobilisation and retention of soluble arsenic, cadmium and zinc by biochar. *Environmental Pollution, 159*, 474–480.

Beesley, L., Moreno-Jiménez, E., Gomez-Eyles, J. L., Harris, E., Robinson, B., & Sizmur, T. (2011). A review of biochars' potential role in the remediation, revegetation and restoration of contaminated soils. *Environmental Pollution, 159*, 3269–3282.

Boehm, H. (1994). Some aspects of the surface chemistry of carbon blacks and other carbons. *Carbon, 32*, 759–769.

Buzier, R., Tusseau-Vuillemin, M.-H., dit Meriadec, C. M., Rousselot, O., & Mouchel, J.-M. (2006). Trace metal speciation and fluxes within a major French wastewater treatment plant: Impact of the successive treatments stages. *Chemosphere, 65*, 2419–2426.

Cantrell, K. B., Hunt, P. G., Uchimiya, M., Novak, J. M., & Ro, K. S. (2012). Impact of pyrolysis temperature and manure source on physicochemical characteristics of biochar. *Bioresource Technology, 107*, 419–428.

Cao, X., Ma, L., Gao, B., & Harris, W. (2009). Dairy-manure derived biochar effectively sorbs lead and atrazine. *Environmental Science & Technology, 43*, 3285–3291.

Chakraborty, P., Zhang, G., Eckhardt, S., Li, J., Breivik, K., Lam, P. K., Tanabe, S., & Jones, K. C. (2013). Atmospheric polychlorinated biphenyls in Indian cities: Levels, emission sources and toxicity equivalents. *Environmental Pollution, 182*, 283–290.

Chakraborty, P., Zhang, G., Li, J., Sivakumar, A., & Jones, K. C. (2015). Occurrence and sources of selected organochlorine pesticides in the soil of seven major Indian cities: Assessment of air–soil exchange. *Environmental Pollution, 204*, 74–80.

Chakraborty, P., Khuman, S. N., Selvaraj, S., Sampath, S., Devi, N. L., Bang, J. J., & Katsoyiannis, A. (2016). Polychlorinated biphenyls and organochlorine pesticides in River Brahmaputra from the outer Himalayan Range and River Hooghly emptying into the Bay of Bengal: Occurrence, sources and ecotoxicological risk assessment. *Environmental Pollution, 219*, 998–1006.

Chakraborty, P., Selvaraj, S., Nakamura, M., Prithiviraj, B., Cincinelli, A., & Bang, J. J. (2018). PCBs and PCDD/Fs in soil from informal e-waste recycling sites and open dumpsites in India: Levels, congener profiles and health risk assessment. Science of the Total Environment 621, 930–938.

Chen, B., Zhou, D., & Zhu, L. (2008). Transitional adsorption and partition of nonpolar and polar aromatic contaminants by biochars of pine needles with different pyrolytic temperatures. *Environmental Science & Technology, 42*, 5137–5143.

Chen, B., & Chen, Z. (2009). Sorption of naphthalene and 1-naphthol by biochars of orange peels with different pyrolytic temperatures. *Chemosphere, 76*, 127–133.

Chen, B., & Yuan, M. (2011). Enhanced sorption of polycyclic aromatic hydrocarbons by soil amended with biochar. *Journal of Soils and Sediments, 11*, 62–71.

Chen, X.-W., Wong, J. T.-F., Ng, C. W.-W., & Wong, M.-H. (2016). Feasibility of biochar application on a landfill final cover—A review on balancing ecology and shallow slope stability. *Environmental Science and Pollution Research, 23*, 7111–7125.

Choppala, G. K., Bolan, N., Megharaj, M., Chen, Z., & Naidu, R. (2012). The influence of biochar and black carbon on reduction and bioavailability of chromate in soils. *Journal of Environmental Quality, 41*, 1175–1184.

Chun, Y., Sheng, G., Chiou, C. T., & Xing, B. (2004). Compositions and sorptive properties of crop residue-derived chars. *Environmental Science & Technology, 38*, 4649–4655.

Debela, F., Thring, R., & Arocena, J. (2012). Immobilization of heavy metals by co-pyrolysis of contaminated soil with woody biomass. *Water, Air, and Soil Pollution, 223*, 1161–1170.

Downie, A., Crosky, A., & Munroe, P. (2009). Physical properties of biochar. In *Biochar for environmental management: Science and technology* (pp. 13–32). London: Earthscan.

Duku, M. H., Gu, S., & Hagan, E. B. (2011). Biochar production potential in Ghana—A review. *Renewable and Sustainable Energy Reviews, 15*, 3539–3551.

Eckel, H., Roth, U., Döhler, H., & Schultheis, U. (2008). Assessment and reduction of heavy metal input into agro-ecosystems. In P. Schlegel & A. W. Durosoy S Jongbloed (Eds.), *Trace elements in animal production systems* (pp. 33–43). Wageningen: Wageningen Academic Publishers.

Fellet, G., Marchiol, L., Delle Vedove, G., & Peressotti, A. (2011). Application of biochar on mine tailings: Effects and perspectives for land reclamation. *Chemosphere, 83*, 1262–1267.

Feng, Y., Dionysiou, D. D., Wu, Y., Zhou, H., Xue, L., He, S., & Yang, L. (2013). Adsorption of dyestuff from aqueous solutions through oxalic acid-modified swede rape straw: Adsorption process and disposal methodology of depleted bioadsorbents. *Bioresource Technology, 138*, 191–197.

Frazzoli, C., Orisakwe, O. E., Dragone, R., & Mantovani, A. (2010). Diagnostic health risk assessment of electronic waste on the general population in developing countries' scenarios. *Environmental Impact Assessment Review, 30*, 388–399.

Gadde, B., Bonnet, S., Menke, C., & Garivait, S. (2009). Air pollutant emissions from rice straw open field burning in India, Thailand and the Philippines. *Environmental Pollution, 157*, 1554–1558.

Glaser, B., Lehmann, J., Steiner, C., Nehls, T., Yousaf, M., & Zech, W. (2002). *Potential of pyrolyzed organic matter in soil amelioration, 12th ISCO conference* (pp. 421–427). Beijing.

Gomez-Eyles, J. L., Sizmur, T., Collins, C. D., & Hodson, M. E. (2011). Effects of biochar and the earthworm Eisenia fetida on the bioavailability of polycyclic aromatic hydrocarbons and potentially toxic elements. *Environmental Pollution, 159*, 616–622.

Hameed, B., & El-Khaiary, M. (2008). Malachite green adsorption by rattan sawdust: Isotherm, kinetic and mechanism modeling. *Journal of Hazardous Materials, 159*, 574–579.

Hartley, W., Dickinson, N. M., Riby, P., & Lepp, N. W. (2009). Arsenic mobility in brownfield soils amended with green waste compost or biochar and planted with Miscanthus. *Environmental Pollution, 157*, 2654–2662.

He, Z. L., Yang, X. E., & Stoffella, P. J. (2005). Trace elements in agroecosystems and impacts on the environment. *Journal of Trace Elements in Medicine and Biology, 19*, 125–140.

Inanc, B., Idris, A., Terazono, A., & Sakai, S.-i. (2004). Development of a database of landfills and dump sites in Asian countries. *Journal of Material Cycles and Waste Management, 6*, 97–103.

Inyang, M., & Dickenson, E. (2015). The potential role of biochar in the removal of organic and microbial contaminants from potable and reuse water: A review. *Chemosphere, 134*, 232–240.

Islam, M. T., Abdullah, A., Shahir, S., Kalam, M., Masjuki, H., Shumon, R., & Rashid, M. H. (2016). A public survey on knowledge, awareness, attitude and willingness to pay for WEEE management: Case study in Bangladesh. *Journal of Cleaner Production, 137*, 728–740.

Jeong, C. Y., Wang, J. J., Dodla, S. K., Eberhardt, T. L., & Groom, L. (2012). Effect of biochar amendment on tylosin adsorption–desorption and transport in two different soils. *Journal of Environmental Quality, 41*, 1185–1192.

Jiang, J., Xu, R.-k., T-y, J., & Li, Z. (2012a). Immobilization of Cu (II), Pb (II) and Cd (II) by the addition of rice straw derived biochar to a simulated polluted Ultisol. *Journal of Hazardous Materials, 229*, 145–150.

Jiang, T.-Y., Jiang, J., Xu, R.-K., & Li, Z. (2012b). Adsorption of Pb (II) on variable charge soils amended with rice-straw derived biochar. *Chemosphere, 89*, 249–256.

Jung, K.-W., Kim, K., Jeong, T.-U., & Ahn, K.-H. (2016). Influence of pyrolysis temperature on characteristics and phosphate adsorption capability of biochar derived from waste-marine macroalgae (Undaria pinnatifida roots). *Bioresource Technology, 200*, 1024–1028.

Kabata-Pendias, A., & Pendias, H. (2001). *Trace elements in soils and plants* (3rd ed.). Boca Raton: CRC Press.

Karami, N., Clemente, R., Moreno-Jiménez, E., Lepp, N. W., & Beesley, L. (2011). Efficiency of green waste compost and biochar soil amendments for reducing lead and copper mobility and uptake to ryegrass. *Journal of Hazardous Materials, 191*, 41–48.

Kloss, S., Zehetner, F., Dellantonio, A., Hamid, R., Ottner, F., Liedtke, V., Schwanninger, M., Gerzabek, M. H., & Soja, G. (2012). Characterization of slow pyrolysis biochars: Effects of

feedstocks and pyrolysis temperature on biochar properties. *Journal of Environmental Quality, 41*, 990–1000.
Korthals, G. W., van de Ende, A., van Megen, H., Lexmond, T. M., Kammenga, J. E., & Bongers, T. (1996). Short-term effects of cadmium, copper, nickel and zinc on soil nematodes from different feeding and life-history strategy groups. *Applied Soil Ecology, 4*, 107–117.
Lehmann, J., da Silva, J. P., Steiner, C., Nehls, T., Zech, W., & Glaser, B. (2003). Nutrient availability and leaching in an archaeological Anthrosol and a Ferralsol of the Central Amazon basin: Fertilizer, manure and charcoal amendments. *Plant and Soil, 249*, 343–357.
Lehmann, J., Rillig, M. C., Thies, J., Masiello, C. A., Hockaday, W. C., & Crowley, D. (2011). Biochar effects on soil biota–a review. *Soil Biology and Biochemistry, 43*, 1812–1836.
Lehmann, J., & Joseph, S. (2015). *Biochar for environmental management: Science, technology and implementation*. Abingdon/New York: Routledge.
Li, R., Wang, J. J., Zhou, B., Awasthi, M. K., Ali, A., Zhang, Z., Lahori, A. H., & Mahar, A. (2016). Recovery of phosphate from aqueous solution by magnesium oxide decorated magnetic biochar and its potential as phosphate-based fertilizer substitute. *Bioresource Technology, 215*, 209–214.
Lu, H., Zhang, W., Yang, Y., Huang, X., Wang, S., & Qiu, R. (2012). Relative distribution of Pb^{2+} sorption mechanisms by sludge-derived biochar. *Water Research, 46*, 854–862.
Luo, L., Ma, Y., Zhang, S., Wei, D., & Zhu, Y.-G. (2009). An inventory of trace element inputs to agricultural soils in China. *Journal of Environmental Management, 90*, 2524–2530.
Ma, J., Wang, H., & Luo, Q. (2007). Movement-adsorption and its mechanism of Cd in soil under combining effects of electrokinetics and a new type of bamboo charcoal. *Huan jing ke xue= Huanjing kexue/[bian ji, Zhongguo ke xue yuan huan jing ke xue wei yuan hui "Huan jing ke xue" bian ji wei yuan hui], 28*, 1829–1834.
McCarl, B. A., Peacocke, C., Chrisman, R., Kung, C.-C., & Sands, R. D. (2009). Economics of biochar production, utilization and greenhouse gas offsets. In *Biochar for environmental management Science and Technology* (pp. 341–358). London: Earthscan.
Méndez, A., Gómez, A., Paz-Ferreiro, J., & Gascó, G. (2012). Effects of sewage sludge biochar on plant metal availability after application to a Mediterranean soil. *Chemosphere, 89*, 1354–1359.
Minh, N. H., Minh, T. B., Watanabe, M., Kunisue, T., Monirith, I., Tanabe, S., Sakai, S., Subramanian, A., Sasikumar, K., Viet, P. H., Tuyen, B. C., Tana, T. S., & Prudente, M. S. (2003). Open dumping site in Asian developing countries: A potential source of polychlorinated dibenzo-dioxins and polychlorinated dibenzofurans. *Environmental Science & Technology 37* (8):1493–1502.
Mohan, D., Rajput, S., Singh, V. K., Steele, P. H., & Pittman, C. U., Jr. (2011). Modeling and evaluation of chromium remediation from water using low-cost bio-char, a green adsorbent. *Journal of Hazardous Materials, 188*, 319–333.
Mohan, D., Sarswat, A., Ok, Y. S., & Pittman, C. U. (2014). Organic and inorganic contaminants removal from water with biochar, a renewable, low-cost and sustainable adsorbent–a critical review. *Bioresource Technology, 160*, 191–202.
Mulligan, C., Yong, R., & Gibbs, B. (2001). Remediation technologies for metal-contaminated soils and groundwater: An evaluation. *Engineering Geology, 60*, 193–207.
Namgay, T., Singh, B., & Singh, B. (2010). *Plant availability of arsenic and cadmium as influenced by biochar application to soil, 19th world congress of soil science*.
Nigam, P., Armour, G., Banat, I., Singh, D., & Marchant, R. (2000). Physical removal of textile dyes from effluents and solid-state fermentation of dye-adsorbed agricultural residues. *Bioresource Technology, 72*, 219–226.
Novak, J. M., Lima, I., Xing, B., Gaskin, J. W., Steiner, C., Das, K., Ahmedna, M., Rehrah, D., Watts, D. W., & Busscher, W. J. (2009). Characterization of designer biochar produced at different temperatures and their effects on a loamy sand. *Annals of Environmental Science, 3*, 195–206.

Olafisoye, O. B., Adefioye, T., & Osibote, O. A. (2013). Heavy metals contamination of water, soil, and plants around an electronic waste dumpsite. *Polish Journal of Environmental Studies, 22*, 1431–1439.

Oleszczuk, P., Hale, S. E., Lehmann, J., & Cornelissen, G. (2012). Activated carbon and biochar amendments decrease pore-water concentrations of polycyclic aromatic hydrocarbons (PAHs) in sewage sludge. *Bioresource Technology, 111*, 84–91.

Park, J. H., Choppala, G. K., Bolan, N. S., Chung, J. W., & Chuasavathi, T. (2011). Biochar reduces the bioavailability and phytotoxicity of heavy metals. *Plant and Soil, 348*, 439.

Pignatello, J. J., Kwon, S., & Lu, Y. (2006). Effect of natural organic substances on the surface and adsorptive properties of environmental black carbon (char): Attenuation of surface activity by humic and fulvic acids. *Environmental Science & Technology, 40*, 7757–7763.

Pratt, K., & Moran, D. (2010). Evaluating the cost-effectiveness of global biochar mitigation potential. *Biomass and Bioenergy, 34*, 1149–1158.

Qiu, Y., Zheng, Z., Zhou, Z., & Sheng, G. D. (2009). Effectiveness and mechanisms of dye adsorption on a straw-based biochar. *Bioresource Technology, 100*, 5348–5351.

Reddy, K. R., Yargicoglu, E. N., Yue, D., & Yaghoubi, P. (2014). Enhanced microbial methane oxidation in landfill cover soil amended with biochar. *Journal of Geotechnical and Geoenviromental Engineering, 140*, 04014047.

Robinson, B. H. (2009). E-waste: An assessment of global production and environmental impacts. *Science of the Total Environment, 408*, 183–191.

Schnoor, J. L., Light, L. A., McCutcheon, S. C., Wolfe, N. L., & Carreia, L. H. (1995). Phytoremediation of organic and nutrient contaminants. *Environmental Science & Technology, 29*, 318A–323A.

Serio, M. A., Chen, Y., Wójtowicz, M., & Suuberg, E. (2000). Pyrolysis processing of mixed solid waste streams. *ACS Division of Fuel Chemistry Preprints, 45*, 466–474.

Shehzad, A., Bashir, M. J., Sethupathi, S., & Lim, J. W. (2016). An insight into the remediation of highly contaminated landfill leachate using sea mango based activated bio-char: Optimization, isothermal and kinetic studies. *Desalination of Water Treatment, 57*, 22244–22257.

Shih, Y.-H., SJe, K., Tseng, C.-H., Wang, H.-C., Chen, L.-L., & Chang, Y.-M. (2016). Health risks and economic costs of exposure to PCDD/Fs from open burning: A case study in Nairobi, Kenya. *Air Quality, Atmosphere and Health, 9*, 201–211.

Someya, M., Ohtake, M., Kunisue, T., Subramanian, A., Takahashi, S., Chakraborty, P., Ramachandran, R., & Tanabe, S. (2010). Persistent organic pollutants in breast milk of mothers residing around an open dumping site in Kolkata, India: Specific dioxin-like PCB levels and fish as a potential source. *Environment International, 36*, 27–35.

Song, Y., Wang, F., Bian, Y., Kengara, F. O., Jia, M., Xie, Z., & Jiang, X. (2012a). Bioavailability assessment of hexachlorobenzene in soil as affected by wheat straw biochar. *Journal of Hazardous Materials, 217*, 391–397.

Song, Y., Wang, F., Yang, X., Bian, Y., Gu, C., Xie, Z., & Jiang, X. (2012b). Influence and assessment of biochar on the bioavailability of chlorobenzenes in soil. *Huan Jing Ke Xue, 33*, 169–174.

Spokas, K. A., & Reicosky, D. C. (2009). Impacts of sixteen different biochars on soil greenhouse gas production. *Annals of Environmental Science, 3*, 179–193.

Sun, K., Gao, B., Ro, K. S., Novak, J. M., Wang, Z., Herbert, S., & Xing, B. (2012). Assessment of herbicide sorption by biochars and organic matter associated with soil and sediment. *Environmental Pollution, 163*, 167–173.

Sun, Y., Gao, B., Yao, Y., Fang, J., Zhang, M., Zhou, Y., Chen, H., & Yang, L. (2014). Effects of feedstock type, production method, and pyrolysis temperature on biochar and hydrochar properties. *Chemical Engineering Journal, 240*, 574–578.

Suppadit, T., Phumkokrak, N., & Poungsuk, P. (2012). The effect of using quail litter biochar on soybean (*Glycine max* [L.] Merr.) production. *Chilean Journal of Agricultural Research, 72*, 244.

Tan, B. H., Teng, T. T., & Omar, A. M. (2000). Removal of dyes and industrial dye wastes by magnesium chloride. *Water Research, 34*, 597–601.

Tang, J., Zhu, W., Kookana, R., & Katayama, A. (2013). Characteristics of biochar and its application in remediation of contaminated soil. *Journal of Bioscience and Bioengineering, 116*, 653–659.

Tao, S., Xu, F., Wang, X., Liu, W., Gong, Z., Fang, J., Zhu, L., & Luo, Y. (2005). Organochlorine pesticides in agricultural soil and vegetables from Tianjin, China. *Environmental Science & Technology, 39*, 2494–2499.

Thanh, N. P., & Matsui, Y. (2011). Municipal solid waste management in Vietnam: Status and the strategic actions. *International Journal of Environmental Research, 5*, 285–296.

Thies, J. E., & Rillig, M. C. (2009). Characteristics of biochar: Biological properties. In *Biochar for environmental management: Science and Technology* (pp. 85–105). London: Earthscan.

Thitame, S. N., Pondhe, G., & Meshram, D. (2010). Characterisation and composition of municipal solid waste (MSW) generated in Sangamner city, district Ahmednagar, Maharashtra, India. *Environmental Monitoring and Assessment, 170*, 1–5.

Tong, X.-j., Li, J.-y., Yuan, J.-h., & R-k, X. (2011). Adsorption of Cu (II) by biochars generated from three crop straws. *Chemical Engineering Journal, 172*, 828–834.

Ubbelohde, A. R., & Lewis, F. A. (1960). *Graphite and its crystal compounds*. Oxford: Clarendon Press.

Uchimiya, M., Lima, I. M., Thomas Klasson, K., Chang, S., Wartelle, L. H., & Rodgers, J. E. (2010). Immobilization of heavy metal ions (CuII, CdII, NiII, and PbII) by broiler litter-derived biochars in water and soil. *Journal of Agricultural and Food Chemistry, 58*, 5538–5544.

Wang, H., Lin, K., Hou, Z., Richardson, B., & Gan, J. (2010). Sorption of the herbicide terbuthylazine in two New Zealand forest soils amended with biosolids and biochars. *Journal of Soils and Sediments, 10*, 283–289.

Wang, S., Gao, B., & Li, Y. (2016). Enhanced arsenic removal by biochar modified with nickel (Ni) and manganese (Mn) oxyhydroxides. *Journal of Industrial and Engineering Chemistry, 37*, 361–365.

Willmott, N., Guthrie, J., & Nelson, G. (1998). The biotechnology approach to colour removal from textile effluent. *Coloration Technology, 114*, 38–41.

World Bank. (2012). https://siteresources.worldbank.org/INTURBANDEVELOPMENT/Resources/336387-1334852610766/What_a_Waste2012_Final.pdf

Xu, T., Lou, L., Luo, L., Cao, R., Duan, D., & Chen, Y. (2012). Effect of bamboo biochar on pentachlorophenol leachability and bioavailability in agricultural soil. *Science of the Total Environment, 414*, 727–731.

Yang, R.-q., Jiang, G.-b., Zhou, Q.-f., Yuan, C.-g., & Shi, J.-b. (2005). Occurrence and distribution of organochlorine pesticides (HCH and DDT) in sediments collected from East China Sea. *Environment International, 31*, 799–804.

Yang, X.-B., Ying, G.-G., Peng, P.-A., Wang, L., Zhao, J.-L., Zhang, L.-J., Yuan, P., & He, H.-P. (2010). Influence of biochars on plant uptake and dissipation of two pesticides in an agricultural soil. *Journal of Agricultural and Food Chemistry, 58*, 7915–7921.

Yip, K., Wu, H., & D-k, Z. (2007). Effect of inherent moisture in collie coal during pyrolysis due to in-situ steam gasification. *Energy & Fuels, 21*, 2883–2891.

Yu, X.-Y., Ying, G.-G., & Kookana, R. S. (2006). Sorption and desorption behaviors of diuron in soils amended with charcoal. *Journal of Agricultural and Food Chemistry, 54*, 8545–8550.

Yu, X.-Y., Ying, G.-G., & Kookana, R. S. (2009). Reduced plant uptake of pesticides with biochar additions to soil. *Chemosphere, 76*, 665–671.

Yu, X., Wang, D., Mu, C., & Liu, X. (2011). Role of biochar in slow sorption and desorption of diuron in soil. *Jiangsu Academy of Agricultural Sciences, 27*, 1011–1015.

Zhang, H., Lin, K., Wang, H., & Gan, J. (2010). Effect of Pinus radiata derived biochars on soil sorption and desorption of phenanthrene. *Environmental Pollution, 158*, 2821–2825.

Zhang, X., Wang, H., He, L., Lu, K., Sarmah, A., Li, J., Bolan, N. S., Pei, J., & Huang, H. (2013). Using biochar for remediation of soils contaminated with heavy metals and organic pollutants. *Environmental Science and Pollution Research, 20*, 8472–8483.

Zhao, L., Cao, X., Mašek, O., & Zimmerman, A. (2013). Heterogeneity of biochar properties as a function of feedstock sources and production temperatures. *Journal of Hazardous Materials, 256*, 1–9.

Zheng, W., Guo, M., Chow, T., Bennett, D. N., & Rajagopalan, N. (2010). Sorption properties of greenwaste biochar for two triazine pesticides. *Journal of Hazardous Materials, 181*, 121–126.

Zhou, J., Deng, C., Chen, J., & Zhang, Q. (2008). Remediation effects of cotton stalk carbon on cadmium (Cd) contaminated soil. *Ecology and Environment, 17*, 1857–1860.

Chapter 2
Scope of Nanoparticles in Environmental Toxicant Remediation

Anupam Dhasmana, Swati Uniyal, Vivek Kumar, Sanjay Gupta, Kavindra Kumar Kesari, Shafiul Haque, Mohtashim Lohani, and Jaya Pandey

Contents

Abstract The requirement and need of novel techniques to speed up the sanitization of polluted and adulterated sites and reduction in the expenses of these methods is a growing concern. The application of nanoparticles, predominantly the iron nanoparticles, as a pioneering and inventive technique to decontaminate the adulterated sites has received attention and consideration in recent times. Though, all over the world, many research studies have been carried on nanoparticles, diminutive level of knowledge is realized about their performance, actions, and

A. Dhasmana (✉) · V. Kumar · S. Gupta
Himalayan School of Biosciences, Swami Rama Himalayan University, Dehradun, Uttarakhand, India
e-mail: anudhas007@gmail.com

S. Uniyal
School of Biotechnology, Gautam Buddh University, Greater Noida, Uttar Pradesh, India

K. K. Kesari
Department of Applied Physics, Aalto University, Espoo, Finland

S. Haque · M. Lohani
Research and Scientific Unit, College of Nursing & Allied Health Sciences, Jazan University, Jazan, Kingdom of Saudi Arabia

J. Pandey
Amity University Uttar Pradesh Campus, Lucknow, Uttar Pradesh, India

R. C. Sobti et al. (eds.), *Environmental Biotechnology: For Sustainable Future*, https://doi.org/10.1007/978-981-10-7284-0_2

conduct in the soil and in aquatic habitats, their adsorption on soil mineral particles, and communication with soil microbes.
Industrial sectors, which are involved in the manufacturing of display, optical and photonic products, semiconductors, memory and storage devices, nano-biotechnology equipment (energy aspects), and health care goods, generate most of the products that contain nanoparticles. On the other hand, nanotechnology is a technique which has been employed as an environmental know-how to guard the nature through prevention, handling, curing, and cleanup of pollution. In this chapter, we have focused on the environmental toxicant cleanup and discuss a background and overview of the existing practices related to the remediation. Thc research findings; social issues; probable environmental, human health, and safety repercussions; and future thoughts for remediation using nanotechnology are also discussed. Here, we also discuss nanoscale zerovalent iron in some detail. The technique of nanoremediation has the capacity to lessen the total costs of decontamination at the bigger polluted sites. Moreover, the purpose of this technique is also to reduce the cleanup duration, eradicate the treatment requirement and dumping of the contaminated soil, and also lessen contaminant amount to almost zero. Further, we believe that suitable evaluation of nanoremediation approaches, especially the large-scale environmental studies, also need to be addressed to avoid and counteract any probable hostile environmental effects.

Keywords Nanoparticles · Remediation · Toxicants · Environment

Introduction

Nanotechnology actually means the exploitation of the substances at their nanometer size and is expected to enhance the quality of life and economic development on the global basis. A decade ago, nanoparticles (NPs) were studied because of their size-dependent physical and chemical properties, but now they have crossed the threshold of commercial exploration. Understanding of biological processes on the nanoscale level is a strong driving force behind development of nanotechnology. Nanotechnology is being employed in imaging, quantifying, modeling, and manipulating matter at nanoscale. Nanoparticle application in commercial sectors involves mostly in the area of semiconductors, nano-memory and storage chip technologies, optical display, and photonic technologies. This technology is also used in electricity, biological sciences, diagnostics, and therapeutic health care. On the other side, the application of nanotechnology in environmental sciences to shield the environment from toxicants and treatment and cleanup of hazardous waste dumping sites is coming up. Nanotechnology could be an effective replacement of current practices for waste remediation. In recently published studies, some of the NPs have been employed in scavenging the high molecular weight polycyclic aromatic hydrocarbons (PAHs) from the contaminated soils and water. Considering this, the review discusses the safeguard properties of NPs against the deleterious effects of toxicants in environmental systems.

Under the American Recuperation and Reinvestment Act (2009), almost $ 1 billion have been granted to the United States Environment Protection Agency (USEPA) for remediation projects. Nanotechnology is the rising technology of current era, which could also be used as an affordable environmental remediation method. NP-associated technology can contribute and may prove as cost-effective and successful remediation strategy to clean up the contaminated sites. In year 2007, the US government allocated $380 million fund to Superfund, for construction and post construction actions for site remediation projects (US EPA 2008). Superfund is a US federal government funding agency, intended to offer the financial grants for cleanup of areas contaminated with hazardous wastes and pollutants. This program was introduced as the legislative act to protect the environment under the Comprehensive Environmental Response, Compensation, and Liability Act of 1980 (CERCLA).

So many other programs were established under Superfund program:

(a) Brownfields Program [under the Small Business Liability Relief and Brownfields Revitalization Act (2002)] and fund allocated was $100 million.
(b) Corrective Action (CA) programs under Subtitle C of the Resource, Conservation, and Recovery Act (RCRA 2002).
(c) Underground Storage Tank program under Subtitle I of the RCRA and fund allocated was $200 million.

USEPA (American Recovery and Reinvestment Act 2009) allocated $600 million for the Superfund remedial. USEPA (2004) estimated that it will take 30–35 years and cost up to $250 billion to clean up the nation's hazardous waste sites. There are other important and major remediation projects such as the remediation of Homebush Bay (New South Wales, Australia), one of the important projects because the pollution directly impacted and affected the food chain as well as local protected and threatened species; Japan-Australia Migratory Bird Agreement (JAMBA) and China-Australia Migratory Bird Agreement (CAMBA) for protected species and the ones which use other Ramsar-protected wetlands in various countries as per Ramsar convention; and Bakar Ex Cokeing Plant Site, Croatia project, European Union contract for cleaning a polluted area of BAKAR. After 3 years of rigorous investigation by the Croatian government, the European Union funded the immobilization project in BAKAR. The site is contaminated with large amounts of total petroleum hydrocarbons, polycyclic aromatic hydrocarbons, and metals. For the cleaning of pollutants, the service provider preferred to apply the mix-in-plant procedure (Operational ProgramME Environment, Ministry of Environmental and Nature Protection, Republic of Croatia 2007–2013).

Nanoremediation

Nanoremediation is an upcoming application; by 2009, nanoremediation technique was documented in almost 44 cleanup sites over the world, largely in the United States of America (Karn et al. 2009; Mueller et al. 2012; Project on Emerging Nanotechnologies. "Nanoremediation Map" 2013). In Europe, this process is being explored by the NanoRem project (Nanotechnology for Contaminated Land Remediation 2014). The objective of NanoRem (Nanotechnological Remediation) is using of nanotechnology for unlocking the potential of remediation process from laboratory level to end user cleanup site like land and water resource for the restoration of environment. This project is funded by the European Commission FP7. This project focuses on the smooth functioning of practical, secured, cost-effective, and usable nanotechnology for in situ remediation. This project helps to understand and develop a comprehensive knowledge of the environmental risk-benefit for the utilization of NPs, current market requirements, sustainability, and observation of users. A report released by the NanoRem consortium has recognized about 70 nanoremediation projects worldwide at pilot or full scale.

In current scenario highly reactive NPs/materials are used in the remediation and detoxification procedures of pollutants. These NPs have the potential to reduce and catalyze the concerned pollutants. In nanoremediation procedures no transportations are required for the treatment of water and soil from site of pollutant to site of treatment (Otto et al. 2008). NPs can be applied in the remediation of several contaminated sites by acting as "super adsorbent" for many pollutants. They also play a significant and deciding role in the development of speedy and accurate environmental sensors which can be applied in the recognition of pollutants at atomic levels and also for deactivating injurious microbes (Khan et al. 2014).

NPs have some specific properties, such as they are tiny in size, they can have innovative surface modifications, and moreover, NPs are able to permeate from very small pores in the subsurface and remain settled down in groundwater. Owing to these properties, NPs travel far and have wide distribution (Tratnyek and Johnson 2006). Various potential nanoparticles have been investigated for remediation processes, like nano-zeolites, metal oxides, carbon nanotubes, carbon fibers, enzymes, bimetallic nanoparticles, and ultrafine titanium dioxide (Ghorbani et al. 2011). Of these, nanoscale zerovalent iron (nZVI) is currently the most widely used. Since the last few years, calcium peroxide NPs are also being used for remediation. There are various physical and chemical methods mostly used for the synthesis of various NPs for remediation applications (Turkevich et al. 1953; Ghorbani et al. 2011). These production methods release toxic by-products; therefore, the utilization of these kinds of NPs, prepared by rigorous methods, creates added environmental issues and hence is not suitable for remediation. Green synthesis of NPs could be the alternative method for the synthesis of these particles which may be used in the remediation procedure as an eco-friendly approach toward the environment. This green synthesis can be performed by the use of plant extracts as well as from microbes like bacteria and fungi. However, this method of synthesis is basically limited to metal oxide NPs. Synthesis of NPs using this technique makes use of intra-/extracellular enzymes produced by the plants and microbes. The bulk metal, truncated to nano-range by organisms apart from using tools for cleaning up environmental resources, is also intensively used in biosensors, bioprobes, and other biosystems. Metal NPs can be designed of desired shape and size and obtained by simply using bacteria to most complicated of eukaryotes (Prashant et al. 2008).

Applications of Nanoremediations

In recently published studies, NPs have been employed in scavenging the high molecular weight polycyclic aromatic hydrocarbons (PAHs) from the contaminated soils (Karnchanasest and Santisukkasaeom 2007). Amphiphilic polymer NPs have been used as nano-absorbent for pollutants in aqueous phase. In a recent report, Bochra et al. (2011) have shown the potential application of titanate nanotubes as solid-phase extraction adsorbents for seven PAHs: acenaphthylene, acenaphthene, anthracene, fluorene, phenanthrene, fluoranthene, and pyrene, from the environmental water samples. The recoveries of PAHs ranged from 90% to 100%. The scavenging capacities of the NPs for PAH and other pollutants could probably be attributed to their higher affinity toward the xenobiotics (Karnchanasest and Santisukkasaem 2007). Recently, titanate nanosheets and titanate nanotubes (TNT) have also been synthesized and used as additives for removing harmful compounds in cigarette smoke (Qixin et al. 2011) including nicotine, tar, ammonia, hydrogen cyanide, selected carbonyls, and phenolic compounds. Interestingly, TNT exhibits highly efficient adsorption capability for most of the harmful compounds. This might be related to the intrinsic properties of NPs (Qixin et al. 2011).

Groundwater remediation is the most popular and common application of nano-based remediation (Bardos et al. 2014; US EPA 2014). Zerovalent metals (ZVMs) are an emerging and promising concept for decontamination of groundwater (Lowry 2007). Nanoremediation process is capable to treat contaminates by passing the excavation or need to pump out the groundwater. This method starts with injecting slurry of NPs along the vertical range of the probe to provide treatment to contaminated groundwater through injection well. NPs are transported to the site of action, where they scavenge the contaminants through absorption process, by immobilizing them, and by converting the harmful contaminates into less harmful compounds. The conversion of contaminants is basically by redox reaction process (Lowry 2007). The transportation of NPs is very important for the successful remediation of environmental toxicants. Direct push well technique is cost-effective than drilling and packing well technique (Pandey and Fulekar 2012). Carbon nanotubes and ultrafine TiO_2 exhibited potential role for treatment, purification, disinfection, and desalination of surface water. NPs act as sorbents, reactive agents (photocatalysts or redox agents), and nanofilters in membrane filtration (Theron et al. 2008).

NPs have the capacity to detect and remediate the trace levels of toxicants from the contaminated sites. Generally, portable kits and instruments are not sensitive enough to detect trace contaminants outside of the laboratory. NPs are highly reactive and, having large surface area, can be utilized as effective sorbents to help remediate the target trace contaminants. This they do by approaching solid-phase micro-extraction process, particularly in the form of self-assembled monolayers on mesoporous matrix. This matrix may be an effective sorbent for heavy/trace metals such as mercury (Hg), lead (Pb), cadmium (Cd), chromium (Cr) and arsenic (As), and radionuclides such as ^{99}Tc, ^{137}CS, uranium, actinides, metal ions, and organic and inorganic compounds to the ppb levels (Addleman et al. 2005; Savage and Diallo 2005). Magnetite nanoparticles (Fe_3O_4) saturated with silica are applied for the removal of a huge amount of toxic elements. The nanostructured silica can also be used in wastewater treatment in order to remove heavy metal ions (Xin et al. 2001). Mechanism of action of remediation, example of NPs used, and contaminates are shown in Table 2.1.

Few research articles demonstrated the remediation properties of TiO_2 NPs in biological systems also. These NPs scavenge the toxicants or carcinogens from the biological systems. TiO_2 NPs have been used to reduce the neurotoxicity of phoxim in the brain of silkworm (Xie et al. 2014). TiO_2 NPs provide protection in cellular model A-549 cells against benzo-alpha-pyrene. Authors used very low or subtoxic concentrations of NPs in cellular system to protect the cells from carcinogens (Dhasmana et al. 2014, 2015).

Nanoparticles Used in Remediation

Metal-Based Nanoparticles

Iron Nanoparticles

Iron (Fe) NPs are extensively applied in the chemical, electronics, and other industries. Recent applications of nano-Fe have been extended to the treatment of toxic and hazardous wastes and for remediation of soil and wastewaters, degradation of dyes, and reduction of aromatic nitro compounds and also dehalogenate organic compounds along with removing metal ions. Nano-iron (nano-Fe) is a striking module of remediation. Nano-Fe usually is synthesized from Fe^{2+} (II) and Fe^{3+} (III), using borohydride (anion BH_4^-) as the reductant. Size range of nanoscale zerovalent (nZVI) nano-Fe particles is from 10 to 100 nm in diameter with typical core shell-like structure. The core consists primarily of nZVI or metal iron, whereas the mixed valent [i.e., Fe^{2+} (II) and Fe^{3+} (III)] oxide shell is formed as a result of oxidation of the Fe. Fe is naturally present in the environment as Fe^{2+} (II) and Fe^{3+} (III) oxides (Li et al. 2006). nZVI are usually ideal for nanoremediation, because of their large surface area and various sites of action which increases the reactivity of NPs (Tratnyek and Johnson 2006) and also holds two kinds of properties: adsorption and reduction (Fig. 2.1). This property provides strength to be used for the in situ remediation of a large range of contaminants. In addition, nZVI nano-Fe efficiently

Table 2.1 Mode of action of NPs for remediation of contaminates

Technique	Procedure	Implications	Disadvantages	Target metals
Precipitation method	Reduction and precipitation of metals, generation of Fe and other metal precipitation	Performance is similar to the natural procedure	Corrosion and clogging of zerovalent irons	Copper, zinc, lead, calcium, nickel, chromium, magnesium, cobalt, arsenic
Membrane filtration technique	Use of 3D structure in arresting of micelle slow electrical charges, complexation, dialysis	High removal efficacy	Filter blockage, restitution of filter materials	Technetium, mercury, copper, lead, chromium, zinc
Denitrification and BSR	Sulfide formation from divalent metals and (OH) hydroxides from trivalent metals	PRBs removal up to 95%	Continuous nutrient supply is required	Copper, zinc, lead, calcium, nickel, magnesium, cobalt, uranium, selenium, arsenic
Absorption method:				
1. Inorganic surfactants	Metal sorption depends upon charge of the surfactant	Complex formation along with surfactants	Aquifer with highest penetrability is needed	Cadmium, lead, zinc, arsenic, copper, nickel
2. Industrial by products (derivatives)	Adsorption at surface site	Industrial raw materials	Field application is needed	Lead, arsenic, cadmium
3. Ferrous/ferric materials	Metal sorption of iron oxide and its derivatives	As(V) with Fe form inner sphere complex	Oxidation process is problematic and materials to be replaced recurrently	Arsenic, chromium, mercury, copper, cadmium, lead
BSR technique	Materials reduced to precipitates, catalyzed by the sulfate-reducing bacteria method	On-site treatment and off-site use of bioreactors and applied in permeable reactive barriers	Limited reaction rate and residence time are needed	Divalent metal cations
Reduction:				
1. Use of dithionites	Precipitation of metals at alkaline pH	Operational over a larger part	Toxic gas formation and handling are problematic	Thorium, uranium, chromium
2. Use of zerovalent iron and colloidal Fe	Precipitation and sorption on zerovalent iron	No toxic exposure in the deep aquifers	No proper scavenging of metals	Chromium, technetium, uranium, arsenic

(continued)

Table 2.1 (continued)

Technique	Procedure	Implications	Disadvantages	Target metals
Flushing of chelate	Stable complexes formation	Ligands action at low dose and regeneration	Expensive, long-lasting, and toxic	Cadmium, copper, iron, arsenic, mercury, zinc, chromium
Biological approach: Subsurface activity	Bioaccumulation, oxidation, and precipitation	Low cost and effective for long duration	Not suitable for aquifers, technique is sluggish, modeling impossible	Lead, cobalt, copper, nickel, chromium, cadmium, zinc
Ion exchange technique	Liquid-liquid extraction and solid-phase separation	Selective in eradicating low level of metals	Not cost-effective and leads to contaminant	Transition and heavy metals
Agricultural wastes and cellulosic materials	Heavy metals adsorbed at pH 4–6 in cellulosic material and agricultural waste	Metals' high concentrations are treated by cellulosic materials	Field studies have not been conducted, therefore doubtful	Zinc, cadmium, copper, nickel, lead
Bio-sorption method	Bacteria, fungi, plants, recover metals from the cytoplasm. Use of DNA has been implicated	Good metal absorption, anionic interference is not present	Owing to high acidic environmental conditions, desorption of metals happens	Cadmium, zinc, arsenic, chromium, nickel, iron

From Kumar and Gopinath (2016)

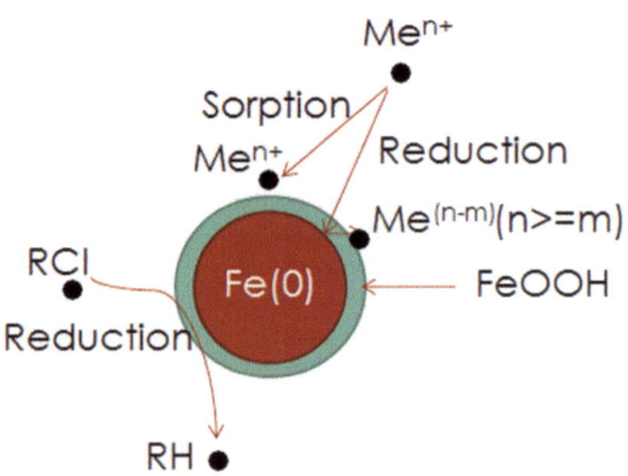

Fig. 2.1 Schematic diagram of zerovalent Fe. (From Rajan 2011)

remediates the contaminants by releasing less amount of hazardous chemicals (Varma and Nadagouda 2009). nZVI nano-Fe can also be customized by using palladium coatings such as polyelectrolyte or triblock polymers according to contaminants present (Saleh et al. 2007). It has been studied that nZVI-Fe very effectively scavenges Cr from soil with reduction of Cr (VI) (Singh et al. 2011). Fe NPs by using super paramagnetic property can be manipulated and able to remove 99% of As from water samples by using 12 nm diameter iron oxide NPs (Rickerby and Morrison 2007). Nano-Fe and Ni are capable to decontaminate the radioactive sites also. Anionic species of uranium (U) sorption are very effective with magnetite NPs in neutral solutions. The amount of Fe^{3+} (III) contents is increased and U (Vl) reduction occurs. Magnetite NPs are very effective in removal of U from the pollutants (Dickinson and Scott 2011; Das et al. 2010).

TiO_2 Nanoparticles

TiO_2 NPs associated with photocatalytic degradation are actively used for the removal of polychlorinated biphenyl (PCB) from contaminated soil samples (Zhu et al. 2012). TiO MCM (mesoporous molecular sieving) has been studied for the photolytic degradation of bisphenol A (2,2-bis(4-hydroxyphenyl)propane, BPA an endocrine disrupting compound) (Tao et al. 2011). Truncation of bulk TiO_2 leads to formation of nano-TiO_2 and nano-TiO_2 (TiO_2-x) holding oxygen vacant sites which have been created using a plasma discharge truncation procedure, and these vacant sites involve in mercury (Hg) removal. Moisture is the competitive inhibitor for the binding of Hg to the oxygen vacant sites of nano-TiO_2 (Tsai et al. 2011). The adsorption capacity of nano-TiO_2 toward contaminant metals like Pb, Cu, Cd, Ni, and Zn is greater than bulk particles (Engates and Shipley 2011).

SiO_2 Nanoparticles

SiO_2 NPs treated with 2,6-pyridine dicarboxylic acid and applied as a sorbent successfully for the remediation of Hg^{2+} ions present in trace levels, from aqueous solutions (Patil et al. 2016). SiO_2 NPs exhibited positive features such as rapid absorption equilibrium, easy elution, good reusability, and increased stability with respect to Hg^{2+} ions as a sorbent (Zhang et al. 2010).

Nano-zirconium Phosphate (ZrP)

Immobilized nano-zirconium phosphate (ZrP) with surface modification groups like -CH_2Cl, -SO^{3-}, and -CH_2N (CH_3)3 is actively used for the removal of lead (Pb) from the water (Silbernage et al. 2014). These fabricated NPs have positively or negatively charged functional groups which can remove different contaminants from the site of action (Zhang et al. 2011).

Carbon-Based Nanoparticles

Recently, nanotechnology has introduced various types of NPs in the water industry with encouraging results. Since the discovery of carbon nanotubes, these have attracted prodigious considerations owing to the exceptional qualities. The carbon nanotubes (CNTs) are NPs that are shaped into a tube and are categorized into two types as single-walled carbon nanotubes (SWCNTs) and multi-walled carbon nanotubes (MWCNTs). This categorization is owing to the layers of carbon atoms in the walls of the nanotubes (Iijima and Ichihashi 1993). CNT have been employed in decontamination of ecosystems. Owing to these properties of detoxification, the nanosorbents are progressively being used since their discovery, due to their extraordinary adsorption potential and their capability to be attached to functional groups to enhance the affinity to the target molecules (Savage and Diallo 2005). Hexagonal arrays of carbon molecules in graphite sheets of CNT surface have a robust communication with other particles or atoms, which makes them an assuring adsorbent material which has replaced the activated carbon in many ways (Liang et al., 2004). These nanotubes are being used for the exclusion of heavy metals like Cr, Pb, and Zn, metalloids such as As compounds (Peng et al. 2005), organic compounds, and natural biological impurities (Savage and Diallo 2005), eliminating several types of inorganic and organic contaminants such as dioxin (Long and Yang 2001) and several volatile organic compounds (VOCs) (Agnihotri et al. 2005). While comparing CNT with other active adsorbents, the scientists proclaim that these are more effective adsorbents for removing pollutants and can be employed in several environmental applications (Rao et al. 2007). The plus point of CNT is that they are exceptional, matchless, and unidimensional macromolecules that retain high thermal and chemical stability (Smart et al. 2006). Owing to this property of the nanomaterials, they have been maneuvered for the management of natural organic matter (NOM) which could produce toxic, carcinogenic agents (Grünwald et al. 2002) and may augment the biofilm formation and bacterial regrowth (Bull et al. 1995). Consequently, thermally handled CNT was used for the management of NOM to attain the applicable absorption. Therefore, CNT possesses all the indispensable and crucial properties to maintain purity of water of high excellence. For the remediation of groundwater, there are certain rigorous regulation approaches. Any augmentation in the release of heavy metals into the aquatic water bodies could prove toxic as they have the capabilities to accumulate in the living tissues of animals and plants. Henceforth, the carbon-based NPs should not be introduced into the water bodies; if introduced, these NPs must be removed from the water, because later they may lead to some aqua toxicity. The remediation efficiency has been achieved by employing multi-walled carbon nanotube (MWCNT) techniques. To enhance the absorption capability of MWCNTs, with the help of nitric oxide, it is oxidized which results into a higher level of adsorption potential. According to Li et al. (2003), the sorption of Pb (II), Cu (II), and Cd (II) onto the MWCNTs was three to four times greater than those of activated powdered carbon and activated granular carbon particles which are the two traditionally and commonly employed sorbents in the water purification methods.

Future Prospectus

Nanotechnology is now playing an important role of maintaining environmental sustainability and also has the potential to enhance and improve the conventional methods. The novel nano-based technologies have the capacity to replace the conventional technologies being much more efficient and cost-effective than conventional methods. Metal- and carbon-based NPs, nanofibers, nanoresins, and nanoenzyme-based portable equipments will be used for the purification and detoxification of the hazardous contaminates from soil, toxic gases such as CO and VOCs from air, and trace heavy metals from water and groundwater. In the future, these technologies will be frequently used as dependable and highly sensitive sensors for toxic and particular substances that are difficult to detect with conventional methods. This kind of nano-based application will be used commonly in the future because NPs are very small in size and highly reactive due to very large surface volume ratio.

Conclusion

Nanoremediation is a cost-effective process of cleaning up large-scale contaminated sites, reduces cleanup time, and trims down the concentrations of pollutants. In order to prevent any serious adverse effect on environment, proper evaluation, including minute studies, of NPs is required before this technique is used on a mass scale. The success of the NP-based remediation technique in field conditions depends upon interdisciplinary factors. The good knowledge of chemistry, material science, atmosphere, and geology is one of the important requirements to face the challenge.

References

Addleman, R. S., Egorov, O. B., O'Hara, M., Zemaninan, T. S., Fryxell, G., & Kuenzi, D. (2005). Nanostructured sorbents for solid phase microextraction and environmental assay. In B. Karn, T. Masciangioli, W. Zhang, V. Colvin, & P. Alivisatos (Eds.), *Nanotechnology and the environment: Applications and implications* (pp. 186–199). Washington, DC: Oxford University Press.

Agnihotri, S., Rood, M. J., & Rostam-Abadi, M. (2005). Adsorption equilibrium of organic vapors on single-walled carbon nanotubes. *Carbon, 43*, 2379–2388.

Bardos, P., Bone, B., Daly, P., Elliott, D., Jones, S., Lowry, G., & Merly, C. (2014). *A risk/benefit appraisal for the application of nano-scale Zero Valent Iron (nZVI) for the remediation of contaminated sites*. WP9 NanoRem. www.nanorem.eu

Bochra, B. K., Latifa, L. E. A., Hafedh, K., & Abdelhamid, G. (2011). TiO_2 nanotubes as solid-phase extraction adsorbent for the determination of polycyclic aromatic hydrocarbons in environmental water samples. *Journal of Environmental Sciences, 23*(5), 860–867.

Bull, R. J., Brinbaum, L. S., Cantor, K. P., Rose, J. B., Butterworth, B. E., Pegram, R., & Tuomisto, J. (1995). Water chlorination: Essential process and cancer hazard. *Fundamental and Applied Toxicology, 28*(1), 155–166.

Das, D., Suresh Kumar, M. K., Koley, S., Mithal, N., & Pillai, C. G. S. (2010). Sorption of uranium on magnetite nanoparticles. *Journal of Radioanalytical and Nuclear Chemistry, 285*, 447–454.

Dhasmana, A., Jamal, M. Q. S., Mir, S. S., Bhatt, M. L. B., Rahman, Q., Gupta, R., Siddiqui, M. H., & Lohani, M. (2014). Titanium dioxide nanoparticles as guardian against environmental carcinogen Benzo [alpha] Pyrene. *PLoS One, 9*(9), e107068.

Dhasmana, A., Mohd, Q., Jamal, S., Gupta, R., Siddiqui, M. H., Kesari, K. K., Wadhwa, G., Khan, S., Lohani, M. (2015). *Titanium dioxide nanoparticles provide protection against polycyclic aromatic hydrocarbon BaP & Chrysene induced perturbation of DNA repair machinery: A Computational Biol Approach. Biotechnology and Applied Biochemistry* (pp. 187–199). Wiley Publication.

Dickinson, M., & Scott, T. B. (2011). The application of zero-valent iron nanoparticles for the remediation of a uranium-contaminated waste effluent. *Journal of Nanoparticle Research, 13*, 3699–3711.

Engates, K. E., & Shipley, H. J. (2011). Adsorption of Pb, Cd, Cu, Zn, and Ni to titanium dioxide nanoparticles: Effect of particle size, solid concentration, and exhaustion. *Environmental Science and Pollution Research, 18*, 386–395.

Ghorbani, H. R., Safekordi, A. A., Attar, H., & Rezayat Sorkhabadi, S. M. (2011). Biological and non-biological methods for silver nanoparticles synthesis. *Chemical and Biochemical Engineering Quarterly, 25*, 317–326.

Grünwald, A., Šť astný, B., Slavíčková, K., & Slavíček, M. (2002). Formation of halo forms during chlorination of natural waters. *Acta Polytechnica, 42*(2), 56–59.

Iijima, S., & Ichihashi, T. (1993). Single shell carbon nanotube of diameter 1 nm. *Nature, 363*, 603–605.

Karn, B., Todd, K., & Martha, O. (2009). Nanotechnology and in situ remediation: A review of the benefits and potential risks. *Environmental Health Perspectives, 117*(12), 1823–1831.

Karnchanasest, B., & Santisukkasaem, O. (2007). A preliminary study for removing Phenanthrene & Benzo(a)Pyrene from soil by nanoparticles. *Journal of Applied Sciences, 7*(21), 3317–3321.

Khan, I., Farhan, M., Singh, P., & Thiagarajan, P. (2014). Nanotechnology for environmental remediation. *Research Journal of Pharmaceutical, Biological and Chemical Sciences, 5*(3), 1916–1927.

Kumar, S. R., & Gopinath, P. (2016). Nano-bioremediation, applications of nanotechnology for bioremediation. In L. K. Wang, M. H. S. Wang, Y. T. Hung, N. K. Shammas, & J. P. Chen (Eds.), *Remediation of heavy metals in the environment* (pp. 46–58). Boca Raton: Taylor & Francis.

Li, Y. H., Dinga, J., Luanb, Z., Dia, Z., Zhua, Y., Xua, C., Wu, D., & Wei, B. (2003). Competitive adsorption of Pb^{2+}, Cu^{2+} and Cd^{2+} ions from aqueous solutions by multiwalled carbon nanotubes. *Carbon, 41*(14), 2787–2792.

Li, X. Q., Elliott, D. W., & Zhang, W. (2006). Zero-valent iron nanoparticles for abatement of environmental pollutants: Materials and engineering aspects. *Critical Reviews in Solid State and Materials Sciences, 31*, 111–122.

Liang, P., Liu, Y., Guo, L., Zeng, J., & Pei, H. L. (2004). Multiwalled carbon nanotubes as solid-phase extraction adsorbent for the pre concentration of trace metal ions and their determination by inductively coupled plasma atomic emission spectrometry. *Journal of Analytical Atomic Spectrometry, 19*, 1489–1492.

Long, R. Q., & Yang, R. T. (2001). Carbon nanotubes as superior sorbent for dioxin removal. *Journal of the American Chemical Society, 123*(9), 2058–2059.

Lowry, G. V. (2007). Nanomaterials for groundwater remediation. In M. R. Wiesner & J. Bottero (Eds.), *Environmental nanotechnology* (pp. 297–336). New York: The McGraw-Hill Companies.

Mueller, N. C., Jürgen, B., Johannes, B., Miroslav, Č., Rissing, P., Rickerby, D., & Nowack, B. (2012). Application of nanoscale zero valent iron (NZVI) for groundwater remediation in Europe. *Environmental Science and Pollution Research, 19*(2), 550–558.

OPERATIONAL PROGRAMME ENVIRONMENT, Ministry of Environmental and Nature Protection, Republic of Croatia. 2007–2013.

Otto, M., Floyd, M., & Bajpai, S. (2008). Nanotechnology for site remediation. *Remediation, 19*(1), 99–108.

Pandey, B., & Fulekar, M. H. (2012). Nanotechnology: Remediation technologies to clean up the environmental pollutants. *Research Journal of Chemical Sciences, 2*(2), 90–96.

Patil, S. S., Shedbalkar, S. U., Truskewycz, A., Chopade, B. A., & Ball, A. S. (2016). Nanoparticles for environmental clean-up: A review of potential risks and emerging solutions. *Environmental Technology & Innovation, 5*, 10–21.

Peng, X., Luan, Z., Ding, J., Di, Z., Li, Y., & Tian, B. (2005). Ceria nanoparticles supported nanotubes for removal of arsenate in water. *Materials Letters, 59*, 399–403.

Prashant, M., Rana, N. K., & Yadav, S. K. (2008). Biosynthesis of nanoparticles: Technological concepts and future applications. *Journal of Nanoparticle Research, 10*, 507–517.

Qixin, D., Chaozhang, H., Wei, X., Jianping, Z., & Yiqiang, Z. (2011). Significant reduction of harmful compounds in tobacco smoke by the use of titanate nanosheets and nanotubes. *Chemical Communications, 47*, 6153–6155.

Rajan, C. S. (2011). Nanotechnology in groundwater remediation. *International Journal of Environmental Science and Technology, 2*(3), 47–53.

Rao, G. P., Lu, C., & Su, F. (2007). Sorption of divalent heavy metal ions from aqueous solutions by carbon nanotubes: A review. *Purification Technology, 58*(1), 224–231.

Rickerby, D. G., & Morrison, M. (2007). Nanotechnology and the environment: A European perspective. *Science and Technology of Advanced Materials, 8*, 19–24.

Saleh, N., Sirk, K., Liu, Y., & Phenrat, T. (2007). Surface modifications enhance nanoiron transport and NAPL targeting in saturated porous media. *Environmental Engineering Science, 24*(1), 45–57.

Savage, N., & Diallo, M. S. (2005). Nanomaterials and water purification: Opportunities and challenges. *Journal of Nanoparticle Research, 7*, 331–342.

Silbernage, R., Díaz, A., Steffensmeier, E., Clearfield, A., & Blüme, J. (2014). Wilkinson-type hydrogenation catalysts immobilized on zirconium phosphate nanoplatelets. *Journal of Molecular Catalysis A: Chemical, 394*(15), 217–223.

Singh, R., Misra, V., & Singh, R. P. (2011). Synthesis, characterization and role of zero-valent iron nanoparticle in removal of hexavalent chromium from chromium-spiked soil. *Journal of Nanoparticle Research, 13*, 4063–4073.

Smart, S. K., Cassidy, A. I., Lu, G. Q., & Martin, D. J. (2006). The biocompatibility of carbon nanotubes. *Carbon, 44*(6), 1034–1047.

Tao, H., Hao, S., Chang, F., Wang, L., Zhang, Y., Cai, X., & Zeng, J. S. D. (2011). Photodegradation of bisphenol a by Titana nanoparticles in mesoporous MCM-41. *Water, Air, & Soil Pollution, 214*, 491–498.

Theron, J. J. A., Walker, T. E., & Cloete, T. E. (2008). Nanotechnology and water treatment: Applications and emerging opportunities. *Critical Reviews in Microbiology, 34*(1), 43–69.

Tratnyek, P. G., & Johnson, R. L. (2006). Nanotechnologies for environmental cleanup. *Nanotoday, 1*(2), 44–48.

Tsai, C. Y., His, H. C., Bai, H., Fan, K. S., & Chen, C. (2011). TiO2-x nanoparticles synthesized using he/ar. Thermal plasma and their effectiveness on low-. Concentration mercury vapor removal. *Journal of Nanoparticle Research, 13*, 4739–4748.

Turkevich, J., Stevenson, P. C., & Hillier, J. (1953). The formation of colloidal gold. *The Journal of Physical Chemistry, 57*, 670–673.
US EPA. (2004). https://nepis.epa.gov/Exe/ZyNET.exe/
US EPA. (2008). https://cfpub.epa.gov/ncea/risk/recordisplay.cfm?deid=190806
US EPA. (2014). *Remediation: Selected sites using or testing nanoparticles for remediation.*
Varma, R., & Nadagouda, M. N. (2009). Risk reduction vie greener synthesis of noble metal nanostructures and nano composites. *Environmental Security, 3*, 209–217.
Xie, Y., Wang, B., Li, F., Ma, L., Ni, M., & Shen, W. (2014). Molecular mechanisms of reduced nerve toxicity by titanium dioxide nanoparticles in the phoxim-exposed brain of *Bombyx mori. PLoS One, 9*(6), e101062.
Xin, R. Z., Ping, Y., & Meng, Y. Z. (2001). Research on photocatalytic degradation of organophosphorus pesticides using TiO2.SiO2/beads. *Industrial Water Treatment, 21*(3), 13–39.
Zhang, L., Chang, X., Hu, Z., Zhang, L., Shi, J., & Gao, R. (2010). Selective solid phase extraction and preconcentration of mercury(II) from environmental and biological samples using nanometer silica functionalized by 2,6-pyridine dicarboxylic acid. *Microchimica Acta, 168*, 79–85.
Zhang, Q., Pan, B., Zhang, S., Wang, J., Zhang, W., & Lv, L. (2011). New insights into nanocomposite adsorbents for water treatment: A case study of polystyrene-supported zirconium phosphate nanoparticles for lead removal. *Journal of Nanoparticle Research, 13*, 5355–5364.
Zhu, X., Zhou, D., Wang, Y., Cang, L., Fang, G., & Fan, J. (2012). Remediation of polychlorinated biphenyl-contaminated soil by soil washing and subsequent TiO2 photocatalytic degradation. *Journal of Soils and Sediments, 12*, 1371–1379.

Chapter 3
Removal of Inorganic and Organic Contaminants from Terrestrial and Aquatic Ecosystems Through Phytoremediation and Biosorption

Dhananjay Kumar, Sangeeta Anand, Poonam, Jaya Tiwari, G. C. Kisku, and Narendra Kumar

Contents

D. Kumar · S. Anand · Poonam · N. Kumar (✉)
Department of Environmental Science, Babasaheb Bhimrao Ambedkar University (A Central University), Lucknow, Uttar Pradesh, India
e-mail: narendrakumar_lko@yahoo.co.in

J. Tiwari
School of Biological Sciences, AIPH University, Bhubaneswar, Odisha, India

G. C. Kisku
Environmental Monitoring Division CSIR-Indian Institute of Toxicology Research
Vishvigyan Bhavan, Lucknow, Uttar Pradesh, India

R. C. Sobti et al. (eds.), *Environmental Biotechnology: For Sustainable Future*,
https://doi.org/10.1007/978-981-10-7284-0_3

Abstract Escalated industrialization, inappropriate waste management practices, mining, landfill operations, and application of sewage sludge have caused excess contamination of aquatic and terrestrial ecosystems. As a consequence, human beings pose serious threats to life-supporting resources, i.e., air, soil, and water. Heavy metals and pesticides are a special class of contaminants having wide variety of effects. When the contaminated lands are used for agriculture practices, contaminants like heavy metals and pesticides get transferred from soil to food chain which leads to bioaccumulation and biomagnification. Phytoremediation (a technique that exploits plants ability to lessen, eradicate, degrade, or immobilize the environmental contaminants, with the aim of restoring the contaminated area) is gaining advantage over other conventional treatment techniques being economical, environmentally sound, and aesthetically acceptable. Conventional approaches for cleanup and restoration of heavy metals and pesticides from contaminated environment have some unavoidable precincts like high cost and creation of secondary pollutants. Many aquatic and terrestrial plants such as *Eichhornia*, *Pistia*, *Lemna*, *Salvinia*, *Typha*, *Hydrilla*, *Ricinus*, *Brassica*, *Arabidopsis*, *Vetiver*, *Solanum*, etc. are capable of accumulating heavy metals and can be used as agents for eco-restoration of degraded ecosystems. Further, biosorption has also emerged as an innovative, eco-friendly, cost-effective, and probable substitute for the removal and/or recovery of inorganic contaminants from aqueous medium. Biosorption can be applicable over wide range of temperature and pH, with rapid kinetics of adsorption and desorption and low capital and operation cost. Even, biological biomass can again be regenerated for reuse.

Keywords Biosorption · Heavy metals · Macrophytes · Pesticides · Phytoremediation

Introduction

Escalated industrialization and urbanization all over the world have generated various luxurious facilities and commodities. However, such rampant industrialization and several anthropogenic deeds like mining, sewage sludge applications, improper waste management, and landfill operations have resulted in excess contamination of soil, air, and water (Ghavari et al. 2013; Ha et al. 2014). As a consequence, many life-forms including humans are facing serious threats to their life-supporting resources. Heavy metals, metalloids, and pesticides are a special class of contaminants having wide variety of effects. Chemically, heavy metals are the transition metals having atomic mass and specific gravity greater than 20 and 5, respectively. Further, some heavy metals like As, Cd, Hg, Pb, Se are nonessential, since they do not play any major role in the physiological functions of plant. Besides these nonessential metals, some heavy metals like Fe, Cu, Zn, Mn are essential for normal growth and development of plants. However, at high concentrations, these metals cause many toxic repercussions and recognized as environmental toxicants (Kumar et al. 2013, 2016). A cumulative discharge of industrial waste comprising of heavy metals to agricultural fields not only degrades the health of soil but also affects the food quality, productivity, and environmental safety (Mapanda et al. 2005; Singh et al. 2010). When the contaminated fields are used for agriculture practices, metals and pesticides easily get transferred from soil to food chain and lead to bioaccumulation and biomagnification (Sakakibara et al. 2011; Kumar and Kumar 2016). Heavy metals and pesticides are primary pollutants in aquatic and terrestrial environment, being recalcitrant in nature. Plants growing in metal-contaminated environment express several disturbances in their physiological and biochemical processes like gaseous exchange, CO_2 fixation, respiration, nutrient absorption, etc. Generally, all the heavy metals cause loss of 25 kDa polypeptide and subsequently cause reduction in cell enlargement and synthesis of heavy metal-binding phytochelatin through inactivation of α-expansin/phytochelatin synthase (Kasim 2005). During the last couple of decades, researchers had explored several remediation techniques to mitigate the problem of contamination of soil, *viz.*, stabilization, solidification, soil washing, encapsulation, vitrification, electrokinetic, and phytoremediation, and water (e.g., coagulation-flocculation, oxidation, chemical precipitation, ion exchange, adsorption, membrane filtration, photocatalytic degradation, electrochemical, and phytoremediation) (Dhir et al. 2009). Among these available techniques, phytoremediation, which exploit plant's ability to reduce, remove, degrade, or immobilize the environmental toxins, with the aim of restoring the contaminated area, has emerged as an efficient, environmentally sound, and aesthetically acceptable technique (Kumar et al. 2012, 2013).

Phytoremediation: An Environmentally Sound and Cost-Effective Approach

Today, phytoremediation is a well-known solar-driven botanical remediation approach. In recent years, the applicability of phytoremediation for the restoration of the soil contaminated with wide varieties of pollutant such as heavy metals (Chen et al. 2010; Kumar et al. 2012), nutrients (Ashworth et al. 2005; Silveira et al. 2013), and organics (Euliss et al. 2008; Nedunuri et al. 2009) has delivered considerable results. Certain plants have unique potential to survive in metal-contaminated matrix. They not only survive but also can significantly extract and accumulate heavy metals/metalloids in their roots/shoots tissues even at concentration greater

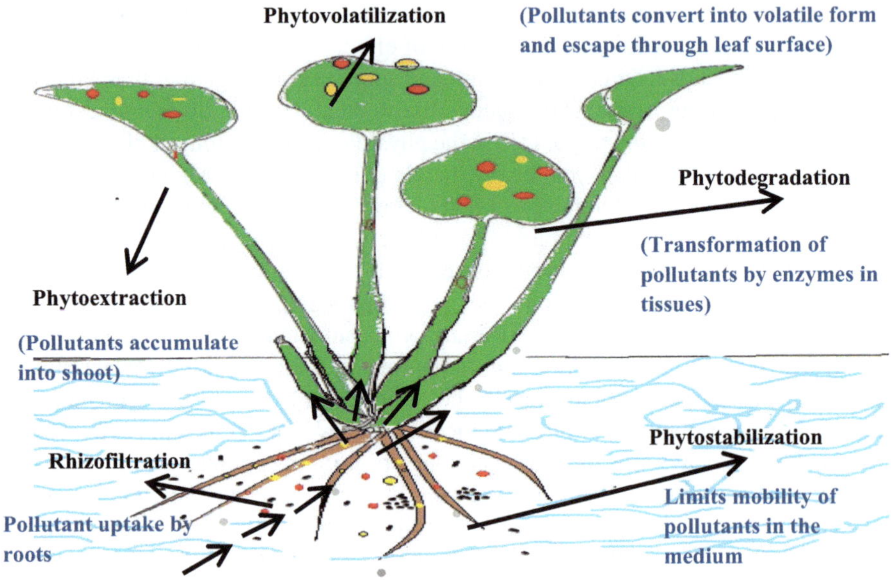

Fig. 3.1 Phytoremediation mechanisms and techniques

than in the contaminated medium (Fig. 3.1) (Lasat 2002). Further, along with its remediation potential, some plants also offer opportunity for the eco-restoration of degraded environment. Till date, over 500 plant species belonging to more than 45 families have been documented for their ability toward metal tolerance and accumulation of potentially toxic elements (PTEs) specially heavy metals (Hemen 2011; Sharma 2011).

Phytoremediation of Contaminated Aquatic Ecosystems

Phytoremediation utilizes natural use of plants to absorb pollutants through roots and their translocation to aerial parts of the plant (Sharma et al. 2015). Several studies have revealed the removal of heavy metals and pesticides from contaminated water by applying phytoremediation (Bhatia and Goyal 2014; Kumar et al. 2013) For remediation of contaminated water, a wide number of macrophytes have been reported (Bhatia and Goyal 2014; Sharma et al. 2015). Macrophytes suitable for phytoremediation should have some special characteristics like rapid growth, extensive root system, high biomass production, diverse adaptations, and high tolerance and potential to accumulate wide range of pollutants in their body parts (Valipour and Ahn 2016). Further, several factors like salinity, temperature, pH, sunlight, and plant growth rate affect the process and success of phytoremediation. Associated limitation like more time requirement can be solved by combining more than one phytoremediation approaches (Parmar and Singh 2015).

Table 3.1 Historical advances in era of wastewater treatment using aquatic macrophytes

Decades	Summary	Scope of works	References
1970s	Exploration of ability of aquatic macrophytes for remediation of wastewater	Research on potential of aquatic and submerged plants	Boyd (1970), Cowgill (1974) and Hutchinson (1975)
	Rooted and rootless submerged macrophytes have potential to uptake contaminants from water rapidly		
	Accumulations of PTEs in the plant tissues were higher than in the growing medium		
1980s	Emergent and surface-floating macrophytes uptake heavy metals via their root systems while submerged uptake metals and nutrients via roots and their leaves	Study on emergent and surface-floating plants	Denny (1980, 1987)
1990s	Uptake and removal of potentially toxic metals by aquatic vascular plants known as "hyperaccumulation"	Study on hyperaccumulation which occurs via root and foliar absorption	Outridge and Noller (1991), Sharma and Gaur (1995) and Rai et al. (1995)
	Nutrients uptake through root, and foliar absorption are also occurred in rooted macrophytes		
2000–2010	Aquatic plants are proficient in remediation of heavy metals/metalloid; an important class of contaminants	Emergence of hyperaccumulator and exploration of effectiveness of macrophytes	Hu et al. (2003), Kamal et al. (2004) and Rai (2009)
	Macrophytes, either living or nonliving, have been used for the treatment and monitoring of metal contamination		
2010–2015	Plants have high potential to accumulate or uptake pollutants by diverse mechanisms from the contaminated medium (Fig. 3.1)	Much emphasis on the mechanism and efficiency improving techniques	Sharma et al. (2015), Ali et al. (2013) and Sasmaz et al. (2015)
	Approaches for phytoremediation of PTEs are natural, economical, and tempted with application of hyperaccumulating plants and chemicals like synthetic chelating ligands		

Classically, three categories of macrophytes, viz., free-floating, submerged, and emergent, have been used for the treatment of aqueous medium (Valipour and Ahn 2016). Historical advancement in the era of wastewater treatment using macrophytes is presented in Table 3.1.

Various investigators have proved that macrophytes, e.g., *Lemna minor*, *Eichhornia crassipes*, *Pistia stratiotes*, *Myriophyllum spicatum*, *Azolla pinnata*, *Hydrilla verticillata*, *Elodea canadensis*, *Salvinia herzogii*, etc., offer promising results for removal of metal from contaminated aquatic medium; Table 3.2

Table 3.2 Metal and pesticides accumulation capacity of some aquatic macrophytes

S. No	Plant species	Contaminants	Accumulation	References
1.	*Eichhornia crassipes*	Cu	6000–7000 mg kg^{-1}	Molisani et al. (2006)
		Cr	4000–6000 mg kg^{-1}.	Hu et al. (2007)
		Ni	1200 mg kg^{-1}	Low et al. (1994)
		Cd	2200 μg kg^{-1}	Zhu et al. (1999)
		Zn	1677 mg g^{-1}	Kamel (2013)
		As	909.58 mg kg^{-1}	Delgado et al. (1993)
		Hg	119 ng g^{-1}	Molisani et al. (2006)
		Mn	300 mg kg^{-1}	Dixit et al. (2011)
		Mevinphos		Wolverton (1975)
		Ethion		Xia and Ma (2006)
2.	*Azolla pinnata*	Ni	16,252 l μg gm^{-1}	Arora et al. (2004)
		Cd	740 μg g^{-1}	Rai (2008)
		Cr	1095 μg g^{-1}	Rai (2010)
		Hg	940 μg g^{-1}	Rai (2009)
		Pb	1383 mg kg^{-1}	Thayaparan et al. (2013)
3.	*Azolla filiculoides*	Cd	2608 μg g^{-1}	Arora et al. (2004)
		Cu	6013 μg g^{-1}	Zhang et al. (2008)
		As	>60 μg g^{-1}	Arora et al. (2006)
		Cr	12,383 μg g^{-1}	Vesely et al. (2011)
		Pb	1607 mg kg^{-1}	
4.	*Azolla caroliniana*	Cr	356 mg kg^{-1}	Bennnicelli et al. (2004)
		Hg	578 mg kg^{-1}	Zhang et al. (2008)
		As	284 mg kg^{-1}	
5.	*Hydrilla verticillata*	As	231 mg kg^{-1}	Srivastava et al. (2010)
		Cu	770–3,0830 mg kg^{-1}	Srivastava et al. (2011)
		Chlordane	1060.95 μg L^{-1}	Chaudhary et al. (2002)
6.	*Typha latifolia*	Ni	295.6 mg kg^{-1}	Afrous et al. (2011)
		Cu	1156.7 mg kg^{-1}	Nguyen et al. (2009)
		Dieldrin	0.60 ng g^{-1}	Guo et al. (2014)
7.	*Pistia stratiotes*	Pb	519 mg kg^{-1}	Vesely et al. (2011)
		Chlorpyrifos	0.036 mg g^{-1}	Prasertsup and Ariyakanon (2011)
8.	*Salvinia minima*	Pb	5469 mg kg^{-1}	Vesely et al. (2011)
9.	*Salvinia natans*	Cr	7.40 mg g^{-1}	Dhir et al. (2009)
10.	*Lemna gibba*	U	896.9 mg kg^{-1}	Mkandawire et al. (2004)
		As	1021.7 mg kg^{-1}	
11.	*Lemna minor*	Pb	561 mg g^{-1}	Leblebici and Aksoy (2011)
		Cu	400 μg g^{-1}	Boule et al. (2009)
		Flazasulfuron	27 μg g^{-1}	Olette et al. (2009)
		Dimethomorph	33 μg g^{-1}	
		Chlorpyrifos	0.23 g^{-10}	

(continued)

Table 3.2 (continued)

S. No	Plant species	Contaminants	Accumulation	References
12.	*Typha angustifolia*	Mn	860 mg kg^{-1}	Sasmaz et al. (2008)
		Cu	50 mg kg^{-1}	
13.	*Myriophyllum spicatum*	Pb	8.94 mg g^{-1}	Kamel (2013)
		Zn	2.66 mg g^{-1}	
14.	*Ceratophyllum submersum*	Pb	258.62 mg kg^{-1}	Sasmaz et al. (2008)
		Zn	1172.8 mg kg^{-1}	Nguyen et al. (2009)
		Aldrin	0.38 n gg^{-1}	
		Endosulfan	0.73 n gg^{-1}	
15.	*Wolffia globosa*	As	> 1000 mg kg^{-1}	Zhange et al. (2009)
16.	*Phragmites communis*	Fe	2813 μg g^{-1}	Chandra and Yadav (2011)
		Mn	814.40 μg g^{-1}	Guo et al. (2014)
		Zn	265.80 μg g^{-1}	
		Pb	92.80 μg g^{-1}	
		γ-HCH	0.97 ng g^{-1}	
		DDTs	0.93 ng g^{-1}	
17.	*Pontederia cordata*	Oryzalin		Fernandez et al. (1999)
18.	*Spirodela oligorrhiza*	o,p′-DDT, p,p′-DDT,	50–66%	Gao et al. (2000)
19	*Schoenoplectus californicus*	DDTs	30.2–45.7 ng g^{-1}	Miglioranza et al. (2004)

(Mahmood et al. 2005; Verma et al. 2005, Aktar et al. 2010; Sasmaz et al. 2015; Rezania et al. 2016). Fritioff et al. (2005) suggested that submerged aquatic macrophytes are completely flooded and have the capability to uptake metals straight from the contaminated water and reducing metal concentrations in microbiologically treated wastewaters. *E. crassipes* possess high remediation potential for a number of heavy metals like copper (Cu), chromium (Cr), iron (Fe), zinc (Zn), manganese (Mn), mercury (Hg), cadmium (Cd), and arsenic (As) from contaminated growing medium and store them in their bladders followed by translocation to stems, leaves, and roots (Mahmood et al. 2005; Jardia and Fulekar 2009; Priya and Selvan 2013; Rizwana et al. 2014; Mohamad and Latif 2010; Rahman et al. 2008). Jafari (2010) observed the high accumulation ability of *A. microphylla* (94%), *A. filiculoides* (96%), *A. pinnata* (71%), and *A. microphylla* (98%), for Pb, Cu, Mn, and Zn, respectively, whereas, *A. caroliniana* have potential to accumulate Hg and Cr. Sasmaza et al. (2009) reported that the *L. gibba* can be applied for the remediation of As from secondary effluents as an alternative way of treatment. Miretzky et al. (2004) reported that *P. stratiotes* is capable of removing the metals, viz., Pb, Cr, Mn, and Zn, almost completely from water in first 24 h of exposure. *Ipomoea aquatica* can remove Cr (III) from contaminated water in the presence of chelating agents' chloride and EDTA (Chen et al. 2009). Further, chloride can also increase the solubility of Cr and enhance its translocation and accumulation in roots and shoots of the plant.

Adhikari et al. (2010) assessed the removal potential of two wetland plant species, i.e., *Typha angustifolia* and *Ipomoea carnea*, and reported that both the plants were able to remove Pb from contaminated wastewaters. Chandra and Yadav (2010) also found that *T. angustifolia* is suitable for the removal of heavy metals (Cu, Pb, Ni, Fe, Mn, and Zn) from wastewater and recommended the application of *Typha* for phytoremediation of heavy metals from industrial wastewater in optimized conditions. Unnikannan et al. (2013) identified *Salvinia natans*, *P. stratiotes*, and *E. crassipes* as hyperaccumulators of Cr. Kumar et al. (2012) evaluated the remediation potential of five macrophytes, viz., *H. verticillata*, *E. crassipes*, *I. aquatica*, *Bacopa monnieri*, and *Melica minuta*, and recommended that *E. crassipes* can be used for removal of Ni and Cu, while *M. minuta* and *H. verticillata* can be applied for remediation of Pb and Cr from the contaminated aqueous medium. Susselan et al. (2006) reported the effectiveness of *Mimosa pudica* for rhizofiltration of Hg, Cd, U, and Zn. Upadhyay et al. (2007) reported that the *E. crassipes*, *L. minor*, *P. stratiotes*, *A. pinnata, and Spirodela polyrhiza* are suitable for remediation of Fe, Cu, Cr, Cd, Zn, and Ni. Comparative heavy metal uptake efficiency of *Salvinia* and *Spirodela* was assessed by Li et al. (2016). They found that *Spirodela* was more efficient than *Salvinia* in removal of Pb and Zn (Li et al. 2016).

Application of Aquatic Plants for Pesticide Removal

Selected macrophytes have developed high potential for accumulation and degradation of pesticides by adopting different cellular mechanisms such as direct root uptake, detoxification by phytotransformation and conjugation with glutathione or sugars, and subsequent accumulation of non-phytotoxic metabolites in plant roots/shoots (Goncalyes and Alpendurada 2005). *E. crassipes* (Xia and Ma 2006), *Juncus effusus*, *Ludwigia peploides* (Bouldin et al. 2006), *L. minor*, *Spirodela polyrhiza*, *E. canadensis*, and *Carya aquatica* have been reported to be efficient in pesticide uptake (Olette et al. 2008, 2009). Olette et al. (2009) examined the remediation potential of *L. minor* and *S. polyrhiza* for remediation of dimethomorph and pyrimethanil from water and noticed high removal efficiency during the 4-day test period (Olette et al. 2009). Prasertsup and Ariyakanon (2011) demonstrated efficient and successful use of *L. minor* and *P. stratiotes* in the remediation of chlorpyrifos from aquatic medium. Both the plants can be utilized for efficient, economical, and ecological alternatives to accelerate the removal of wastewater contaminated with a relatively low concentration of chlorpyrifos (0.5 mg/L). Miglioranza et al. (2004) observed significant variances in the dichlorodiphenyltrichloroethane (DDT) levels between root and shoot of *Typha*, indicating its ability to uptake and translocate DDT from water. Phytoremediation potential of some macrophytes are presented in Table 3.2.

Application of Terrestrial Plants in Phytoremediation

For terrestrial plants, roots are the main contact site for exposure to metal contaminants. Metal contamination in growing medium causes disturbances in normal physiological functioning of plant. In an endeavor to survive and safeguard the susceptible cellular component, plants have developed some effective and specific mechanisms to cope up with the metal stress. Adaptive mechanisms evolved by plants to mitigate the metal toxicity include immobilization, plasma membrane exclusion, restriction of uptake and transport and sequestration by specific ligands (Dalcorso et al. 2008; Hossain et al. 2009, 2012; Sharma and Dietz 2009; Hossain and Fujita 2009; Cobbett 2000; Clemens 2006). Cellular defense mechanism for metal tolerance consists of two basic approaches, to retain low level of toxic metal ions into the cytoplasm by preventing them from being transported through the plasma membrane. This can be achieved either by increasing the binding of metal ions to cell wall or by pumping out the metal by active efflux pumps from cell. The other approach is detoxification of toxic metal ions by inactivation via chelation of the metal cation by specific ligands or organic acids or conversion of toxic metal ion into less toxic form or vacuolar/cell wall sequestration away from metabolic sites (Zhu et al. 2004; Chaney et al. 2007; Mejare and Bulow 2001).

Hyperaccumulator of Heavy Metals

Plant species having ability to accumulate 100 mg Cd; 1000 mg Ni, Cu, Co, Cr, and Pb; and 10,000 mg kg^{-1} Zn and Mn (dry weight) are called as hyperaccumulators (Baker and Brooks 1989; Kumar et al. 2013; Brooks 1998; Yuan et al. 2016). Further, translocation of metals from roots to aboveground plant parts is also an essential feature of phytoremediation plants and is deduced by calculating by the translocation factor (TF) (Kumar et al. 2013). TF higher and lower than unity is a key feature of metal accumulator and excluder plant species, respectively (Singh et al. 2010). Identification and use of hyperaccumulator plants for soil remediation can help up to a great extent to remediate the metal-contaminated matrix (Lasat 2002). Till date more than 500 hyperaccumulator plants have been reported belonging to the family Asteraceae, Brassicaceae, Caryophyllaceae, Cyperaceae, Cunouniceae, Fabaceae, Flacourtiaceae, Lamiaceae, Poaceae, Violaceae, and Euphorbiaceae. Some of the hyperaccumulator plants having high phytoremediation potential and TF of >1 for selected metals are mentioned in Table 3.3. And the key important factors affecting the phytoremediation potential are mentioned in Table 3.4.

Table 3.3 Hyperaccumulators of various metals

Plant name	Family	Contaminants	References
Ricinus communis	Euphorbiaceae	Cd	Bauddh and Singh (2012)
Brassica juncea	Brassicaceae	Cd, Se	Qadir et al. (2004) and Banuelos et al. (2005)
Vetiveria zizanioides	Poaceae	Cd, Pb	Danh et al. (2009)
Cyperus rotundus	Cyperaceae	Cr, Cu, Ni, Pb, Cd, As, Sn	Asharaf et al. (2011), Kumar et al. (2013), Jaison and Muthukmar (2016) and Yuan et al. (2016)
Parthenium hysterophorus	Asteraceae	Mg Fe, Pb, Zn, Cd	Mazumdar and Das (2015), Sanghamitra et al. (2011) and Ahmad and Al-Othman (2014)
Amaranthus cruentus	Amaranthaceae	Mg, Fe, Pb, Zn, Cr	Mazumdar and Das (2015) and Liu et al. (2008)
Populus tremula	Salicaceae	Zn, Cd, Cu	Ruiz et al. (2011) and Pierre et al. (2011)
Solanum americanum	Solanaceae	Mg, Fe, Pb, Zn	Mazumdar and Das (2015)
Croton bonplandianum	Euphorbiaceae	Cr, Cu, Ni, Pb, Cd	Kumar et al. (2013)
Arabidopsis thaliana	Brassicaceae	Cd, As	Guo et al. (2012) and Kiyono et al. (2012)
Solanum nigrum	Solanaceae	Zn, Mn, Cu, Cr, Ni, Co, Cd, Pb	Malik et al. (2010) and Varun et al. (2012)
Cannabis sativa	Cannabaceae	Pb, Cu, Zn, Ni, Co, Cr	Malik et al. (2010)

Table 3.4 Factors affecting the phytoremediation potential of aquatic macrophytes

S. No.	Parameter	Effects
1	Temperature	Metal uptake/toxicity decreases at lower temperature
2	Light	Sometime depends upon light
3	pH	Generally metal uptake/accumulation potential decreases at higher pH
4	Salinity	High salinity decreases the content/toxicity
5	Monovalent cation	Enhances metal uptake in the presence of lower concentration of monovalent cation in growing medium
6	Divalent cation	Enhances metal uptake in the presence of lower concentration of divalent cation in growing medium
7	Nitrate	Decreases metal toxicity significantly
8	Sulfate	Decreases metal uptake and toxicity
9	Polysaccharides	Chelate metals, reduces uptake/toxicity
10	Organic acid/ selenite (Se)	Reduces metal uptake and toxicity
11	Heavy metals	Reduces metal uptake/toxicity by binding metal complexes; Zn/Cd, Cu, and Ni combination is antagonistic, while Fe can encourage Cu uptake

Sood et al. (2012), De Fillippis (1979), Rai et al. (1981), Trevor et al. (1986), Wallen (1990), Kelly and Whitton (1989), Whitton et al. (1989), Mallick et al. (1990), Reed and Gadd (1990), Wong and Chau (1990), Mukherjee et al. (2004)

Biosorption of Contaminants Using Plant Biowastes

Generally, conventional treatment techniques, viz., precipitation, ion exchange, membrane filtration, electroplating, adsorption, etc., have considerable drawbacks, such as high consumption of chemical and energy, hazardous sludge generation, low efficiency when metals concentration is below 100 mgL^{-1}, and high-cost involvement (Marin-Rangel et al. 2012). In the recent years, biosorption has attracted considerable interest for the removal of inorganic and organic pollutants. Biosorption can be defined as the use of inactive, dead, biological waste to bind and concentrate contaminants from the medium. It comprises of a solid phase (sorbent or biosorbent; biowaste) and a liquid phase (solvent; water or wastewater) having varieties of dissolved species to be sorbed on sorbent (sorbate; metal ions, organic and inorganic pollutants). Due to high affinity of sorbent, the sorbate is attracted and adheres on sorbent by various mechanisms. The biosorption process continues till equilibrium is established between the amount of solid-bound sorbate species and its remaining portion present in the solution. The degree of affinity of sorbent determines the distribution of sorbate between solid and liquid phases (Ahalya et al. 2003). Biowastes are a form of biomass which can be decomposed under anaerobic or aerobic conditions into simplest forms, i.e., carbon dioxide and water. The major advantages of biosorption over other conventional treatment methods include low cost, greater efficiency even with low metal concentration, minimization of chemical or biological sludge, regeneration of biosorbents, no additional nutrients requirements, easy operation, possibility of metal recovery, and no detrimental effects to the environment (Jiménez-Cedillo et al. 2013).

Potential Biowastes for Biosorption of Contaminants

Biowastes are composed of food scraps (vegetable peels, eggshells), algae, yeast, fungi, manure; plant's parts such as bark, stem, wood, leaves, root, husk, skin, shell, and bran; and paper and cardboard produced by industries, agriculture, and individuals. Some biowastes which have been applied for the removal of different organic and inorganic pollutants are presented in Table 3.5. Sludge from wastewater treatment plants is also considered as biowaste. Further, biowaste can be classified into two categories (Fig. 3.2).

Naturally Occurring

These include various microorganisms: algae (*Chaetomorpha* sp., *Ulva* sp., *Cladophora* sp., *Chlorella* sp., *Fucus* sp., *E. crassipes*, etc.), fungi (*Aspergillus* sp., *Rhizopus* sp., *Saccharomyces* sp., *Lentinus* sp., *Mucor* sp., *Trichoderma*, *Fusarium*, etc.), and agricultural wastes (orange peel, lemon peel, pine leaves, rice husk, wheat straw, potato peel, gram husk, bagasse, coir, cashew nutshell, almond shell, cassava seeds, flower parts, sawdust, etc).

Table 3.5 Biowastes for removal of different organic and inorganic pollutants

S. No.	Biowaste	Pollutant	q_e (mg/g)	t_e (hrs)	Reference
1	*Parthenium hysterophorus*	Methylene blue	98.06		Chatterjee et al. (2012)
2	Coconut coir (activated carbon)	Malachite green	27.44		Banerjee et al. (2012)
3	Animal bone meal	Rhodamine B	65		Haddad et al. (2014)
4	Rutile TiO_2	Direct fast blue B2RL	56		El-Mekkawi and Galal (2013)
	Degussa P25		144		
5	$CuFe_2O_4$ nanocomposite	Cyanine acid blue (CAB)	178.56		Hashemian and Salimi (2012)
	Sawdust nanocomposite		151.46		
6	Olive stone	Fe^{2+}	62.5		Alslaibi et al. (2013)
		Pb^{2+}	23.47		
		Cu^{2+}	22.73		
		Zn^{2}	15.1		
		Ni^{2+}	12		
		Cd^{2+}	11.72		
7	Peanut husk	Cu^{2+}	0.345		Abdel et al. (2011)
			0.182		
	Fly ash	Zn^{2+}	0.368		
			0.180		
8	Peanut husk	Cu^{2+}	25.39		Witek-Krowiak et al. (2011)
		Cr^{3+}	27.86		
9	Rapeseed oil cake (activated by Na_2CO_3)	Pb^{2+}	129.87		Uçar et al. (2014)
		Ni^{2+}	133.33		
10	Rose petal waste	Pb^{2+}	119.92		Manzoor et al. 2013
11	Bentonite clay	Zn^{2+}	4.49		Araujo et al. (2013)
12	Parsley (iron-modified non-pyrolyzed)	As^{5+}	0.19		Jiménez-Cedillo et al. (2013)
13	Cortex banana waste	Cd^{2+}	67.2		Kelly et al. (2012)
14	Coir pith	Cr^{6+}	165		Suksabye and Thiravetyan (2012)
15	Orange peel (NaOH $CaCl_2$ modified)	Cu^{2+}	70.73		Feng and Guo (2012)
16	*Litchi chinensis* seeds	Ni^{2+}	66.62		Flores-Garnica et al. (2013)
17	*Schizosaccharomyces pombe* (marine algae)	Cu^{2+}	70%	96	Prakash and Kumar (2013)

(continued)

Table 3.5 (continued)

S. No.	Biowaste	Pollutant	q_e (mg/g)	t_e (hrs)	Reference
18	*Ascophyllum nodosum* (algae)	Cd^{2+}	114.9		Kumar and Oommen (2012)
		Ni^{2+}	50		
		Zn^{2+}	53.2		
		Cu2+	70.9		
		Pb^{2+}	204.1		
19	Immobilized *Bacillus subtilis* beads (IBSB)	Cd^{2+}	251.91	3	Ahmad et al. (2014)
20	Coconut (cationic polymer-modified granular activated carbon)	Nitrate (NO_3^-)	27.56	1.5	Cho et al. (2011)
		Cr(VI)	80.56	2	
21	Aleppo pine sawdust (activated by ethanol, urea and H_2SO_4)	Phosphate (PO_4^{3-})	29.67	0.6	Benyoucef and Amrani (2011)
22	Wheat stalk resin (amine-cross-linked biopolymer-based resin)	Perchlorate (ClO_4^-)	170.4		Song et al. (2015)
23	Cotton stalk based resin (CS-resin)	Perchlorate (ClO_4^-)	138.9		Ren et al. (2015)
24	Cotton stalk	Nitrate (NO_3^-)	33.35		Xu et al. (2013b)
	Wheat stalk		43.18		
25	Green algae (*Cladophora hutchinsiae*)	Se(IV)	74.9	1	Tuzen and Sarı (2010)
26	Cotton stalk	Perchlorate (ClO_4^-)	38.1		Xu et al. (2013a)
27	Cotton stalk (amine-impregnated)	Perchlorate (ClO_4^-)	83.8		Xu et al. (2015)
28	Sugarcane bagasse (activated by $ZnCl_2$)	Cr(VI)	>87%		Cronje et al. (2011)
29	Sugarcane bagasse	Phosphate (PO_4^{3-})	152		Carvalho et al. (2011)

t_e equilibrium time of adsorption, q_e equilibrium adsorption capacity

Synthesized Products

These include various waste products produced by industries (sludge, fly ash, red mud, fertilizer wastes, paper and pulp waste, tea wastes, etc.) and domestic wastes (food and vegetable wastes, etc.).

Biosorption Mechanisms of Biowaste

It is very important to explore the mechanisms followed by different biowastes for the process of biosorption to concentrate, remove, and recover the contaminants from aqueous solutions (Ahemad and Malik 2011; Abas et al. 2013). Several factors

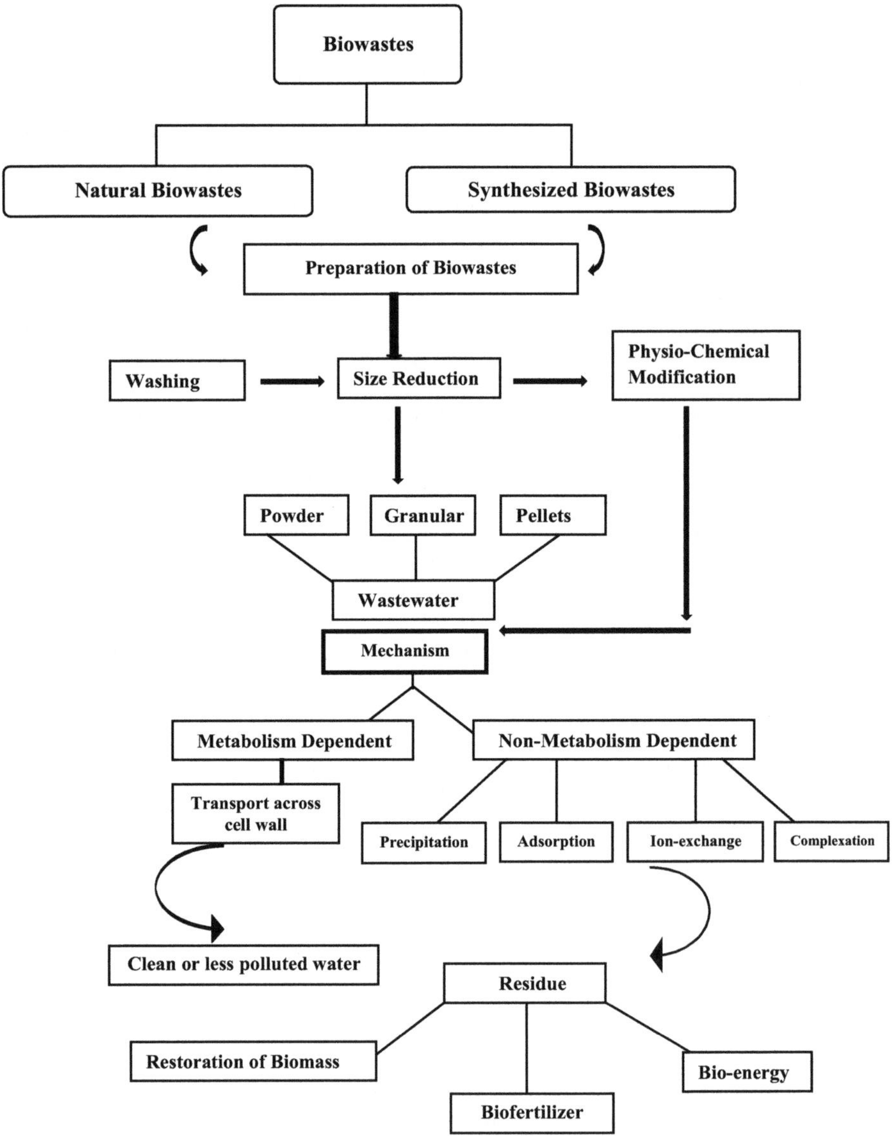

Fig. 3.2 Schematic description on application of biowastes for removal of pollutants from water

are found to affect working of biowastes for biosorption of pollutants, including types of biomass (living or nonliving), properties of metal solution chemistry, and environmental conditions (pH, temperature, humidity, etc.). Hence, the actual mechanism of the metal biosorption is not well understood, though a number of mechanisms have been suggested for the removal of different pollutants from the medium (Abbas et al. 2014; Nguyen et al. 2013). Broadly, the process of biosorption mechanisms can be categorized as mentioned in Fig 3.2.

Cell Metabolism

This category can be further divided into metabolism and non-metabolism dependent.

Metabolism Dependent

As the name indicates, this mechanism involves living cells and transport of contaminants (metals) across the cell membrane involving intracellular accumulation. It is also linked with the defense system of the living cells which participate in different reactions with toxic contaminants. Uptake of contaminants includes metabolism-dependent binding of metals to the cell walls and intracellular uptake and transportation of metal ions across the cell membrane. Further, the same mechanism will be used to transport metabolically important ions like potassium, magnesium, and sodium into the living cell (Ahalya et al. 2004).

Non-metabolism Dependent

This mechanism works on the physicochemical interaction between pollutants and the functional groups present in the biosorbents which depends on the physisorption, ion exchange, and chemisorption. Cell walls of the biomass are chiefly composed of polysaccharides, proteins, lipids, cellulose, hemicellulose, and lignin and have numerous metal-binding groups such as hydroxyl, carboxyl, hydroxyl, sulfate, phosphate, and amino groups. Due to abundant binding groups, biowastes have an enormous potential for decontaminating organic and inorganic pollutants from the contaminated medium (Nguyen et al. 2013).

Location Based

This depends upon the location where biosorption takes place. It is further divided into intra- and extracellular accumulation/precipitation and cell surface sorption/precipitation.

Other Mechanisms

Other kind of interactions between the biowastes and the pollutants can be further categorized into the following types.

Physical Adsorption

It takes place with the help of weak van der Waals' forces (Babak et al. 2012). Babak et al. (2012) hypothesized that biosorption of Cr, Cu, and Zn by *Pseudomonas aeruginosa* takes place through electrostatic forces of interaction between the metal ions present in the aqueous solutions and microbial cell walls. Electrostatic interactions have been found to be accountable for metal biosorption by bacterium, e.g., Cr and Fe biosorption by sulfate-reducing bacteria (SRB) and Cu biosorption by *Bacillus licheniformis* (Karakagh et al. 2012).

Ion Exchange

Cell walls of biological wastes contain polysaccharides, and metal ions get exchanged with the counterions of the polysaccharides. Ofomaja et al. (2010) proposed that ion exchange was the prevailing mechanism for the removal of Cu(II) and Pb(II) using KOH treated pine cone powder (Ofomaja et al. 2010). Vázquez et al. (2012) explained the role of functional groups in the biosorption of Cu(II), Zn(II), Pb(II), and Cd(II) onto chestnut (*Castanea sativa*) shell by ion exchange (Vázquez et al. 2012).

Complexation

Removal of metal from aqueous medium also takes place by complex formation, i.e., chemical reaction between functional groups present on the surface of biowaste and pollutants. The functional groups make biowastes capable for complexing metals and other pollutants by exchanging their hydrogen ions with metal ions or giving an electron pair to form metal complexes (Kumar et al. 2011). Microorganisms also produce organic acids such as citric, oxalic, fumaric, lactic, and malic acids which chelate with toxic metals and form metallo-organic molecules. These organic acids help in solubilization of metal compounds and their leaching from the surfaces. Metals may also be biosorbed or complexed by carboxyl groups found in polysaccharides and other biopolymers (Ahalya et al. 2004).

Precipitation

In precipitation, soluble ions present in separate solutions mixed together to form an insoluble compound that settles out as a solid. Aman et al. (2008) reported that precipitation of $Cu(OH)_2$ took place at pH greater than 6.0 in case of Cu(II) removal by potato peels (Aman et al. 2008).

Factors Affecting the Process of Biosorption

Biosorption of inorganic and inorganic contaminants from wastewater is influenced by several physical and chemical factors such as pH, temperature, initial metal concentration, biosorbent dose and size, ionic strength, coexisting ions. These factors determine the overall biosorption by affecting the uptake rate, selectivity, and concentration of contaminants to be removed (Nguyen et al. 2013).

pH

pH is a most important factor which governs the heavy metal adsorption. Rajoriya and Kaur (2014) reported that the uptake of Zn increased on increasing pH of solution till the equilibrium was achieved by using lemon and banana peel as adsorbent (Rajoriya and Kaur 2014). Tomar et al. (2014) also supported the pattern of increased removal percentage of Fe with increasing pH by using Zr-Mn composite material as adsorbent, and maximum adsorption, i.e., 90%, was found at pH 7. Kılıç et al. (2013) also reported the same pattern of increased amount of adsorption on increasing pH for Ni and Co, by using biochar of by-product of almond shell pyrolysis.

Specific Surface Area

It can be defined as the total available area on the surface for the purpose of the biosorption (Naeem et al. 2010). Ahmady-Asbchin et al. (2008) reported that the surface area of *Ficus serratus* biosorbents was increased by the water immersion, and increase of the surface area enhanced the adsorption efficiency of adsorbent.

Particle Size

This factor is directly correlated with adsorbent surface area as finer the particle of the adsorbent, the larger the surface area and higher will be the adsorption of contaminants on the surface.

Temperature

Temperature of the solution affects the solubility and diffusion rate of contaminants and also damages the active binding sites and, thus, affects the adsorption rate too (Park et al. 2013).

Dosage of the Adsorbents

Several researchers have taken this factor in account to analyze the biosorption of different contaminants and reported that as the dosages of the adsorbent increase, the adsorption rates have also been found to improve (Wasewar et al. 2008; Naeem et al. 2010; Tomar et al. 2014).

Contact Time

It was found that initially the rate of adsorption has been found to be rapid as there are numerous vacant active sites available for adsorption. Agitation and contact time provides energy and enough time to bring contaminant from the bulk of the solution to the available sites by reducing the resistance (Carvalho et al. 2011).

Conclusion

Contamination of environment by toxic metals and pesticides is a precarious threat; therefore, effective remediation techniques are necessary. Conventional approaches for cleanup and restoration of heavy metals and pesticides from contaminated environment have serious limitations like cost and generation of secondary pollutants. Being economical, eco-friendly, and aesthetically acceptable phytoremediation is gaining lead over other conventional remediation approaches. However, the information on the physiological and molecular mechanisms involved in phytoremediation is still scanty and needs further exploration for attainment of goal of remediation. The development of new fast-growing transgenic plant species through genetic modification process, which have high biomass, deep root system, high tolerance, and high accumulation potential and are easily harvestable, can help to mitigate the problem of metal contamination. Many aquatic and terrestrial plants are capable of accumulating heavy metals and can be used as agents for eco-restoration of degraded ecosystems.

Further, biosorption has emerged as an innovative, eco-friendly, cost-effective, and efficient alternative for the removal and/or recovery of contaminants from aqueous solutions. Biosorption can be performed over wide range of pH and temperature, with rapid kinetics of adsorption and desorption with low capital and operation cost. Even biological biomass can again be regenerated for reuse. However, further research work is required with multiple pollutant systems and real wastewater to make industrial usage of biowastes more feasible.

References

Abas, S. N. A., Ismail, M. H. S., Kamal, L., & Izhar, S. (2013). Adsorption process of heavy metals by low-cost adsorbent: A review. *World Applied Sciences Journal, 28*, 1518–1530.

Abbas, S. H., Ismail, I. M., Mostafa, T. M., & Sulaymon, A. H. (2014). Biosorption of heavy metals: A review. *Journal of Chemical Science and Technology, 3*(4), 74–102.

Abdel Salam, O. E., Reiad, N. A., & ElShafei, M. M. (2011). A study of the removal characteristics of heavy metals from wastewater by low-cost adsorbents. *Journal of Advanced Research, 2*(4), 297–303.

Adhikari, T., Kumar, R., Singh, M. V., & Rao, A. S. (2010). Phytoaccumulation of lead by selected wetland plant species. *Communications in Soil Science and Plant Analysis, 41*, 2623–2632.

Afrous, A., Manshouri, M., Liaghat, A., Pazira, E., & Sedghi, H. (2011). Mercury and arsenic accumulation by three species of aquatic plants in Dezful, Iran. *African Journal of Agricultural Research, 6*(24), 5391–5397.

Ahalya, N., Ramachandra, T. V., & Kanamadi, R. D. (2003). Biosorption of heavy metals. *Research Journal of Chemistry and Environment, 7*, 71–78.

Ahalya, N., Ramachandra, T. V., & Kanamadi, R. D. (2004). Biosorption of heavy metals. *Journal of Chemistry and Environment, 7*(4), 71–79.

Ahemad, M., & Malik, A. (2011). Bioaccumulation of heavy metals by zinc resistant bacteria isolated from agricultural soils irrigated with wastewater. *Bacteriology Journal, 2*, 12–21.

Ahmad, A., & Al-Othman, A. A. S. (2014). Remediation rates and translocation of heavymetals from contaminated soil through *Parthenium hysterophorus*. *Chemistry and Ecology, 30*(4), 317–327.

Ahmad, M. F., Haydar, S., Bhatti, A. A., & Baria, A. J. (2014). Application of artificial neural network for the prediction of biosorption capacity of immobilized *Bacillus subtilis* for the removal of cadmium ions from aqueous solution. *Biochemical Engineering Journal, 84*, 83–90.

Ahmady-Asbchin, S., Andre's, Y., Ge'rente, C., & Cloirec, P. L. (2008). Biosorption of Cu(II) from aqueous solution by *Fucus serratus*: Surface characterization and sorption mechanisms. *Bioresource Technology, 99*, 6150–6155.

Aktar, M. W., Paramasivam, M., Ganguly, M., Purkait, S., & Sengupta, D. (2010). Assessment and occurrence of various heavy metals in surface water of Ganga river around Kolkata: A study for toxicity and ecological impact. *Environmental Monitoring and Assessment, 160*, 207–213.

Ali, H., Khan, E., & Sajad, M. A. (2013). Phytoremediation of heavy metals – Concepts and applications. *Chemosphere, 91*, 869–881.

Alslaibi, T. M., Abustan, I., Ahmad, M. A., & Abu Foul, A. (2013). Application of response surface methodology (rsm) for optimization of Cu^{2+}, Cd^{2+}, Ni^{2+}, Pb^{2+}, Fe^{2+}, and Zn^{2+} removal from aqueous solution using microwaved olive stone activated carbon. *Journal of Chemical Technology and Biotechnology, 88*(12), 141–151.

Aman, T., Kazi, A. A., Sabri, M. U., & Bano, Q. (2008). Potato peels as solid waste for the removal of heavy metal copper(II) from waste water/industrial effluent. *Colloids and Surfaces, B: Biointerfaces, 63*, 116–121.

Araujo, A. L. P., Bertagnolli, C., Silva, M. G. C., Gmenes, M. L., & Barros, M. A. S. D. (2013). Zinc adsorption in bentonite clay: Influence of pH and initial concentration. *Acta Scientiarum – Technology, 35*, 325–332.

Arora, A., Sood, A., & Singh, P. K. (2004). Hyperaccumulation of cadmium and nickel by *Azolla* species. *Indian Journal of Plant Physiology, 3*, 302–304.

Arora, A., Saxena, S., & Sharma, D. K. (2006). Tolerance and phytoaccumulation of chromium by three *Azolla* species. *World Journal of Microbiology and Biotechnology, 22*, 97–100.

Ashraf, M. A., Maah, M. J., & Yusof, I. (2011). Heavy metals accumulation in plants growing in tin mining catchment. *International Journal of Environmental Science and Technology, 8*(2), 401–416.

Ashworth, J., Barnes, C., Oates, P., & Schaw, A. (2005). *Indicators for land contaminants science*. Environment agency report SC030039/SR. Bristol Environment Agency.

Babak, L., Šupinova, P., Zichova, M., Burdychova, R., & Vitova, E. (2012). *Biosorption of Cu, Zn and Pb by thermophilic bacteria–effect of biomass concentration on biosorption capacity*. Acta Universitatis Agriculturae Et Silviculturae Mendelianae Brunensis LX (5).

Baker, A. J. M., & Brooks, R. R. (1989). Terrestrial higher plants which hyperaccumulate metallic elements. A review of their distribution, ecology and phytochemistry. *Biorecovery, 1*(2), 81–126.

Banerjee, K., Ramesh, S. T., Nidheesh, P. V., & Bharathi, K. S. (2012). A novel agricultural waste adsorbent, watermelon shell for the removal of copper from aqueous solutions. *Iranica Journal of Energy & Environment, 3*, 143–156.

Banuelos, G., Terry, N., Leduc, D. L., Pilon-Smits, E. A. H., & Mackey, B. (2005). Field trial of transgenic Indian mustard plants shows enhanced phytoremediation of selenium-contaminated sediment. *Environmental Science & Technology, 39*, 1771–1777.

Bauddh, K., & Singh, R. P. (2012). Growth, tolerance efficiency and phytoremediation potential of *Ricinus communis* (L.) and *Brassica juncea* (L.) in salinity and drought affected cadmium contaminated soil. *Ecotoxicology and Environmental Safety, 85*, 13–22.

Bennicelli, R., Stezpniewska, Z., Banach, A., Szajnocha, K., & Ostrowski, J. (2004). The ability of *Azolla caroliniana* to remove heavy metals (Hg(II), Cr(III), Cr(VI)) from municipal waste water. *Chemosphere, 55*, 141–146.

Benyoucef, S., & Amrani, M. (2011). Adsorption of phosphate ions onto low cost Aleppo pine adsorbent. *Desalination, 275*, 231–236.

Bhatia, M., & Goyal, D. (2014). Analyzing remediation potential of wastewater through wetland plants: A review. *Environmental Progress & Sustainable Energy, 33*, 9–27.

Bouldin, J. L., Farris, J. L., Moore, M. T., Smith, J. S., & Cooper, C. M. (2006). Hydroponic uptake of atrazine and lambda-cyhalothrin in *Juncus effusus* and *Ludwigia peploides*. *Chemosphere, 65*, 1049–1057.

Boule, K. M., Vicente, J. A. F., Nabais, C., Prasad, M. N. V., & Freitas, H. (2009). Ecophysiological tolerance of duckweeds exposed to copper. *Aquatic Toxicology, 91*, 1–9.

Boyd, C. E. (1970). Vascular aquatic plants for mineral nutrient removal from polluted waters. *Economic Botany, 24*, 95–103.

Brooks, R. R. (1998). Phytochemistry of hyperaccumulators. In R. R. Brooks (Ed.), *Plants that hyperaccumulate heavy metals: Their role in phytoremediation, microbiology, archaeology, mineral exploration and phytomining* (pp. 15–54). Wallingford: CAB International.

Carvalho, W. S., Martins, D. F., Gomes, F. R., Leite, I. R., Gustavo da Silva, L., Ruggiero, R., & Richter, E. M. (2011). Phosphate adsorption on chemically modified sugarcane bagasse fibres. *Biomass and Bioenergy, 35*, 3913–3919.

Chandra, R., & Yadav, S. (2010). Potential of Typha angustifolia for phytoremediation of heavy metals from aqueous solution of phenol and melanoidin. *Ecological Engineering, 36*, 1277–1284.

Chandra, R., & Yadav, S. (2011). Phytoremediation of Cd, Cr, Cu, Mn, Fe, Ni, Pb and Zn from aqueous solution using *Phragmites communis, Typha angustifolia* and *Cyperus esculentus*. *International Journal of Phytoremediation, 13*, 580–591.

Chaney, R. L., Angle, J. S., Broadhurst, C. L., Peters, C. A., Tappero, R. V., & Sparks, D. L. (2007). Improved understanding of hyperaccumulation yields commercial phytoextraction and phytomining technologies. *Journal of Environmental Quality, 36*, 1429–1443.

Chatterjee, S., Kumar, A., Basu, S., & Dutta, S. (2012). Application of response surface methodology for methylene blue dye removal from aqueous solution using low cost adsorbent. *Chemical Engineering Journal, 181–182*, 289–299.

Chaudhry, Q., Schroder, P., Werck-Reichhart, D., Grajek, W., & Marecik, R. (2002). Prospects and limitations of phytoremediation for the removal of persistent pesticides in the environment. *Environmental Science and Pollution Research, 9*, 4–17.

Chen, J. C., Wang, K. S., Chen, H., Lu, C. Y., Huang, L. C., Li, H. C., Peng, T. H., & Chang, S. H. (2009). Phytoremediation of Cr(III) by *Ipomoea aquatica* (water spinach) from water in the presence of EDTA and chloride: Effects of Cr speciation. *Bioresource Technology, 101*, 3033–3039.

Chen, J.-C., Wang, K.-S., Chen, H., Lu, C.-Y., Huang, L.-C., Li, L.-C., Peng, T.-H., & Chang, C.-H. (2010). Phytoremediation of Cr(III) by Ipomonea aquatica (water spinach) from water in the presence of EDTA and chloride: Effects of Cr speciation. *Bioresource Technology, 101*(9), 3033–3039.

Cho, D., Chon, C., Kim, Y., Jeon, B., Schwartz, F. W., Lee, E., & Song, H. (2011). Adsorption of nitrate and Cr (VI) by cationic polymer-modified granular activated carbon. *Chemical Engineering Journal, 175*, 298–305.

Clemens, S. (2006). Toxic metal accumulation, responses to exposure and mechanisms of tolerance in plants. *Biochimie, 88*(11), 1707–1719.

Cobbett, C. S. (2000). Phytochelatins and their roles in heavy metal detoxification. *Plant Physiology, 123*(3), 825–832.

Cowgill, V. M. (1974). The hydro geochemical of Linsley Pond, North Braford. Part 2. The chemical composition of the aquatic macrophytes. *Archiv für Hydrobiologie, Supplement, 45*, 1–119.

Cronje, K. J., Chetty, K., Carsky, M., Sahu, J. N., & Meikap, B. C. (2011). Optimization of chromium(VI) sorption potential using developed activated carbon from sugarcane bagasse with chemical activation by zinc chloride. *Desalination, 275*, 276–284.

Dalcorso, G., Farinati, S., Maistri, S., & Furini, A. (2008). How plants cope with cadmium: Staking all on metabolism and gene expression. *Journal of Integrative Plant Biology, 50*(10), 1268–1280.

Danh, L. T., Truong, P., Mammucari, R., Tran, T., & Foster, N. (2009). Vetiver grass, *Vetiveria zizanioides*: A choice plant for phytoremediation of heavy metals and organic wastes. *International Journal of Phytoremediation, 11*, 664–691.

De Fillippis, L. F. (1979). The effect of sub-lethal concentrations of mercury and zinc in Chlorella: The counteraction of metal toxicity by selenium and sulphhydryl compounds. *Zeitschrift für Pflanzenphysiologie, 93*, 63–68.

Delgado, M., Bigeriego, M., & Guardiola, E. (1993). Uptake of Zn, Cr and Cd by water hyacinth. *Water Research, 27*, 269.

Denny, P. (1980). Solute movement in submerged angiosperms. *Biological Reviews, 55*, 65–92.

Denny, P. (1987). Mineral cycling by wetland plants a review. *Archiv fur Hydrobiologie Beith, 27*, 1–25.

Dhir, B., Sharmila, P., & Saradhi, P. P. (2009). Potential of aquatic macrophytes for removing contaminants from the environment. *Critical Reviews in Environmental Science and Technology, 39*, 754–781.

Dixit, A., Dixit, S., & Goswami, S. (2011). Process and plants for wastewater remediation: A review. *Scientific Reviews and Chemical Communications, 1*(1), 71–77.

El-Mekkawi, D., & Galal, H. R. (2013). Removal of a synthetic dye "Direct Fast Blue B2RL" via adsorption and photocatalytic degradation using low cost rutile and Degussa P25 titanium dioxide. *Journal of Hydro-Environment Research, 7*, 219–226.

Euliss, K., Ho, C. H., Schwab, A. P., Rock, S., & Banks, M. K. (2008). Greenhouse and field assessment of phytoremediation for petroleum contaminants in a Riparian zone. *Bioresource Technology, 99*, 1961–1971.

Feng, N., & Guo, X. (2012). Characterization of adsorptive capacity and mechanisms on adsorption of copper, lead and zinc by modified orange peel. *Transactions of the Nonferrous Metals Society of China, 22*, 1224–1231.

Fernandez, R. T., Whitwell, T., Riley, M. B., & Bernard, C. R. (1999). Evaluating semiaquatic herbaceous perennials for use in herbicide phytoremediation. *Journal of the American Society for Horticultural Science, 124*, 539.

Flores-Garnica, J. G., Morales-Barrera, L., Pineda-Camacho, G., & Cristiani-Urbina, E. (2013). Biosorption of Ni(II) from aqueous solutions by Litchi chinensis seeds. *Bioresource Technology, 136*, 635–643.

Fritioff, A., Kautsky, L., & Greger, M. (2005). Influence of temperature and salinity on heavy metal uptake by submersed plants. *Environmental Pollution, 133*, 265–274.

Gao, J., Garrison, A. W., Mazur, C. S., Wolfe, N. L., & Hoehamer, C. F. (2000). Uptake and phytotransformation of o, p′-DDT and p, p′-DDT by axenically cultivated aquatic plants. *Journal of Agricultural and Food Chemistry, 48*(12), 6121–6127.

Ghavri, S. V., Bauddh, K., Kumar, S., & Singh, R. P. (2013). Bioaccumulation and translocation potential of Na+ and K+ in native weeds grown on industrially contaminated soil. *International Journal of ChemTech Research, 5*(4), 1869–1875.

Goncalves, C., & Alpendurada, M. F. (2005). Assessment of pesticide contamination in soil samples from an intensive horticulture area, using ultrasonic extraction and gas chromatography-mass spectrometry. *Talanta, 65*, 1179–1189.

Guo, J., Xu, W., & Ma, M. (2012). The assembly of metals chelation by thiols and vacuolar compartmentalization conferred increased tolerance to and accumulation of cadmium and arsenic in transgenic *Arabidopsis thaliana*. *Journal of Hazardous Materials, 199–200*, 309–313.

Guo, W., Zhang, H., & Huo, S. (2014). Organochlorine pesticides in aquatic hydrophyte tissues and surrounding sediments in Baiyangdian wetland, China. *Ecological Engineering, 67*, 150–155.

Ha, H., Olson, J., Bian, L., & Rogerson, P. A. (2014). Analysis of heavy metals sources in soil using kriging interpolation on principal components. *Environmental Science and Technology, 48*, 4999–5007.

Haddad, M. E., Mamouni, R., Saffaj, N., & Lazar, S. (2014). Evaluation of performance of animal bone meal as a new low cost adsorbent for the removal of a cationic dye Rhodamine B from aqueous solutions. *Journal of Saudi Chemical Society*. in Press.

Hashemian, S., & Salimi, M. (2012). Nano composite a potential low cost adsorbent for removal of cyanine acid. *Chemical Engineering Journal, 188*, 57–63.

Hemen, S. (2011). Metal hyperaccumulation in plants: A review focusing on phytoremediation technology. *Journal of Environmental Science and Technology, 4*(2), 118–138.

Hossain, M. A., & Fujita, M. (2009). Purification of glyoxalase I from onion bulbs and molecular cloning of its cDNA. *Bioscience Biotechnology and Biochemistry, 73*(9), 2007–2013.

Hossain, M. A., Hossain, M. Z., & Fujita, M. (2009). Stress-induced changes of methylglyoxal level and glyoxalase I activity in pumpkin seedlings and cDNA cloning of glyoxalase I gene. *Australian Journal of Crop Science, 3*(2), 53–64.

Hossain, M. A., Piyatida, P., Teixeira da Silva, J. A., & Fujita, M. (2012). Molecular mechanism of heavy metal toxicity and tolerance in plants: Central role of glutathione in detoxification of reactive oxygen species and methylglyoxal and in heavy metal chelation. *Journal of Botany, 2012*, 1–37. https://doi.org/10.1155/2012/872875.

Hu, M. J., Wei, Y. L., Yang, Y. W., & Lee, J. F. (2003). Immobilization of chromium (VI) with debris of aquatic plants. *Environmental Contamination and Toxicology, 71*, 0840–0847.

Hu, C., Zhang, L., Hamilton, D., Zhou, W., Yang, T., & Zhu, D. (2007). Physiological responses induced by copper bioaccumulation in Eichhornia crassipes (Mart.). *Hydrobiologia, 579*(1), 211–218.

Hutchinson, G. E. (1975). *A treatise on limnology*. London: Wiley.

Jadia, C. D., & Fulekar, M. H. (2009). Review on phytoremediation of heavy metals: Recent techniques. *African Journal of Biotechnology, 8*(6), 921.

Jafari, N. (2010). Ecological and socio-economic utilization of water hyacinth *(E. crassipes Mart Solms)*. *Journal of Applied Sciences and Environmental Management, 14*, 2.

Jaison, S., & Muthukmar, T. (2016). Chromium accumulation in medicinal plants growing naturally on tannery contaminated and non-contaminated soils. *Biological Trace Element Research*. https://doi.org/10.1007/s12011-016-0740-1.

Jiménez-Cedillo, M. J., Olguín, M. T., Fall, C., & Colin-Cruz, A. (2013). As(III) and As(V) sorption on iron-modified non-pyrolyzed and pyrolyzed biomass from *Petroselinum crispum* (parsley). *Journal of Environmental Management, 117*, 242–252.

Kamal, M., Ghaly, A. E., Mahmoud, N., & Cote, R. (2004). Phytoaccumulation of heavy metals by aquatic plants. *Environment International, 29*, 1029–1039.

Kamel, A. K. (2013). Phytoremediation potentiality of aquatic macrophytes in heavy metal contaminated water of El-Temsah Lake, Ismailia, Egypt. *Middle-East Journal of Scientific Research, 14*(12), 1555–1568.

Karakagh, R. M., Chorom, M., Motamedi, H., YuseKalkhajeh, Y. K., & Oustan, S. (2012). Biosorption of Cd and Ni by inactivated bacteria isolated from agricultural soil treated with sewage sludge. *Ecohydrology and Hydrobiology, 12*(3), 191–198.

Kasim, W. A. (2005). The correlation between physiological and structural alterations induced by copper and cadmium stress in broad beans (*Vicia faba* L.). *Egyptian Journal of Biology, 7*, 20–32.

Kelly, M. G., & Whitton, B. A. (1989). Interspecific differences in Zn, Cd and Pb accumulation by freshwater algae and bryophytes. *Hydrobiologia, 175*(1), 12.

Kelly-Vargas, K., Cerro-Lopez, M., Reyna-Tellez, S., Bandala, E. R., & Sanchez-Salas, J. L. (2012). Biosorption of heavy metals in polluted water, using different waste fruit cortex. *Physics and Chemistry of the Earth, 37–39*, 26–29.

Kılıc, M., Kırbıyık, Ç., Çepelioˇgullar, Ö., & Pütüna, A. E. (2013). Adsorption of heavy metal ions from aqueous solutions by bio-char, aby-product of pyrolysis. *Applied Surface Science, 283*, 856–862.

Kiyono, M., Oka, Y., Sone, Y., Tanaka, M., Nakamura, R., Sato, M. H., Pan-Hou, H., Sakabe, K., & Inoue, K. (2012). Expression of bacterial heavy metal transporter MerC fused with a plant SNARE, SYP121 in *Arabidopsis thaliana* increases cadmium accumulation and tolerance. *Planta, 235*, 841–850.

Kumar, D., & Kumar, N. (2016). Tannery effluent toxicity assessment on the growth and germination of phaseolus vulgaris L (Bean). *International Journal of Green and Herbal Chemistry, 5*(2), 139–144.

Kumar, I. N., & Oommen, C. (2012). Removal of heavy metals by biosorption using freshwater algae *Spirogyra hyaline*. *Journal of Environmental Biology, 33*, 27–31.

Kumar, J., Balomajumder, C., & Mondal, P. (2011). Application of agro-based biomasses for Zinc removal from wastewater–a review. *Clean: Soil, Air, Water, 39*, 641–652.

Kumar, N., Bauddh, K., Barman, S. C., & Singh, D. P. (2012). Accumulation of metals in selected macrophytes grown in mixture of drain water and tannery effluent and their phytoremediation potential. *Journal of Environmental Biology, 33*, 323–327.

Kumar, N., Bauddh, K., Kumar, S., Dwivedi, N., Singh, D. P., & Barman, S. C. (2013). Extractability and phytotoxicity of heavy metals present in petrochemical industry sludge. *Clean Technologies and Environmental Policy, 15*, 1033–1039.

Kumar, D., Singh, D. P., Barman, S. C., & Kumar, N. (2016). Heavy metal and their regulation in plant system: An overview. In *Plant responses to xenobiotics* (pp. 19–38). New York: Springer.

Lasat, M. M. (2002). Phytoextraction of toxic metals: A review of biological mechanisms. *Journal of Environmental Quality, 31*, 109–120.

Leblebici, Z., & Aksoy, A. (2011). Growth and lead accumulation capacity of *L. minor* and *Spirodela polyrhiza* (Lemnaceae): Interactions with nutrient enrichment. *Water, Air, and Soil Pollution, 214*, 175–184.

Li, Q., Chen, B., lin, P., Zhou, J., Zhan, J., Shen, Q., & Pan, X. (2016). Adsorption of heavy metal from aqueous solution by dehydrated root powder of long-root *E. crassipes*. *International Journal of Phytoremediation, 18*, 103–109.

Liu, D., Zou, J., Wang, M., & Jiang, W. (2008). Hexavalent chromium uptake and its effects on mineral uptake, antioxidant defence system and photosynthesis in *Amaranthus viridis* L. *Bioresource Technology, 99*(7), 2628–2636.

Low, K. S., Lee, C. K., & Tai, C. H. (1994). Biosorption of copper by water hyacinth roots. *Journal of Environmental Science and Health, Part A, 29*(1), 171.

Mahmood, Q., Zheng, P., Islam, E., Hayet, Y., Hassan, M. J., Jilani, G., & Jin, R. C. (2005). Lab scale studies on water Hyacinth (*E. crassipes* Marte Sloms) for biotreatment of textile waste water. *Caspian Journal of Environmental Sciences, 3*, 83–88.

Malik, R. N., Husain, S. Z., & Nazir, I. (2010). Heavy metal contamination and accumulation in soil and wild plant species from industrial area of Islamabad. *Pakistan Journal of Botany, 42*(1), 291–301.

Mallick, N., Singh, A. K., & Rai, L. C. (1990). Impact of bimetallic combinations of Cu, Ni and Fe on growth rate, uptake of nitrate and ammonium, $14CO_2$ fixations, nitrate reductase and urease activity of *Chlorella vulgaris*. *Biology of Metals, 2*, 223–228.

Manzoor, Q., Nadeem, R., Iqbal, M., Saeed, R., & Ansari, T. M. (2013). Organic acids pre-treatment effect on *Rosa bourboniaphyto* biomass for removal of Pb(II) and Cu(II) from aqueous media. *Bioresource Technology, 132*, 446–452.

Mapanda, F., Mangwayana, E. N., Nyanangara, J., & Giller, K. E. (2005). The effect of long-term irrigation using wastewater on the heavymetal contents of soils under vegetables in Harare, Zimbabwe. *Agriculture, Ecosystems & Environment, 107*, 151–165.

Marin-Rangel, V. M., Cortes-Martines, R., Villanueva, R. A. C., Garnica-Romo, M. G., & Martinez-Flores, H. E. (2012). As(V) biosorption in an aqueous solution using chemically treated lemon (*Citrus aurantifolia* Swingle) residues. *Journal of Food Science, 71*, 10–14.

Mazumdar, K., & Das, S. (2015). Phytoremediation of Pb, Zn, Fe, and Mg with 25 wetland plant species from a paper mill contaminated site in North East India. *Environmental Science and Pollution Research, 22*, 701–710.

Mejare, M., & Bulow, L. (2001). Metal-binding proteins and peptides in bioremediation and phytoremediation of heavy metals. *Trends in Biotechnology, 19*, 67–73.

Miglioranza, K. S. B., De Moreno, J. E. A., & Moreno, V. J. (2004). Organochlorine pesticides sequestered in the aquatic hydrophyte Schoenoplectus californicus (C. A. Meyer) Soják from a shallow lake in Argentina. *Water Research, 38*, 1765–1772.

Miretzky, P., Saralegui, A., & Cirelli, A. F. (2004). Aquatic macrophytes potential for the simultaneous removal of heavy metals (Buenos Aires, Argentina). *Chemosphere, 57*, 997–1005.

Mkandawire, M., Taubert, B., & Dude, E. G. (2004). Capacity of *Lemna gibba* L. (Duckweed) for uranium and arsenic phytoremediation in mine tailing waters. *International Journal of Phytoremediation, 6*(4), 347–362.

Mohamad, H. H., & Latif, P. A. (2010). Uptake of Cadmium and Zinc from synthetic effluent by water hyacinth (*E. crassipes*). *Environment Asia, 3*, 36–42.

Molisani, M. M., Rocha, R., Machado, W., Barreto, R. C., & Lacerda, L. D. (2006). Mercury contents in aquatic macrophytes from two reservoirs in the paraiba do sul: Guandu river system, Se Brazil. *Brazilian Journal of Biology, 66*, 101–107.

Mukherjee, S., Bhattacharya, P., & Duttagupta, A. K. (2004). Heavy metal levels and esterase variations between metal-exposed and unexposed duckweed *L. minor*: Field and laboratory studies. *Environment International, 30*, 811–814.

Naeem, K., Yawar, W., Akhter, P., & Rehana, I. (2010). Atomic absorption spectrometric determination of cadmium and lead in soil after total digestion. *Asia-Pacific Journal of Chemical Engineering, 7*, 295–301. https://doi.org/10.1002/apj.535.

Nedunuri, K. V., Lowell, C., Meade, W., Vonderheide, A. P., & Shann, J. R. (2009). Management practices and phytoremediation by native grasses. *International Journal of Phytoremediation, 12*(2), 200–214.

Nguyen, T. T. T., Davy, F. B., Rimmer, M., & De Silva, S. (2009). *Use and exchange of genetic resources of emerging species for aquaculture and other purposes*. FAO/ NACA expert meeting on the use and exchange of aquatic genetic resources relevant for food and agriculture, Chonburi, Thailand.

Nguyen, T. A. H., Ngo, H. H., Guo, W. S., Zhang, J., Liang, S., Yue, Q. Y., Li, Q., & Nguyen, T. V. (2013). Applicability of agricultural waste and by-products for adsorptive removal of heavy metals from wastewater: Review. *Bioresource Technology, 148*, 574–585.

Ofomaja, A. E., Naidoo, E. B., & Modise, S. J. (2010). Biosorption of Cu(II) and Pb(II) onto potassium hydroxide treated pine cone powder. *Journal of Environmental Management, 91*, 1674–1685.

Olette, R., Couderchet, M., Biagianti, S., & Eullaffroy, P. (2008). Toxicity and removal of pesticides by selected aquatic plants. *Chemosphere, 70*, 1414–1421.

Olette, R., Couderchet, M., & Eullaffroy, P. (2009). Phytoremediation of fungicides by aquatic macrophytes: Toxicity and removal rate. *Ecotoxicology and Environmental Safety, 72*, 2096–2101.

Outridge, P. M., & Noller, B. N. (1991). Accumulation of toxic trace elements by freshwater vascular plants. In *Reviews of environmental contamination and toxicology* (pp. 1–63). New York: Springer.

Park, J., Hung, I., Gan, Z., Rojas, O. J., Lim, K. M., & Park, S. (2013). Activated carbon from biochar: Influence of its physicochemical properties on the sorption characteristics of phenanthrene. *Bioresource Technology, 149*, 383–389.

Parmar, S., & Singh, V. (2015). Phytoremediation approaches for heavy metal pollution: A review. *Journal of Plant Science & Research, 2*, 135.

Pierre, V., Terry, M., & Madeleine, S. G. (2011). Compartmentation of metals in foliage of *Populus tremula* grown on soil with mixed contamination from the tree crown to leaf cell level. *Environmental Pollution, 159*, 324–336.

Prakash, B. S., & Kumar, S. V. (2013). Batch removal of heavy metals by biosorption onto marine algae-equilibrium and kinetic studies. *International Journal of ChemTech Research, 5*(3), 1254–1262.

Prasertsup, P., & Ariyakanon, N. (2011). Removal of Chlorpyrifos by Water Lettuce (*P. stratiotes* L.) and Duckweed (*L. minor* L.). *International Journal of Phytoremediation, 13*(4), 383–395.

Priya, E. S., & Selvan, P. S. (2013). Water hyacinth (*E. crassipes*) – An efficient and economic adsorbent for textile effluent treatment – A review. *Arabian Journal of Chemistry*. https://doi.org/10.1016/j.arabjc.2014.03.002.

Qadir, S., Qureshi, M. I., Javed, S., & Abdin, M. Z. (2004). Genotypic variation in phytoremediation potential of *Brassica juncea* cultivars exposed to Cd stress. *Plant Science, 167*, 1171–1181.

Rahman, S. M. B., Kumar, S., Sayeed, M. A. B., Sabbir, M. W., Hasanuzzaman, A. F. M., Alam, M. I., & Sarower, M. G. (2008). Ecological diversity and distribution of aquatic and semi-aquatic weeds in Khulna district, Bangladesh. *South Asian Journal of Agriculture, 3*(1&2), 163–168.

Rai, P. K. (2008). Phytoremediation of Hg and Cd from industrial effluents using an aquatic free floating macrophyte *Azolla pinnata*. *International Journal of Phytoremediation, 10*, 430–439.

Rai, P. K. (2009). Heavy metal phytoremediation from aquatic ecosystems with special reference to macrophytes. *Environmental Science & Technology, 39*, 697–753.

Rai, P. K. (2010). Micrococosom investigation of phytoremediation of Cr using *Azolla pinnata*. *International Journal of Phytoremediation, 12*, 96–104.

Rai, L. C., Gaur, J. P., & Kumar, H. D. (1981). Phycology and heavy metal pollution. *Biological Reviews of the Cambridge Philosophical Society, 56*, 99–151.

Rai, U. N., Sinha, S., Tripathi, R. D., & Chandra, P. (1995). Waste water treatability potential of some aquatic macrophytes: Removal of heavy metals. *Ecological Engineering, 5*, 5–12.

Rajoriya, S., & Kaur, B. (2014). Adsorptive removal of Zinc from waste water by natural biosorbents. *International Journal of Engineering and Science Invention, 3*(6), 60–80.

Reed, R. H., & Gadd, G. M. (1990). Metal tolerance in eukaryotic and prokaryotic algae. In A. J. Shaw (Ed.), *Heavy metal tolerance in plants: Evolutionary aspects* (pp. 105–118). Boca Raton: CRC Press.

Ren, Z., Xu, X., Gao, B., Yue, Q., & Song, W. (2015). Integration of adsorption and direct bio-reduction of perchlorate on surface of cotton stalk based resin. *Journal of Colloid and Interface Science, 459*, 127–135.

Rezania, S., Ponraj, M., Din, M. F. M., Chelliapan, S., & Sairan, F. M. (2016). Effectiveness of *E. crassipes* in nutrient removal from domestic wastewater based on its optimal growth rate. *Desalination and Water Treatment, 57*, 360–365.

Rizwana, M., Darshan, M., & Nilesh, D. (2014). Phytoremediation of textile waste water using potential wetland plant: Eco sustainable approach. *International Journal of Interdisciplinary and Multidisciplinary Studies, 1*(4), 130–138.

Ruiz, O. N., Alvarez, D., Gonzalez-Ruiz, G., & Torres, C. (2011). Characterization of mercury bioremediation by transgenic bacteria expressing metallothionein and polyphosphate kinase. *BMC Biotechnology, 11*, 82–89.

Sakakibara, M., Ohmoril, Y., Ha, N. T. H., Sano, S., & Sera, K. (2011). Phytoremediation of heavy metal-contaminated water and sediment by *Eleocharis acicularis*. *CLEAN – Soil, Air, Water, 39*(8), 735–741.

Sanghamitra, K., Prasada, R. P. V. V., & Naidu, G. R. K. (2011). Heavy metal tolerance of weed species and their accumulations by phytoextraction. *Indian Journal of Science and Technology, 4*(3), 285–290.

Sasmaz, A., Obek, E., & Hasar, H. (2008). The accumulation of heavy metals in *Typha latifolia* L. grown in a stream carrying secondary effluent. *Ecological Engineering, 33*, 278–284.
Sasmaz, M., Topal, E. I. A., Obek, E., & Sasmaz, A. (2015). The potential of *Lemna gibba* L. and *L. minor* L. to remove Cu, Pb, Zn, and As in gallery water in a mining area in Keban, Turkey. *Journal of Environmental Management, 163*, 246–253.
Sasmaza, A., Obekb, E., & Hasarb, H. (2009). The accumulation of heavy metals in Typha latifolia L. Grown in a stream carrying secondary effluent. *Ecological Engineering, 33*, 278–284.
Sharma, H. (2011). Metal hyperaccumulation in plants: A review focusing on phytoremediation technology. *Journal of Environmental Science and Technology, 4*(2), 118–138.
Sharma, S. S., & Dietz, K. J. (2009). The relationship between metal toxicity and cellular redox imbalance. *Trends in Plant Science, 14*(1), 43–50.
Sharma, S. S., & Gaur, J. P. (1995). Potential of *Lemna polyrrhiza* for removal of heavy metals. *Ecological Engineering, 4*, 37–43.
Sharma, S., Singh, B., & Manchanda, V. K. (2015). Phytoremediation: Role of terrestrial plants and aquatic macrophytes in the remediation of radionuclides and heavy metal contaminated soil and water. *Environmental Science and Pollution Research, 22*, 946–962.
Silveira, M. L., Vendramini, J. M. B., Sui, X. L., Sollenberger, L., & O'Connor, G. A. (2013). Screening perennial warm-season bioenergy crops as an alternative for phytoremediation of excess soil P. *Bioenergy Research, 6*, 469–475.
Singh, R., Singh, D. P., Kumar, N., Bhargava, S. K., & Barman, S. C. (2010). Accumulation and translocation of heavy metals in soil and plants from fly ash contaminated area. *Journal of Environmental Biology, 31*, 421–430.
Song, W., Xu, X., Tan, X., Wang, Y., Ling, J., Gao, B., & Yue, Q. (2015). Column adsorption of perchlorate by amine-crosslinked biopolymer based resin and its biological, chemical regeneration properties. *Carbohydrate Polymers, 115*, 432–438.
Sood, A., Uniyal, P. L., Prasanna, R., & Ahluwalia, S. A. (2012). Phytoremediation potential of aquatic macrophyte, Azolla. *Ambio, 41*, 122–137.
Srivastava, S., Mishra, S., Dwivedi, S., & Tripathi, R. (2010). Role of thio-metabolism in arsenic detoxification in *Hydrilla verticillata* (L.f.) Royle. *Water, Air, and Soil Pollution, 212*, 155–165.
Srivastava, S., Srivastava, M., Suprasanna, S., & D'Souza, F. (2011). Phytofiltration of arsenic from simulated contaminated water using *Hydrilla verticillata* in field conditions. *Ecological Engineering, 37*, 1937–1941.
Suksabye, P., & Thiravetyan, P. (2012). Cr(VI) adsorption from electroplating plating wastewater by chemically modified coir pith. *Journal of Environmental Management, 102*, 1–8.
Susselan, K. N., Salskar, D. A., Suvarna, S., Udas, A., & Bhagawat, A. (2006). Uptake of mercury, cadmium, uranium and zinc by *Mimosa pudica*. *Indian Journal of Plant Physiology, 11*, 432–436.
Thayaparan, M., Iqbal, S. S., Chathuranga, P. K. D., & Iqbal, M. C. M. (2013). Rhizofiltration of Pb by *Azolla pinnata*. *International Journal of Environmental Sciences, 3*, 6.
Tomar, V., Prasad, S., & Kumar, D. (2014). Adsorptive removal of fluoride from water samples using Zr–Mn composite material. *Microchemical Journal, 111*, 116–124.
Trevors, J. T., Stratton, G. W., & Gadd, G. M. (1986). Cadmium transport, resistance and toxicity in bacteria, algae and fungi. *Journal of Microbiology, 32*, 447–464.
Tuzen, M., & Sarı, A. (2010). Biosorption of selenium from aqueous solution by green algae (*Cladophora hutchinsiae*) biomass: Equilibrium, thermodynamic and kinetic studies. *Chemical Engineering Journal, 158*, 200–206.
Uçar, S., Erdem, M., Tay, T., & Karagöz, S. (2014). Removal of lead (II) and nickel (II) ions from aqueous solution using activated carbon prepared from rapeseed oil cake by Na_2CO_3 activation. *Clean Technologies and Environmental Policy, 17*, 747–756.
Unnikannan, P., Baskaran, L., Chidambaram, A. L. A., & Sundaramoorthy, P. (2013). Chromium phytotoxicity in tree species and its role on phytoremediation. *Insight Botany, 3*, 15–25.
Upadhyay, A. R., Mishra, V. K., Pandey, S. K., & Tripathi, B. D. (2007). Biofiltration of secondary treated municipal wastewater in a tropical city. *Ecological Engineering, 30*, 9–15.

Valipour, A., & Ahn, Y. H. (2016). Constructed wetlands as sustainable ecotechnologies in decentralization practices: A review. *Environmental Science and Pollution Research, 23*, 180–197.

Varun, M., D'Souza, P. J., & Paul, M. S. (2012). Metal contamination of soils and plants associated with the glass industry in North Central India: Prospects of phytoremediation. *Environmental Science and Pollution Research, 19*, 269–281.

Vázquez, G., Mosquera, O., Freire, M. S., Antorrena, G., & González-álvarez, J. (2012). Alkaline pre-treatment of waste chestnut shell from a food industry to enhance cadmium, copper, lead and zinc ions removal. *Chemical Engineering Journal, 184*, 147–155.

Verma, V. K., Gupta, R. K., & Rai, J. P. N. (2005). Biosorption of Pb and Zn from pulp and paper industry effluent by water hyacinth. *Journal of Scientific and Industrial Research, 64*, 778–781.

Vesely, T., Tlustos, P., & Szakova, J. (2011). The use of water lettuce (*P. stratiotes*) for rhizofiltration of a highly polluted solution by cadmium and lead. *International Journal of Phytoremediation, 13*, 859–872.

Wallen, D. G. (1990). The toxicity of chromium (VI) to photosynthesis of the phytoplankton assemblage of Lake Erie and the diatom *Fragilaria crotonensis*. *Aquatic Botany, 38*, 331–340.

Wasewar, K. L., Mohammad, A., Prasad, B., & Mishra, I. M. (2008). Adsorption of Zn using factory tea waste: Kinetics, equilibrium and thermodynamics. *Clean: Soil, Air, Water, 36*(3), 320–329.

Whitton, B. A., Burrows, I. G., & Kelly, M. G. (1989). Use of *Cladophora glomerata* to monitor heavy metals in rivers. *Journal of Applied Phycology, 1*, 293–299.

Witek-Krowiak, A., Szafran, R. G., & Modelski, S. (2011). Biosorption of heavy metals from aqueous solutions onto peanut shell as a low-cost biosorbent. *Desalination, 265*(1–3), 126–134.

Wolverton, B. C. A. (1975). Water hyacinth for removal of phenols from polluted waters, NASA Tech. Memo. (TM-X-72722), 18p. *Science Technology Aerospace Report 13*(7), 79.

Wong, P. T. S., & Chau, Y. K. (1990). Zinc toxicity to fresh water algae. *Toxicity Assess, 5*, 167–177.

Xia, H., & Ma, X. (2006). Phytoremediation of ethion by water hyacinth (*E. crassipes*) from water. *Bioresource Technology, 97*, 1050–1054.

Xu, X., Gao, B., Tan, X., Zhang, X., Yue, D., & Yue, Q. (2013a). Uptake of perchlorate from aqueous solutions by amine-crosslinked cotton stalk. *Carbohydrate Polymers, 98*, 132–138.

Xu, X., Gao, B., Yue, Q., Li, Q., & Wang, Y. (2013b). Nitrate adsorption by multiple biomaterial based resins: Application of pilot-scale and lab-scale products. *Chemical Engineering Journal, 234*, 397–405.

Xu, X., Gao, B., Huang, X., Ling, J., Song, W., & Yue, Q. (2015). Physicochemical characteristics of epichlorohydrin, pyridine and trimethylamine functionalized cotton stalk and its adsorption/desorption properties for perchlorate. *Journal of Colloid and Interface Science, 440*, 219–228.

Yuan, Y., Yu, S., Banuelos, G. S., & He, Y. (2016). Accumulation of Cr, Cd, Pb, Cu, and Zn by plants in tanning sludge storage sites: Opportunities for contamination bioindication and phytoremediation. *Environmental Science and Pollution Research, 23*, 22477–22487.

Zhang, X., Lin, A. J., Zhao, F. J., Xu, G. Z., Duan, G. L., & Zhu, Y. G. (2008). Arsenic accumulation by aquatic fern Azolla: Comparison of arsenate uptake, speciation and efflux by *A. caroliniana* and *A. filiculoides*. *Environmental Pollution, 156*, 1149–1155.

Zhang, X., Zhao, F. J., Huang, Q., Williams, P. N., Sun, G. X., & Zhu, Y. G. (2009). Arsenic uptake and speciation in the rootless duckweed *Wolffia globosa*. *New Phytologist, 182*, 421–428.

Zhu, Y. L., Zayed, A. M., Quian, J.-H., de Souza, M., & Terry, N. (1999). Phytoaccumulation of trace elements by wetland plants: II Water hyacinth. *Journal of environmental quality, 28*, 339–344.

Zhu, Y.-G., Ralf, K., & Tong, Y.-P. (2004). Vacuolar compartmentalization: A second-generation approach to engineering plants for phytoremediation. *Trends in Plant Science, 9*(1), 7–9.

Chapter 4
Environmental Health Hazards of Post-Methanated Distillery Effluent and Its Biodegradation and Decolorization

Sangeeta Yadav and Ram Chandra

Contents

Abstract Anaerobically digested distillery effluent is a mixture of complex organic and inorganic pollutants which is composed of several plant sterols which do not only affect the water quality but also aquatic flora and fauna. Research has revealed the adverse effects of post-methanated distillery effluent (PMDE) on the seed germination and plant growth of *Phaseolus mungo* even at lower concentrations. Studies have also showed the adverse effect on soil fertility by inhibiting the nitrogen-fixing bacteria and root nodulation. The major colorant of distillery effluent is melanoidin, reaction product of amino-carbonyl compounds at elevated temperature in the sugar industries and distilleries due to condensation reaction. Due to its high solubility in aquatic ecosystem and negative charge, it makes complexation with all the humic substances and heavy metals in the environment. Therefore, the decolorization and degradation of PMDE is still a global challenge due to its complexity. The physical, chemical, and biological techniques have been attempted for

S. Yadav · R. Chandra (✉)
Department of Environmental Microbiology, Babasaheb Bhimrao Ambedkar University (A Central University), Lucknow, Uttar Pradesh, India
e-mail: rc_microitrc@yahoo.co.in; prof.chandrabbau@gmail.com

R. C. Sobti et al. (eds.), *Environmental Biotechnology: For Sustainable Future*,
https://doi.org/10.1007/978-981-10-7284-0_4

its detoxification and color removal but still warranted for its feasible application. Manganese peroxidase (MnP) and laccase have been reported as key enzymes from fungi and bacteria. During the degradation process of PMDE, different metabolic products through GC-MS/MS analysis have also been characterized. The integration of bacterial treatment with constructed wetland plant treatment (*Phragmites communis*, *Typha angustifolia*, and *Cyperus esculentus*) technique has been reported recently as an effective approach for decolorization and degradation of PMDE. The major challenge of PMDE biodegradation and decolorization is its high total dissolved solids (TDS) containing complex organic pollutants including heavy metals. The high TDS is a result of precipitation of metal sulfides during anaerobic digestion of distillery spentwash due to complexation of heavy metals and sulfates which impose inhibitory effects on the microorganisms, consequently inhibiting the biodegradation process. Several complex organic pollutants present in PMDE have been also reported as endocrine-disrupting chemicals (EDCs) which directly affect the aquatic and terrestrial ecosystems.

Keywords Degradation · Decolorization · Ligninolytic enzymes · Metabolites · Phytoremediation · Post-methanated distillery effluent

Introduction

The disposal of wastewater generated from sugarcane molasses-based alcohol industries is a global challenge and environmental threat for sustainable development. There is production of huge amount of wastewater, i.e., 90 m^3 of effluent from per kilo liter of alcohol production. However, wastewater discharge standard fixed by Environmental Protection Agency (EPA) is 12 m^3 per kilo liter of alcohol produced (EPA 2008). Hence it is more than seven times higher discharge in comparison to the prescribed standard. Moreover, the wastewater remains very dark, viscous, and chemically complex even after secondary treatment which causes water and soil pollution due to its indiscriminate disposal in the environment. Therefore, alcohol industries are rated among the 17 most polluting industries in India. Currently around 397 distilleries are operating in India. In international scenario two countries, i.e., the United States and Brazil, are playing leading role for production of ethanol as fuel. However, the countries following them and rising very fast are India, China, Canada, European, Argentina, Colombia, and Thailand. Moreover, the world's total production of alcohol from cane molasses is more than 13 million m^3/annum (Khairnar et al. 2013)

In alcohol industries, major wastewater is generated during distillation process, where only 5–12% of ethanol is recovered, and 88–95% volume of wastewater is generated from fermented molasses which is known as spentwash. The spentwash remains acidic in nature and contains a variety of recalcitrant coloring compounds such as melanoidins, phenolics, metal sulfide, and methylated complex organics that contribute toxicity and color to distillery effluent. The pH of spentwash increases from 4.5 to 8.5 during the anaerobic treatment process which is finally known as post-methanated distillery effluent (PMDE). Sugarcane molasses contains high level of caramelized sugar, heavy metals, sulfates, and phenolics along with the fermentable sugar. But, during the fermentation process, the fermentable sugar gets converted into ethyl alcohol, leaving the complex, nondegradable organic residues as spent after distillation process. The water soluble hemicelluloses, proteins, gums, nonsugars, and organic compounds present in sugarcane juice also come into the spentwash in their original or transformed forms. Besides this, there is a high chemical oxygen demand (COD) and biological oxygen demand (BOD) of the spentwash (Chandra et al. 2008a). During conventional secondary treatment, i.e., anaerobic digestion of spentwash, phenolics and butyric acid are produced by reducing sugars, and methane is emitted. Phenolics, sulfides, and heavy metals are left which acted as inhibitors for biodegradation of PMDE during the tertiary treatment process (Borja et al. 1993). The heavy metals also form complexes with sulfides, and these complexes result in toxicity of the effluent (Krishnanand et al. 1993; Kumar and Chandra 2004). Apart from this, distillery wastewater also contains nitrogen (1660–4200 mg/l), phosphorus (225–3038 mg/l), and potassium (9600–17,475 mg/l) as nutrients that can lead to the eutrophication of water bodies. Indiscriminate disposal or application of this effluent on crop plants causes health hazards for environment, humans, and animals. Chandra et al. (2009) have reported accumulation and distribution pattern of metals in wheat (*Triticum aestivum* L.) and Indian mustard (*Brassica campestris* L.) irrigated with mixed effluent of distilleries and tanneries. The study revealed that these plants have accumulated metals in different parts beyond the permissible limit as suggested by Food and Agriculture Organization of the United Nations (FAO)/World Health Organization (WHO) when treated with mixed effluent of distillery and tannery. Maximum accumulation was of Fe, i.e., 340 mg/kg in wheat's root and 560 mg/kg in mustard's leaves, followed by Mn and Zn in the order root > shoot > leaves > seeds. Maximum increase in photosynthetic pigment and protein content was observed between 30–60 days and 60–90 days of plant growth, respectively. Another study with distillery sludge-amended soil (10, 20, 40, 60, 80, and 100%) showed 10% (w/w) sludge was favorable for *Phaseolus mungo*, while above 10% sludge concentration showed inhibitory effect. Soil with 10% (w/w) distillery sludge induced the growth of root, shoot, number of leaves, biomass, photosynthetic pigment, protein, and starch, while 20% (w/w) sludge-amended soil had variable effects on *P. mungo*. Concentrations >40% reduced all the growth parameters, viz., root length, shoot length, number of leaves, biomass, photosynthetic pigment, and protein content, of *P. mungo* (Chandra et al. 2008b). Hence, prior to discharge of effluent into the environment, its treatment is mandatory. Though various wastewater treatment approaches including physical, chemical, and biological have been reported, but the feasible and economic technology has to be developed. Recently the Central Pollution Control Board (CPCB) has recommended two technologies, namely, (1) concentration and incineration and (2) concentration and biocomposting, to achieve zero discharge. But, these technologies are still not operative due to cost feasibility. Complete chemical properties of PMDE and their degradation process are still to be understood. In the present review the available knowledge on the subject for the awareness and ignition of scientific society to manage this challenging problem has been discussed.

Physico-chemical Properties of PMDE

The spentwash generated after the distillation in industries is highly acidic in nature and has a variety of recalcitrant coloring compounds such as melanoidins, phenolics, and metal sulfides that are mainly responsible for the dark color of this effluent. The bad smell of effluent might be due to the presence of indole, skatole, and other sulfur compounds. Moreover, due to the presence of high amount of organic pollutants, most of the distilleries generate biogas from these by anaerobic digestion. In this process organic substances are converted into methane-rich biogas, which is utilized as fuel. This biogas normally contains 55–65% methane gas, which is well-recognized fuel gas with minimum air pollution potential. The main steps involved for methane production from spentwash are (1) hydrolysis, (2) acidogenesis, (3) acetogenesis, and (4) methanation (Fig. 4.1). In anaerobic treatment process hydrogen sulfide (H_2S) gas is also produced as a result of the reduction of oxidized sulfur compounds, and the resulting wastewater is known as PMDE. Sulfides bind

Spentwash
Polysaccharides, Proteins & Fats

Hydrolysis ↓
1. $C_6H_{10}O_5 + 2H_2O \rightarrow C_6H_{12}O_6 + 2H_2$
Fermentable bacteria *Bactoroids&Clostridim spp*

Sugar, amino acids, Fatty acids, Glycerol

Acedogenesis ↓
2. $C_6H_{12}O_6 \leftrightarrow 2CH_3CH_2OH + 2CO_2$
3. $C_6H_{12}O_6 + 2H_2 \leftrightarrow 2CH_3CH_2COOH + 2H_2O$
4. $C_6H_{12}O_6 \rightarrow 3CH_3COOH$

Production of volatile fatty acid
Propionic acid, Butyric acid, acetic acid, formic acid, lactic acid, ethanol, methanol

Acetogenesis ↓
5: $CH_3CH_2COO^- + 3H_2O \leftrightarrow CH_3COO^- + H^+ + HCO_3^- + 3H_2$
6: $C_6H_{12}O_6 + 2H_2O \leftrightarrow 2CH_3COOH + 2CO_2 + 4H_2$
7: $CH_3CH_2OH + 2H_2O \leftrightarrow CH_3COO^- + 2H_2 + H^+$

Acetic acid, Hydrogen and CO_2

Methanogenesis ↓
8: $CO_2 + 4H_2 \rightarrow CH_4 + 2H_2O$
9: $2C_2H_5OH + CO_2 \rightarrow CH_4 + 2CH_3COOH$

Methane and CO_2

Fig. 4.1 Steps of biomethanation for generation of methane from spentwash

Table 4.1 Physico-chemical properties of spentwash and post-methanated distillery effluent (PMDE)

Physico-chemical properties	Spentwash	PMDE	Permissible limits
Color (Co-pt)	Brown (1,50,000)	Dark brown (70,000)	Not clear
Odor	Jaggery smell	Mild sulfur smell	Not clear
pH	3.90 ± 0.25	8.3 ± 0.31	5–9
TS	103,084 ± 5.50	34,317 ± 455	2200
TDS	77,776 ± 3768	20,022 ± 438	2100
TSS	25,308 ± 1201	14,276 ± 16.00	100
COD	90,000 ± 231	58,018 ± 185	120
BOD	42,000 ± 123	29,120 ± 265	40
Chloride	2200 ± 105	1300 ± 60.50	1500
Phenol	4.20 ± 1.80	1.65 ± 0.76	0.50
Sulfate	5760 ± 260	13,656 ± 21.23	1500
Phosphate	5.36 ± 0.17	1.16 ± 0.15	Not clear
Total nitrogen	2800 ± 130	568 ± 23.00	25
Total organic carbon	25,368 ± 1.06	10,904 ± 0.34	Not clear
Metals			
Cd	0.020 ± 0.00	2.281 ± 0.07	0.010
Cr	0.192 ± 0.01	0.40 ± 0.01	0.050
Fe	6.312 ± 0.21	84.01 ± 1.98	2.000
Ni	0.171 ± 0.01	1.241 ± 0.04	0.100
Cu	0.961 ± 0.00	0.955 ± 0.02	0.500
Pb	0.945 ± 0.00	4.446 ± 0.06	0.050
Zn	2.012 ± 0.00	4.631 ± 0.11	2.000
Mg	0.214 ± 0.00	2.112 ± 0.05	0.200

Chandra et al. (2004) and Chandra and Srivastava (2004)
All the values are in mg/l, except color, odor, and pH

with heavy metals present in effluent and form a colloidal solution of metal sulfides, which results in dark black color of PMDE (Chandra et al. 2008a). The PMDE is a complex, cumbersome, and highly toxic waste, which has very high BOD, COD, color, sulfate, phenol, total suspended solids (TSS), TDS, and various heavy metals (Table 4.1). The dark color of effluent is deleterious to aquatic life because it hinders the photosynthesis by blocking sunlight. The increasingly stringent environmental laws are forcing the distilleries to improve the existing treatment processes and explore suitable methods for effluent management.

Major Colorants in PMDE

Color-contributing substances present in distilleries are generated and concentrated during the processing of sugarcane juice, while some are added during the processing of sugar. The wastewater generated from alcoholic fermentation has a large

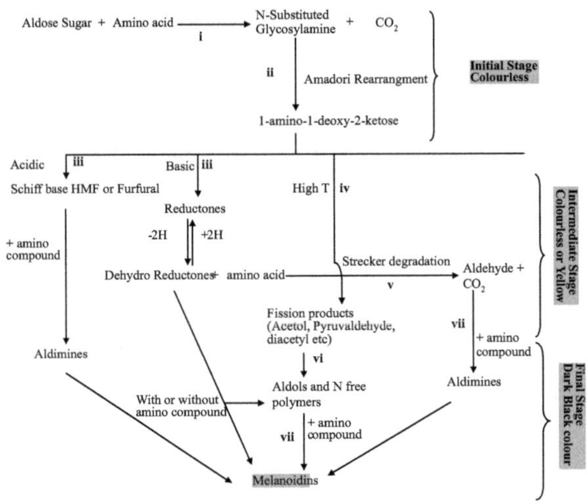

Fig. 4.2 Nonenzymatic browning reaction. (Based on Hodge 1953)

amount of melanoidin (i.e., brown pigment). The major colorant is synthesized as per Maillard reaction due to reaction of sucrose or reducing sugar with mixture of amino acids present in sugarcane juice (glycine, aspartic acid, histidine, lysine, tryptophan, and arginine) at high temperature during concentration of sugar juice. Though the reaction is complicated which forms mixture of compounds at various stages, but to understand the process it can be simplified into three stages as given in Fig. 4.2 (Hodge 1953).

1. *Initial stage*: Products formed during this stage are colorless which show no absorption in ultraviolet range. This stage involves the following two reactions:

   Reaction i: Sugar-amine condensation
   Reaction ii: Amadori rearrangement

2. *Intermediate stage*: Products formed during this stage are colorless or yellow which show strong absorption in the ultraviolet range. This stage involves the following reactions:

   Reaction iii: Sugar dehydration
   Reaction iv: Sugar fragmentation
   Reaction v: Amino acid degradation (Strecker degradation)

3. *Final Stage*: Products formed during this stage are highly colored. This stage involves the following reactions:

   Reaction vi: Aldol condensation
   Reaction vii: Aldehyde-amine condensation and formation of heterocyclic nitrogen compounds

Table 4.2 Major colorants present in post-methanated distillery effluent (PMDE) and their characteristics

Colorant	Characteristics
Melanoidin	Browning reaction products of sugar and amino acids with high molecular weight
Caramel	Process colorant, thermal degradation products of sugar, high molecular weight with low net charge
Phenolics	Plant pigment, low molecular weight may be attached to polysaccharides, pH sensitive, darker at high pH, pale yellow to orange color, react with iron to produce very dark color, may dimerize or oxidize to form darker color
Alkaline degradation products of fructose	Process colorants, reddish to dark brown in color, low molecular weight up to polymeric depending on degree of degradation
Sulfides	Process colorant, toxic to microorganisms and creating foul smell, and have strong binding tendency with metals
Heavy metals	Process colorants, toxic to microorganisms, animals, and humans

The coloring compounds present in spentwash are not degraded by the conventional treatment process and even be increased during the anaerobic treatment process in methane reactors due to the re-polymerization of compound. Melanoidins from overheated sugars, furfurals from acid hydrolysis, and humic and tannic acids from feedstock mainly contribute to the color of the effluent (Chandra et al. 2008a). However, heating of glucose and fructose at acidic or basic conditions leads to the degradation reaction forming the highly reactive intermediate compounds, which undergo condensation and polymerization reactions forming more complex and colored polymers known as melanoidins (Table 4.2). The colorants can be divided into enzymatic colorants such as melanins and nonenzymatic colorants such as melanoidins. Around 6% of melanoidins are present in sugarcane molasses that impart colors to the effluent, and molasses is raw material of distilleries. High concentration of sulfate is also present in sugarcane molasses, which is added during the cleaning of sugar crystals. The high level of sulfate can lead to production of sulfides during the anaerobic digestion of distillery spentwash, which is precipitated out along with the existing metals as metallic sulfides in PMDE, consequently increasing the total solids (TS) contents of the distillery effluent. In addition to these color-contributing components, during the heat treatment process, the Maillard reaction is accompanied by the formation of a class of compounds known as Maillard reaction products (MRPs) in the different fractions of distillery wastewater. The suspended solid fraction of distillery effluent comprises hydroxypropanone, methylbenzene phenol, methylphenol, p-chloroanisole, methylbenzaldehyde, indole, and methylindole. Most of the compounds have been reported as pyrolysis products of different amino acids as methylbenzene from phenylalanine, indole and methylindole from tryptophan, and finally phenol and methylphenol from the tyrosine amino acid (Bharagava and Chandra 2010; Gonzalez et al. 2000).

The presence of all these compounds in distillery wastewater could be derived from the amino acids present in original molasses or from the rest of the yeast cells that remain in distillery effluent after alcoholic distillation process. Despite hydroxypropanone has been reported as a dehydration product of glycerol and typical carbohydrate pyrolysis product, glycerol represents 5–6% of the dry weight of vinasse which is also a raw material of distilleries industries. The existence of carbohydrate-derived compounds such as furfuryl alcohol and 2,3-dihydro-5-methylfuran-2-one or lignin-derived moieties such as 4-vinylguaiacol and syringol could be ascribed to residual sugarcane fragments present in the suspended solids fraction of distillery effluent. Moreover, 4-vinylguaiacol has been described as a pyrolysis product of ferulic acid, one of the two cinnamic acids present in the vegetal cell wall of gramineous plants as is the case of sugarcane (*Saccharum officinarum* L.). The polysaccharide-free fraction of distillery effluent contains hydroxypropane and 2,5-dimethylfuran along with some other minor carbohydrate pyrolysis products such as 2-hydroxymethylfuran and 2,4-dimethylfuran, 1-hexadecanol, and palmitic acid. Moreover, the polysaccharide-rich fraction of distillery effluent has major portion of hydroxypropanone, 2,5-dimethylfuran, 2-hydroxymethylfuran, 2,4-dimethylfuran, 2,3-dihydro-5-methylfuran-2-one, and 5-methyl-2-furfuraldehyde (Gonzalez et al. 2000; Bharagava and Chandra 2010). The presence of all these types of toxic compounds in distillery wastewater indicated that there is an urgent need to develop an environment-friendly and cost-effective treatment technique, which can remove/degrade/detoxify all these toxic compounds from distillery effluent, so that it can be safely disposed into environment.

Environmental Pollution and Toxicity Profile of PMDE

PMDE turns more dark and viscous with alkaline pH than original spentwash during the anaerobic treatment process in industries. The increase in pH from 4 to 8.5 in methane reactor during the anaerobic digestion process is mainly due to the oxidation of organic acids to carbon dioxide (CO_2), which react with basic compounds present in PMDE and generate carbonates and bicarbonates resulting in the increase of pH. The highly colored PMDE is the major source of soil as well as water pollution. PMDE causes water pollution in two ways. First, the highly colored (1,50,0000–1,80,000 co-pt) nature of PMDE blocks out the sunlight penetration in rivers and streams, reducing the photosynthesis leading to the reduction in oxygenation of which becomes detrimental for aquatic life. Secondly, due to the high organic load, i.e., high COD and BOD, it causes eutrophication. In addition, due to the presence of putriciable organics like indole, skatole, and other sulfur compounds, the PMDE that is disposed in aquatic resources also produce bad smell.

The organic and inorganic ions added in water and agricultural land through PMDE pose a serious threat to the soil and groundwater (Table 4.3), if applied without proper dilution/treatment (Jain et al. 2005; Kaushik et al. 2005). The agricultural crops growing on such metal-contaminated soil may accumulate high

Table 4.3 Effect of higher concentration of PMDE on soil and water properties

Properties	Toxicity effect
Soil	
pH	Acidify the soil with spentwash application and decrease the productivity
Electric conductivity	Increase electricity and results in salinization of the soil which is important cause of lower productivity
Organic carbon	Heavy infield transport machinery is most commonly associated with soil compaction problems. Soil compaction decreases porosity and water infiltration rate, restricting the rooting ability of the crop
Total phosphorous	Stunted plant growth decreases the plant's ability to uptake zinc, decreases the plant's ability to uptake iron
Total nitrogen	Increase acidity of soil
Total potassium	Deleterious effects on soil as well as on crops, since very high exchangeable K can also cause dispersion of soil. Higher anionic and cationic concentration can decrease the bulk density as well as water holding capacity of soil by reducing the porosity in clay soil due to deflocculation of clay particles which occurs in the presence of higher Na content, and the consequent effects are seen in the cation exchange capacity in the soil which adversely affects the seed germination and plant growth
Total sodium	Change the electric conductivity, soil dispersion affecting soil permeability and aeration, which causes hindrance to seed germination
Total calcium	Change the electric conductivity
Heavy metals (iron zinc, cadmium, copper)	Inhibit the biodegradation of organic contaminants, adverse effects on the soil microbial diversity, plant growth, and genetic variation
Microbial parameters	*Staphylococcus aureus*, *Pseudomonas aeruginosa*, *Escherichia coli*, *Enterobacter* spp., *Klebsiella* spp., *Proteus* spp., and *Bacillus* spp., while the fungi isolated were *Aspergillus* spp., *Rhizopus* spp., and *Penicillium* spp.
Soil enzyme (phosphatase, dehydrogenase, urease)	Soil enzyme activity is increased and found to be more than the recommended NPK and farmyard manure
Water	
pH	The pH of water affects the solubility of chemicals. Therefore, the availability of these chemicals to aquatic organisms is affected. As acidity of water increases by application of spentwash, most metals and other toxic compounds become more soluble and more toxic which decrease the productivity
COD	Increase organic content leading to depletion of oxygen and reduction of dissolve oxygen (DO) which creates anaerobic conditions. Anaerobic condition is deleterious to higher aquatic life
BOD	Fish and other aquatic organisms may not survive under very high level of BOD
Total phosphorous	Eutrophication
Total nitrogen	Increase eutrophication
Total potassium	Increase eutrophication
Heavy metals	Act as inhibitor for many biological treatment processes, recalcitrant and persistence in nature

amount of toxic metals which is excessive enough to cause clinical problems to both humans and animals (Rattan et al. 2005). Wastewater treatment inhibited seed germination, reduced soil alkalinity and soil manganese deficiency leading to damage to agricultural crops. Bharagava et al. (2008) have reported the effect of different concentrations of distillery effluent irrigation on morphological and physiological parameters of Indian mustard (*Brassica nigra* L) at different time intervals. During this study it was found that distillery effluent beyond 50% (v/v) concentration significantly decreases the plant growth parameters (root, shoot length, number of leaves, chlorophyll, and protein content) in tested plants in respect to control. However, the maximum increase in root length, shoot length, number of leaves, and total dry weight were recorded in mustard plants treated with 50% (v/v) distillery effluent. Moreover, the mustard plants treated with 75% (v/v) distillery effluent have shown the vigorous growth initially. But later on reduced growth and delayed flowering and fruiting were also observed as compared to control plants. Lower concentration of distillery effluent which supported the plant growth parameters might be due to the induction of plant growth hormones. While higher concentration of effluent (>50%) suppresses the hormone activities, it has been also reported that higher distillery effluent content acts as inhibitor for plant growth hormone(s) (auxin and gibberellins), which are mainly required for the growth and development of plants (Subramani et al. 1997)

The reduction in plant growth parameters at higher concentration of distillery effluent might be also due to the entrance of toxic heavy metals into the protoplasm resulting in the loss of intermediate metabolites, which are essential for further growth and development of plants. Further, Chandra et al. (2008b) have also observed the effect of distillery sludge amendments in soil (10, 20, 40, 60, 80, and 100%) and found that 10% (w/w) sludge was favorable for *P. mungo*, while >10% was inhibitory for plant growth (Fig. 4.3). Greater than 40% concentrations of sludge-amended soil reduced all the growth parameters, viz., root length, shoot length, number of leaves, biomass, photosynthetic pigment, protein, and starch, of *P. mungo*.

Fig. 4.3 Morphological effect of distillery sludge amendment soils at different concentrations (10–80%) on the growth of root, shoot, and leaves of *P. mungo* after 60 days vs. the control

Disposal of undiluted and untreated PMDE in water bodies has toxic effects on fishes and other aquatic organisms. Ramakritinan et al. (2005) have estimated LC_{50} for distillery spentwash at 0.5% using a bio-toxicity study on freshwater fish *Cyprinus carpio* var. communis. They reported that the respiratory process in *C. carpio* under the PMDE stress is severely affected resulting in a shift toward the anaerobiosis at organ level during sublethal intoxication. The sulfide toxicity of PMDE toward microorganisms is also well documented as sulfide has strong tendency of metal binding which contributed more toxicity and color to the PMDE (Krishnanand et al. 1993). Several studies have reported that PMDE inhibits the microbial growth by sequestering the ammonia, amino acid, and peptide, while melanoidins inhibit growth by cross-linking with polypeptide chain and sequestering essential multivalent metal cations (Silvan et al. 2006). In vitro studies have also revealed that melanoidins (the major color-contributing component in distillery effluent) have mutagenic, carcinogenic, and cytotoxic effects in humans and animals (Silvan et al. 2006). The excessive glycation has also been reported to cause destruction of essential amino acids, decreased digestibility, inactivation of enzymes, inhibition of regulatory molecule binding sites, cross-linking of the glycated extracellular matrix, decreased susceptibility to proteolysis, abnormality of nucleic acid functions, altered macromolecular recognition, and endocytosis as well as increased immunogenicity (Silvan et al. 2006), which are involved in the development of diseases such as diabetes mellitus, cardiovascular complications, and Alzheimer's disease (Silvan et al. 2006; Chandra et al. 2008a).

Since melanoidins are equally harmful to plants, animals, and humans, hence the nutritional and physiological effects of melanoidins have been widely investigated. Lee et al. (1977) showed the effect of non-dialyzable melanoidins prepared from reduction sugars and amino acids on the rats. Results showed that growth was depressed and diarrhea was induced in rats feeding on diets containing melanoidin. The high molecular weight part of non-dialyzable melanoidins was hardly digested or absorbed, and long-term feeding on it did not affect the growth of rats, while the low molecular weight components of melanoidins have depressing effects on protein digestion and absorption.

Over the last decades, toxicity profile of distillery effluent was mainly focused on the persistent organic pollutants (POPs) present in effluent which are not removed during the secondary treatment by the industries. Among these POPs most of the compounds come under the endocrine-disrupting chemicals (EDCs) on which there is need for further research. Table 4.4 presents some EDCs on the basis of authoritative reports (TEDX 2011). EDCs are complex mixtures of compounds that are persistent and bioaccumulative or less persistent and not bioaccumulative. EDCs were mainly associated with nuclear hormone receptors and affected their functions. Due to the EDCs mostly androgen, estrogen, thyroid, and progesterone are affected. Androgen receptors are proteins which are found in the cytoplasm that specifically bind to androgens, which are responsible for male characteristics. Estrogen receptors are also cytoplasm proteins that bind to estrogens, which are most responsible for female characteristics. While thyroid receptors which are usually found in the nucleus regulate DNA transcription and regulate the metabolism. Similarly,

Table 4.4 List of detected organic pollutants present in PMDE screened under EDC compounds

S.No.	Organic compounds present in PMDE	Toxicity	References
1.	4-Methylphenol (p-cresol)	Irritation and burning of the skin, eyes, mouth, and throat; abdominal pain and vomiting, heart damage, anemia, liver and kidney damage, facial paralysis, coma and death	Camarero et al. (1999) and Gonzalez et al. (2000)
2.	4-Methoxyvinylbenzene	Not clear	Camarero et al. (1999)
3.	1,2-Dimethoxybenzene (veratrole)	Pneumoconiosis	Camarero et al. (1999)
4.	Nitroacetone phenone	Group D carcinogenic	Bharagava and Chandra (2010) and Sangeeta et al. (2011)
5.	3,4-Dimethoxytoluene	Carcinogenic	Gonzalez et al. (2000)
6.	p-Anisaldehyde	Neurotoxicant	Camarero et al. (1999)
7.	1-Methoxy-4-(2-propenyl) benzene	Not clear	Camarero et al. (1999)
8.	1,2,3-Trimethoxybenzene	Carcinogenic effects, hazardous in case of inhalation (lung irritant)	Camarero et al. (1999)
9.	1,2-Dimethoxy-4-ethylbenzene	Allergic reactions in certain sensitive individuals, carcinogenic effects	Camarero et al. (1999)
10.	1,2-Dimethoxy-4-vinylbenzene		Camarero et al. (1999)
11.	p-Chloroanisole	Liver and kidney damage, carcinogenic, reproductive damage	Bharagava and Chandra (2010) and Sangeeta et al. (2011)
12.	1,2,3-Trimethoxy-5-methylbenzene s, 1,3,5-tri-tert-Butylbenzene (standard)	Skin or sensory organ toxicants	Gonzalez et al. (2000)
13.	Cis-1,2-Dimethoxy-4-(1-propenyl)benzene	Cause cancer or mutations, inhalation of vapors may cause drowsiness and dizziness, hepatotoxic	Gonzalez et al. (2000) and Sangeeta et al. (2011)
14.	Trans-1,2-Dimethoxy-4-(1-propenyl)benzene	Do	Camarero et al. (1999)
15.	1,2,3-Trimethoxy-5-vinylbenzene	Do	Camarero et al. (1999)
16.	1,2,3-Trimethoxy-5-(2-propenyl) benzene	Do	Camarero et al. (1999)
17.	1,2-Dimethoxy-4-(2-ketopropyl) benzene	Do	Camarero et al. (1999)
18.	Cis-1,2,3-Trimethoxy-5-(1-propenyl)benzene	Do	Camarero et al. (1999)
19.	2,3-dimethyl pyrazine	Irritating to the eyes, respiratory system, and skin	Bharagava and Chandra (2010) and Sangeeta et al. (2011)

(continued)

Table 4.4 (continued)

S.No.	Organic compounds present in PMDE	Toxicity	References
20.	Trans-1,2,3-Trimethoxy-4-(1-propenyl)benzene	Anticoagulant effect	Gonzalez et al. (2000)
21.	Methyl 4-methoxycinnamate	Not clear	Camarero et al. (1999)
22.	3,4,5-Trimethoxyacetophenone	Group D carcinogenic	Camarero et al. (1999)
23.	Methyl 3,4,5-trimethoxybenzoate	Carcinogenic, reproductive and developmental toxicity, neurotoxicity and acute toxicity	Camarero et al. (1999)
24.	Methyl 3,4-dimethoxycinnamate	Not clear	Camarero et al. (1999)
25.	4-Vinylphenol	Pneumotoxicity and hepatotoxicity	Camarero et al. (1999)
26.	1,2-Dimethoxy-4-(2-propenyl) benzene	Persistence and bioaccumulation potential	Camarero et al. (1999)
27.	3,4,5-Trimethoxybenzaldehyde	Carcinogenic, reproductive and developmental toxicity, neurotoxicity, and acute toxicity	Camarero et al. (1999)
28.	2,3-Dihydro-5-methylfuran		Bharagava and Chandra (2010) and Sangeeta et al. (2011)

TEDX (2011)

progesterone receptors are specific proteins found in or on cells of progesterone target tissues that specifically combine with progesterone, which control the menstrual cycle, pregnancy, and embryogenesis; the cytosol progesterone-receptor complex then associates with the nucleic acids to initiate protein synthesis. EDCs can mimic the natural hormone, fooling the body to give overresponse against stimulus and responding at inappropriate times; block the effects of the hormone from certain receptors or directly stimulate or inhibit the endocrine system; and cause overproduction or underproduction of hormones (Kandarakis et al. 2009). Several pollutants of distillery waste act as EDCs, and their exposure may have a much wider impact on the body's endocrine system, including health problems, i.e., nonreproductive cancers, immune effects, metabolic effects, obesity, diabetes, brain development, cardiovascular disease, and behavior. Many metals and metalloids also work as endocrine disruptors. Metal exposure can target all five steroid receptor pathways (progesterone, estrogen, testosterone, corticosteroids, and mineralocorticoids) and also receptors for retinoic acid, thyroid, and peroxisome proliferators. Some EDCs are persistent in the environment and bioaccumulate through food webs to high concentrations in wildlife and humans and can be transferred to the developing fetus and the newborn through the placenta or breast milk, respectively, and cause abnormalities. Hence, their remediation from the distillery wastewater is mandatory and very challenging for the researchers.

Bacterial Treatment of PMDE

Microorganisms including bacteria, fungi, and actinomycetes due to their versatile inherent properties to metabolize a variety of complex compounds have been widely utilized for biodegradation purposes. Degradation and detoxification of industrial wastes by microbes is an environment-friendly and cost-competitive method for industrial waste cleanup. During industrial wastewater treatment process, the utility of microbes largely depends on the nutrient requirement, enzymatic setup, environment conditions, as well as the nature and chemical structure of recalcitrant compounds (Chandra et al. 2007; Mohana et al. 2007). However, most of the researchers have reported the microbial degradation of distillery effluent by using the various fungal strains at laboratory scale, but the application of fungal strains for the treatment of wastewater at industrial scale is not practically possible because of slow growth rate of fungus and their less stability under stress environment such as higher pH, low oxygen, and high pollution load. Hence, bacteria appear to be more effective than fungi for the bioremediation of environmental pollutants due to their broad environmental adaptability and biochemical versatility (Chandra et al. 2007; Bharagava et al. 2009).

Many workers have reported several aerobic bacterial strains for the degradation and decolorization of PMDE (Table 4.5). Bharagava et al. (2009) isolated *Bacillus* sp., *Bacillus licheniformis*, and *Alcaligenes* sp. from distillery effluent-contaminated soil and found that these bacterial strains were capable for 52, 49, and 59% decolorization of natural melanoidins, respectively, in axenic condition, while in mixed condition these showed 70% decolorization in the presence of glucose (1.0%) and peptone (0.1%) at pH 7.0 and temperature 37 °C. Further, Ghosh et al. (2004) have also isolated *Acinetobacter*, *Enterobacter*, *Pseudomonas*, *Klebsiella*, and *Stenotrophomonas* from the distillery effluent contaminated soil which showed degradation of recalcitrant compounds from PMDE with 44% COD reduction. Sangeeta et al. (2011) have also isolated and characterized the potential MnP producing *Bacillus*, *Raoultella planticola*, and *Enterobacter sakazakii*. The consortium of these bacteria showed maximum 60% decolorization of synthetic melanoidin (SAA-MP; 2400 mg/l) at optimized nutrient, pH (7.0 ± 0.2), shaking speed (180 rpm), and temperature (35 ± 2 °C) after 144 h incubation. The addition of D-xylose at stationary stage enhanced the decolorization from 60 to 75% along with reduction of BOD and COD. Jain et al. (2002) isolated three bacterial strains, *B. megaterium*, *B. cereus*, and *B. fragariae*, from the activated sludge of distillery effluent, and these strains showed 38–58% color and 55–68% COD reduction from distillery effluent. Sirianuntapiboon et al. (2004) isolated acetogenic bacterial strains from juice and vegetable samples that were found effective to decolorize 76% molasses pigment medium and 73–76% anaerobically treated distillery effluent within 5 days when media was supplemented with glucose and nitrogen sources. Patel et al. (2001) have reported *Oscillatoria* sp., *Lyngbya* sp., and *Synechocystis* sp. for 96, 81, and 26% decolorization of distillery waste, respectively. Dahiya et al. (2001) have also reported *Pseudomonas fluorescens* from the reactor liquid and

Table 4.5 Different bacterial strains reported for the decolorization of PMDE

Bacterial strains	Decolorization conditions	Decolorization efficiency (%)	References
Lactobacillus hilgardii	Immobilized cells of the heterofermentative lactic acid bacterium decolorized 40% of the melanoidins solution within 4 days aerobically	40	Ohmomo et al. (1988)
Bacillus cereus, *B. megaterium*, and *B. smithii*	Decolorization occurred at 55 °C in 20 days under anaerobic conditions in the presence of peptone or yeast extract as supplemental nutrient. Strains could not use MWW as sole carbon source	76, 82 and 35.5, respectively	Kambe et al. (1999)
Acinetobacter sp., *Bacillus* sp., *P. paucimobilis*, *Pseudomonas* sp., *Acinetobacter* sp., and *Agrobacterium radiobacter*	These entire organisms were isolated from air bubble column reactor belonged to species of the genus *Bacillus* were used for treating winery wastewater after 6 months of operation	Not checked in this study	Petruccioli et al. (2002)
Acetobacter aceti	The organism required sugar, especially, glucose and fructose, for decolorization of MWWs	76.40	Sirianuntapiboon et al. (2004)
Pseudomonas fluorescens	Decolorization was obtained with cellulose carrier coated with collagen. Reuse of decolorization cells reduced the color	94	Dahiya et al. (2001)
Pseudomonas putida	The organism used glucose as a carbon source and produce hydrogen peroxide with reduction of color	60	Ghosh et al. (2002)
Xanthomonas fragariae, *B. megaterium*, and *B. cereus*	These three strains needed glucose as carbon and NH_4Cl as nitrogen source. Decolorization efficiency was found better in free cells as compared to immobilized cells	76	Jain et al. (2002)
B. brevis	The three strains were part of a consortium which decolorized the anaerobically digested spentwash in the presence of basal salts and glucose	22	Kumar and Chandra (2004)
Pseudomonas aeruginosa		27.4	
B. thuringiensis		67	Kumar and Chandra (2004)
Enterobacter, *Pseudomonas*, *Stenotrophomonas*, *Klebsiella*, and *Acinetobacter*	All were isolated from the soil of effluent discharge site, and addition of 1% glucose was necessary for decolorization	44	Ghosh et al. (2004)

(continued)

Table 4.5 (continued)

Bacterial strains	Decolorization conditions	Decolorization efficiency (%)	References
B. licheniformis, *Bacillus* sp., and *Alcaligenes faecalis*	All require the presence of glucose (1.0%) and peptone (0.1%) in medium and pH 7.0 and temperature 37 °C	70	Bharagava et al. (2009)
Bacillus sp., *Raoultella planticola*, and *Enterobacter sakazakii*	The three strains have potential to decolorized the synthetic melanoidin, and decolorization was enhanced by addition of D-xylose at stationary stage of growth	60 and 75, respectively	Sangeeta et al. (2011)
Proteus mirabilis, *Bacillus* sp., *Raoultella planticola*, and *Enterobacter sakazakii*	They decolorized the molasses melanoidin and PMDE after extraction of melanoidin	75 and 70, respectively	Sangeeta and Chandra (2012, 2013)

found that this bacterial strain was capable to decolorize melanoidin-containing wastewater up to 90 and 76% under sterile and non-sterile conditions, respectively. Marine cyanobacteria such as *Oscillatoria boryna* have potential to produce H_2O_2, hydroxyl, perhydroxyl, and active oxygen radicals for degradation of melanoidin (Kalavathi et al. 2001). The decolorization of molasses wastewater by immobilized cells of *P. fluorescence* on porous cellular carrier was undertaken achieving 76% decolorization in 24 hrs incubation at optimized conditions. *Bacillus* sp. has been reported to decolorize molasses wastewater by up to 36% under aerobic conditions within 20 days at 55 °C (Kambe et al. 1999). Kumar and Viswanathan (1991) isolated bacterial strains from sewage and acclimatized on increasing concentration of distillery waste. These strains reduced 80% COD within 4–5 days without any aeration, and the major products left after the effluent treatment were biomass, carbon dioxide, and volatile acids. Ohmomo et al. (1988) have used *Lactobacillus hilgardii* immobilized cells on calcium alginate for the decolorization of melanoidin under continuous supply of small amount of oxygen resulting in 40% decolorization of PMDE. Moreover, the differences in decolorization of PMDE might be due to the fact that melanoidin's stability varies with pH and temperature. At higher temperature, during the sterilization process, melanoidin pigment decomposes into low molecular weight compounds, while during the anaerobic treatment process, the pH of distillery spentwash increases from 4.5 to 8.5 during which melanoidins undergo complexation process forming complex which can't be easily degraded during the bacterial treatment of distillery effluents.

Recent Approaches for the Treatment of PMDE

The sequential bioreactor treatment of wastewater has become promising over the conventional treatment techniques to remove the complex mixture of pollutants from distillery wastewater. The bacterial pre-treatment of distillery effluent enhanced the phytoremediation process of distillery effluent (Kumar and Chandra 2004). In this study it was found that the recalcitrant compounds present in effluent that were ameliorated by *Bacillus thuringiensis* is through utilization of complex compounds. Therefore, foster phytoremediation of heavy metals was observed from the PMDE. Enzymatic pre-treatment for the enhancement of biodegradability of distillery wastewater has also been reported by Sangave and Pandit (2006a). The cellulase (enzymatic hydrolysis) followed by aerobic oxidation was investigated in this study for the treatment of distillery spentwash. The rate of anaerobic oxidation enhanced by two- to three fold in the pre-treatment sample as compared to the untreated sample when the pH during the pre-treatment step was maintained at a value of 4.8. Similarly, two fold increase in the aerobic oxidation rate was also observed when effluent was pre-treated with the enzyme without any pH control (effluent pH 3.8). This might be due to the enzymatic pre-treatment transforming the complex and large pollutant molecules into simpler biologically assailable smaller molecules. Sangave and Pandit (2006b) have also reported the role of the ultrasound and enzyme assisted biodegradation of distillery wastewater. Efficiency of these two techniques was analyzed by subjecting the effluent to subsequent aerobic biological oxidation (AO). The distillery effluent when treated with ultrasound (US) alone for 30 min, COD reduction was found best after 2 h during the aerobic oxidation step (US+AO). However, the 50 U of enzyme was used alone at pH 4.8 for 24 h, which yielded better COD removal as compared to untreated effluent, while US followed by enzymatic pre-treatment resulted in maximum COD removal efficiency during the aerobic oxidation (US+E + AO) process as compared to the other combinations tested for the treatment of distillery effluent. Moreover, the application of constructed wetland at pilot scale for the treatment of industrial wastewaters was also reported by various authors. Chandra et al. (2008c) reported that the integration of bacterial pre-treatment of PMDE with *Typha angustata* enhanced the reduction of heavy metals, i.e., Cr (36–58%), Cd (34–62%), Fe (33–51%),Cu (33–54%), Ni (36–60%), Mn (36–83%), Zn (32–54%), and Pb (33–60%) in comparison with control lacking this pre-treatment step. In addition to the removal of heavy metals, other physico-chemical parameters, viz, color, COD, BOD, total nitrogen, and phenol, were also reduced significantly by 98%, 99%, 99%, 94%, and 82%, respectively, at 30% effluent after 7 days of free water surface flow treatment process. This indicated that the bacterial pre-treatment of PMDE followed by phytoremediation will significantly boast the treatment process of PMDE. Chen et al. (2006) used four different wetland plant species including floating plants (*Pistia stratiotes, Ipomea aquatica*) and emergent plants (*Phragmites communis, Typha orientalis*) for the removal of COD, BOD, and suspended solids from industrial wastewater and found that the system with a 50 HRT (hydraulic retention time) (feed rate 0.4 m^3/day) and vesicles ceramic bioballs as the media was effective to remove 61% COD, 89% BOD and 81% suspended solid (SS), 35% total phosphate (TP), and 56% NH_3-N.

The use of distillery effluent for agriculture has been advocated due to the presences of high amount of nitrogen and phosphorous. But its indiscriminate application may create adverse effects on crop and contamination of groundwater. Kaushik et al. (2005) reported that long-term application of PMDE causes significant increase in total organic carbon (TOC), total kjeldahl nitrogen (TKN), potassium, and phosphorus along with accumulation of high concentration of sodium ions (Na^+) in soil. Moreover, the short-term application of 50% PMDE along with bioamendments proved most useful in improving the properties of sodic soil and also favored the successful germination and improving seedling growth of pearl millet. Similar observations were recorded by Chandra et al. (2004) during a study on the impact of pre-treatment and PMDE irrigation on soil microflora and growth of *Phaseolus aureus* in pot-grown experiment. They concluded that lower concentration of raw distillery effluent (1–5%) and PMDE (1–10%) stimulated the growth of *P. aureus* and soil microflora, except soil bacteria (inhibited by all the concentrations of raw effluent). However, higher concentration (raw effluent, 10–20%, and PMDE, 15–20%) showed toxicity. Similar toxicity pattern was also observed by Sangeeta and Chandra (2006) on *Vicia faba* where they reported that the sludge of PMDE also contains high quantity of organic and inorganic pollutants. The sludge-amended soil above 10% (W/V) concentration causes adverse effects on legume crops. The sludge concentration above 50% causes toxicity to plants and even inhibits the seed germination. Further, the continuous irrigation with PMDE affects the underground water quality by increasing 40% TDS with salt (Jain et al., 2005). A study by Hati et al. (2007) on soil properties and crop yield on a vertisol in India shows increase in organic carbon and salinity. The study supported the judicious application of distillery effluent as an amendment to the agricultural field can be considered as an available option for its safe disposal into the environment.

PMDE Treatment with Consortium

During the last two decades experiments were restricted only to use the axenic culture and immobilized culture on solid supports for the degradation of toxic compounds. Recently, immobilized consortia of two or more selected bacterial strains were employed (Kowalska et al. 1998). The use of bacterial consortium for the treatment of distillery wastewater is highly effective because mixed bacterial culture comprise of many strains which are very helpful in co-metabolism employing different sets of enzymes. The pH and temperature play a very important role for optimum activity of enzymes. In co-metabolism, the metabolites or intermediates produced by one bacterial strain are utilized by other bacterial strains as source of nutrient and energy (Chandra et al. 2007; Bharagava and Chandra 2010). Some authors have also reported the use of jet loop reactors (JLR) for the aerobic treatment of winery wastewater. Petruccioli et al. (2002) used a JLR with working volume of 15 dm^3 for the aerobic treatment of winery wastewater, and they recorded more than 90% COD removal efficiency with an organic load of the final effluents ranging between 0.11 and 0.3 kg COD m^{-3} at much diluted concentration of effluent. Eusibio et al. (2004) have also reported the operation of a JLR for more than 1 year treating winery wastewater collected in different seasons and achieved 80% COD removal efficiency at diluted concentration of distillery effluent. They also reported that JLR have higher oxygen transfer rate at lower energy costs. Hence the use of mixed bacterial culture for the treatment of industrial wastewater is more effective compared to a single pure bacterial strain.

Phytoremediation Approach for Treatment of PMDE

Phytoremediation is a low-cost technique which is very useful to remediate sites contaminated with recalcitrant toxic compounds and heavy metals. Phytoremediation takes advantage of plant nutrient utilization processes, transpires water through leaves, and acts as transformation system to metabolize organic compounds. They may also absorb, sequester, and bioaccumulate toxic compounds. Billore et al. (2001) used *Phragmites kharka* in a constructed wetland for the treatment of wastewater from the same industry and obtained 36% removal of TKN and 48% removal of TSS. Detoxification of distillery effluent through *B. thuringiensis* (MTCC 4714) enhanced phytoremediation potential of *Spirodela polyrrhiza* (Kumar and Chandra 2004).

Recently, a comparative heavy metal accumulation pattern and ultrastructural changes were studied in *P. communis*, *T. angustifolia*, and *C. esculentus* in both in situ and ex situ conditions (Chandra and Sangeeta 2011; Sangeeta and Chandra 2011). In situ study showed direct correlation of bioaccumulation with the metal content in sediments. Both macrophytes *T. angustifolia* and *C. esculentus* were observed as root accumulators for Fe, Cr, Pb, Cu, and Cd, while ex situ condition showed the maximum accumulation of all tested heavy metals in *P. communis* followed by *T. angustifolia* and *C. esculentus*. Results revealed that *P. communis* and *T. angustifolia* had more potential for tested metals than *C. esculentus*. Similarly, *Potamogeton pectinatus* has also been reported for the bioaccumulation of heavy metals from distillery effluent (Singh et al. 2005), and an increase in effluent concentration greatly reduced the biomass of plant with maximum accumulation of Fe being recorded in plants growing at 100% effluent concentration. In another study, Trivedy and Nikate (2000) have employed *Typha latifolia* for the treatment of distillery effluent in a constructed wetland treatment system resulting in 78% and 47% reduction in COD and BOD, respectively, within a period of 10 days. Using a combined treatment with *Lemna minuscula* and *Chlorella vulgaris*, Valderrama et al. (2002) have achieved 52% color removal from distillery effluent. The microalgal treatment removed nutrients and organic matter from wastewater and produced oxygen for other organisms. The microphyte removed organic matter and eliminated the microalgae from treated wastewater. However, despite the potential of aquatic macrophytes in cleaning wastewater, the use of these plants in designing a low-cost treatment system is still at experimental stage and is considered to be a potentially important area of environmental management.

Role of Enzymes in PMDE Decolorization

Although the enzymatic process related with detoxification and decolorization of melanoidins and PMDE is yet to be completely understood, it seems greatly connected with fungal ligninolytic mechanisms. The white-rot fungi have complex enzymatic systems which are nonspecific, extracellular, and work under nutrient-limiting conditions. A large number of enzymes from different microorganisms and plants have been reported for the treatment of wastewater discharge from various industries. Important amongst these are ligninolytic enzymes, i.e., manganese peroxidase (MnP), lignin peroxidase (LiP), and laccase reported by several workers for degradation and detoxification of PMDE and synthetic and molasses melanoidin (Ghosh et al. 2002; Petruccioli et al. 2002; Sangeeta et al. 2011; Sangeeta and Chandra, 2012; Sangeeta and Chandra 2013). The catalytic cycle of MnP also resembles with other heme peroxidases, i.e., horseradish peroxidase (HRP) and lignin peroxidase and includes the native ferric enzyme as well as the reactive intermediates compound I and compound II (Fig. 4.4). In contrast to other peroxidases, MnP uses Mn^{2+} as the preferred substrate as electron donor. The catalytic cycle of peroxidases is starting with the reaction of native ferric enzyme (Fe^{3+}-P (porphyrin)) with H_2O_2 yielding compound 1. Compound 1 is a complex of high-valent oxo-iron and porphyrin cation radical (Fe^{4+} = O, P). Compound 1 is reducing a substrate (RH) and yielding a radical cation (R·) and one-electron-oxidized enzyme intermediate compound II (Fe^{4+} = O, P). A single one-electron oxidation of a substrate molecule returns the enzyme to the native ferric state, completing the catalytic cycle. Similarly MnP activity is initiated by binding of H_2O_2 or organic peroxide to the native ferric enzyme and formation of an iron-peroxide complex. Subsequent cleavage of the peroxide oxygen-oxygen bond requires a two-electron transfer from the heme resulting in formation of MnP compound I, which is a Fe^{4+}-oxo-porphyrin-radical complex. Afterwards, the

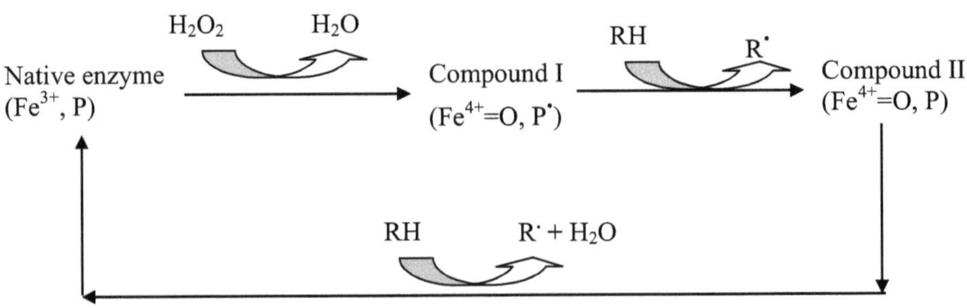

Fig. 4.4 Catalytic cycle of peroxidases

dioxygen bond is heterolytically cleaved and one water molecule expelled. Subsequent reduction proceeds through MnP compound II (Fe^{4+}-oxo-porphyrin complex). A monochelated Mn^{2+} ion acts as one-electron donor for this porphyrin intermediate and is oxidized to Mn^{3+}. The reduction of compound II proceeds in a similar way, and another Mn^{3+} is formed from Mn^{2+} thereby leading to generation of native enzyme and release of the second water molecule. The Mn^{3+} formed is stabilized by organic acids such as oxalate and acts in turn as a low-molecular mass, diffusible redox mediator that attacks organic molecules nonspecifically via hydrogen and one-electron abstraction.

Laccases are common enzymes in nature and are found widely in plants and fungi as well as in some bacteria and insects. Different types of laccases are secreted by different organisms, but they all catalyze polymerization or depolymerization processes including phenols, polyphenols, anilines, and even certain inorganic compounds by a one-electron transfer mechanism. Laccase is a multicopper blue oxidase which reduces oxygen molecules leading to oxidation of broad range of organic substrates. The laccase is a dimeric or tetrameric glycoprotein usually containing per monomer four copper (Cu) atoms bound to three redox sites (Type 1, Type 2 and Type 3 Cu pair). The Type 2 and Type 3 Cu sites are close together and form a trinuclear center that is involved in the catalytic mechanism of the enzyme. The molecular mass of the monomer ranges from about 50 to 100 kDa with acidic isoelectric point around pH 4.0. For the catalytic activity a minimum of four Cu atoms per active protein unit are needed. To function, laccase depends on Cu atoms distributed among the three different binding sites. Cu atoms play an essential role in the catalytic mechanism. There are three major steps in laccase catalysis. The Type 1 Cu is reduced by a reducing substrate, which therefore is oxidized. The electron is then transferred internally from Type 1 Cu to a trinuclear cluster made up of the Type 2 and Type 3 Cu atoms. The O_2 molecule is reduced to water at the trinuclear cluster. The O_2 molecule binds to the trinuclear cluster for asymmetric activation, and it is postulated that the O_2 binding pocket appears to restrict the access of oxidizing agents other than O_2. H_2O_2 is not detected outside of laccase during steady-state laccase catalysis indicating that a four-electron reduction of O_2 to water is occurring. A one-electron substrate oxidation is coupled to a four-electron reduction of oxygen.

Several reports claimed that intracellular sugar-oxidase-type enzymes (sorbose-oxidase or glucose-oxidase) also had melanoidins-decolorizing activities. It was suggested that melanoidins were decolorized by the active oxygen (O_2; H_2O_2) produced by the reaction with sugar oxidases. Ohmomo et al. (1985) used *Coriolus versicolor* Ps4a, which decolorized molasses wastewater by 80% in darkness under optimum conditions. Decolorization activity involved two types of intracellular enzymes, i.e., sugar-dependent and independent. One of these enzymes required no sugar and oxygen for their activity and could decolorize molasses wastewater up to 20% in darkness and 11–17% of synthetic melanoidins. Thus, the participation of these H_2O_2 producing enzymes as a part of the complex enzymatic system for melanoidin degradation by fungi should be taken into account while designing any treatment strategy. Color removal of synthetic melanoidin by *Coriolus hirsutus* involved the participation of peroxidases (MnP and MIP) and the extracellular H_2O_2

produced by glucose-oxidase, without disregard of a partial participation of fungal laccase. Mansur et al. (1997) obtained a maximum decolorization of around 60% by fungus *Trametes* sp. I-62 after 8th day incubation. Here effluent was added at a final concentration of 20% (v/v) after 5 days of fungal growth, the time at which high levels of laccase activity were detected in the extracellular mycelium. The white-rot basidiomycete *Trametes versicolor* is an active degrader of humic acids as well as of melanoidins. Melanoidins mineralizing 47 KDa extracellular proteins corresponding to the major mineralizing enzyme system from *T. versicolor* was isolated by Dehorter and Blondeau (1993). This Mn^{2+}-dependent enzyme system required oxygen and was described to be as peroxidase. Some studies have identified the lignin degradation-related enzymes participating in the melanoidin decolorization. Intracellular H_2O_2 producing sugar oxidases have been isolated from *Coriolus* strains. Also, *C. hirsutus* have been reported to produce enzymes that catalyze melanoidins decolorization directly without additions of sugar and O_2. Miyata et al. (1998) used *C. hirsutus* pellets to decolorize a melanoidin-containing medium. It was elucidated that extracellular H_2O_2 and two extracellular peroxidases, a manganese-independent peroxidase (MIP) and manganese peroxidase (MnP), were involved in decolorization activity. Lee et al. (2000) investigated the dye-decolorizing peroxidase by cultivating *Geotrichum candidum* Dec1 using molasses as a carbon source. Components in the molasses medium stimulated the production of decolorizing peroxidase but inhibited the decolorizing activity of the purified enzyme. D'souza et al. (2006) reported 100% decolorization of 10% spentwash by a marine fungal isolate whose laccase production increased several folds in the presence of phenolic and non-phenolic compounds. A combined treatment technique consisting of enzyme followed by aerobic biological oxidation was investigated by Sangave and Pandit (2006a) for the treatment of spentwash. It was suggested that enzymatic pre-treatment of the distillery effluent leads to in situ formation of the hydrolysis products, which have different physical properties and are easier to assimilate than the parent pollutant molecules by the microorganisms, leading to faster aerobic oxidation even at low biomass. In another study, Sangave and Pandit (2006b) used ultrasound combined with the use of an enzyme as pre-treatment technique for treatment of distillery wastewater. The combination of the ultrasound and enzyme also reported significant COD removal efficiencies as compared to the processes when they were used as stand-alone treatment techniques. Panicker et al. (2015) have showed the possibility of using immobilized enzyme to enhance the biodegradability of the distillery spentwash. The investigations so far can be seen as an initial step toward solving the problem. Moreover, most of these microbial decolorization studies required effluent dilution for optimal activity. While, using microorganisms use of media supplement pose extra burden on overall effluent treatment process. Further, the emerging treatment methods like enzymatic treatment have technological advantages and yet are in its infancy, requiring economic considerations in order to apply on the plant scale. Hence, there is need of suitable technique to treat distillery effluent. Furthermore, capital and operating costs of the available physico-chemical and biological treatment processes of distillery waste stream are inevitably high thus making these processes less lucrative to the industry.

Metabolites Produced During the Bacterial Treatment of PMDE

The study of metabolites produced during the degradation process is very essential to explore the degradation pathway of any pollutant as well as to develop the effective bioremediation techniques. The structural characterization of Maillard reaction products (MRPs) has proved to be rather challenging due to extreme complexity of MRPs that are known to range from small molecules to extremely large polymers. There are many techniques available which are used for characterization of large, complex pollutants and metabolites/intermediate compounds, produced during degradation process (Table 4.6). By using these techniques, some authors have studied the chemical nature of melanoidins and reported that these are complex polymers consisting of repeating units of furans and/or paroles, which are produced during the advance stages of Maillard reaction

Table 4.6 Analytical techniques used for the analysis of Maillard reaction products (MRPs)/ melanoidin

Maillard reaction products (MRPs)	Analytical techniques
Amadori compounds (direct analysis)	Column chromatography, HPLC differential refractometry detection, HPLC involving derivatization, HPAEC coupled electrochemical and/or DAD, FAB-MS, ESI coupled HPLC and EC, EC coupled MS, MALDI-TOF, NMR, LC-MS, NBT, ELISA
Indirect analysis (2-FM-AA)	Immunoblotting (lactosylated proteins), ion-pair RP-HPLC, CEC UV detection
Unreactive lysine	Colorimetric and fluorimetric methods
Advanced Maillard reaction products	FAST
General AGEs	HPLC-DAD
CML	RP-HPLC, RP-HPLC o-phthalaldehyde pre-column derivatization, GC-MS, ELISA
Pyrraline	Amino acid analysis with PAD, RP-HPLC
Cross-linking products lysine dimmers	LC-MS with ESI
Arginine-lysine	LC-MS with ESI, ion-exchange chromatography
Other amino acid derivatives, argpyrimidine	HPLC-coupled GC-MS
OMA	ELISA
PIO	RP-HPLC/LC-ESI-TOF-MS/NMR
Pyrazinones	HPLC with UV and fluorescence detection
Lysine amnioreduction	HPLC-DAD
Final stage MRPs	
General melanoidins	HPLC, NMR, MS, UV, IR spectrometry
Pronyl-L-lysine	GC-MS chemical ionization

Jones et al. (1998), Silvan et al. (2006) and Shen et al. (2007)

and remain linked by polycondensation reactions (Tressl and Wondrak 1998; Gonzalez et al. 2000; Bharagava and Chandra 2010). Hofmann (1998) also reported that MRPs are low molecular weight colored compounds, which cross-link to proteins via amino groups of lysine or arginine and polymerized through aldol-type condensation reactions to produce high molecular weight colored MRPs. The chemical structure of these MRPs is mainly built with sugar degradation products, which largely depends on the starting materials as well as on the reaction conditions (Cammerer and Kroh 1995). Some authors have also reported that the electro spray ionization mass spectrum (ESI-MS) infrared (IR)^{1}H NMR and LC-MS-MS analysis of untreated distillery wastewater has shown the presence of dihydroxyconiferyl alcohol ($C_{10}H_{14}O_3$), 2,2′-bifuran-5-carboxylic acid ($C_9H_6O_4$), 2-nitroacetophenone ($C_8H_7NO_3$), p-chloroanisole (C_7H_7ClO), 2,3-dimethyl pyrazine ($C_6H_8N_2$), 2-methylhexane (C_7H_{16}), methyl benzene (C_7H_8),1,2- dihydro-5 methylfuran (C_5H_8O),3-pyrroline (C_4H_7N), and acetic acid ($C_2H_4O_2$), while in treated distillery wastewater, compounds 2-nitroacetophenone ($C_8H_7NO_3$), p-chloroanisole (C_7H_7ClO), 2,2-bifuran($C_8H_6O_2$), indole (C_8H_7N), 2-methylhexane (C_7H_{16}), and 2,3-dihydro-5-methylfuran (C_5H_8O) were reported (Gonzalez et al. 2000; Bharagava and Chandra 2010). Further, most of the detected metabolic products were reported to as degradation products of pyrroles/furans, which remains present in distillery wastewater.

Challenges for the Degradation and Decolorization of PMDE

The major problem of PMDE is its complex nature due to the presence of recalcitrant constituents that inhibit the growth of microbes during the effluent treatment process in industries. The major coloring constituent, i.e., melanoidins, has a net negative charge and, thus, forms large complex molecules with different heavy metal ions (Cu^{2+}, Cr^{3+}, Fe^{3+}, Zn^{2+}, Pb^{2+}, etc.) in acidic medium and gets precipitated. Besides melanoidins, distillery effluent contains various phenolic compounds such as gallic acids, p-coumaric acid, gentisic acid, and 3–4 dimethoxy 2 (1,1-hydroxyethyl) α-naphthol (Borja et al. 1993; Bharagava et al. 2006). These compounds have inhibitory and antimicrobial activity reducing the anaerobic digestion process of distillery effluent in methane reactors. Further, the sulfates and heavy metals present in PMDE get reduced into black-colored precipitate of metal sulfides, which also act as competitive inhibitor for sulfate-reducing bacteria (SRB) and non-SRB (McCartney and Oleszkiewicz 1991) leading to the inhibition of methanogenesis process. In addition, the presence of EDCs makes effluent more toxic, and complete knowledge of these is lacking.

Thus, it becomes very essential to study the effect of sulfides and metals along with phenolics as well as other compounds on decolorization and the detoxification of PMDE. Similarly complete knowledge of EDCs is mandatory before biological treatment and discharge of effluent in the environment.

Future Prospects

Alcohol production and distilleries are one important commodity and industry world over. But, the huge generation of complex wastewater is still a challenge for its safe disposal and pollution prevention for sustainable development. Therefore, the detailed characterization and biodegradation of complex pollutants present in PMDE is a priority need of researches and industries. The potential microorganisms including bacteria, fungi, and cyanobacteria may be optimized for the attenuation of the complex pollutants with the enzymatic activity (MnP and laccase). Further, the potential wetland plants which enhance the biodegradation may be integrated for degradation of PMDE. The detoxified and decolorized PMDE may be recycled for agricultural practices and green vegetation development around the industrial and urban areas. In addition, generated sludge and biomass in developed treatment technology can be formulated for biocomposting as organic manure for the sustainable development and pollution prevention of aquatic resources.

Conclusion

PMDE is generated after methanogenesis of sugarcane molasses-based spentwash which contains dark color, high TDS due to the presence of complex mixture of organic and inorganic pollutants. The complex organic and inorganic pollutants inhibit biodegradation. Therefore, the degradation and decolorization of PMDE is still a challenge for researchers and industries. Thus the complete knowledge regarding the nature of organic and inorganic pollutants present in PMDE is essential. Further, its biodegradation in valuable products is also essential. The reported research knowledge is still inadequate. The intergradations of bacterial treatment with phytoremediation may be effective and feasible techniques for treatment of this complex wastewater.

Acknowledgments The authors are grateful to the Department of Biotechnology, Government of India, for the financial assistance to project vide letter no BT/PR13922/BCE/8/1129/2015.

References

Bharagava, R. N., & Chandra, R. (2010). Biodegradation of the major colour containing compounds in distillery wastewater by an aerobic bacterial culture and characterization of their metabolites. *Biodegradation, 21*, 703–711.

Bharagava, R. N., Chandra, R., & Singh, S. K. (2006). Elucidation of chemical structure of phenolic compounds by H^1NMR and GC-mass spectrometry present in anaerobically digested distillery effluent. *Indian Journal of Environmental Protection, 26*(11), 1015–1018.

Bharagava, R. N., Chandra, R., & Rai, V. (2008). Phytoextraction of trace elements and physiological changes in Indian mustard plants (*Brassica nigra* L.) grown in post methanated distillery effluent (PMDE) irrigated soil. *Bioresource Technology, 99*, 8316–8324.

Bharagava, R. N., Chandra, R., & Rai, V. (2009). Isolation and characterization of aerobic bacteria capable of the degradation of synthetic and natural melanoidins from distillery wastewater. *World Journal of Microbiology and Biotechnology, 25*, 737–744.

Billore, S. K., Singh, N., Ram, H. K., et al. (2001). Treatment of a molasses based distillery effluent in a constructed wetland in central India. *Water Science and Technology, 44*, 441–448.

Borja, R., Martin, A., Maestro, R., et al. (1993). Enhancement of the anaerobic digestion of wine distillery wastewater by the removal of phenolic inhibitors. *Bioresource Technology, 45*(2), 99–104.

Camarero, S., Bocchini, P., Galletti, G. C., et al. (1999). Pyrolysis-gas chromatography/mass spectrometry analysis of phenolic and etherified units in natural and industrial lignins. *Rapid Communications in Mass Spectrometry, 13*, 630–636.

Cammerer, B., & Kroh, L. W. (1995). Investigation of the influence of reaction conditions on the elementary composition of melanoidins. *Food Chemistry, 53*, 55–59.

Chandra, R., & Sangeeta, Y. (2011). Phytoremediation of Cd, Cr, Cu, Mn, Fe, Ni, Pb and Zn from aqueous solution using *Phragmites communis, Typha angustifolia* and *Cyperus esculentus*. *International Journal of Phytoremediation, 13*, 580–591.

Chandra, R., & Srivastava, A. (2004). Toxicological evaluation of bacteria decolourised anaerobically treated distillery effluent with common duckweed (*Lemna minor*). *Journal of Environmental Biology, 25*, 93–98.

Chandra, R., Kumar, K., & Singh, J. (2004). Impact of anaerobically treated and untreated (raw) distillery effluent irrigation on soil microflora, growth, total chlorophyll and protein contents of *Phaseolus aureus* L. *Journal of Environmental Biology, 25*(4), 381–385.

Chandra, R., Raj, A., Purohit, H. J., et al. (2007). Characterization and optimization of three potential aerobic bacterial strains for kraft lignin degradation from pulp paper waste. *Chemosphere, 67*, 839–846.

Chandra, R., Bharagava, R. N., & Rai, V. (2008a). Melanoidins as major colorant in sugarcane molasses based distillery wastewater and its degradation. *Bioresource Technology, 99*, 4648–4660.

Chandra, R., Yadav, S., & Mohan, D. (2008b). Effect of distillery sludge on seed germination and growth parameters of green gram (*Phaseolus mungo* L.). *Journal of Hazardous Materials, 152*, 431–439.

Chandra, R., Sangeeta, Y., Bharagava, R. N., et al. (2008c). Bacterial pretreatment enhances removal of heavy metals during treatment of post-methanated distillery effluent by *Typha angustata* L. *Journal of Environmental Management, 88*, 1016–1024.

Chandra, R., Bharagava, R. N., Yadav, S., & Mohan, D. (2009). Accumulation and distribution of toxic metals in wheat (*Triticum aestivum* L.) and Indian mustard (*Brassica campestris* L.) irrigated with distillery and tannery effluents. *Journal of Hazardous Materials, 162*, 1514–1521.

Chen, T. Y., Kao, C. M., Yeh, T. Y., et al. (2006). Application of a constructed wetland for industrial wastewater treatment: A pilot-scale study. *Chemosphere, 64*, 497–502.

D'Souza, D. T., Tiwari, R., Sah, A. K., et al. (2006). Enhanced production of laccase by a marine fungus during treatment of coloured effluents and synthetic dyes. *Enzyme and Microbial Technology, 38*, 504–511.

Dahiya, J., Singh, D., & Nigam, P. (2001). Decolourisation of molasses wastewater by cells of *Pseudomonas fluorescens* immobilized on porous cellulose carrier. *Bioresource Technology, 78*, 111–114.

Dehorter, B., & Blondeau, R. (1993). Isolation of an extracellular Mn dependent enzyme mineralizing melanoidins from the white rot fungus *Trametes versicolour*. *FEMS Microbiology Letters, 109*, 117–122.

Environment (Protection) Amendment Rules. (2008). On waste water generation standards Substituted by Rule 2(ii) (a) of the notified by G.S.R.186 (E), dated 18 Mar 2008.

Eusibio, A., Petruccioli, M., Lageiro, M., et al. (2004). Microbial characterization of activated sludge in jet-loop bioreactors treating winery wastewater. *Journal of Industrial Microbiology and Biotechnology, 31*, 29–34.

Ghosh, M., Ganguli, A., & Tripathi, A. K. (2002). Treatment of anaerobically digested distillery spentwash in a two-stage bioreactor using *Pseudomonas putida* and *Aeromonas sp. Process Biochemistry, 37*, 857–862.

Ghosh, M., Verma, S. C., Mengoni, A., et al. (2004). Enrichment and identification of bacteria capable of reducing chemical oxygen demand of anaerobically treated molasses spentwash. *Journal of Applied Microbiology, 96*, 1278–1286.

Gonzalez, T., Terron, M. C., Yague, S., et al. (2000). Pyrolysis/gas chromatography/mass spectrometry monitoring of fungal-biotreated distillery wastewater using *Trametes sp.* I-62 (CECT 20197). *Rapid Communications in Mass Spectrometry, 14*, 1417–1424.

Hati, K. M., Biswas, A. K., Bandyopadhyay, K. K., et al. (2007). Soil properties and crop yields on a vertisol in India with application of distillery effluent. *Soil and Tillage Research, 92*, 60–68.

Hodge, J. E. (1953). Chemistry of browning reactions in model systems. *Journal of Agricultural and Food Chemistry, 1*, 928–943.

Hofmann, T. (1998). Studies on melanoidin-type colorants generated from the Maillard reaction of protein-bound lysine and furan-2-carboxaldehyde-chemical characterisation of a red coloured domaine. *European Food Research and Technology, 206*, 251–258.

Jain, N., Minocha, A. K., & Verma, C. L. (2002). Degradation of predigested distillery effluent by isolated bacterial strains. *Indian Journal of Experimental Biology, 40*, 101–105.

Jain, N., Bhatia, A., Kausik, R., et al. (2005). Impact of post methanation distillery effluent irrigation on ground water quality. *Environmental Monitoring and Assessment, 110*, 243–255.

Jones, A. D., Tier, C. M., & Wilkins, J. P. G. (1998). Analysis of the Maillard reaction products of b-lactoglobulin and lactose in skimmed milk powder by capillary electrophoresis and electrospray mass spectrometry. *Journal of Chromatography, 822*, 147–154.

Kalavathi, D. F., Uma, L., & Subramanian, G. (2001). Degradation and metabolization of the pigment-melanoidin in distillery effluent by marine cyanobacterium *Oscillatoria boryana* BDU. *Enzyme and Microbial Technology, 29*, 246–251.

Kambe, T. N., Shimomura, M., Nomura, N., et al. (1999). Decolourization of molasses wastewater by *Bacillus sp.* under thermophilic and anaerobic conditions. *Journal of Bioscience and Bioengineering, 87*, 119–121.

Kandarakis, E. D., Bourguignon, J. P., Giudice, L. C., et al. (2009). Endocrine-disrupting chemicals: An Endocrine Society scientific statement. *Endocrine Reviews, 30*(4), 293–342.

Kaushik, A., Nisha, R., Jagjeeta, K., et al. (2005). Impact of long and short term irrigation of a sodic soil with distillery effluent in combination with bioamendments. *Bioresource Technology, 96*, 1860–1866.

Khairnar, P., Chavan, F., & Diware, V. R. (2013). Generation of energy from distillery waste water. *Pratibha: International Journal of Science, Spirituality, Business and Technology, 2*(1), 29–35.

Kowalska, A., Bodzek, M., & Bohdziewicz, J. (1998). Biodegradation of phenols and cyanides with immobilized microorganisms. *Process Biochemistry, 33*, 189–197.

Krishnanand, Y., Maillacheruvu, G. F. P., et al. (1993). Sulfide toxicity in anaerobic systems fed sulfate and various organics. *Water Environment Research, 65*(2), 100–109.

Kumar, P., & Chandra, R. (2004). Detoxification of distillery effluent through Bacillus thuringiensis (MTCC 4714) enhanced phytoremediation potential of *Spirodela polyrrhiza* (L.) Schliden. *Bulletin of Environmental Contamination and Toxicology, 73*, 903–910.

Kumar, S., & Viswanathan, L. (1991). Production of biomass, carbon dioxide, volatile acids, and their interrelationship with decrease in chemical oxygen demand, during distillery waste treatment by bacterial strains. *Enzyme and Microbial Technology, 13*, 179–186.

Lee, C. M., Chichester, C., & Lee, T. C. (1977). *Physiological consequences of browned food products*. Proceedings of the IVth international congress of food science and technology.

Lee, T. H., Aoki, H., Sugano, Y., et al. (2000). Effect of molasses on the production and activity of dye-decolourizing peroxidase from *Geotrichum candidum* Dec 1. *Journal of Bioscience and Bioengineering, 89*, 545–549.

Mansur, M., Suarez, T., Fernandez-Larrea, J., et al. (1997). Identification of a Laccase gene family in the new lignindegrading basidiomycete CECT 20197. *Applied and Environmental Microbiology, 63*, 2637–2646.
McCartney, D. M., & Oleszkiewicz, J. A. (1991). Competition between methanogens and sulfate reducers: Effect of COD: Sulfate ratio and acclimation. *Water Environment Research, 65*(5), 655–664.
Miyata, N., Iwahori, K., & Fujita, M. (1998). Manganese independent and dependent decolourisation of melanoidin by extracellular hydrogen peroxide and peroxidases from *Coriolus hirsutus* pellets. *Journal of Fermentation and Bioengineering, 85*, 550–553.
Mohana, S., Desai, C., & Madamwar, D. (2007). Biodegradation and decolourization of anaerobically treated distillery spentwash by a novel bacterial consortium. *Bioresource Technology, 98*, 333–339.
Ohmomo, S., Itoh, N., Wantanabe, Y., et al. (1985). Continuous decolorization of molasses wastewater with mycelia of *Coriolus versicolor* Ps4a. *Agricultural and Biological Chemistry, 49*, 2551–2555.
Ohmomo, S., Daengsabha, W., Yoshikawa, H., et al. (1988). Screening of anaerobic bacteria with the ability to decolourize molasses melanoidin. *Agricultural and Biological Chemistry, 57*, 2429–2435.
Panicker, S., Singh, A., & Agnihotri, S. (2015). Decolorization of biomethanated distillery effluent by immobilized enzymes. *International Journal of Bioassays, 4*(11), 4518–4522.
Patel, A., Pawar, P., Mishra, S., et al. (2001). Exploitation of marine cyanobacteria for removal of colour from distillery effluent. *Indian Journal of Environmental Protection, 21*, 1118–1121.
Petruccioli, M., Duarte, J. C., Eusibio, A., et al. (2002). Aerobic treatment of winery wastewater using a jet –loop activated sludge reactor. *Process Biochemistry, 37*, 821–829.
Ramakritinan, C. M., Kumaraguru, A. K., & Balasubramanian, M. P. (2005). Impact of distillery effluent on carbohydrate metabolism of freshwater fish, *Cyprinus carpio*. *Ecotoxicology, 14*, 693–707.
Rattan, R. K., Datta, S. P., Chhonkar, P. K., et al. (2005). Long-term impact of irrigation with sewage effluents on heavy metal content in soils, crops and ground water-A case study. *Agriculture, Ecosystems and Environment, 109*, 310–322.
Sangave, P. C., & Pandit, A. B. (2006a). Enhancement in biodegradability of distillery wastewater using enzymatic pretreatment. *Journal of Environmental Management, 78*, 77–85.
Sangave, P. C., & Pandit, A. B. (2006b). Ultrasound and enzyme assisted biodegradation of distillery wastewater. *Journal of Environmental Management, 80*, 36–46.
Sangeeta, Y., & Chandra, R. (2006). Effect of post methanated distillery effluent on various morphological, physiological and biochemical parameters of *Vicia fabae* after two step biological treatment. Presented in 26th annual session of the academy of environmental biology (pp. 40–41).
Sangeeta, Y., & Chandra, R. (2011). Heavy metals accumulation and ecophysiological effect on *Typha angustifolia* L. and *Cyperus esculentus* L. growing in distillery and tannery effluent polluted natural wetland site, Unnao, India. *Environment and Earth Science, 62*, 1235–1243.
Sangeeta, Y., & Chandra, R. (2012). Biodegradation of organic compounds of molasses melanoidin (MM) from biomethanated distillery spentwash (BMDS) during the decolourisation by a potential bacterial consortium. *Biodegradation, 23*(4), 609–620.
Sangeeta, Y., & Chandra, R. (2013). Effect of pH on melanoidin extraction from post methanated distillery effluent (PMDE) and its decolorization by potential bacterial consortium. *International Journal of Recent Scientific Research, 4*(10), 1492–1496.
Sangeeta, Y., Chandra, R., & Vibhuti, R. (2011). Characterization of potential MnP producing bacteria and its metabolic products during decolourisation of synthetic melanoidins due to biostimulatory effect of d-xylose at stationary phase. *Process Biochemistry, 46*, 1774–1784.
Shen, S. C., Tseng, K. C., & Wu, J. S. B. (2007). An analysis of Maillard reaction products in ethanolic glucose-glycine solution. *Food Chemistry, 102*, 281–287.

Silvan, J. M., Lagemaat, J. V. D., Olano, A., et al. (2006). Analysis and biological properties of amino acid derivates formed by Maillard reaction in foods. *Journal of Pharmaceutical and Biomedical Analysis, 41*, 1543–1551.

Singh, N. K., Pandey, G. C., Rai, U. N., et al. (2005). Metal accumulation and ecophysiological effects of distillery effluent on *Potamogeton pectinatus* L. *Bulletin of Environmental Contamination and Toxicology, 74*, 857–863.

Sirianuntapiboon, S., Phothilangka, P., & Ohmomo, S. (2004). Decolourisation of molasses wastewater by a strain no. BP 103 of acetogenic bacteria. *Bioresource Technology, 92*, 31–39.

Subramani, A., Saravanan, S., Tamizhiniyan, P., et al. (1997). Influence of heavy metals on germination and early seedling growth of *Vigna mungo* L. *Pollution Research, 16*(1), 29–31.

TEDX. (2011). The Endocrine Disruption Exchange, Inc. (TEDX), a 501(c)(3) organization, is based in Paonia, Colorado, and is incorporated as a business under the laws of that state.

Tressl, R., & Wondrak, G. T. (1998). New melanoidin like Maillard polymer from 2-deoxypentoses. *Journal of Agricultural and Food Chemistry, 46*, 104–110.

Trivedy, R. K., & Nakate, S. S. (2000). Treatment of diluted distillery waste by constructed wetlands. *Indian Journal of Environmental Protection, 20*, 749–753.

Valderrama, L. T., Del Campo, C. M., Rodriguez, C. M., et al. (2002). Treatment of recalcitrant wastewater from ethanol and citric acid using the microalga *Chlorella vulgaris* and the macrophyte *Lemna minuscule*. *Water Research, 36*, 4185–4192.

Chapter 5
Heavy Metal Contamination: An Alarming Threat to Environment and Human Health

Sandhya Mishra, Ram Naresh Bharagava, Nandkishor More, Ashutosh Yadav, Surabhi Zainith, Sujata Mani, and Pankaj Chowdhary

Contents

S. Mishra · R. N. Bharagava (✉) · A. Yadav · S. Zainith · P. Chowdhary
Laboratory for Bioremediation and Metagenomics Research (LBMR), Department of Environmental Microbiology, Babasaheb Bhimrao Ambedkar University (A Central University), Lucknow, Uttar Pradesh, India
e-mail: bharagavarnbbau11@gmail.com

N. More
Department of Environmental Sciences (DES), Babasaheb Bhimrao Ambedkar University (A Central University), Lucknow, Uttar Pradesh, India

S. Mani
Department of Biochemistry, Gramin Science (Vocational) College, Vishnupuri, Nanded, Maharashtra, India

R. C. Sobti et al. (eds.), *Environmental Biotechnology: For Sustainable Future*,
https://doi.org/10.1007/978-981-10-7284-0_5

Abstract Metals naturally exist in the crust of the earth, and their compositions vary among different localities, resulting in structural disparity of surrounding concentrations. Some heavy metals are much important in trace amounts in respect to living organisms related to their metabolic activities. High solubility of various heavy metals changes them into extremely toxic and perilous contaminant of water and soil when discharged by many industrial activities. When these metals are released into the environment, they can be leached into the underground waters, depositing in the aquifers, or run off into surface waters and soils thereby resulting in water and soil pollution. Thus, heavy metals become a potential contaminant for environment that can partake in trophic transfer in food chains. The toxicity of heavy metals mainly depends upon its relative oxidation state, which is responsible for physiological bio-toxic effects. When these metals enter into the living organisms, they, combine with proteins, enzymes, and DNA molecules, form highly stable biotoxic compounds, thus altering their proper functioning and obstructing them from the bioreactions. Arsenic, chromium, cadmium, and lead are highly toxic and produce mutagenic, carcinogenic, and genotoxic effects. Hence, this chapter is focused on occurrence and allocation of heavy metals, their toxicological impact on environment, and their possible eco-friendly remedies for green and healthy environment.

Keywords Heavy metals · Contamination · Environment · Toxicity · Bioremediation

Introduction

Heavy metal pollution is a major environmental problem faced by the modern world. Rapid industrialization leads to manufacturing expansion which causes continuous increase in metal concentration and serious pollution problems due to improper dumping and discarding of industrialized waste products directly into water bodies and land areas (Dixit et al. 2015). Heavy metals are the main elements of the earth's crust which exist as persistent environmental contaminants that cannot be purely biodegraded but can only be transformed into nontoxic forms (Coral et al. 2005).

Generally, heavy metal is a collective term used for a group of metals having higher atomic number (above 20) and greater density (5 g/cm^3) such as cadmium, lead, mercury, nickel, chromium, arsenic, copper, and zinc. These metals are directly related with environmental pollution and biological toxicity problems because of their powerful inhibitory actions on biodegradation activities (Nagajyoti et al. 2010; Gupta et al. 2014).

Mostly, metals are essential in a small quantity as nutrients, involved in a range of enzymatic and metabolic pathways, and act as cofactors. However, large quantities of such metals can become strongly inhibitory/deadly to all forms of lives including microorganisms, plants, animals, and humans. Although few metals like cadmium (Cd), arsenic (As), and mercury (Hg) are extremely poisonous at a very low concentrations (Atlas and Bartha 1993; Nies 1999). High concentration of metals, when accumulated in soil, causes detrimental effects on productivity and fertility. These can also go into the system through food, water, or air and bioaccumulate for a long time (Lenntech 2004; Deeb and Altahali 2009). Accumulation of lethal metals into the human body creates severe health consequences such as growth and developmental abnormalities, carcinogenesis, neuromuscular defects, mental illnesses, and failure of metabolic activities (Thiele 1995; Chandra et al. 2011).

The most significant way of releasing toxic metals into the environment occurs through a range of practices and pathways, together with combustion and extraction of toxic waste to the atmosphere, and excess emanations of wastewaters/effluents containing harmful noxious chemicals and contaminants, polluting surface waters and soils. The continuous use of hazardous metals in anthropogenic activities from industrial sectors like electroplating, painting, tanning, textiles and dyes, papermaking, mining, and others has increased enormously and has become detrimental for diversity of life on earth (Dhal et al. 2013; Yadav et al. 2017). The indiscriminate discharge of hazardous metals from these industrial processes often at elevated concentrations (above permissible limits) creates a toxicity risk for environmental contamination. Hence, their removal/remediation has become necessary for environmental safety (Dhal et al. 2013; Velma et al. 2009).

Thus, biological remediation using microbes and plants is generally considered as environment-friendly, safe, cost-effective, and sustainable approach for the treatment of toxic metals in comparison with conventional treatment technologies. Nevertheless, conventional technologies have some disadvantages, regarding their expensive charge, maintenance, and generation of hazardous by-products or inefficiency, whereas biological treatment technologies elucidate such problems of cost and operating system since they are easy to operate and do not produce secondary pollutants (Vargas et al. 2012; Chandra et al. 2011). The practices of involvement of microorganisms and plants for the reduction and detoxification of highly toxic pollutants into innocuous and less harmful forms are known as bioremediation and phytoremediation, respectively (Jain et al. 2014). This chapter is focused on heavy metal contamination into the environment and their occurrence, applications, and toxic impact related to their accumulation in plants, animals, and humans. The chapter also discusses the possible remediation of heavy metal-contaminated sites by biological means.

Distribution of Heavy Metals into the Environment

The environment (soil, water, and air) has been severely contaminated with various toxic metals. Metals are introduced into the environment through different anthropogenic and natural activities. Naturally, rock formation, weathering, erosion, and volcanoes are particularly responsible for the exposure and emission of huge amount of Al, Cu, Hg, Mn, Ni, and Zn into the environment (Seaward and Richardson

1990). Usually metals like aluminum, copper, and silver get deposited in a variety of rocks including igneous, hornblende, olivine, augite, and sedimentary as minerals, and the rocks having sufficient quantity of metals for economic extraction are known as ores, contributing considerable amount of metals in their pure extracted forms (Nagajyoti et al. 2010).

Anthropogenic deeds of electroplating, mining, smelting, pesticides and fertilizer discharges, bio-solids, municipal sewage/sludge, and textiles and paint industries have become most significant sources of metal contamination (Barkat 2011; Yadav et al. 2017; Mani and Bharagava 2016). Prevalently, heavy metals are present in toxic concentrations in terrestrial and aquatic water bodies attributed to the direct discharge of solid and liquid waste products emanating from these industries. This rapid and modern industrialization increased the level of heavy metal's contamination in the environment. Metalliferous mining operations such as coal mines are mainly responsible for the extraction of metals such as As, Cd, Fe, etc., and such operations are ultimately causing nonstop soil contamination because of regular practices around the coalfield. Transportation of metals from mining areas to various industries for their extensive applications in manufacturing of various products is also an important source of metal contamination into the environment.

Leather industries are primarily responsible for Cr contamination into the environment because of extensive applications of Cr salts during the chrome tanning process (Mishra and Bharagava 2016). Electroplating and paint industries contribute Pb-based contamination to the environment. The widespread use of fertilizers, pesticides, and fungicides in agricultural activities is also contributing to metal pollution into soils. The extensive use of chemical-based fertilizers in agricultural practices to enhance fertility of soil for production of crop yields has become one of the important sources of metal contamination (Yadav et al. 2016). These fertilizers have significant amount of As, Cr, Cd, Pb, Zn, Ni, Fe, Mo, and Mn that play a vital role in growth improvement and healthy progression of plants, but excess concentration of such metals results in toxic effects in soil as well as in aquatic resources (Reeves and Baker 2000; Blaylock and Huang 2000).

Arsenic (As)

It belongs to the VA group of periodic table and mainly presents as As_2O_3 in variability of mineral forms. It exists in several oxidation states (ranging from -III, 0, III, V) (Smith et al. 1995). It is conspicuously lethal, cancer causing, and expansively existing in the semimetallic form of oxides/sulfides or as salts of Fe, Na, Ca, Cu, etc. (Singh et al. 2007). It is predominantly used as coloring agent in textiles, wallpaper, and toy-making industries. It is also used as insecticide and rat killer to protect timber and hide from termites and rats. Furthermore, As is present in ashes from coal combustion. Elemental As and its simplest organic compound AsH_3 (arsine), which is highly toxic and flammable, usually occur under extremely reduced environments. Several compounds of arsine, dimethyl arsine $HAs(CH_3)_2$

and trimethylarsine As $(CH_3)_3$, which are generated via methylation of As, are also extremely dangerous and toxic in nature. Conversion of As to arsenite via methylation into mono-methyl (MMA) or dimethyl (DMA) compounds is also deadly poisonous in comparison with other arsenicals. MMA is also responsible for causing As-induced carcinogenesis (Singh et al. 2007). Contamination with high concentrations of As is a matter of great concern because As causes numerous human health disorders. As compounds cause skin damage, cancer, and marked problems with circulatory system (Scragg 2006). The presence of As in drinking water more than its permissible limit is also responsible for severe clinicopathological diseases, developmental abnormalities, neurobehavioral sicknesses, cardiovascular diseases, and hearing sickness, together with anemia, leukopenia, eosinophilia, and carcinoma (Tchounwou et al. 2004; ATSDR 2000; Centeno et al. 2005; NRC 2001). Damage of cellular respiration due to the inhibition of various mitochondrial enzymes is one of the main mechanisms which is mainly responsible for As associated with uncoupling of oxidative phosphorylation, which results in As toxicity from its capability to interact with sulfhydryl groups of proteins and enzymes and to substitute P in a variety of biochemical reactions (Wang and Rossman 1996).

Cadmium (Cd)

Predominantly, Cd is available in the associated forms with Pb and Zn ores in natural environments. It is a by-product of Zn refining. Zn production constitutes prevalent source of Cd contamination into the environment (Ziemacki et al. 1989). Cd is considerably applicable in numerous industrial processes involving electroplating, stabilizing agent in plastic industry, and vending machines of soft drinks, cigarette smoke, and batteries, and it provides protection against corrosion. Industrial wastewater contains higher amounts of Cd from 10 mM to 100 mM (Shuttleworth and Unz 1993).

Cd is the seventh most hazardous and noxious metal as per Agency for Toxic Substances and Disease Registry (ATSDR) ranking. Extensive exposure of Cd contamination severely affects human and animals more probably by inhalation or ingestion via a number of sources including metal industries, spoiled and wasted food, cigarettes, and Cd products related to factories and work areas (IARC 1990; Paschal et al. 2000). The chemical similarity of Cd to Zn accounts for its toxicity, located just below the Zn in the periodic table. Therefore, the substitution of Zn by Cd causes disruption of metabolic activities (Campbell 2006).

Cd causes an indirect oxidative stress that might be relative to its carcinogenic and mutagenic nature resulting in severe health hazards such as kidney damage, prostate dysfunction, bone diseases, and cancer (Admis et al. 2003). The prolonged exposure of Cd in the kidneys adversely affects several enzymes which are accountable for reabsorption of proteins in kidney tubules, resulting in kidney dysfunction and proteinuria.

Cd is highly toxic and nonessential heavy metal that is well known for its severe and inhibitory effects on enzymatic reactions inducing oxidative stress and nutritive deficiency in plants (Irfan et al. 2013). Cd toxicity is a matter of concern due to its strong cancer-causing behavior with a half-life of 20 years (Giaginis et al. 2006). Cd causes Fe deficiencies as it can easily bind with histidine, cysteine, aspartate, and glutamate ligands (Castagnetto et al. 2002). It is also responsible for renal and hepatic damages and coma due to after stark absorption and acts as severe gastrointestinal irritant (Baselt 2000).

Chromium (Cr)

Cr is a d-block transition metal located in the VIB group of periodic table. Elemental or pure form of Cr cannot be found into the environments; it can exist only in Cr compounds. Mostly Cr metal is held inside the sediments and rocks as ores mainly in the compounds of chromite ($FeOCr_2O_3$) and is commonly introduced into the natural waters, soils, air, and food products (Smith et al. 1995).

Different forms of Cr exist in nature ranging from divalent to hexavalent in which Cr (VI) and Cr (III) forms are highly stable in relation to human exposure and toxicity (Zhitkovich 2005). Cr (VI) is extremely harmful and reported as priority pollutant and human carcinogen and causes severe health implications and clinical disorders sometimes leading to death (IARC 1990; ATSDR 2008). Cr is extensively used in numerous industrial processes of leather tanning, metal refining, textile dyeing, pharmaceutical drugs, inks and pigment, refractories, as a wood preserver, etc. (Patra et al. 2010; Bharagava and Mishra 2018). Both Cr forms, Cr (III) and Cr (VI), are unlike in relation to their mobility, solubility, bioavailability, and toxicity. Cr (VI) is highly toxic due to prevailing stability and cellular permeability to cross membranes via sulfate transport system and causes denaturation and mutation of nucleic acids and proteins. However, trivalent Cr is less harmful and vital for humans as nutrient supplement, primarily bound to organic substances present in water and soil environments. Cr (VI) is very well known to cause life-threatening harmful toxic effects, resulting in cancer mutations and genotoxic abnormalities in humans and animals (Turpeinen et al. 2004; Mishra and Bharagava 2016). Cr (VI) contamination in soil changes the organization of native microbial diversity by reducing their growth and metabolic activities. Moreover, Cr disturbs the human physiology by accumulating in the food chain and creates critical health issues of skin problems, nasal irritation, hear impairment, and lung carcinoma (Verkleji 1993; Stern 2010). Cr compounds and its several intermediates are also responsible for chromosomal aberration alteration in replication and DNA damage by forming DNA-protein complexes (Chang and Gu 2007; Xu et al. 2005; Mastsumoto et al. 2006).

Lead (Pb)

Pb is a bluish gray metal belonging to IV group and period 6 of the periodic table. It occurs naturally as mineral in association with other components such as sulfides with S (PbS, $PbSO_4$) or oxides with oxygen ($PbCO_3$). It exists in the earth's crust in the concentrations between 10 and 30 mg kg^{-1} (USDHHS 1999). Pb is commonly used in auto exhaust, hair dyes, paints, sindoor (vermillion), surma (collyrium), glazing of pottery, and enamelware. Road transports and vehicles are principally responsible for high amounts of Pb emissions into the environment. Anthropogenic activities of coal combustion, leaded fuels, and production of pyrometallurgical nonferrous metals are noticeably contributed for Pb-related pollution of water bodies and soil sources (Monterroso et al. 2003; Andrews and Sutherland 2004).

Generally, Pb can be present in ionic, oxide, hydroxide, and oxyanion forms into the environment. Among which, Pb (II) is predominantly common and highly reactive and ionic form of Pb, whereas Pb (VI) compounds tend to be covalent and strong oxidants. Industrial exposure to Pb causes decrease in intelligent quotient, memory loss, infertility, mood swing, and sterility. The most significant way of Pb exposure is through direct intake of polluted soil or dust by eating. Inhalation is an important cause of Pb poisoning. Pb is highly hazardous and nonessential, but it can enter into the human body through contaminated food uptake and causes severe and evident illnesses, weakness of joints, nausea, insomnia, anorexia, or even death (NSC 2009). Young and infants are more sensitive to lead poisoning in comparison with adults.

Pb is potentially toxic and causes significant modifications and alterations in cell signaling, protein folding, enzymatic reactions, and other physiological processes due to interaction with proteins and inhibitory or mimicking the action of bivalent cations (Mg^{2+}, Fe^{2+}, Ca^{2+}) and monovalent cations (ATSDR 1999; Flora et al. 2007). The presence of high quantity of Pb in plants produces reactive oxygen species (ROS), causing lipid membrane damage that results in damage and disruption of chlorophyll and photosynthetic processes and suppresses the overall growth of the plant (Najeeb et al. 2017).

Mercury (Hg)

Hg is a heavy silver white transition metal belonging to d-block and period 6 of the periodic table. Hg residues are frequently used in scientific tools and equipments such as in thermometers and barometers, in instruments used to measure blood pressure, in amalgam for dental restoration, and also in fluorescent lighting. It is largely involved in the electrochemical process of chlorine manufacturing and as an electrode in chlor-alkali industry. Hg has been involved in the production of caustic soda and preservation of pharmaceutical products, nuclear reactors, and antifungal agents for wood processing and acts as solvent for reactive and precious metals (Tchounwou 2003).

Hg is a well-reported hazardous metal that usually exists in mercuric (Hg^{2+}), mercurous (Hg_2^{2+}), elemental (Hg0), or alkylated form (methyl/ethyl Hg) in nature. Redox potential and pH of the system are accountable for the steadiness of Hg. Alkylated form of Hg is extremely toxic and easily soluble in water and volatile in air (Smith et al. 1995). Hg is primarily found in metallic elements, inorganic salts, and organic compounds, which possess diverse toxicity and bioavailability (Bodek et al. 1988). All these forms are exceedingly poisonous and produce gastrointestinal toxicity, neurotoxicity, and nephrotoxicity (Tchounwou et al. 2003). It can also alter the structure of tertiary and quaternary protein-damaging cellular function by interacting with selenohydryl and sulfhydryl groups. Hg complexes increase MDA levels in the kidneys, liver, lungs, and testes (Lash et al. 2007). Methyl Hg causes mitochondrial damage, lipid peroxidation, accumulation of neurotoxic molecules, and microtubule destruction (Patrick 2002).

Nickel (Ni)

Ni is a lustrous, silvery, white metal with a slight golden tinge belonging to the transition metals. It is hard and ductile, existing in minute quantity usually in ultramorphic rocks. Oral consumption is a primary mode of Ni contamination. Ni is categorized as borderline metal ion because it has both soft and hard metal properties and can bind to Sr, N, and O_2 (Costa and Klein 1999). It naturally occurs in food products and water and concentrations may increase by pollution. Metal plating industries, combustion of fossil fuels, Ni mining, and electroplating industries are main causes of Ni pollution into the environment (Khodadoust et al. 2004). Ni is an essential element for various metabolic reactions used as catalyst required in small amounts and becomes more hazardous when exceeding the maximum permissible limits. Ni is not a cumulative toxin, but higher concentration and industrial exposure make it toxic and even carcinogenic and also result in occupational hazards. Nickel carbonyl [Ni $(CO)_4$] is an extremely toxic gas. It has been implicated as an embryo toxin teratogenic (Chen and Lin 1998; WHO 1991). The atmosphere is continuously being contaminated with the increasing application of Ni metal refining and fossil fuel combustion. Humans can engross nickel directly from tobacco smoke, jewelry, shampoos, detergents, and coins. Residues of nickel are also absorbed into the plasma from the chelating action of albumin through hemodialysis (Cempel and Nikel 2006).

Toxicological Influence of Heavy Metals in Environment and on Human Health

Toxicological impacts of heavy metals into the environment involve the contamination of soils, groundwater, sediments, natural water, and air. Metal contamination can affect human health through several exposure routes (Bade et al. 2013; Yadav et al. 2016). There are three major pathways: inhalation, diet, and dermal contact or

Table 5.1 Various toxic metals used in industries and their health hazards in living organisms

Metals	Applications	Health hazards
Cr	Tanning, paints pigment, fungicide	Cancer, nephritis, ulceration, and hair loss
Hg	Coal vinyl chlorides, electrical batteries, thermometers	Autoimmune disease, depression, drowsiness, fatigue, hair loss, insomnia, loss of memory, restlessness, disturbance of vision, tremors, temper outbursts, brain damage, lung and kidney failure
Pb	Plastic, paint, pipe, batteries, gasoline, auto exhaust	Neurotoxic and risk of cardiovascular disease
Cd	Fertilizer, plastic, pigments	Carcinogenic, mutagenic, endocrine disrupter, kidney damage, lung damage, and fragile bones, affect calcium regulation in biological systems
Zn	Fertilizer	Dizziness, fatigue, vomiting, renal damage, and cramps
Co	Vitamin B-12, wood preservative	Diarrhea, low blood pressure, and paralysis
Se	Coal, sulfur	Dietary exposure of around 300 μg/day affects endocrine function, impairment of natural killer cell activity, hepatotoxicity, gastrointestinal disturbances, damage of the liver, kidney, spleen, and nervousness
Ni	Electroplating	Cancer of the lungs, allergic disease such as itching, immunotoxic, neurotoxic, teratogenic, carcinogenic, genotoxic, and mutagenic, affects fertility and hair loss
Be	Coal, rocket fuel	Carcinogenic, acute, and chronic poison
Cu	Electronics, wood preservative, architecture	Brain and kidney damage, elevated levels result in liver cirrhosis, chronic anemia, stomach and intestine irritation
As	Pesticides, treated wood products, herbicides	Affects essential cellular processes such as oxidative phosphorylation and ATP synthesis, arsenicosis, carcinogen, and cancer
Ba	Appropriate dust control equipment and industrial controls	Cause cardiac arrhythmias, respiratory failure, gastrointestinal dysfunction, muscle twitching, and elevated blood pressure

Source: Adopted from Yadav et al. (2017)

via manual handling of pollutants (Luo et al. 2012). Metals are extremely hazardous, lethal, and non-biodegradable unlike organic pollutants. Heavy metals can severely inhibit or interfere in the degradation and reduction of organic substances. Soils are polluted by excessive depositions or accumulation of toxic metals and metalloids discharged by numerous harmful human activities (Maslin and Maier 2000; Kirpichtchikova et al. 2006). Heavy metal contamination in environment (soil, water, and air) may pose dangerous toxicological risk and problems to humans and animals as shown in Table 5.1 (Yadav et al. 2017). Natural environment gets severely contaminated by direct or indirect exposure of heavy metals present in drinking and natural water supplies (Table 5.2) and results in the reduction in productivity and fertility of soil, air quality, and food quality (Ling et al. 2007; Wuana and Okieimen 2011).

Table 5.2 Guideline values and limits of detection of some heavy metals present in drinking water

Metal	Guideline value	Occurrence in drinking water resources	Limit of detection
As	0.01 mg/L	Generally range between 1 and 2 mg/L in natural waters, although concentrations may be elevated (up to 12 mg/L) in areas containing natural sources	0.1 mg/L by ICP/MS, 2 mg/L by hydride generation AAS or FAAS
Cr	0.05 mg/L for total chromium	Total chromium concentrations in drinking water are usually less than 2 mg/L, although concentrations as high as 120 mg/L have been reported	0.05–0.2 mg/L for total chromium by AAS
Cu	2 mg/L	Concentrations in drinking water range from 0.005 to > mg/L, primarily as a result of the corrosion of interior copper plumbing	0.02–0.1 mg/L by ICP/MS, 0.3 mg/L by ICP/optical emission spectroscopy, 0.5 mg/L by FAAS
Pb	0.01 mg/L	Concentrations in drinking water are generally below 5 mg/L, although much higher concentrations (above 100 mg/L) have been measured where lead fittings are present	1 mg/L by AAS
Hg	0.006 mg/L for inorganic mercury	Mercury is present in the inorganic form in surface water and groundwater at concentrations usually below 0.5 mg/L, although local mineral deposits may produce higher levels in groundwater	0.05 mg/L by cold vapor AAS, 0.6 mg/L by ICP, 5 mg/liter by FAAS
Se	0.01 mg/L	Levels in drinking water vary greatly in different geographical areas but are usually much less than 0.01 mg/L	0.5 mg/L by AAS with hydride generation
Zn		Levels of zinc in surface water and groundwater normally do not exceed 0.01 and 0.05 mg/L	
Ni	0.07 mg/L	The concentration of nickel in drinking water is normally less than 0.02 mg/L, although nickel released from taps and fittings may contribute up to 1 mg/liter	0.1 mg/liter by ICP-MS, 0.5 mg/liter by FAAS, 10 mg/liter by ICP-AES

Source: World Health Organization (WHO) (2003)

Impacts of Heavy Metals on Humans

Heavy metals greatly affect human health when they are exceeding from their particular recommended limit of dietary intake and show various toxicological effects that have been well studied and documented (Dong et al. 2010). Occurrence of excess concentrations of metals in cultivated soils changes food quality and affects safety, which in turn causes increased risks of kidney and liver failure, infertility and reproductive disorders, cancers, nervous breakdown, leukemia, mental illness, and other toxicity problems (Khan et al. 2011).

Heavy metals in urban soils change ecological quality by polluting the food, air level, water supply, and surrounding environment which directly damage the population specifically kids and youngsters through dermal contact and inhalation (Nagajyoti et al. 2010). ATSDR Committee has listed Cd as the sixth most toxic substance since it damages or affects metabolic rate of calcium, leading to Ca deficiency and resulting in cartilage disease, bone fractures, etc. (ATSDR 2008).

Pb mainly enters human body via gastrointestinal and respiratory tract and then circulates into the blood in soluble salts, protein complexes or ions, etc., in which 95% of the Pb accumulates in the bones in insoluble phosphate form. It also damages the body organs and systems such as the kidney, liver, reproductive system, nervous system, urinary system, and immune system and the basic physiological processes and genetic expressions (Chao et al. 2014).

Some vital trace metals like Ni, Cu, Zn, and Mo are crucial for enzymatic and physiological activities, but higher quantity of these metals might cause destruction and injuries to human health if they are taken in excessive amount from outside environment. Ni and Cu are tumor-promoting factors, and their carcinogenic effect is of global concern. Direct exposure of Ni-manufacturing products to industrial workers is responsible for respiratory cancer and nasopharyngeal carcinoma (Chen 2011; Chao et al. 2014). Inhalation of Cr (VI) is very dangerous and toxic, creates severe symptoms of irritation and itching of the nose and skin, and damages nasal septum and ulcers, whereas absorption and ingestion of high Cr (VI) dose can cause kidney and liver damage, nausea, irritation of the gastrointestinal tract, stomach ulcers, convulsions, and death (Mishra and Bharagava 2016).

Impact of Heavy Metals on Plants

Metal contamination shows detrimental, chronic, and acute toxicity symptoms on developmental activities, yielding capacity and growth pattern of plants. Heavy metal exposure causes cellular damage, ionic homeostasis, and oxidative stress and generates higher amount of reactive oxygen; inhibition of necessary microelements, enzymes, and pigments; and disruption of respiratory and photosynthetic activity (Reddy et al. 2005; Hennery 2000). Cd and Pb are considered as nonessential elements for plants. But excess accumulation of such metals in plant severely harms the plant growth and reproduction by damaging ion channels and disrupting metabolic reactions and absorption of essential elements (Xu and Shi 2000). Pb is highly reactive with sulfhydryl substances in the cells, and therefore, it easily inhibits cellular enzyme actions, changes membrane permeability, causes water imbalance, and reduces nutritive quality (Sinha et al. 1988; Sharma and Dubey 2005). It has been reported that in early stage, Cd inhibits the photosynthesis and growth of rice, then inhibits the reproductive organs and differentiation, and finally disturbs the nutrient transport and mobilization. Cd and Zn reduce catalytic efficiency of enzymes, cause chlorosis, and adversely affect both root and shoot growth (Somasekharaiah et al.

1992; Fontes and Cox 1998). Cd toxicity results in browning of root tips, chlorosis, disrupting chlorophyll synthesis, induces lipid peroxidation, altering membrane permeability and death of plant (Guo et al. 2008). Cr complexes are noxious pollutants that adversely affect seed germination and damage plant growth due to inhibitory action on amylase activity and sugar transport, which easily reduces germination process (Peralta et al. 2001; Zeid 2001).

Impact of Heavy Metals on Soil

Human activities and industrial development have greatly degraded and deteriorated the soil quality due to the unnecessary and increased use of metals. Metals are highly potent and noxious contaminants, which ultimately lead to severe soil pollution and affect soil fertility and productivity (Raju et al. 2013; Prajapati and Meravi 2014). Metals can be found in very smaller to higher concentrations (~10,000 mg Kg^{-1}) in contaminated soil (Long et al. 2002). The regular use of metals, their salts, and residues in fertilizers, pesticides, compost, mines, smelting and textile industries, and agricultural practices increases risks of soil contamination, and once soil gets contaminated or polluted with heavy metals, it is difficult to remediate and is harmful for farming (Zhang et al. 2011). Metal-contaminated soil is considered as chemical time bombs, which may cause serious ecological damage (Wood 1974). Wastewaters from tannery, textile, pigments, transportation, and automotive vehicles and electroplating industries contain high quantity of metals as their waste products, and direct disposal and dumping of such wastewaters into open soil areas is one of the major causes of soil pollution. These metals are accumulating in soil and affecting plant and aquatic life and severely disrupt ecological balance (Falahiardakani 1984; Chen 2002). Continuous long-term exposure of metal-contaminated soil is highly dangerous for plant diversity and aquatic and terrestrial organisms. It negatively affects the species richness and diversity of plants and microorganisms (Zhang et al. 2011; Arao et al. 2010).

Remediation Technologies of Heavy Metal Contamination for Eco-friendly Environment

The contamination of soil, groundwater, sediments, surface water, and air with toxic metals, organic and inorganic pollutants, phenols, dyes, and other xenobiotic compounds are the major threats facing the world today, as these are persistent in the environment or take several years to be broken down into the nontoxic forms and, therefore, have long-lasting effects on the ecosystem. According to recent study, the urgent need to remediate natural resources has led to the development of new technologies that emphasized on the destruction of pollutants rather than on the conventional approaches to prevent their entry into the food chain for the safe disposal of pollutants into the environment (Fulekar 2010; Dixit et al. 2015).

Bioremediation

Bioremediation is a viable alternative in comparison with the use of costly conventional physicochemical methods for metal decontamination. Bioremediation is a process where naturally occurring microorganisms are used to convert harmful toxic substances to either less or nontoxic compounds (Asha and Sandeep 2013). It generally involves the naturally occurring bacteria, fungi, plants, and even earthworm to remove or degrade/detoxify hazardous pollutants and waste material in an eco-friendly manner for environmental safety (Venkateshkumar et al. 2015). For the remediation of contaminated sites, the microorganisms native to that area or isolated from elsewhere and transferred to the contaminated site can be used (Kumar et al. 2011).

There are various environmental factors such as pH, temperature, soil type, and texture, nutrient amendments and O_2, which influence the bioremediation process. Bioremediation has the ability to detoxify inorganic pollutants like metals by adsorption, uptake, or accumulation by microbes. Bioremediation can be used in situ and ex situ for the removal of contaminates. In in situ bioremediation, the contaminants are treated through direct contact of microorganisms with the dissolved and sorbed contaminants and used as substrates in comparison with other methods for cleaning up of contaminated water as well as of soil (Bharagava et al. 2017).

A precise knowledge of heavy metal concentrations, the form in which they are found and their dependence on soil physicochemical properties, lends careful management of soil, thus limiting the adverse impacts on the ecosystem. Since metals cannot be destroyed like organic contaminates, therefore, these must be either removed or transformed to a stable form through biotransformation (Chandra et al. 2011). Mechanisms including biosorption, bioleaching, biomineralization, intracellular accumulation, and enzyme-catalyzed transformation (redox reactions) can be utilized by microorganisms for bioremediation of metals (Lloyd and Lovely 2001). Interactions of microbes with heavy metals can have several implications on the environment. Microbes play a very important role in heavy metal's biogeochemical cycling as well as in cleaning up or treating metal-contaminated environments (Cheung and Gu 2007). Several metal-resistant bacteria have been reported to survive under metal-stressed condition through certain mechanisms including accumulation and complexation of the metal ions into less toxic state, which helps them to tolerate the uptake of these ions (Diaz et al. 2008). Other than the said mechanisms, microorganisms have developed certain other ways such as uptake, adsorption, oxidation, reduction, and methylation through which they protect themselves from toxicity of heavy metals. In order to survive at extreme environmental conditions, bacteria exhibit certain morphological changes at cellular as well as colony level (shape, color, texture, opacity, convexity, margin appearance, etc.) when exposed to some environmental stress, for example, temperature, aeration, and presence of heavy or toxic metals. As a result of these morphological changes, bacterial colonies show phenomenon of phase variation (Tamara et al. 2006).

Many studies have demonstrated the microbe's ability to remove/transform the toxic metals into nontoxic form. Many microorganisms such as *Pseudomonas* spp., *Alcaligenes* spp., *Arthrobacter* spp., *Bacillus* spp., *Corynebacterium* spp., *Flavobacterium* spp., *Azotobacter* spp., *Rhodococcus* spp., *Microbacterium* spp., *Nocardia* spp., *Methosinus* spp., methanogens, *Apergillus niger*, *Pleurotus ostreatus*, *Rhizopus arrhizus*, *Stereum hirsutum*, *Ganoderma applanatum*, etc. are reported as metal-resistant microbes that cause biotransformation of highly toxic metal into less toxic state (Katz and Salem 1993; Zaki and Farag 2010; Rajendran et al. 2003; Megharaj et al. 2003).

For enzymatic functions and bacterial growth, some heavy metal ions are very necessary, which are drawn into the cells through some existing uptake mechanisms, which are of two types. One system is quick and unspecific driven by a chemiosmotic gradient across the cell membrane with no ATP requirement, and the second one is time-consuming, more substrate-specific, and driven by energy from ATP hydrolysis (Cervantes and Campos-Gracia 2007). Being quick and more energy efficient, the first mechanism results in an influx of wider variety of heavy metals and, when these metals are accumulated in higher concentration once inside the cell, cause more toxic effects.

Phytoremediation

The majority of the conventional technologies are too costly to implement and further can cause disturbances to the already damaged environment (Alloway and Jackson 1991). Phytoremediation term is collectively used for all plant-based bioremediation technologies involving the use of green plant and their associated microbiota for in situ remediation of contaminated soil and groundwater (Sadowsky 1999; Lal et al. 2013).

The concept of using metal-accumulating plants for removing heavy metals and other compounds from contaminated soil and plants was actually implemented 30 years ago, introduced in 1983 (Henry 2000). For developing countries, phytoremediation is an eco-friendly, aesthetically pleasing, and cost-effective approach, but despite its potential, it is yet to become a commercially viable technology in India (Ghosh and Singh 2005).

Depending on the mechanism of remediation, phytoremediation may take place through phytoextraction, phytofiltration, phytostabilization, phytovolatilization, and phytodegradation (Fig. 5.1). In the phytoextraction, the metal ions are accumulated in the aerial parts of plants, which can be further removed by disposing off or burning to recover metals. In phytofiltration, plant roots or seedlings are

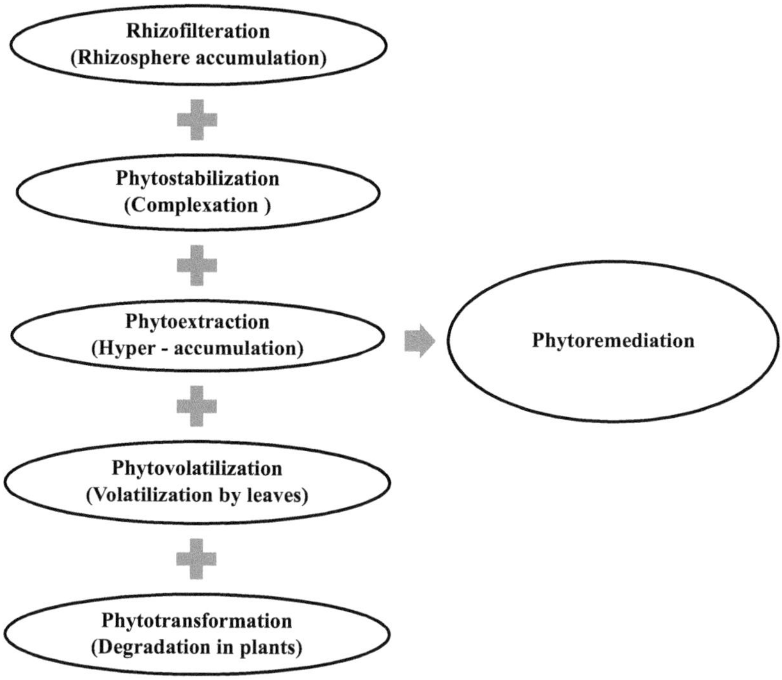

Fig. 5.1 Various phytoremediation processes used by plants to remove toxic metals

involved for removing metals from aqueous wastes, whereas in the phytostabilization, the pollutants from soil are absorbed through plant root and are stored in the rhizosphere, rendering them harmless by preventing them from leaching. The pollutants such as Se and Hg are volatilized from foliage in phytovolatilization, whereas plants and its associated microorganisms are involved to degrade organic pollutants in the phytodegradation (Garbisu and Alkorta 2001).

Several researchers have investigated and reported the hyperaccumulation of heavy metals in various plant species making clear that the different mechanisms of metal accumulation, exclusion, and compartmentation might exist in these plant species (as presented in Table 5.3). For example, in *Thlaspi caerulescens*, Zn is a requisite preferentially in soluble forms in vacuoles of epidermal cells (Frey et al. 2000). Kidd and Monterroso (2005) successfully grew the plant *Alyssum serpyllifolium* sp. *lusitanicum* (Brassicaceae) on two mine-spoil soils to investigate its efficiency in phytoextraction of polymetal-contaminated soils. For the study, they planted on one contaminated with only Cr (283 mg/kg) and other contaminated with Cr (263 mg/kg), Cu (264 mg/kg), Pb (1433 mg/kg), and Zn (377 mg/kg).

Among all the ferns, *Pteris vittata*, also known as brake fern, has been well identified as hyperaccumulator, which can accumulate up to 7500 mg/kg As from As-contaminated soils and waters without showing toxicity symptoms (Ma et al. 2001). Other than ferns, many other species including sharp dock (*Polygonum amphibium*), duckweed (*Lemna minor*), water hyacinth (*Eichhornia crassipes*),

Table 5.3 Some hyperaccumulating plants and aquatic macrophytes used for the phytoremediation of various heavy metals

Plants species	Heavy metal
Brassica juncea	As, Ni, Cr
Pistia stratiotes	Zn, Pb, Ni, Hg, Cu, Cd, and Cr
Solanum nigrum	Cd
Thlaspi caerulescens	Cd, Zn, and Pb
Jatropha curcas	Cd, Cu, Ni, Pb
Zea mays	Cd, Pb, Zn
Salix fragilis	Cd, Cu, Pb, Zn
Populus trichocarpa	Cd, Cu, Pb, Zn
Aquatic macrophytes	
Azolla pinnata	Fe, Cu, Hg, Cr, Cd
Eichhornia crassipes	Zn, As, Cd, Ni, Cr, Se, Cu
Lemna minor	Pb, Ni, Fe, Cu, As, Hg, Ti
Vallisneria spiralis	Cr, Cd, Ni
Nasturtium officinale	Cu, Zn, Ni
Cyperus alternifolius	Cu, Zn, Pb, Cd
Eleocharis acicularis	Fe, Pb, Mn, Cu, Cr, Ni, Zn

Source: Modified from Yadav et al. (2017) and Kumar and Singh (2017)

water lettuce (*Pistia stratiotes*), water dropwort (*Oenanthe javanica*), calamus (*Lepironia articulata*), dollarweed (*Hydrocotyle umbellata*), etc. have been identified and tested for the phytoremediation of heavy metals from the polluted water (Prasad and Freitas 2003; Verma et al. 2016). Further, the Indian mustard roots were found effective in removing Cd, Cr, Cu, Ni, Pb, and Zn and sunflower in removing Pb, U, Cs-137, and Sr-90 hydroponic solutions (Wang et al. 2002).

Challenges and Future Prospects

Heavy metal pollution has become a major environmental issue as well as major health concern worldwide. The contamination of heavy metal in soil and water leads to the bioaccumulation and bio-magnifications of the available heavy metals. Heavy metals cannot be degraded into any harmless by-product through any physical, chemical, and biological means; instead they can only be transformed into less toxic forms by means of plants and microbes. Bioremediation/phytoremediation is an emerging advanced green technology that holds promise for safe and healthy environment. However, the future of phytoremediation and bioremediation is still in developing phase regarding their technical limitations which need to be addressed. Many metal-resistant microorganisms and hyperaccumulator plants remain to be discovered. However, there is a need to know more about microbial diversity as well as metal-accumulating plants and their physiology to survive in metal-contaminated environments.

Conclusion

Due to the potential impact on human and animal health, the contamination of heavy metals in the environment is of great concern, and its treatment from the soil and water around industrial plant has been a challenge since years. Thus, for protecting the precious natural resources and biological diversity, cheaper and effective technologies are needed. During the past few years, the application of microorganisms for the recovery of metals from waste streams as well as the employment of plants for landfill practice has gained attention. Therefore, the cost-effective and eco-friendly newer biotechnological processes, viz., bioremediation and phytoremediation, through microbial and hyperaccumulator plants may become the most promising approach for green environment in the future.

Acknowledgment The financial support received from "Science and Engineering Research Board" (SERB), Department of Science & Technology (DST), Government of India (GOI), New Delhi, India, as "Major Research Project" (Grant No.:EEQ/2017/000407) and University Grant Commission (UGC) Fellowship received for doctoral research work is duly acknowledged.

References

Adamis, P. D. B., Panek, A. D., Leite, S. G. F., & Eleuthero, E. C. A. (2003). Factors involved with cadmium absorption by a wild-type strain of *Saccharomyces cerevisiae*. *Brazilian Journal of Microbiology, 34*(1), 55–60.

Alloway, B. J., & Jackson, A. P. (1991). The behaviour of heavy metals in sewage sludge amended soils. *Science of the Total Environment, 100*, 151–176.

Andrews, S., & Sutherland, R. A. (2004). Cu, Pb and Zn contamination in Nuuanu watershed, Oahu, Hawaii. *Science of the Total Environment, 324*, 173–182.

Arao, T., Ishikawa, S., Murakam, I. M., Abe, K., Maejima, Y., & Makino, T. (2010). Heavy metal contamination of agricultural soil and counter, measures in Japan. *Paddy and Water Environment, 8*(3), 247–257.

Asha, L. P., & Sandeep, R. S. (2013). Review on bioremediation – Potential tool for removing environmental pollution. *International Journal of Basic and Applied Chemical Sciences, 3*, 2277–2073.

Atlas, R. M., & Bartha, R. (1993). *Microbial ecology, fundamentals and applications* (Vol. 410, pp. 237–238). Hamilton: The Benjamin/Cummings.

ATSDR. (1999). *Agency for toxic substances and disease registry.* Public Health Service Atlanta: U.S. Department of Health and Human Services; Toxicological Profile for Lead.

ATSDR. (2000). *Agency for toxic substances and disease registry toxicological profile for Arsenic TP-92/09*. Atlanta: Center for Disease Control.

ATSDR. (2008). *Agency for toxic substances and disease registry*. Atlanta: U.S. Department of Health and Human Services, Public Health Service. Toxicological profile for chromium.

Bade, R., Oh, S., Shin, W. S., & Hwang, I. (2013). Human health risk assessment of soils contaminated with metal (loid)s by using DGT uptake: A case study of a former Korean metal refinery site. *Human and Ecological Risk Assessment, 19*, 767–777.

Barakat, M. A. (2011). New trends in removing heavy metals from industrial wastewater. *Arabian Journal of Chemistry, 4*, 361–377.

Baselt, R. C. (2000). *Disposition of toxic drugs and chemicals in man* (5th ed.). Foster City: Chemical Toxicology Institute.

Bharagava, R. N., & Mishra, S. (2018). Hexavalent chromium reduction potential of Cellulosimicrobium sp. isolated from common effluent treatment industries. *Ecotoxicology and Environmental Safety, 147*, 102–109.

Bharagava, R. N., Chowdhary, P., & Saxena, G. (2017). Bioremediation: An eco-sustainable green technology, its application and limitations. In *Environmental pollutants and their bioremediation approaches*. Boca Raton: CRC press.

Blaylock, M. J., & Huang, J. W. (2000). Phytoextraction of metals. In I. Raskin & B. D. Ensley (Eds.), *Phytoremediation of toxic metals: Using plants to clean up the environment* (pp. 53–70). New York: Wiley.

Bodek, I., Lyman, W. J., & Reehl, W. F. (1988). *Environmental inorganic chemistry: Properties, processes and estimation methods*. Elmsford: Pergamon Press.

Campbell, P. G. C. (2006). Cadmium – A priority pollutant. *Environment and Chemistry, 3*(6), 387–388.

Castagnetto, J. M., Hennessy, S. W., Roberts, V. A., Elizabeth, D. G., Tainer, J. A., & Pique, M. E. (2002). MDB: The metalloprotein database and browser at the Scripps Research Institute. *Nucleic Acids Research, 30*(1), 379–382.

Cempel, M., & Nikel, G. (2006). Nickel: A review of its sources and environmental toxicology. *Polish Journal of Environmental Studies, 15*(3), 375–382.

Centeno, J. A., Tchounwou, P. B., & Patlolla, A. K. (2005). Environmental pathology and health effects of arsenic poisoning: A critical review. In R. Naidu, E. Smith, J. Smith, & P. Bhattacharya (Eds.), *Managing arsenic in the environment: From soil to human health*. Adelaide: Australia CSIRO Publishing Corp.

Cervantes, C., & Campos-Gracia, J. (2007). Reduction and efflux of chromate by bacteria. In D. H. Nies & S. Silver (Eds.), *Molecular microbiology of heavy metals* (pp. 407–420). Berlin: Springer.

Chandra, R., Bharagava, R. N., Kapley, A., & Purohit, H. J. (2011). Bacterial diversity, organic pollutants and their metabolites in two aeration lagoons of common effluent treatment plant (CETP) during the degradation and detoxification of tannery wastewater. *Bioresource Technology, 102*, 2333–2341.

Chao, S., LiQin, J., & WenJun, Z. (2014). A review on heavy metal contamination in the soil worldwide: Situation, impact and remediation techniques. *Environmental Skeptics and Critics, 3*(2), 24–38.

Chen, H. M. (2002). *Behaviours and environmental quality of chemical substances in the soil*. Beijing: Science Press.

Chen, Y. F. (2011). Review of the research on heavy metal contamination of China's city soil and its treatment method. *China Population, Resources and Environment, 21*(3), 536–539.

Chen, C. Y., & Lin, T. H. (1998). Nickel toxicity to human term placenta: In vitro study on lipid per oxidation. *Journal of Toxicology & Environmental Health Part A: Current Issues, 54*, 37–47.

Cheung, K. H., & Gu, J. D. (2007). Mechanism of hexavalent chromium detoxification by microorganism and bioremediation application potential: A review. *International Biodeterioration and Biodegradation, 59*(1), 8–15.

Coral, M. N. U., Korkmaz, H., & Arikan, B. (2005). Plasmid mediated heavy metal resistance in *Enterobacter* spp. isolated from Sofulu landfill, in Adana, Turkey. *Annales de Microbiologie, 55*(3), 175–179.

Costa, M., & Klein, C. B. (1999). Nickel carcinogenesis, mutation, epigenetics or selection. *Environmental Health Perspectives, Part A, 107*, 438–439.

Deeb, B. E., & Altalhi, A. D. (2009). Degradative plasmid and heavy metal resistance plasmid naturally co-exist in phenol and cyanide assimilating bacteria. *American Journal of Biochemistry and Biotechnology, 5*(2), 84–93.

Dhal, B., Thatoi, H. N., Das, N. N., & Pandey, B. D. (2013). Chemical and microbial remediation of hexavalent chromium from contaminated soil and mining/metallurgical solid waste: A review. *Journal of Hazardous Materials, 15*, 250–251.

Diaz, R. M., Diaz-Perez, & Vergas, E. (2008). Mechanism of bacterial resistance to chromium compounds. *Biometals, 21*, 321–332.

Dixit, R., Wasiullah, Malaviya, D., Pandiyan, K., Singh, B. U., Sahu, A., Shukla, R., Singh, B. P., Rai, J. P., Sharma, P. K., Lade, H., & Paul, D. (2015). Bioremediation of heavy metals from soil and aquatic environment: An overview of principles and criteria of fundamental processes. *Sustainability, 7*, 2189–2212.

Dong, X., Li, C., Li, J., Wang, J., Liu, S., & Ye, B. (2010). A novel approach for soil contamination assessment from heavy metal pollution: A linkage between discharge and adsorption. *Journal of Hazardous Materials, 175*, 1022–1030.

Falahiardakani, A. (1984). Contamination of environment with heavy metals emitted from automotives. *Ecotoxicology and Environmental Safety, 8*, 152–161.

Flora, S. J. S., Saxena, G., Gautam, P., Kaur, P., & Gill, K. D. (2007). Lead induced oxidative stress and alterations in biogenic amines in different rat brain regions and their response to combined administration of DMSA and MiADMSA. *Chemico-Biological Interactions, 170*, 209–220.

Fontes, R. L. S., & Cox, F. R. (1998). Zinc toxicity in soybean grown at high iron concentration in nutrient solution. *Journal of Plant Nutrition, 21*, 1723–1730.

Frey, B., Keller, C., & Zierold, K. (2000). Distribution of Zn in functionally different leaf epidermal cells of the hyperaccumulator *Thlaspi caerulescens. Plant, Cell & Environment, 23*(7), 675–687.

Fulekar, M. H. (2010). *Bioremediation technology: Recent advances*. Dordrecht: Springer.

Garbisu, C., & Alkorta, I. (2001). Phytoextraction: A cost-effective plant-based technology for the removal of metals from the environment. *Bioresource Technology, 77*, 229–236.

Ghosh, M., & Singh, S. P. (2005). A review of phytoremediation of heavy metal and utilization of its by-products. *Applied Ecology and Environmental Research, 3*(1), 1–8.

Giaginis, C., Gatzidou, E., & Theocharis, S. (2006). DNA repair systems as targets of cadmium toxicity. *Toxicology and Applied Pharmacology, 213*, 282–290.

Guo, J., Dai, X., & Xu W Ma, M. (2008). Over expressing GSHI and AsPCSI simultaneously increases the tolerance and accumulation of cadmium and arsenic in *Arabidopsis thaliana. Chemosphere, 72*, 1020–1026.

Gupta, M. K., Kumari, K., Srivastava, A., & Shikha, G. (2014). Bioremediation of heavy metal polluted environment using resistant bacteria. *Journal of Environmental Research and Development, 8*(4), 883–889.

Henry, J. R. (2000). *An overview of phytoremediation of lead and mercury – NNEMS report*, pp. 3–9. Washington, DC.

IARC. (1990). International Agency for Research on Cancer, IARC Monographs on the Evaluation of Carcinogenic Risks to Humans: IARC Scientific Publications, IARC; Chromium, nickel and welding. Lyon, France, p. 49.

Irfan, M., Hayat, S., Ahmad, A., & Alyemeni, M. N. (2013). Soil cadmium enrichment: Allocation and plant physiological manifestations. *Saudi Journal of Biological Sciences, 20*(1), 1–10.

Jain, A. N., Udayashankara, T. H., & Lokesh, K. S. (2014). Review on bioremediation of heavy metals with microbial isolates and amendments on soil residue. *International Journal of Science and Research, 3*(8), 2319–7064.

Katz, S. A., & Salem, H. (1993). The toxicology of chromium with respect to its chemical speciation: A review. *Journal of Applied Toxicology, 13*, 217–224.

Khan, M. N., Wasim, A. A., Sarwar, A., & Rasheed, M. F. (2011). Assessment of heavy metal toxicants in the roadside soil along the N-5, National Highway, Pakistan. *Environmental Monitoring and Assessment, 182*, 587–595.

Khodadoust, A. P., Reddy, K. R., & Maturi, K. (2004). Removal of nickel and phenanthrene from kaolin soil using different extractants. *Environmental Engineering Science, 21*(6), 691–704.

Kidd, P. S., & Monterroso, C. (2005). Metal extraction by *Alyssum serpyllifolium ssp.lusitanicum* on mine-spoil soils from Spain. *Science of the Total Environment, 336*(1–3), 1–11.

Kirpichtchikova, T. A., Manceau, A., Spadini, L., Panfili, F., Marcus, M. A., & Jacquet, T. (2006). Speciation and solubility of heavy metals in contaminated soil using X-ray microfluorescence, EXAFS spectroscopy, chemical extraction, and thermodynamic modelling. *Geochimica et Cosmochimica Acta, 70*(9), 2163–2190.

Kumar, M., & Singh, H. (2017). Phytoremediation: A green technology for remediation metal contaminated site. In *Environmental pollutants and their bioremediation approaches* (pp. 297–328). Boca Raton: CRC press.

Kumar, A., Bisht, B. S., Joshi, V. D., & Dhewa, T. (2011). Review on bioremediation of polluted environment: A management tool. *International Journal of Environmental Sciences, 1*(6), 1079–1093.

Lal, S., Singh, R., & Kumar, R. (2013). Heavy metal concentration and bacterial load on Spinach (*Spinacia oleracea L.*) phyllosphere under different regions in Lucknow, Uttar Pradesh. *International Journal of Pharmacy & Life Sciences, 4*(7), 1–7.

Lash, L. H., Putt, D. A., Hueni, S. E., Payton, S. G., & Zwickl, J. (2007). Interactive toxicity of inorganic mercury and trichloroethylene in rat and human proximal tubules (Effects of apoptosis, necrosis, and glutathione status). *Toxicology and Applied Pharmacology, 221*(3), 349–362.

Lenntech Water Treatment and Air Purification. (2004). *Water treatment Lenntech, Rotterdamseweg, Netherlands*. http://www.excelwater.com/thp/filters/Water-Purification.htm

Ling, W., Shen, Q., Gao, Y., Gu, X., & Yang, Z. (2007). Use of bentonite to control the release of copper from contaminated soils. *Australian Journal of Soil Research, 45*(8), 618–623.

Lloyd, J. R., & Lovley, D. R. (2001). Microbial detoxification of metals and radionuclides. *Current Opinion in Biotechnology, 12*, 248–253.

Long, X. X., Yang, X. E., & Ni, W. Z. (2002). Current status and prospective on phytoremediation of heavy metal polluted soils. *Journal of Applied Ecology, 13*, 757–756.

Luo, X. S., Ding, J., Xu, B., Wang, Y. J., Li, H. B., & Yu, S. (2012). Incorporating bioaccessibility into human health risk assessments of heavy metals in urban park soils. *Science of the Total Environment, 424*, 88–96.

Ma, L. Q., Komar, K. M., & Tu, C. (2001). Bioremediation: A fern that hyperaccumulates arsenic. *Nature, 409*, 579.

Mani, S., & Bharagava, R. N. (2016). Exposure to Crystal Violet, its toxic, genotoxic and carcinogenic effects on environmental and its degradation and detoxification for environmental safety. *Reviews of Environmental Contamination and Toxicology, 237*, 71–104.

Maslin, P., & Maier, R. M. (2000). Rhamnolipid-enhanced mineralization of phenanthrene in organic-metal co-contaminated soils. *Bioremediation Journal, 4*(4), 295–308.

Matsumoto, S. T., Mantovani, M. S., Malaguttii, M. I. A., Dias, A. L., Fonseca, I. C., & Marin-Morales, M. A. (2006). Genotoxicity and mutagenicity of water contaminated with tannery effluents, as evaluated by the micronucleus test and comet assay using the fish *Oreochromis niloticus* and chromosome aberrations in onion root-tips. *Genetics and Molecular Biology, 29*(1), 148–158.

Megharaj, M., Avudainayagam, S., & Naidu, R. (2003). Toxicity of hexavalent chromium and its reduction by bacteria isolated from soil contaminated with tannery waste. *Current Microbiology, 47*, 51–54.

Mishra, S., & Bharagava, R. N. (2016). Toxic and genotoxic effects of hexavalent chromium in environment and its bioremediation strategies. *Journal of Environmental Science and Health, Part C*. https://doi.org/10.1080/10590501.2015.1096883.

Monterroso, P., Pato, P., & Pereira, E. (2003). Distribution and accumulation of metals (Cu, Cd, Zn and Pb) in sediments of a lagoon on the North-western coast of Portugal. *Marine Pollution Bulletin, 46*, 1200–1205.

Nagajyoti, P. C., Lee, K. D., & Sreekanth, T. V. M. (2010). Heavy metals, occurrence and toxicity for plants: A review. *Environmental Chemistry Letters, 8*, 199–216.

Najeeb, U., Ahmad, W., Zia, M. H., Malik, Z., & Zhou, W. (2017). Enhancing the lead phytostabilization in wetland plant *Juncus effusus* L. through somaclonal manipulation and EDTA enrichment. *Arabian Journal of Chemistry*. https://doi.org/10.1016/j.arabjc.2014.01.009.

Nies, D. H. (1999). Microbial heavy-metal resistance. *Applied Microbiology and Biotechnology, 51*, 730–750.

NRC. (2001). *National Research Council. Arsenic in drinking water*. On line at: http://www.nap.edu/books/0309076293/html/

NSC. (2009). *Lead poisoning*. National Safety council, 2009. http:// www.nsc.org/news resources/Resources/Documents/Lead Poisoning.pdf

Paschal, D. C., Burt, V., Caudill, S. P., Gunter, E. W., Pirkle, J. L., Sampson, E. J., Miller, D. T., & Jackson, R. J. (2000). Exposure of the U.S. population aged 6 years and older to cadmium: 1988–1994. *Archives of Environmental Contamination and Toxicology, 38*, 377–383.

Patra, R. C., Malik, B., Beer, M., Megharaj, M., & Naidu, N. (2010). Molecular characterization of chromium (VI) reducing potential in gram positive bacteria isolated from contaminated sites. *Soil Biology and Biochemistry, 42*, 1857–1863.

Patrick, L. (2002). Mercury toxicity and antioxidants: Part 1: Role of glutathione and alpha-lipoic acid in the treatment of mercury toxicity. *Alternative Medicine Review, 7*(6), 456–471.

Peralta, J. R., Gardea-Torresdey, J. L., & Tiemann, K. J. (2001). Uptake and effects of five heavy metals on seed germination and plant growth in alfalfa (*Medicago sativa* L.). *Bulletin of Environmental Contamination and Toxicology, 66*(6), 727–734.

Prajapati, S. K., & Meravi, N. (2014). Heavy metal speciation of soil and *Calotropis procera* from thermal power plant area. *Proceedings of the International Academy of Ecology and Environmental Sciences, 4*(2), 68–71.

Prasad, M. N. V., & Freitas, H. M. D. (2003). Metal hyperaccumulation in plants-biodiversity prospecting for phytoremediation technology. *Electronic Journal of Biotechnology, 93*(1), 285–321.

Rajendran, P., Muthukrishnan, J., & Gunasekaran, P. (2003). Microbes in heavy metal remediation. *Indian Journal of Experimental Biology, 41*, 935–944.

Raju, K. V., Somashekar, R. K., & Prakash, K. L. (2013). Spatio-temporal variation of heavy metals in Cauvery River basin. *Proceedings of the International Academy of Ecology and Environmental Sciences, 3*(1), 59–75.

Reddy, A. M., Kumar, S. G., Jyotsnakumari, G., et al. (2005). Lead induced changes in antioxidant metabolism of horse gram (*Macrotyloma uniflorum* (Lam.) Verdc.) and bengal gram (*Cicer arietinum* L.). *Chemosphere, 60*, 97–104.

Reeves, R. D., & Baker, A. J. M. (2000). Metal accumulating plants. In I. Raskin & B. Ensley (Eds.), *Phytoremediation of toxic metals: Using plants to clean up the environment* (pp. 193–229). New York: Wiley.

Sadowsky, M. J. (1999). *Phytoremediation: Past promises and future practices*. In: Proceedings of the 8th international symposium on microbial ecology. Halifax, Canada, pp. 1–7.

Scragg, A. (2006). *Environmental biotechnology* (2nd ed.). Oxford: Oxford University Press.

Seaward, M. R. D., & Richardson, D. H. S. (1990). Atmospheric sources of metal pollution and effects on vegetation. In A. J. Shaw (Ed.), *Heavy metal tolerance in plants evolutionary aspects* (pp. 75–94). Boca Raton: CRC Press.

Sharma, P., & Dubey, R. S. (2005). Lead toxicity in plants. *Brazilian Journal of Plant Physiology, 17*(1), 35–52.

Shuttleworth, K. L., & Unz, R. F. (1993). Growth of Filamentous bacteria in the presence of heavy metals. In D. Jenkins & B. H. Olson (Eds.), *Water and wastewater microbiology* (pp. 485–487). Oxford: Pergamon.

Singh, N., Kumar, D., & Sahu, A. (2007). Arsenic in the environment: Effects on human health and possible prevention. *Journal of Environmental Biology, 28*(2), 359–365.

Sinha, S. K., Srinivastava, H. S., & Mishra, S. N. (1988). Effect of lead on nitrate reductase activity and nitrate assimilation in pea leaves. *Acta Societatis Botanicorum Poloniae, 57*, 457–463.

Smith, L. A., Means, J. L., & Chen, A. (1995). *Remedial options for metals-contaminated sites*. Boca Raton: Lewis Publishers.

Somasekharaiah, B. V., Padmaja, K., & Prasad, A. R. K. (1992). Phytotoxicity of cadmium ions on germinating seedlings of mung bean (*Phaseolus vulgaris*): Involvement of lipid peroxidase in chlorophyll degradation. *Physiologia Plantarum, 85*, 85–89.

Stern, B. R. (2010). Essentiality and toxicity in copper health risk assessment: Overview, update and regulatory considerations. *Toxicology and Environmental Health, 73*(2), 114–127.

Tamara, H., Tom, R., Brett, F., Smith, B., & Oliver, J. (2006). Capsular polysaccharide phase variation in *Vibrio vulnificus*. *Applied and Environmental Microbiology, 72*(11), 6986–6993.

Tchounwou, P. B., Patlolla, A. K., & Centeno, J. A. (2003). Carcinogenic and systemic health effects associated with arsenic exposure-a critical review. *Toxicologic Pathology, 31*(6), 575–588.

Tchounwou, P. B., Centeno, J. A., & Patlolla, A. K. (2004). Arsenic toxicity, mutagenesis and carcinogenesis – A health risk assessment and management approach. *Molecular and Cellular Biochemistry, 255*, 47–55.

Thiele, D. J. (1995). *Metal detoxification in eukaryotic cells*. Washington, DC: CRISP database of National Institute of Health.

Turpeinen, R., Kairesalo, T., & Haggblom, M. M. (2004). Microbial community structure and activity in arsenic-, chromium- and copper contaminated soils. *FEMS Microbiology Ecology, 47*, 39–50.

USDHHS. (1999). *Toxicological profile for lead*. Atlanta: United States Department of Health and Human Services.

Vargas-Garcia, M. C., Suarez-Estrella, F., Lopez, M. J., & Moreno, J. (2012). *Bioremediation of heavy metals with microbial isolates*. Universidad de Almeria, Crta. Sacramento s/n, La Canada de Urbano04008Almeria, Espana. https://doi.org/10.1016/j.scitotenv.2012.05.026.

Velma, V., Vutukuru, S. S., & Tchounwou, P. B. (2009). Ecotoxicology of hexavalent chromium in freshwater fish: A critical review. *Reviews on Environmental Health, 24*(2), 129–145.

Venkatesh, K. R., More, N., & Kanoujia, S. (2015). Chromium accumulation in *Eisenia fetida* in modified vermi compost supplemented with tannery sludge. *Journal of Agroecology and Natural Resource Management, 1*(4), 135–141.

Verkleji, J. A. S. (1993). The effects of heavy metals stress on higher plants and their use as biomonitors. In B. Markert (Ed.), *Plant as bioindicators: Indicators of heavy metals in the terrestrial environment* (pp. 415–424). New York: VCH.

Verma, A., Bharagava, R. N., Kumar, V., Singh, A., Dhusia, N., & More, N. K. (2016). Role of macrophytes in heavy metal removal through rhizo-filtration in aquatic system. *European Journal of Biotechnology and Bioscience, 4*(10), 15–20.

Wang, Z., & Rossman, T. G. (1996). In L. W. Cheng (Ed.), *The toxicology of metals* (Vol. 1, pp. 221–243). Boca Raton: CRC Press.

Wang, Q., Cui, Y., & Dong, Y. (2002). Phytoremediation of polluted waters potential and prospects of wetland plants. *Acta Biotechnologica, 22*(1–2), 199–208.

WHO. (1991). World Health Organization Environmental. Health Criteria 108 Nickel. WHO, Geneva.

WHO. (2003). *World Health Organization. Chromium in drinking-water. Background document for preparation of WHO guidelines for drinking-water quality*. Geneva: World Health Organization (WHO/SDE/WSH/03.04/4).

Wood, J. M. (1974). Biological cycles for toxic elements in the environment. *Science, 183*, 1049–1052.

Wuana, R. A., & Okieimen, F. E. (2011). Heavy metals in contaminated soils: A review of sources, chemistry, risks and best available strategies for remediation. *International Scholarly Research Network (ISRN) Ecology, 2011*, 1–20.

Xu, Q., & Shi, G. (2000). The toxic effects of single Cd and interaction of Cd with Zn on some physiological index of [*Oenanthe javanica* (Blume) DC]. *Journal of Nanjing Normal University (Natural Science), 23*(4), 97–100.

Xu, X. R., Li, H. B., Gu, J. D., & Li, X. Y. (2005). Kinetics of the reduction of chromium (VI) by vitamin C. *Environmental Toxicology and Chemistry, 24*, 1310–1131.

Yadav, A., Raj, A., & Bharagava, R. N. (2016). Detection and characterization of a multi-drug and multi-metal resistant *Enterobacterium Pantoea sp.* from tannery wastewater after secondary treatment process. *International Journal of Plant and Environment, 2*(1–2), 37–42.

Yadav, A., Chowdhary, P., Kaithwas, G., & Bharagava, R. N. (2017). Toxic metals in environment, threats on ecosystem and bioremediation approaches. In S. Das & Singh (Eds.), *Handbook of metal-microbe interactions and bioremediation*. Boca Raton: CRC Press/Taylor & Francis Group.

Zaki, S., & Farag, S. (2010). Identification of bacterial strains from tannery effluent andreduction of hexavalent chromium. *Journal of Environmental Biology, 31*(5), 877–882.

Zeid, I. M. (2001). Responses of *Phaseolus vulgaris* to chromium and cobalt treatments. *Biologia Plantarum, 44*, 111–115.

Zhang, W. J., Jiang, F. B., & Ou, J. F. (2011). Global pesticide consumption and pollution: With China as a focus. *Proceedings of the International Academy of Ecology and Environmental Sciences, 1*(2), 125–144.

Zhitkovich, A. (2005). Importance of chromium-DNA adducts in mutagenicity and toxicity of chromium (VI). *Chemical Research in Toxicology, 18*(1), 3–11.

Ziemackei, G., Viviano, G., & Merli, F. (1989). Heavy metal source and environmental presence. *Annali dell'Istituto Superiore di Sanità, 25*(3), 531–536.

Part II
Sustainable Agriculture

Chapter 6
Plant Growth-Promoting Rhizobacteria: Diversity and Applications

Maya Verma, Jitendra Mishra, and Naveen Kumar Arora

Contents

Abstract The rhizosphere is the region around plant roots where maximum microbial activities occur. In the rhizosphere both beneficial and harmful activities of microorganisms affect plant growth and development. The mutualistic rhizospheric bacteria which improve the plant growth and health are known as plant growth-promoting rhizobacteria (PGPR). They are of much importance due to their ability to help the plant in diverse manners. PGPR such as *Pseudomonas*, *Bacillus*, *Azospirillum*, *Azotobacter*, *Arthrobacter*, *Achromobacter*, *Micrococcus*, *Enterobacter*, *Rhizobium*, *Agrobacterium*, *Pantoea*, and *Serratia* are now very well known. Application of PGPR as bioinoculants/bioformulations is found to be very

M. Verma · J. Mishra · N. K. Arora (✉)
Department of Environmental Microbiology, School for Environmental Sciences, Babasaheb Bhimrao Ambedkar University, Lucknow, Uttar Pradesh, India
e-mail: nkarora_net@rediffmail.com

R. C. Sobti et al. (eds.), *Environmental Biotechnology: For Sustainable Future*,
https://doi.org/10.1007/978-981-10-7284-0_6

effective in enhancing crop productivity in a sustainable way. The use of PGPR in agriculture is also ecologically important as the synthetic chemicals used in agriculture are a severe threat to agroecosystems.

Keywords Rhizosphere · PGPR · Bioinoculants · Agriculture · Agroecosystem

Introduction

In soil, there is a region around plant's root known as rhizosphere (Hiltner 1904). Rhizosphere has diverse number of microorganisms which are 10–100 times higher in comparison with bulk soil (Weller and Thomashow 1994). These rhizospheric microbes have been found to have beneficial, harmful, or neutral impact on plant's health (Whipps 2001; Bais et al. 2006). However PGPR, which show beneficial effect on their host plant, are of prime importance. The term PGPR was coined by Kloepper and Schroth (1978). Among all the rhizospheric bacteria, about 2–5% are considered as PGPR (Antoun and Kloepper 2001). PGPR have three main features: (i) root colonization ability, (ii) high survivability and multiplicity in root surroundings helping in plant growth promotion, and (iii) inhibition of phytopathogens (Lugtenberg et al. 2001; Gamalero et al. 2004). Currently diverse types of bacterial genera are considered as PGPR (Lucy et al. 2004; Adesemoye et al. 2008; Saharan and Nehra 2011; Tailor and Joshi 2014). On the basis of their relationship with the plants, PGPR have been divided into two major groups: symbiotic and free-living (Khan 2005; Hayat et al. 2010). *Pseudomonas* and *Bacillus* are widely reported as free-living whereas rhizobia as symbiotic plant growth promoters (Podile and Kishore 2006).

As the PGPR live in close association with plant roots, they also improve soil quality (Haghighi et al. 2011). Plant growth stimulation by PGPR involves multiple mechanisms (Martinez-Viveros et al. 2010). However, few activities are very common among certain bacteria, and others might be species specific (Schwachtje et al. 2012). PGPR play fundamental roles in physiology and development of plants by influencing various activities (Arora et al. 2013; Arora 2015; Kundan et al. 2015; Gouda et al. 2018). They directly enhance plant growth by the process of nitrogen fixation (Islam et al. 2013), phytohormone production (auxins, cytokinins, gibberellins, ethylene, abscisic acid) (Glick et al. 1995; Vacheron et al. 2013; Maheshwari et al. 2015), mineral (phosphorus, iron, zinc, potassium, and sulfur) solubilization (Rodríguez et al. 2006; Delvasto et al. 2009), and siderophore, enzyme, and organic acid production (Sayyed et al. 2005; Ahemad and Kibret 2014). Indirect growth promotion is mainly due to their biocontrol activities attributed to antibiotic production, iron chelation, cyanide production, induced resistance, synthesis of extracellular enzymes, and competition for niches within the rhizosphere (Beneduzi et al. 2012; Ramadan et al. 2016). In addition to these activities, PGPR also have a role in the management of abiotic stress conditions such as nutrient deficiency, salinity, drought, floods, and extreme temperatures (Da Mota et al. 2008; Arzanesh et al. 2011; Arora et al. 2012; Asari 2015; Tewari and Arora 2016) and removal of various pollutants (heavy metals, organic pollutants) by the process of phyto- and rhizoremediation (Wani et al. 2007; Ma et al. 2011; Kong and Glick 2017; Mishra et al. 2017a). PGPR thus can be used in the enhancement of crop productivity in stressed soils as well as for the bioremediation of such habitats.

Application of PGPR on different crops has shown a notable success. These crops include cereals (Lucas et al. 2009; Qaisrani et al. 2014; Majeed et al. 2015; Karnwal 2017), oil crops (Shahid et al. 2012; Tewari and Arora 2014; Sharifi 2017), pulses (Khare and Arora 2010; Pérez-Montaño et al. 2013; Kumari et al. 2018), and vegetables (Loganathan et al. 2014; Moustaine et al. 2017). Studies indicate that PGPR increase seed germination, seedling vigor, seed emergence, root and shoot growth, total biomass of the plants, seed weight, flowering, and yields (Van Loon et al. 1998; Spaepen et al. 2009; Ahmad et al. 2013; Korir et al. 2017).

Plant productivity is governed by various climatic conditions and increases under suitable conditions, but unfortunately various types of negative changes are occurring in global climate due to anthropogenic activities. Various adverse effects are generated due to climate change which cause degenerative impacts on global vegetation including crop productivity. Climate change is also a threat causing reduction in soil microbial diversity (Bradford et al. 2008; Ladau et al. 2017). Due to climate change, various abiotic stresses have developed which reduce the crop production. PGPR can work as suitable agents to combat the impact of abiotic stresses caused by climate changes. Hence, PGPR inoculation is emerging as a very effective method for enhancing crop production not only for our current food requirements but for future needs as well. In this review we have discussed about the diversity of rhizospheric bacteria and the plant growth-promoting (PGP) mechanisms along with their important roles in agriculture. Some future potential uses of PGPR in enhancing crop productivity and combating global warming are also projected.

PGPR Diversity

The rhizospheric region of plants harbors diverse types of PGPR. Different plants can have specific dominant PGPR genera and species. Root microbiome of plants depends on various environmental (biotic and abiotic) factors such as root type, plant species, plant age, soil type (Campbell 1985), and type of plant species (genotypes) (Lareen et al. 2016). Exudates secreted by plant roots are the most important factors responsible for high microbial diversity in rhizospheric region. On the basis of their functions and taxonomical status, PGPR have been categorized in many groups. According to Tilak et al. (2005), PGPR strains widely belong to five main taxa: Actinobacteria, Bacteroidetes, Cyanobacteria, Firmicutes, and Proteobacteria. It is reported that most commonly studied PGPR of these taxa are *Azospirillum*, *Azotobacter*, *Bacillus*, *Enterobacter*, *Gluconacetobacter*, *Paenibacillus*, *Pseudomonas*, *Rhizobium*, and *Serratia* (Bhattacharyya and Jha 2012; Arora 2015). According to Gray and Smith (2005), PGPR are divided into four categories based on their association level with plant root: (i) those living in the root vicinity, (ii) living on the root surface (rhizoplane-colonizing bacteria), (iii) residing in root tissue (spaces between cortical cells), and (iv) living inside cells of specialized root structures or nodules. Broadly on the basis of these associations, PGPR can be separated into two major groups: (i) extracellular (ePGPR) or rhizospheric and (ii) intracellular or endophytic (iPGPR) (Vessey 2003; Gray and Smith 2005; Martinez-Viveros et al. 2010; Santoyo et al. 2016).

ePGPR cause enhancement in plant growth by various mechanisms and are known as efficient producers of various secondary metabolites which work as potent PGP agents (Gray and Smith 2005). These free-living rhizobacteria, e.g., *Achromobacter*, *Acetobacter*, *Alcaligenes*, *Arthrobacter*, *Azospirillum*, *Azotobacter*, *Azomonas*, *Bacillus*, *Beijerinckia*, *Clostridium*, *Corynebacterium*, *Derxia*, *Enterobacter*, *Herbaspirillum*, *Klebsiella*, *Pseudomonas*, *Rhodospirillum*, *Rhodopseudomonas*, *Serratia*, and *Xanthobacter*, have great agricultural importance (Prithiviraj et al. 2003; Bhattacharyya and Jha 2012).

Among all ePGPR, *Bacillus* and *Pseudomonas* are more abundant because of their outstanding root-colonizing and PGP abilities (Sivasakthi et al. 2014). *Bacillus* is known as very versatile and an important PGPR having many physiological characters to survive in extreme environmental conditions (Shafi et al. 2017). Their ability to form endospore, multilayered cell wall, stress resistance, and secretion of diverse secondary metabolites (peptide antibiotics, peptide signal molecules, extracellular enzymes) are of importance (Gutiérrez-Mañero et al. 2001; Kumar et al. 2011). Many members of *Bacillus* and *Bacillus*-derived genera (BBDG) of phylum Firmicutes are associated with different plants and show various PGP attributes (Yadav et al. 2017). Among all the reported bacilli, *Bacillus* and *Paenibacillus* are highly explored members of the PGPR group (Choudhary and Johri 2008) followed by *Alicyclobacillus*, *Aneurinibacillus*, *Virgibacillus*, *Solibacillus*, and *Gracilibacillus* isolated from different crop plants (Yadav et al. 2016). *Pseudomonas* is also ubiquitous in soil and rhizosphere and one of the dominant PGPR genus with diverse traits (Sivasakthi et al. 2014). Fluorescent pseudomonads are considered as most effective and metabolically and physiologically diverse group of bacteria showing fast growth rate with simple and diverse nutrients (Lugtenberg and Dekkers 1999; Lata et al. 2002). Fluorescent pseudomonads are known as efficient biocontrol agents because of their ability to produce various types of metabolites against phytopathogens (Khare and Arora 2011; Mishra et al. 2012; Tewari and Arora 2014; Mishra and Arora 2018). The most studied members of this group are *P. fluorescens*, *P. putida*, *P. aeruginosa*, *P. chlororaphis*, *P. aureofaciens*, and *P. syringe* (Tewari and Arora 2016; Dorjey et al. 2017).

Actinomycetes are also known as potential group of PGPRs and mainly reported as biocontrol agents against diverse phytopathogens (Merzaeva and Shirokikh 2006; Franco-Correa et al. 2010). Actinobacteria are well known for their ability to produce secondary metabolites and plant growth regulators (Sathya et al. 2017). Besides biocontrol Actinobacteria are also able to mobilize minerals and metals in various crops (He et al. 2010; Sathya et al. 2016). *Streptomyces*, *Actinomadura*, *Microbispora*, *Micromonospora*, *Nocardia*, *Nonomuraea*, *Mycobacterium*, *Frankia*,

Actinoplanes, *Saccharopolyspora*, and *Verrucosispora* are common genera of actinomycetes reported in soil and rhizosphere (Vijayabharathi et al. 2016). Some Actinobacteria such as *Arthrobacter*, *Agromyces*, *Corynebacterium*, *Mycobacterium*, *Micromonospora*, *Propionibacteria*, and *Streptomyces* are also reported as nitrogen-fixing endophytes (Sellstedt and Richau 2013). Some other important ePGPR are from among Cyanobacteria, Proteobacteria, and Firmicutes which include *Anabaena*, *Nostoc*, *Azotobacter*, *Azospirillum*, *Beijerinckia*, and *Clostridium* (Steenhoudt and Vanderleyden 2000; Duc et al. 2009). These are mainly free-living nitrogen fixers present in soil (Zhan and Sun 2012).

iPGPR exist inside plant cells, in specialized nodular structures. Rhizobia and *Frankia* are known as the most explored members of this group (Figueiredo et al. 2011; Gopalakrishnan et al. 2015). These two groups of bacteria can symbiotically fix atmospheric nitrogen with higher plants and are widely studied as symbiotic plant growth promoters (Ahemad and Kibret 2014). Rhizobia are a very broad group of phenotypically heterogeneous Gram-negative, aerobic, non-sporulating, rod-shaped bacteria (Tak et al. 2017; Rao et al. 2018). Rhizobia show various PGP activities, but biological nitrogen fixation (BNF) is most important (Dardanelli et al. 2010; Arora et al. 2017). Rhizobia are known as legume symbionts with 238 species and 18 genera, belonging to 3 different classes: α- proteobacteria (*Rhizobium*, *Mesorhizobium*, *Bradyrhizobium*, *Azorhizobium*, *Ensifer* (formerly *Sinorhizobium*), *Methylobacterium*, *Devosia*, *Microvirga*, *Ochrobactrum*, *Phyllobacterium*, *Shinella*, *Allorhizobium*, *Pararhizobium*, *Aminobacter*, *Blastobacter*, *Photorhizobium*), ß-proteobacteria (*Burkholderia* and *Cupriavidus* (formerly *Ralstonia*)), and ϒ-proteobacteria (*Pseudomonas*) (Berrada and Fikri-Benbrahim 2014; Shamseldin et al. 2017). *Frankia* forms root nodules in 200 species of non-leguminous woody plants (24 genera of angiosperms in 8 different families) (Welsh et al. 2009; Franche et al. 2011). Nitrogen-fixing cyanobacteria such as *Nostoc*, *Anabaena*, *Calothrix*, *Aulosira*, *Tolypothrix*, *Fischerella*, and *Chlorogloeopsis* also have the ability to colonize different tissues of plants including Bryophyta, Pteridophyta, gymnosperms, and angiosperms (Bergman et al. 2007; Santi et al. 2013; Mus et al. 2016).

Besides these, some rhizobacteria live inside plant roots with more intimacy and are categorized as root endophytes (Schulz and Boyle 2006). Endophytic bacteria are an important group of PGPR and now considered as more effective in comparison with rhizospheric bacteria (Coutinho et al. 2015; Asaf et al. 2017). Endophytes belong to various bacterial phyla (Acidobacteria, Actinobacteria, Ascomycota, Bacteroidetes, Basidiomycota, Deinococcus-Thermus, and Firmicutes) (Suman et al. 2016). Dominant bacterial genera which are endophytes are *Achromobacter*, *Bacillus*, *Burkholderia*, *Enterobacter*, *Herbaspirillum*, *Micrococcus*, *Microbacterium*, *Pantoea*, *Pseudomonas*, and *Streptomyces* (Ryan et al. 2008; Verma et al. 2013; Suman et al. 2016). Some researchers also consider rhizobia as endophytes. Diversity of various PGP bacterial endophytes are explored and applied for higher crop production (Rosenblueth and Martínez-Romero 2006; Liu et al. 2017).

Roles of PGPR in Rhizosphere

PGPR perform various beneficial activities to promote plant growth and health (Fig. 6.1). These mechanisms can be divided into four major activities: (i) growth-enhancing activities, (ii) biocontrol activities, (iii) management of abiotic stresses, and (iv) soil renovations by rhizoremediation (García-Fraile et al. 2015).

Growth-Enhancing Activities

PGPR enhance the plant growth by several mechanisms, and these work as direct benefits to the plant (Table 6.1) (Bhattacharyya and Jha 2012; Arora 2015).

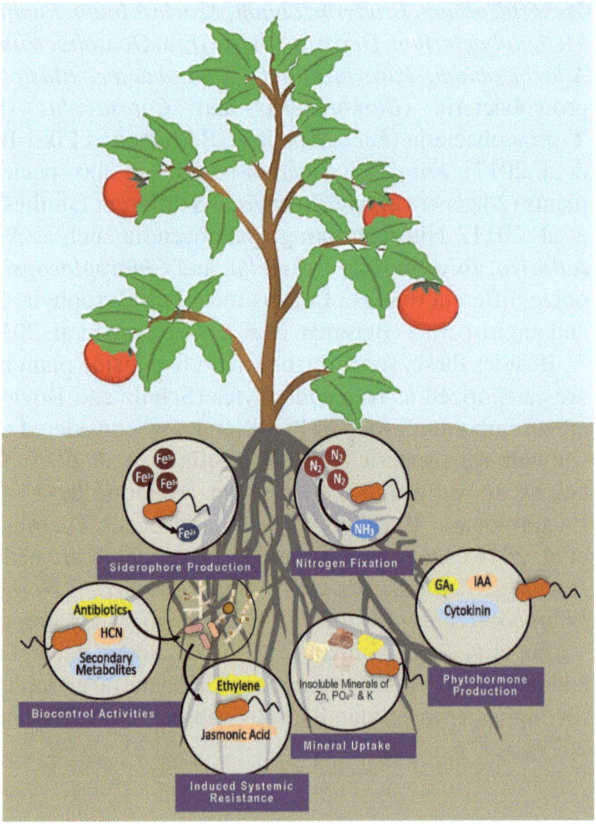

Fig. 6.1 Plant growth promoting activities of PGPR

Table 6.1 Direct mechanisms and PGPR used

Mechanism		PGPR	Crops	References
Nitrogen fixation	Symbiotic	*Rhizobium* and allied genera	Legumes, e.g., Soybeans, Peanut, Chickpea etc.	Lucas-Garcia et al. (2004), Vargas et al. (2010), Laranjo et al. (2014), and Abd-Alla et al. (2017)
		Frankia	Higher Agiospermic plants (Actinorhizal plants), e.g., *Alnus, Casurina*	Crannell et al. (1994), Santi et al. (2013), Diagne et al. (2013), and Ballhorn et al. (2017)
	Free-living	Cyanobacteria, *Azotobacter, Azospirillum, Beijerinckia*	Cereals, e.g., Wheat, Rice, Maize	Steenhoudt and Vanderleyden (2000), Cassán et al. (2009), and Shariatmadari et al. (2013)
Phosphate solubilisation		*Pseudomonas, Bacillus, Rhizobium*	Any crops	Mamta et al. (2010), Schoebitz et al. (2013), and Oteino et al. (2015)
Iron sequestration		*Alcaligenes, Pseudomonas, Bacillus*	Any crops	Gamit and Tank (2014) and Aznar and Dellagi (2015)
Zinc solubilisation		*Burkholderia, Pseudomonas, Bacillus*	Any crops	Goteti et al. (2013), Vaid et al. (2014), and Sunithakumari et al. (2016)
Potassium solubilisation		*Bacillus, Pseudomonas*	Any crops	Bagyalakshmi et al. (2012), Parmar and Sindhu (2013), and Prajapati and Modi (2016)
Phytohormone production		*Bacillus, Rhizobium, Pseudomonas*	Any crops	Khare and Arora (2010), Reetha et al. (2014), and Pandya and Desai (2014)

Nutrient Uptake

Nitrogen (N) is among the most vital nutrients essential for plant growth. In the atmosphere, nitrogen is present in very large amounts (78%) but remains unavailable for the direct uptake by the plant. Diazotrophic microorganisms, particularly bacteria and archaea, can fix atmospheric dinitrogen (N_2) through the BNF process (Dixon and Kahn 2004). Nitrogen-fixing microorganisms change nitrogen to ammonia by using a complex enzyme system known as nitrogenase (Postgate 1998; Leigh 2002). Nitrogen-fixing bacteria are of two types: symbiotic (rhizobia and *Frankia*) (Ahemad and Khan 2012) and nonsymbiotic (Cyanobacteria, *Azospirillum, Azotobacter, Gluconacetobacter, Azoarcus*) (Bhattacharyya and Jha 2012; Roper and Gupta 2016). Among all nitrogen fixers, root nodule-associated rhizobia are most studied and exploited. It is reported that in agricultural systems rhizobia are annually able to fix 180×10^6 tonnes of nitrogen through BNF process at global level (Postgate 1998; Sahgal and Johri 2003) which causes efficient increment in the productivity and quality of crops (Herridge et al. 2008; Krapp et al. 2011). It is also estimated that the cost of total nitrogen fixed by BNF process is equivalent to US $ 160–180 billion (Rajwar et al. 2013). Due to this (BNF) outstanding property, rhizobia are being used as biofertilizers to enhance crop production in several countries (Mia and Shamsuddin 2010; Arora et al. 2017). In this regard, Yadegari et al. (2010) reported that on inoculation of *Rhizobium phaseoli*, growth and yield of *Phaseolus vulgaris* are increased due to high nitrogen fixation. Similarly Sarr et al. (2015) also worked on the enhanced growth, nitrogen fixation, and increased nodulation of cowpea by application of *Bradyrhizobium* strains.

Phosphorus (P) is the second major nutrient required for plant growth and development and present at levels of 400–1200 mg/kg of soil (Begon et al. 1990; Khan et al. 2007). Most of the soils are deficient in P due to its high fixation rate in insoluble forms (Batjes 1997; Kravchenko et al. 2004), and very less amount (1 mg or less) of P is in soluble form hence in general is not available for plant uptake (Khan et al. 2009). It is reported that only 5% or less of total P present in soil is available for plant uptake. PGPR solubilize these insoluble compounds by acidification, chelation, exchange reactions, and gluconic acid production which make the P uptake process by plants very easy (Rodriguez et al. 2004; Chung et al. 2005; Khan et al. 2009; Coutinho et al. 2012; Li et al. 2017). Phosphate-solubilizing bacteria (PSB) help in phosphate uptake by the plants due to one or more aforesaid mechanisms. PSB belong to various genera such as *Azotobacter*, *Bacillus*, *Beijerinckia*, *Burkholderia*, *Enterobacter*, *Erwinia*, *Flavobacterium*, *Microbacterium*, *Pseudomonas*, *Rhizobium*, and *Serratia* (Bhattacharyya and Jha 2012). PSB enhance the plant growth and yield such as Oteino et al. (2015) reported that phosphate-solubilizing *Pseudomonas* strain increased the growth of *Pisum sativum* L. by producing gluconic acid. Similarly, Demissie et al. (2013) and Gusain et al. (2015) reported the enhanced plant growth and phosphate uptake in faba bean and rice, respectively. Recently, Li et al. (2017) also worked on the growth promotion of maize plants by inoculation of PSB isolates (*Paenibacillus*, *Pseudomonas*, and *Sphingomonas*). Besides P uptake, PSB also enhance the plant growth by stimulating BNF (Ahmad et al. 2008; Zaidi et al. 2009).

Apart from N and P, PGPR also influence the uptake of other nutrients such as potassium (K), zinc (Zn), iron (Fe), and sulfur (S) (Peix et al. 2001; Wu et al. 2005; Adesemoye et al. 2008; Lukkani and Reddy 2014). K is also a very important nutrient, essential for the growth and development of plants. However, more than 90% of K in the soil exists in insoluble forms (micas, illite, and orthoclases) (Li et al. 2006; Bahadur et al. 2016). PGPR such as *Pseudomonas*, *Burkholderia*, *Acidithiobacillus*, *Bacillus*, and *Paenibacillus* are known to solubilize K for plant uptake mainly by excreting organic acids (Rogers et al. 1998; Parmar and Sindhu 2013). Han and Lee (2005) reported the increased K uptake, photosynthesis, and yield of eggplant grown on K-limited soils by inoculation of K-solubilizing bacteria. According to Bagyalakshmi et al. (2012), K-solubilizing bacteria caused high productivity of tea plants by enhancing nutrient uptake efficiency. Shanware et al. (2014) and Etesami et al. (2017) reported the detail about the mechanism, occurrence, and functions of potassium-solubilizing rhizobacteria.

Zn is an essential micronutrient for plant growth, but according to FAO it is estimated that about 50% of lands are Zn deficient due to fixed forms of Zn compounds (FAO 2002; Review 2008). PGPR help plants to mobilize the fixed form of Zn compound and its easy uptake (Iqbal et al. 2010; Hussain et al. 2015). Several bacterial genera are now considered as Zn solubilizers which include *Bacillus*, *Pseudomonas*, *Rhizobium*, *Burkholderia*, *Acinetobacter*, *Mycobacterium*, *Stenotrophomonas*, *Enterobacter*, and *Xanthomonas* (Vaid et al. 2014; Naz et al. 2016). PGPR solubilize Zn by various mechanisms such as excretion of organic acids (2-ketogluconic acid and gluconic acid), protein extrusion, and production of chelating agents (Nahas 1996; Seshadre et al. 2002). Goteti et al. (2013) observed enhanced nutrient uptake and growth of maize plants when inoculated with Zn-solubilizing bacteria. Shakeel et al. (2015) also recorded that root-associated *Bacillus* sp. enhanced Zn translocation toward grains of rice varieties, basmati-385 and super basmati, and also increased total yields by 22–49% and 18–47%, respectively. Zinc-solubilizing PGPR have a very important role in agriculture and food security (Shaikh and Saraf 2017).

Fe is the fourth most abundant and essential element present in the soil and works as an essential nutrient for all forms of life on earth. Iron is spatially distributed in soils with 0.2% to 55% (20,000–550,000 mg/kg), but in the aerobic environment, Fe occurs in insoluble hydroxides and oxyhydroxide form, which is generally inaccessible to both plants and microorganisms (Rajkumar et al. 2010). In low Fe conditions, certain bacteria produce siderophores as low-molecular mass Fe chelators (Miethke and Marahiel 2007). Siderophore-producing PGPR help in Fe uptake by the plants through formation of a ferric-siderophore complex even at very low concentrations (Podile and Kishore 2006; Dimkpa et al. 2009; Boiteau et al. 2016). Bacterial genera *Pseudomonas*, *Burkholderia*, *Enterobacter*, and *Grimontella* are reported as efficient siderophore producers, while *Klebsiella*, *Stenotrophomonas*, *Rhizobium*, *Herbaspirillum*, and *Citrobacter* are also known to be siderophore producers (De Souza et al. 2015; Arora and Verma 2017). In this context, Sharma et al. (2013) reported that iron content of rice grains is increased on treatments with strain of *P. putida*. In another study, Wang et al. (2017) observed the enhanced iron uptake by *Arabidopsis thaliana* plants on inoculation with *B. amyloliquefaciens*. Detailed information about the application of bacterial siderophores is described by Ali and Vidhale (2013) and Saha et al. (2016).

S is also an essential nutrient for plant growth and taken up by plants mainly in the form of sulfate and sulfur dioxide (Marschner 2012). A major part of S (95%) is unavailable (bound with organic complex) for plant uptake, and only 5% is available (Kertesz and Mirleau 2004). Sulfur-oxidizing PGPR can be used to convert unavailable S into available form for direct uptake by the plants (Gahan and Schmalenberger 2014). Currently PGPR have gained much attention in agriculture to fulfil the sulfur requirements of plants (Salimpour et al. 2010; Anandham et al. 2014). Grayston and Germida (1991) reported that sulfur-oxidizing PGP isolates stimulate canola growth by enhancing mineral nutrient uptake. Awad et al. (2011) reported that onion growth, yield, and nutrient content were increased on inoculation with sulfur-oxidizing bacteria. Similar work was also done by Eslamyan et al. (2013) who reported that on inoculation with *P. fluorescens*, growth of rapeseed cultivars is increased due to high S uptake.

Phytohormone Production

Phytohormones act as chemical messengers and enhance plant growth by affecting all the major activities such as formation and development of various parts of plants mainly leaves and flowers and ripening of fruits (Khalid et al. 2006). There are several PGPR which show phytohormone production ability and influence the physiological processes of plants to facilitate their growth by regulating hormonal balance (Patten and Glick 2002; Asghar et al. 2004; Boiero et al. 2007; Maheshwari et al. 2015). PGPR can control the production of hormones such as indole acetic acid (IAA), abscisic acid (ABA), cytokinins, gibberellins, and ethylene (Kudoyarova et al. 2015). Among all, IAA is reported as quantitatively most abundant and secreted by several bacterial genera including *Azospirillum, Bacillus, Pseudomonas*, and *Rhizobium* (Arora 2013; Islam et al. 2015). It is reported that more than 80% of the rhizospheric bacterial strains are able to produce auxins (Khalid et al. 2004; Hayat et al. 2010). The biosynthesis of IAA depends on various environmental factors such as pH, carbon, and precursor concentration (Spaepen et al. 2009). IAA plays a very important role in rhizobacteria-plant interactions (Spaepen and Vanderleyden 2011) and promotes the plant growth mainly by stimulating the development of root system (Khare and Arora 2010; Tewari and Arora 2013). Mohite (2013) reported the enhanced growth of wheat plant due to inoculation of IAA-producing strains of *Bacillus* and *Lactobacillus*. Etesami et al. (2015) observed the enhanced growth of rice crop on inoculation with IAA producing rhizospheric and endophytic bacteria. Recently Pérez-Fernández and Alexander (2017) also reported that IAA-producing strains significantly enhanced the plant biomass, flower, pod, and seed production along with nitrogen content in *Cicer arietinum*. Vidhyasekaran (2015) reported that auxin also controls various plant defense-signaling pathways by affecting other plant hormone activities, e.g., cytokinin, abscisic acid, ethylene, jasmonate, and salicylic acid.

PGPR are also known to produce cytokinins which have major role in root initiation, cell division, cell enlargement and increase the root surface area of plants by enhanced formation of adventitious roots (Werner et al. 2003; Salamone et al. 2005). Cytokinins mainly affect the leaf growth by delaying the senescence or aging of plant tissues (Kundan et al. 2015). Some bacterial genera, e.g., *Azotobacter, Rhizobium, Pantoea, Rhodospirillum, Pseudomonas, Bacillus*, and *Paenibacillus*, are reported to produce cytokinins (Salamone et al. 2001; Glick 2012). In a study

Ortíz-Castro et al. (2009) reported that cytokinin-producing *Bacillus megaterium* promote biomass production of *Arabidopsis thaliana* plant. Hussain and Hasnain (2009) reported enhanced growth in cucumber on inoculation with cytokinin-producing bacterial extract. Selvakumar et al. (2016) reported that cytokinin-producing osmotolerant strains (*Citricoccus zhacaiensis* and *B. amyloliquefaciens*) enhance the growth of tomato under irrigation-deficit conditions.

Gibberellins (GA) are a class of phytohormones that commonly result in modifications of plant morphology by extension of plant tissues, particularly stem tissues (Salisbury 1994). In bacteria, they act as signaling factors toward the host plant (Bottini et al. 2004). GA production by PGPR induces growth promotion in plants (Piccoli et al. 1997; Gutiérrez-Mañero et al. 2001; Kang et al. 2014). It is also reported that GA interact with other phytohormones which modify the plant hormonal balance and affect plant growth (Bömke and Tudzynski 2009). Bottini et al. (2004) stated that gibberellin-producing bacteria have potential to enhance the growth and yields of crops. Kang et al. (2014) reported that gibberellin-producing PGP strain *Leifsonia soli* (SE134) has efficient potential to enhance the biomass, hypocotyl, and root lengths of cucumber seeds in comparison with control.

Ethylene is a gaseous phytohormone that controls the plant growth and development (Abeles et al. 1992; Khalid et al. 2006). It is also important to induce different physiological changes in plants by performing various activities such as inhibition of root elongation and auxin transport which promotes senescence and abscission of various organs and fruit ripening (Glick et al. 2007; Cassán et al. 2014). Besides regulating plant growth, ethylene has also been recognized as a stress hormone (Saleem et al. 2007). Ethylene production also regulates other important plant growth functions such as seed germination, seedling establishment, development of root hair, formation of adventitious roots, nodulation, flower and leaf senescence, and leaf and fruit abscission (Kucera et al. 2005; Bakshi et al. 2015). Corbineau et al. (2014) reported that ethylene production breaks seed dormancy and regulates seed germination. Ribaudo et al. (2006) reported that ethylene produced by *Azospirillum brasilense* promotes root hair development in tomato plants.

Production of Volatile Organic Compounds (VOCs)

It has recently been reported that some rhizobacteria promote plant growth by releasing VOCs (Ping and Boland 2004; Ortíz-Castro et al. 2009; Tahir et al. 2017). These compounds are of diverse types and very helpful for PGPR to communicate with each other and their host plants (Peñuelas et al. 2014). VOCs include various compounds such as 2, 3-butanediol, acetoin, terpenes, and jasmonates. The synthesis of bioactive VOCs is considered to be a strain-specific phenomenon. These VOCs can act as signaling molecules in plant-microbe interactions and elicit the plant responses mainly when generated at sufficient concentrations (Ryu et al. 2003; Reddy et al. 2014). Gutierrez-Luna et al. (2010) observed that VOC emission by

PGPR can modulate both plant growth promotion and root system architecture of *A. thaliana* plant. Similarly Tahir et al. (2017) reported that volatile compounds produced by *Bacillus subtilis* play a very significant role in growth promotion of tomato plants by regulating biosynthesis of phytohormones.

Biocontrol Activities

Several phytopathogens cause severe diseases in crops and reduce crop yields. Contamination of food grains with phytopathogens is also considered to diminish quality of food and responsible for high economic losses (Guo et al. 2013). Biocontrol of phytopathogens by using natural organisms is an effective method to treat plant diseases in an eco-friendly manner without causing harm to the environment (Compant et al. 2005; Fatima et al. 2009). Hence application of biocontrol appears as a more beneficial and profitable approach in comparison with synthetic chemicals. Biocontrol PGPR are reported to control plant diseases caused by various types of pathogens and pests such as bacteria, fungi, protozoa, viruses, nematodes, and insects (de Bruijn et al. 2007; Raaijmakers et al. 2010).

Biological control of phytopathogens by PGPR is boon to modern as well as conventional agriculture. Currently various strains of PGPR in genera *Agrobacterium*, *Arthrobacter*, *Azoarcus*, *Azotobacter*, *Azospirillum*, *Bacillus*, *Burkholderia*, *Caulobacter*, *Chromobacterium*, *Enterobacter*, *Erwinia*, *Flavobacterium*, *Klebsiella*, *Micrococcus*, *Rhizobium*, *Pantoea*, *Pseudomonas*, and *Serratia* are being used to control the diseases of agriculturally important crops (Ahemad and Kibret 2014; Figueroa-López et al. 2016). *Pseudomonas* and *Bacillus* are very preferential among them because they are aggressive colonizers of the rhizosphere and have broad-spectrum antagonistic activities (Weller et al. 2002; Ahmad et al. 2008; Khare and Arora 2011; García-Gutiérrez et al. 2013; Arora and Mishra 2016; Mishra and Arora 2018). PGPR show various biocontrol activities against phytopathogens such as competition for nutrients, antagonism by production of various metabolites (antibiotics, hydrolytic enzymes, cyanide, siderophores), and induction of systemic resistance (Table 6.2) (Siddiqui 2006; Lugtenberg and Kamilova 2009; Ramyasmruthi et al. 2012; Mishra et al. 2017b). These aforesaid mechanisms are not only very efficient to control phytopathogens, but they also act in synergistic manner to resist the plants against pathogens (Jha et al. 2011).

Competition

PGPR control the growth of pathogens by creating competitive environments for nutrient uptake by them and cause the reduction in nutrient availability around the host plant. This competition between pathogens and non-pathogens is responsible to control the phytopathogen population (Pal and Gardener 2006). Rabosto et al.

Table 6.2 Indirect mechanisms and PGPR used

Mechanisms	PGPR	Crops	References
Competition	*Enterobacter, Pseudomonas, Bacillus*	Any crops, e.g., Wheat, Rice, Tomato	Raaijmakers and Weller (2001), Kageyama and Nelson (2003), and Liu et al. (2013)
Enzyme production	*Bacillus, Pseudomonas, Serratia*	Any crops, e.g., Cucumber, Grape, Peanut	Singh et al. (1999), Kishore et al. (2005), Arora et al. (2008), and Kejela et al. (2016)
Antibiotic production	*Pseudomonas, Bacillus, Agrobcaterium, Pantoea*	Any crops, e.g., Tobacco, Cotton, Cabbage, Mung bean	Hill et al. (1994), Mishra and Arora (2012), and Dhanya and Adeline (2014)
HCN production	*Pseudomonas, Bacillus, Burkholderia*	Any crops e. g. Wheat, Barley	Reetha et al. (2014), and Nandi et al. (2015)
Siderophore production	*Rhizobium, Pseudomonas, Bacillus*	Any crops, e.g., Groundnut, Chickpea, Capsicum	Arora et al. (2001), Omidvari et al. (2010), Mishra and Arora (2011), and Rais et al. (2017)
ISR and SAR	*Pseudomonas, Serratia, Bacillus, Rhizobium*	Any crops, e.g., Cucumber, Tobacco Bean, Tomato	Press et al. (1997), Reitz et al. (2002), Meziane et al. (2005), Pal and Gardener (2006), and Tahir et al. (2017)

(2006) studied that *Bacillus* sp. showed antagonistic activity against *Botrytis cinerea* creating competition for nutrients. Similarly, Haidar et al. (2016) reported the biocontrol activity of antagonistic bacteria by creating competitive environments for fungal pathogens. Siderophore production is very important for this type of interaction in which siderophore-producing bacteria suppress the growth of pathogens by creating iron-limiting conditions in soil (Arora et al. 2001; Sayyed et al. 2008; Verma et al. 2011). Siderophores work as antagonistic metabolites for biocontrol of phytopathogens which helps bacteria to rapidly colonize around the plant root (Keel et al. 1989; Loper and Buyer 1991; Haas and Défago 2005). According to Hofte et al. (1992), siderophore (salicylic acid, pyochelin, and pyoverdine) production causes disease suppression by limiting the supply of essential trace minerals for pathogens in rhizosphere. Yu et al. (2011) reported that the siderophore-producing bacterium, *Bacillus subtilis* CAS15, has a biocontrol effect on *Fusarium* wilt and promotes the growth of pepper plant. Recently Kotasthane et al. (2017) also reported that siderophore-producing *Pseudomonas* strain controls the collar rot disease of chickpea plant.

Production of Metabolites

Production of metabolites is one of the potent and broad-spectrum mechanisms of biocontrol (Mishra and Arora 2018). Besides siderophore, PGPR also produce many other antagonistic compounds against pathogens such as (i) volatile compounds, e.g., hydrogen cyanide, aldehydes, alcohols, ketones, and sulfides, and (ii) nonvolatile compounds, e.g., polyketides (diacetylphloroglucinol (DAPG) and mupirocin) and heterocyclic nitrogenous compounds such as derivatives of phenazines (pyocyanin, phenazine-1-carboxylic acid (PCA), phenazine-1-carboxamide (PCN), and hydroxy phenazines) (De Souza et al. 2003). Meyer et al. (2016) reported that DAPG-producing *P. fluorescens* is used for biocontrol of *Meloidogyne incognita* and *Fusarium oxysporum*. In this context, recently Mishra and Arora (2018) and Shahid et al. (2018) reported the detail information about the secondary metabolites produced by pseudomonads for effective biocontrol activities. Exopolysaccharide (EPS) production has also very promising role in growth enhancement of plants by performing biocontrol activities. EPS control plant diseases by protecting their root against phytopathogenic attacks (Noumavo et al. 2016). In this regard, Tewari and Arora (2014) reported that EPS-producing *Pseudomonas* strain enhances the yield of sunflower crop by protecting it from the attack of a pathogen, *Macrophomina phaseolina*. Recently, Tewari and Arora (2018) reported role of salicylic acid (SA) produced by fluorescent pseudomonads in biocontrol of fungal phytopathogens. Khare et al. (2018) suggested the role of biologically active peptides known as peptaibols, in inhibition of phytopathogen, *Fusarium oxysporum*.

Several biocontrol PGPR show antagonistic activity against phytopathogens by causing lysis of their cell wall which is performed by secreting various hydrolytic enzymes such as chitinases, glucanases, proteases, and lipases (Maksimov et al. 2011; Jadhav and Sayyed 2016). These enzymes are extracellular and hydrolytic and cause the digestion or deformation of cell wall components of fungi (Aeron et al. 2011). Hydrolytic enzymes producing bacteria are also used in combination with other biocontrol agents, leading to a synergistic inhibitory effect against phytopathogens (Someya et al. 2007). Arora et al. (2007) and Saraf et al. (2014) reported chitinase and glucanase production by *Bacillus* and *Pseudomonas* sp. as growth suppressor of filamentous fungi both in laboratory and field conditions. Figueroa-López et al. (2016) also observed the antagonistic activity of rhizospheric bacteria against *Fusarium verticillioides* by producing various hydrolytic enzymes (glucanases, proteases, or chitinases) for disease control in maize plant.

Induced Systemic Resistance (ISR) and Systemic Acquired Resistance (SAR)

ISR and SAR activate chemical and physical defense mechanisms of the host plant by an inducer (chemical or microorganism) (Pieterse et al. 2014). However, induction of host defenses can be local and/or systemic in nature, depending on the type, source, and number of stimuli or signaling compounds (Van Loon 2007). ISR is triggered by PGPR and starts in the root and then spreads to the other plant parts (Ramos-Solano et al. 2008). Various chemical elicitors are produced by PGPR strains such as lipopolysaccharides (LPS), siderophores, cyclic lipopeptides, DAPG, homoserine lactones, and volatiles (acetoin and 2, 3-butanediol) (Lugtenberg and

Kamilova 2009; Pieterse et al. 2014). ISR is mainly elicited by ethylene and jasmonic acid signaling in the plant (Van Loon 2007). Several PGPR strains among *Pseudomonas*, *Bacillus*, *Serratia*, and *Azospirillum* have been reported as ISR inducers (Bakker et al. 2013). Zhang et al. (2002) observed the role of SA in ISR elicited by PGPR against blue mold of tobacco. Boukerma et al. (2017) evaluated that *P. fluorescens* (PF15) and *P. putida* (PP27) have potency to protect tomato plants against *Fusarium* wilt by ISR. In contrast, SAR is typically activated by necrotic pathogenic bacteria and the molecule that typically leads to the expression of pathogenesis-related (PR) proteins such as SA. These PR proteins include enzymes some of which may act directly to lyse invading cells, reinforce cell wall boundaries to resist infections, or induce localized cell death (Beneduzi et al. 2012).

Abiotic Stress Management

Abiotic stresses are the major constraints that are affecting the soil quality as well as productivity of the crops (Grayson 2013). According to the Food and Agriculture Organization (FAO), it is estimated that due to abiotic stresses, 30% land will degrade in the next 25 years and it will reach up to 50% by the year 2050 (Wang et al. 2003). PGPR can support plants to resist against various types of stresses to ameliorate their effect (Lugtenberg and Kamilova 2009; Egamberdieva 2012). Stress-tolerant PGP inoculants work as cost-effective tools to improve production of crops in a hostile environment (Nadeem et al. 2010; Tewari and Arora 2015). PGPR can induce abiotic stress tolerance in plants by imposing physical and chemical changes which is known as "induced systemic tolerance" (IST) (Yang et al. 2009). PGPR like *Achromobacter*, *Azotobacter*, *Azospirillum*, *Acetobacter*, *Bacillus*, *Chryseobacterium*, *Flavobacterium*, *Enterococcus*, *Klebsiella*, *Pseudomonas*, *Rhizobium*, *Serratia*, and *Paenibacillus* are being used for ameliorating various stresses in diverse habitats (Rabie and Almadini 2005; Egamberdieva 2012). Arora et al. (2012) recommend that the use of such bioproducts is helpful to combat the soil stress with high crop production. It is reported by various workers that application of PGPR in stressed soils reduces their harmful effects by regulating the production of various metabolites such as phytohormones, antioxidants, enzymes, and EPS (Raghavendra et al. 2010; Kong et al. 2014; Arora and Mishra 2016).

Salinity stress is a major issue, affecting more than 900 million hectares (ha) land around the globe (Khan and Panda 2008). Salinity causes alkalization of soil; hence soil nutrients become unavailable for plant leading to nutrient stress (Maheshwari et al. 2012a). Application of PGPR in saline soil can be helpful for reducing the impact of salinity along with plant growth promotion (Ilangumaran and Smith 2017). EPS production is known to be a potential trait for salinity stress amelioration. EPS production by rhizobacteria is helpful for effective survival of plant under salinity stress in rhizospheric soil by inducing osmotolerance activity (Grover et al. 2010; Upadhyay et al. 2011). EPS production by salt-tolerant PGPR causes high accumulation of soil particles around plant roots with high water retention activity along with nutrients and facilitates the plant growth in saline soil (Roberson and Firestone 1992; Paul and Lade 2014). Due to this activity, physiological properties of soils and metabolic properties of crops increase, leading to higher crop productivity with improved quality. In this regard, Qurashi and Sabri (2012) reported that salt-tolerant PGP strains have the ability to enhance plant growth by secreting large amount of EPS under high salinity. Tewari and Arora (2016) also reported the enhanced yield of sunflower under salinity stress on inoculation with EPS-producing *Pseudomonas* sp. An enzyme, 1-aminocyclopropane-1-carboxylate (ACC) deaminase produced by PGPR, plays an important role in stress management by acting as regulator for phytohormone ethylene (Glick 2014; Saikia et al. 2018). It cleaves ACC (immediate precursor of ethylene) in the biosynthetic pathway for ethylene in plants (Glick et al. 1998). In this context Saleem et al. (2007) and Glick (2014) reported that ACC producing PGPR can contribute to growth promotion efficiently in stressful soil conditions when inoculated on plants. In another study Habib et al. (2016) worked on ACC deaminase producing PGPR and their effect on reduction of salinity stress in okra plant through reactive oxygen species (ROS)-scavenging enzyme activity.

Temperature is an important abiotic factor for plant growth, but exposure at very high or very low temperatures is harmful. Cold-tolerant and heat-tolerant PGPR are recently being used to ameliorate the adverse effects of temperature on plants (Hoflich and Kuhn 1996; Meena et al. 2015). Ali et al. (2011) reported the effect of thermotolerant PGPR inoculation on the growth of wheat under heat stress. According to Mishra et al. (2012), cold-tolerant bioinoculants are also useful in stress management. Drought is also reported as a major stress for crop production (Glick 2004). Drought stress can be ameliorated by using various drought-tolerant PGPR including *Azotobacter*, *Azospirillum*, *Bacillus*, *Rhizobium*, and *Serratia* (Vurukonda et al. 2016). Sandhya et al. (2009) used the EPS-producing PGPR to promote the growth of sunflower crop under drought stress. Kaushal and Wani (2016) recommended the use of PGPR to combat the drought stress problems in drylands. Flooding is also one of the major abiotic stresses that works as a limiting factor for crop growth (Normile 2008). It is reported that on an average flooding affects 140 million people per year around the globe (WDR 2003). Chakraborty et al. (2013) reported the water stress amelioration and high yield of wheat crop by using osmotic stress-tolerant bacteria. Similarly, Tewari and Arora (2016) reviewed the effect of flooding stress on soybean production and their amelioration by using PGPR. According to Tewari and Arora (2013), application of microbes to ameliorate soil abiotic stresses is easier and beneficial in comparison with developing stress-tolerant crops. PGPR are now being used as efficient and inexpensive tools to ameliorate the soil stresses (Batool et al. 2014; Gontia-Mishra et al. 2016).

Soil Renovation

Presently it is estimated that a large portion of the land around the globe is wasted due to soil erosion and degradation problems. According to Riadh et al. (2010), out of the world's 5.2 billion ha of dryland agriculture, 3.6 billion ha is affected by the problems of erosion and soil degradation. Soil properties and fertility depend on the combination of various biotic and abiotic factors. Richness and diversity of microbial communities in soil and their activity work as a major indicator of soil health (fertility) (Griffiths and Philippot 2012; De Souza et al. 2015). In low-quality pollutant-contaminated soil, microbial load and nutrients are very less, and PGPR inoculation improves physicochemical as well as biological properties (Sahoo et al. 2013; Bhardwaj et al. 2014). Several methods have been developed for pollutant removal, but microbes have emerged as successful solution for bioremediation due to their high sensitivity, tolerance, and the sequestration ability (Burd et al. 2000; Amora-Lazcano et al. 2010). The role of PGPR in pollutant degradation is very useful for plants to grow as natural vegetation at a contaminated site (Pérez-Montaño et al. 2014). It is reported that use of PGPR can be extensively applied for crop improvement along with contaminant removal in deteriorating soils (Huang et al. 2005; Yang et al. 2009). Several researchers reported that on application of PGPR, high crop production and metal removal takes place simultaneously in polluted environment (Zhuang et al. 2007; Romeh and Hendawi 2014; Khan and Bano 2016). This bioremediation technology (rhizoremediation) is now being used for the removal of pollutants such as pesticides, polyaromatic hydrocarbons, heavy metals, and other toxic wastes (Bhalerao 2012; Hansda et al. 2014; Bisht et al. 2015). Efficiency of rhizoremediation process is controlled by various factors such as pollutant type and their bioavailability, plant variety and diversity and activity of microbes, and environmental conditions (Wenzel 2009). In this regard, Okon et al. (2014) worked on the bioremediation of palm oil mill effluent polluted soil and their effect on the growth of *Amaranthus hybridus* L. They reported that the crop growth in polluted soil can be upgraded by using bioproducts. It is also studied that PGPR are used for rhizoremediation of petroleum compounds in soil along with the growth of various plants such as cotton, ryegrass, tall fescue, and alfalfa (Tang et al. 2010).

Heavy metals are present in soils as contaminants (Gadd 2010) with concentration ranging from 1 to 100,000 mg/kg in typical soil (Long et al. 2002). Metal contamination from soil can be removed mainly by geo-active action of soil microbes (Mishra et al. 2017a). Metal-mobilizing PGPR remediate heavy metals from soil and perform multiple functions, e.g., soil quality improvement, plant growth enhancement, detoxification, and metal removal from soil (Hassan et al. 2016; Mishra et al. 2017a). PGPR enhance rhizoremediation of metals through various mechanisms such as acidification, chelation, complexation, precipitation, and redox reactions by producing various metabolites, e.g., organic acids, siderophores, exopolymers, and biosurfactants (Ma et al. 2011; Ma et al. 2016). According to Pires et al. (2017) *Bacillus*, *Pseudomonas*, and *Arthrobacter* are considered as predominant genera of the bacterial population in metal-contaminated sites. Abou-Shanab et al. (2006) reported that application of certain rhizobacteria can increase the uptake of nickel (Ni) from soils by changing its phase. Wani and Khan (2010) worked on chromium (Cr) toxicity and reported its reduced uptake in roots, shoots, and grains of *Cicer* plant on application of *Bacillus* strain. In the same year, Dary et al. (2010) worked on the application of PGPR for remediation of copper (Cu), cadmium (Cd), and lead (Pb) with increased biomass of *Lupinus* plant. Pinter et al.

(2017) observed that on the application of arsenic (As)-tolerant PGPR (*B. licheniformis*, *Micrococcus luteus*, and *P. fluorescens*), grapevine biomass is increased. Zn-tolerant rhizobia from Zn mining soil is reported as growth promoter of *Leucaena leucocephala* in contaminated soil (Rangel et al. 2017). Cu-resistant *Kocuria* sp. is also reported as potential PGPR isolated from the dry tailing of copper mine (Hansda et al. 2017). Ma et al. (2015) studied that PGP strains of *Bacillus* sp. are able to mobilize high amount of metals. Rhizoremediation, thus, can be a very important and low input biotechnology of the future for cleaning contaminated soils off pollutants and simultaneously enhancing their productivity.

Diverse Applications of PGPR

PGPR are essential components of soil and for crop management. PGPR play significant role in solving the problems related to soil stress, soil fertility, soil degradation, and plant growth (Glick 2010). Generally PGPR have been used as biofertilizers or biopesticides (Vessey 2003; Zahir et al. 2004; Arora et al. 2016a; Vandenberghe et al. 2017), but in the future these useful microbes can be used to play various other roles in sustaining the agroecosystems.

Biofertilizers

Biofertilizers are playing crucial role in enhancing crop productivity and quality in present-day agriculture (Mahanty et al. 2016). Currently biofertilizers are also termed as biostimulants. Biostimulants are considered as any substance (of biological origin) or microorganism or their combinations applied to plants, seeds, or soil to enhance nutrient uptake efficiency, stress tolerance activation, and crop quality improvement (du Jardin 2015). Biofertilizers/biostimulants are formulated products of one or more microorganisms which promote plant growth by colonizing the rhizosphere or inner parts of plants (Vessey 2003; Malusá and Vassilev 2014). Biofertilizers work as important components of integrated nutrient management in soil and nourish the plant by playing a very active role in nutrient cycling between soil, plant roots, and microorganisms (Abdel Ghany et al. 2013; Vejan et al. 2016).

Globally biofertilizers have been commercially used in agriculture since more than 120 years when "Nitragin" was launched (Nobbe and Hiltner 1896). Rhizobia-based bioinoculants are commercially available since more than 100 years now (Bashan 1998; Arora et al. 2017). According to BCC research (2014), it is estimated that total fertilizer market has 5% share of biofertilizers and more than 150 microbe-based products are registered for agricultural purposes. Currently various types of biofertilizers are used in agriculture: nitrogen fixers, phosphate solubilizers, potash solubilizers, and others (Zn, boron, S solubilizers). Among all the biofertilizers, nitrogen fixers (*Rhizobium*, *Azospirillum*, *Azotobacter*, *Azoarcus*, and cyanobacteria)

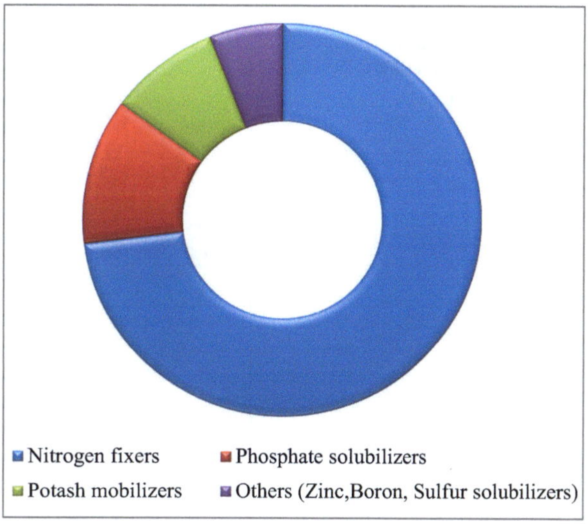

Fig. 6.2 Market shares of various types of biofertilizers at global level. (Research Nester 2018)

and phosphate solubilizers (*B. megaterium*, *Pseudomonas* sp.) have wide commercial applications at global level (Mishra and Das 2014) (Fig. 6.2). Zn-, K-, and S-based biofertilizers are also emerging as important bioinoculants to overcome their deficiency in plants (Khatibi 2011; Shaikh and Saraf 2017). It is reported that biofertilizers increase the crop yield by up to 10–40% by improving uptake of various nutrients such as proteins, amino acids, vitamins, nitrogen, etc. (Bhardwaj et al. 2014). Nitrogenous biofertilizers can increases up to 15–25% of total crop yield by adding 20–200 kg N/ha/year, while PSB-based biofertilizers can increase 10–20% crop yield by solubilizing about 30–50 kg phosphate/ha/year (AgriInfo.in 2015). Although various types of biofertilizers are commercially available in the global market, rhizobia-based inoculants have the largest market share estimated at approximately 78%, while phosphate solubilizers and other bioinoculants have 15% and 7% shares respectively (Owen et al. 2015; Transparency Market Research 2017). It is reported that various other nitrogen-fixing products are also used at large scale which are based on *Azospirillum*, *Azotobacter*, and *Gluconacetobacter* species (Parnell et al. 2016). Phytohormone-producing biofertilizers are also globally used for sustainable agriculture mainly as growth regulators of plants (Narula et al. 2006; Maheshwari et al. 2015). According to Nadeem et al. (2014), biofertilizers for high nutrient uptake have wide applicability in global market, but growth regulator-based biofertilizers need more concern for their commercialization. Currently, the requirement and growth of fertilizers, mainly biofertilizer market in all regions of the world (Fig. 6.3), are increasing to fulfil the food requirements at global level. It is reported that the biofertilizer market was valued at USD 946.6 million in 2015 (marketsandmarkets.com 2016). This market is projected to grow at a cumulative annual growth rate (CAGR) of 14.08% from 2016 to 2022 to reach USD 2305.5 million (marketsandmarkets.com 2016).

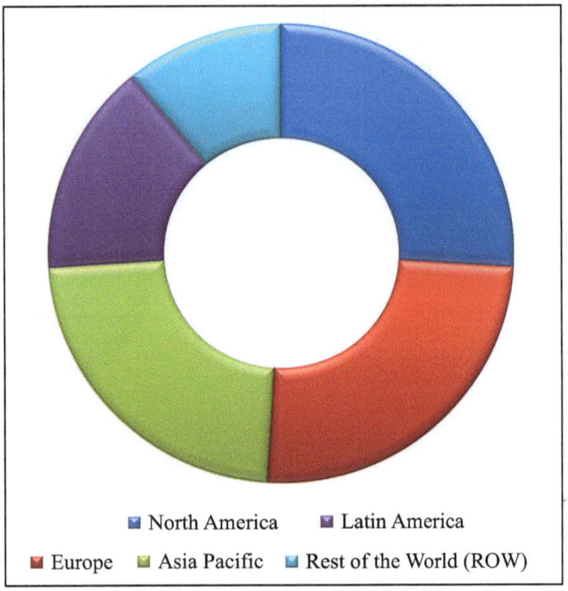

Fig. 6.3 Market share of biofertilizers in various regions of world. (BCC Research 2014)

Biopesticides

Various plant diseases are responsible for the loss of approximately one-third of the crop yields at global level (Lugtenberg et al. 2002). According to Lugtenberg (2015), approximately 25% of the crop yield is lost every year due to diseases caused by phytopathogens around the globe. To overcome this problem, PGPR with biocontrol traits are used as biopesticides. Biopesticides are also attracting attention to manage several pests and weeds (Kumar and Singh 2015). The global market for pesticides has 2.5% contribution of biopesticides (marketsandmarkets 2014). Biopesticides are of various types including living organisms, their products (phytochemicals, microbial products), and byproducts (semiochemicals) (Dutta 2015). Microbial biopesticides contribute in global biopesticide market with 30% share (Thakore 2006; Cawoy et al. 2011). Although first commercial biopesticide "Sporeine" based on *Bacillus thuringiensis* (Bt) was used in France in 1938, effective development and commercialization of microbial biopesticides was started by the development and registration of "Thuricide," as first registered commercial product of *B. thuringiensis* in 1961. Biopesticides are target-specific, eco-friendly, and effective solutions for the eradication of phytopathogenic diseases (Gupta and Dikshit 2010; Kumar and Singh 2015). Biopesticides are used to protect various crops (cereals, legumes, fruits, flowers, and ornamentals) from phytopathogenic diseases. Presently biopesticides are gaining much attention and used as better alternatives of chemicals all over the world (Mishra et al. 2015).

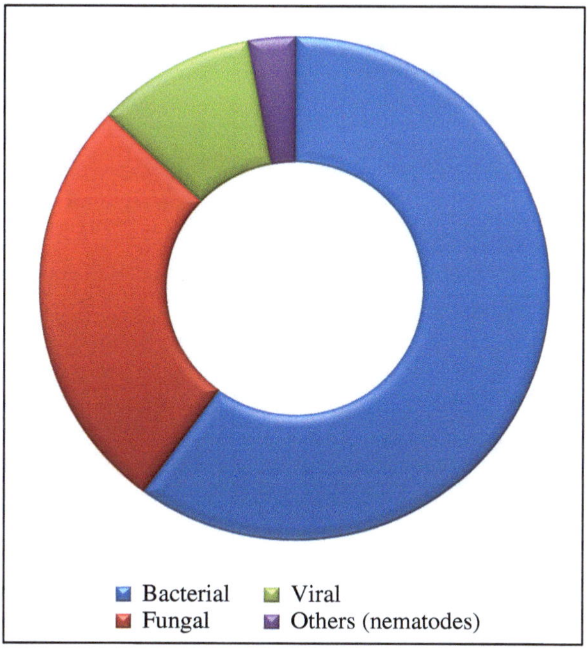

Fig. 6.4 Market share of various types of biopesticides. (CPL Business Consultants 2010)

In the global market, bacterial biopesticides have higher share in comparison with other microbe-based products (Fig. 6.4). Most of the bacterial strains exploited as biopesticides belong to the genera *Bacillus* and *Pseudomonas* (Fravel 2005). Among various biopesticides, *Bacillus*-based products are considered as very popular and have been widely used (Schallmey et al. 2004). The most successful bacterial biopesticides are *B. thuringiensis* (Bt) based which represent about 95% of total market of biocontrol products (Bravo et al. 2011). Bt is specifically used for insect pest control (Sanchis and Bourguet 2008). Other species of *Bacillus* used as biopesticides are *B. subtilis*, *B. licheniformis*, and *B. pumilus* (Ongena and Jacques 2008).

Pseudomonas is also used as efficient and popular biopesticide (Ganeshan and Kumar 2006). Among all *Pseudomonas* sp., fluorescent pseudomonads are reported as important biopesticides in crop protection (Wilson et al. 1992; Fravel 2005). Besides *Bacillus* and *Pseudomonas*, some other bacterial genera being used as biopesticides are *Agrobacterium*, *Azospirillum*, *Burkholderia*, and *Streptomyces*, which are commercially available in the market (Nakkeeran et al. 2005). Currently biopesticides are being used all over the world, and their market share by geographical region is shown in Fig. 6.5. According to global market report of BCC research, the total sale of biopesticide was $1.2 billion in 2008 and $1.6 billion in 2009 (BCC Research 2010). According to PR Newswire (2017), the global biopesticide market reached $3.7 billion in 2015 and nearly $4.0 billion in 2016. The market is projected to gain at a CAGR of 14.1% from 2016 to 2021 and should reach $7.7 billion by 2021 (PR Newswire 2017).

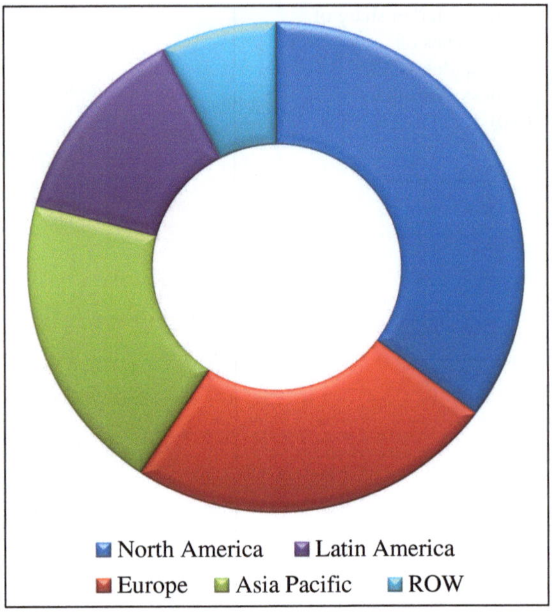

Fig. 6.5 Market share of biopesticides in various regions of world. (BCC Research 2014)

Future Roles of PGPR

To fulfil the food requirements of the growing world population, there is a need of a suitable, economic, and beneficial approach which should not have side effects on the environment. For the safe and sustainable agricultural future, there is a need to produce sufficient amount of food crops with three main parameters: disease resistance, stress tolerance, and high nutrient content (Arora et al. 2016a). PGPR application can be an effective method for approaching this goal. PGPR impart a very big role in increasing crop productivity by various beneficial processes, e.g., biofertilization, biocontrol, bioremediation, and biofortification, which are important to achieve agricultural sustainability for a better future (Rana et al. 2015; Shaikh and Saraf 2017). Although PGPR are being used for crop production, their contribution either as biofertilizers or biopesticides is very less at global level, hence by applying various advanced techniques such as genomics, proteomics, metabolomics, and nanotechnology, we can improve upon bioinoculant technology for better and reliable products (Vejan et al. 2016; Gouda et al. 2018). Various issues related with plant-microbe interactions require more concern for better future approaches such as advanced culture techniques for microbes (independent culture techniques), diversity analysis of microbes, study of mechanism and genetic constructions of plants and microbes, and the interaction of PGPR with plants at metabolic and molecular levels. Some advanced inoculation methods can also be used for improving crop productivity, and in this regard Arora and Mishra (2016) suggested the application of secondary metabolite-based formulations for high crop productivity. Similarly, Timmusk et al. (2017) suggested that implementation of metabolomics of microbiomes has potent capability to advance the quality and applications of bioinoculants in agriculture. Recently, Mitter et al. (2017) reported a new technique to inoculate bioinoculants in plants: inoculating flowers to develop progeny seeds which modify plant microbiomes and characters as per requirements. It is also suggested that the integrated use of PGPR with other organic farming techniques (Maheshwari et al. 2012b; Nazir et al. 2017) and seed biopriming techniques (Mahmood et al. 2016) can also be focused for high crop productivity. Only exploration and selection of efficient PGPR is not sufficient but also their proper registration, regulation, and effective delivery system are required (Arora et al. 2016b). There is necessity to attract the farmers toward bioinoculants by selecting suitable and reliable products; hence we can gain their faith in agrobiologicals.

Climate change is responsible for various problems and global warming is one of them. Global warming causes reduction in crop production by causing hormonal imbalances which has negative impacts on agriculture and global food security (Lobell and Gourdji 2012). Global warming is increasing due to high emission of carbon dioxide (CO_2) in the atmosphere (Anderson et al. 2016). Soil systems work as the largest sink of CO_2, but due to high human interference, their deterioration takes place which leads to the high release of CO_2 in the atmosphere causing enhanced global warming (Salam and Noguchi 2005). It is suggested by various researchers that applications of microbial inoculants in agricultural crop productions can be used as sustainable tool to overcome the negative effect of climate change as well as global warming (Dimkpa et al. 2009; Tewari and Arora 2013). It is reported that maximum parts (90%) of plant biomass are derived from CO_2 assimilation by photosynthetic activity (Long et al. 2006) and photosynthetic rate of plants is increased upon the application of biofertilizers (PGPR) (Mia and Shamsuddin 2010; Mahanty et al. 2016). Hence it can be concluded that due to PGPR inoculation, plants consume more CO_2 which reduce the atmospheric CO_2 level and global warming (Pendall et al. 2004; Nie et al. 2015). According to Nie et al. (2015), the rate of microbial respiratory carbon loss is also reduced on the application of PGPR which decreases global warming. Higher application of microbial inoculants in agriculture minimizes the chemical load which also reduces the risk of global warming. It is also reported that stress-tolerant PGPR significantly help to maintain the soil fertility and reclamation of wastelands (Hrynkiewicz and Baum 2011; Mishra et al. 2017b). It can be considered that due to the reclamation of wastelands, sinks of CO_2 are increased that cause reduction of atmospheric CO_2 and global warming.

Conclusion

Extension of agriculture services in an eco-friendly manner is a key issue in present-day era of increasing population and climate change. To attain surplus agricultural production, farmers are still dependent on synthetic chemicals. However, indiscriminate use of these harmful chemicals has only raised the problems. Soil microbes are always considered as chief components of soil fertility. In this context the use of PGPR for enhancing plant growth, preventing deadly diseases, alleviating abiotic and biotic stresses, and restoring soil health can be very useful. PGPR-based products such as biofertilizers and biopesticides are already being used at the global level and found to be suitable alternatives to dangerous chemicals. There is a need to explore PGPR now for purposes such as stress management, bioremediation, and combating climate change (global warming). In recent years, considerable developments are visible in the field of plant-microbe interactions affirming their role in solving key environmental problems. However, in agroecosystems, contribution of PGPR is not fully exploited in accordance with economic and social needs. The possibilities for their wide application in agriculture can increase by the advent of newer techniques. A number of PGPR strains showing multifarious plant beneficial activities are now known, but very few have been formulated in the form of bioinoculants. Hence a strong perspective focused on their extended use in remediation of soil problems and diverse applicability in agroecosystems is required.

References

Abd-Alla, M. H., Elenany, A. E., Ramadan, T., Zohri, E. M., & Nafady, I. M. (2017). Nodulation and nitrogen fixation of some wild legumes from differing habitats in Egypt. *European Journal of Biological Research, 7*(1), 9–21.

Abdel Ghany, T. M., Alawlaqi, M. M., & Al Abboud, M. A. (2013). Role of biofertilizers in agriculture: A brief review. *Mycopathology, 11*(2), 95–101.

Abeles, F. B., Morgan, P. W., & Saltveit, M. E. J. (1992). *Ethylene in plant biology* (pp. 26–55). San Diego: Academic.

Abou-Shanab, R. A. I., Angle, J. S., & Chaney, R. L. (2006). Bacterial inoculants affecting nickel uptake by *Alyssum murale* from low, moderate and high Ni soils. *Soil Biology and Biochemistry, 38*, 2882–2889.

Adesemoye, A. O., Obini, M., & Ugoji, E. O. (2008). Comparison of plant growth promotion with *Pseudomonas aeruginosa* and *Bacillus subtilis* in three vegetables. *Brazilian Journal of Microbiology, 39*, 423–442.

Aeron, A., Pandey, P., Kumar, S., & Maheshwari, D. K. (2011). Emerging role of plant growth promoting rhizobacteria. In D. K. Maheshwari (Ed.), *Bacteria in agrobiology: Crop ecosystem* (pp. 1–26). Berlin/Heidelberg: Springer.

AgriInfo.in. (2015). *Role of biofertilizers in soil fertility and agriculture.* http://agriinfo.in/?page=topic&superid=5&topicid=176

Ahemad, M., & Khan, M. S. (2012). Productivity of greengram in tebuconazole-stressed soil, by using a tolerant and plant growth-promoting *Bradyrhizobium* sp. MRM6 strain. *Acta Physiologiae Plantarum, 34*, 245–254.

Ahemad, M., & Kibret, M. (2014). Mechanisms and applications of plant growth promoting rhizobacteria: Current perspective. *Journal of King Saud University, 26*(1), 1–20.

Ahmad, F., Ahmad, I., & Khan, M. S. (2008). Screening of free-living rhizospheric bacteria for their multiple plant growth promoting activities. *Microbiological Research, 163*, 173–181.

Ahmad, M., Zahir, Z. A., Khalid, M., Nazli, F., & Arshad, M. (2013). Efficacy of *Rhizobium* and *Pseudomonas* strains to improve physiology, ionic balance and quality of Mung bean under salt-affected conditions on farmer's fields. *Plant Physiology and Biochemistry, 63*, 170–176.

Ali, S. S., & Vidhale, N. N. (2013). Bacterial siderophore and their application: A review. *International Journal of Current Microbiology and Applied Sciences, 2*, 303–312.

Ali, S. Z., Sandhya, V., Grover, M., Linga, V. R., & Bandi, V. (2011). Effect of inoculation with a thermotolerant plant growth promoting *Pseudomonas putida* strain AKMP7 on growth of wheat (*Triticum* spp.) under heat stress. *Journal of Plant Interactions, 6*(4), 239–246.
Amora-Lazcano, E., Guerrero-Zúñiga, L. A., Rodriguez-Tovar, A., Rodriguez-Dorantes, A., & Vasquez-Murrieta, M. S. (2010). Rhizospheric plant-microbe interactions that enhance the remediation of contaminated soils. In A. Méndez-Vilas (Ed.), *Current research, technology and education topics in applied microbiology and microbial biotechnology* (pp. 251–256). Badajoz: FORMATEX.
Anandham, R., Janahiraman, V., Gandhi, P. I., Kwon, S. W., Chung, K. Y., Han, G. H., Choi, J. H., & Sa, T. M. (2014). Early plant growth promotion of maize by various sulfur oxidizing bacteria that uses different thiosulfate oxidation pathway. *African Journal of Microbiology Research, 8*(1), 19–27.
Anderson, T. R., Hawkins, E., & Jones, P. D. (2016). CO2, the greenhouse effect and global warming: From the pioneering work of Arrhenius and Callendar to today's earth system models. *Endeavour, 40*, 178–187.
Antoun, H., & Kloepper, J. W. (2001). Plant growth promoting rhizobacteria. In S. Brenner & J. F. Miller (Eds.), *Encyclopedia of genetics* (pp. 1477–1480). New York: Academic.
Arora, N. K. (Ed.). (2013). *Plant microbe symbiosis: Fundamental and advances* (Vol. 459). New Delhi: Springer.
Arora, N. K. (Ed.). (2015). *Plant microbe symbiosis: Applied facets* (p. 381). New Delhi: Springer.
Arora, N. K., & Mishra, J. (2016). Prospecting the roles of metabolites and additives in future bioformulations for sustainable agriculture. *Applied Soil Ecology, 107*, 405–407.
Arora, N. K., & Verma, M. (2017). Modified microplate method for rapid and efficient estimation of siderophore produced by bacteria. *3 Biotech, 7*, 381.
Arora, N. K., Kang, S. C., & Maheshwari, D. K. (2001). Isolation of siderophore producing strains of Rhizobium meliloti and their biocontrol potential against *Macrophomina phaseolina* that causes charcoal rot of groundnut. *Current Science, 81*, 673–677.
Arora, N. K., Kim, M. J., Kang, S. C., & Maheshwari, D. K. (2007). Role of chitinase and β-1,3-glucanase activities produced by a fluorescent pseudomonad and in vitro inhibition of *Phytophthora capsici* and *Rhizoctonia solani*. *Canadian Journal of Microbiology, 53*, 207–212.
Arora, N. K., Khare, E., Verma, A., & Sahu, R. K. (2008). *In vivo* control of *Macrophomina phaseolina* by a chitinase and β-1,3-glucanase-producing pseudomonad NDN1. *Symbiosis, 46*, 129–135.
Arora, N. K., Tewari, S., Singh, S., & Lal, N. (2012). PGPR for protection of plant health under saline conditions. In D. K. Maheshwari (Ed.), *Bacteria in agrobiology* (pp. 239–258). Dordrecht: Springer.
Arora, N. K., Tewari, S., & Singh, R. (2013). Multifaceted plant-associated microbes and their mechanisms diminish the concept of direct and indirect PGPRs. In N. K. Arora (Ed.), *Plant microbe symbiosis: Fundamentals and advances* (pp. 411–449). New Delhi: Springer.
Arora, N. K., Mehnaz, S., & Balestrini, R. (Eds.). (2016a). *Bioformulations: For sustainable agriculture* (p. 299). Lucknow: Springer.
Arora, N. K., Verma, M., Prakash, J., & Mishra, J. (2016b). Regulation of biopesticides: Global concerns and policies. In N. K. Arora, S. Mehnaz, & R. Balestrini (Eds.), *Bioformulations: For sustainable agriculture* (pp. 283–299). New Delhi: Springer.
Arora, N. K., Verma, M., & Mishra, J. (2017). Rhizobial bioformulation: Past, present and future. In S. Mehnaz (Ed.), *Rhizotrophs: Plant growth promotion to bioremediation* (pp. 69–99). Singapore: Springer.
Arzanesh, M. H., Alikhani, H. A., Khavazi, K., Rahimian, H. A., & Miransari, M. (2011). Wheat (*Triticum aestivum* L.) growth enhancement by *Azospirillum* sp. under drought stress. *World Journal of Microbiology and Biotechnology, 27*, 197–205.
Asaf, S., Khan, M. A., Khan, A. L., Waqas, M., Shahzad, R., Kim, A. Y., Kang, S. M., & Lee, I. J. (2017). Bacterial endophytes from arid land plants regulate endogenous hormone content and promote growth in crop plants: An example of *Sphingomonas* sp. and *Serratia marcescens*. *Journal of Plant Interactions, 12*, 31–38.

Asari, S. Y. (2015). Studies on plant-microbe interaction to improve stress tolerance in plants for sustainable agriculture. *Diss. (sammanfattning/summary) Uppsala: Sveriges lantbruksuniv, Acta Universitatis agriculturae Sueciae, 76*, 1652–6880.

Asghar, H. N., Zahir, Z. A., Arshad, M., & Khaliq, A. (2004). Relationship between in-vitro production of auxins by rhizobacteria and their growth-promoting activities in *Brassica juncea* L. *Biology and Fertility of Soils, 35*, 231–237.

Awad, N. M., Abd El-Kader, M. A., Alva, A. K., & Narale, S. H. (2011). Effects of nitrogen fertilization and soil inoculation of sulfur oxidizing or nitrogen-fixing bacteria on onion plant growth and yield. *International Journal of Agronomy, 2011*, 1–6.

Aznar, A., & Dellagi, A. (2015). New insights into the role of siderophores as triggers of plant immunity: What can we learn from animals? *Journal of Experimental Botany, 66*(11), 3001–3010.

Bagyalakshmi, T. A., Ramesh, V., Arivudainambi, U. S. E., & Rajendran, A. (2012). A novel endophytic fungus *Pestalotiopsis* sp. inhibiting *Pinus canariensis* with antibacterial and antifungal potential. *International Journal of Advanced Life Sciences, 1*, 1–7.

Bahadur, I., Maurya, B. R., Kumar, A., Meena, V. S., & Raghuwanshi, R. (2016). Towards the soil sustainability and potassium-solubilizing microorganisms. In V. S. Meena, B. R. Maurya, J. P. Verma, & R. S. Meena (Eds.), *Potassium solubilizing microorganisms for sustainable agriculture* (pp. 255–266). New Delhi: Springer.

Bais, H. P., Weir, T. L., Perry, L. G., Gilroy, S., & Vivanco, J. M. (2006). The role of root exudates in rhizosphere interactions with plants and other organisms. *Annual Review of Plant Biology, 57*, 233–266.

Bakker, P. A. H. M., Doornbos, R. F., Zamioudis, C., Berendsen, R. L., & Pieterse, C. M. J. (2013). Induced systemic resistance and the rhizosphere microbiome. *Plant Pathology Journal, 29*, 136–143.

Bakshi, A., Shemansky, J. M., Chang, C., & Binder, B. M. (2015). History of research on the plant hormone ethylene. *Journal of Plant Growth Regulation, 34*, 809–827.

Ballhorn, D. J., Elias, J. D., Balkan, M. A., Fordyce, R. F., & Kennedy, P. G. (2017). Colonization by nitrogen-fixing Frankia bacteria causes short-term increases in herbivore susceptibility in red alder (*Alnus rubra*) seedlings. *Oecologia, 184*(2), 497–506.

Bashan, Y. (1998). Inoculants of plant growth-promoting bacteria for use in agriculture. *Biotechnology Advances, 16*(4), 729–770.

Batjes, N. H. A. (1997). World data set of derived soil properties by FAO UNESCO soil unit for global modeling. *Soil Use and Management, 13*, 9–16.

Batool, N., Ilyas, N., & Shahzad, A. (2014). Role of plant growth promoting rhizobacteria as ameliorating agent in saline soil. *Pure and Applied Biology, 3*(4), 167.

BCC Research. (2010). *Biopesticides: The global market*. Available at: http://www.bccresearch.com/market-research/chemicals/biopesticides-marketchm029c.html

BCC Research. (2014). *Global market for biopesticides*. Wellesley: Market Research Reports. Available at: https://www.bccresearch.com/market-research/chemicals/biopesticides-market-chm029c.html

Begon, M., Harper, J. L., & Townsend, C. R. (1990). *Ecology: Individuals, populations and communities* (2nd ed.p. 945). Boston: Blackwell Scientific Publications.

Beneduzi, A., Ambrosini, A., & Passaglia, L. M. P. (2012). Plant growth-promoting rhizobacteria (PGPR): Their potential as antagonists and biocontrol agents. *Genetics and Molecular Biology, 35*(4), 1044–1051.

Bergman, B., Rai, A. N., Rasmussen, U., Elmerich, C., & Newton, W. E. (2007). *Cyanobacterial associations, associative and endophytic nitrogen-fixing bacteria and cyanobacterial associations* (pp. 257–301). Dordrecht: Springer.

Berrada, H., & Fikri-Benbrahim, K. (2014). Taxonomy of the rhizobia: Current perspectives. *British Microbiology Research Journal, 4*, 616–639.

Bhalerao, T. S. (2012). Bioremediation of endosulfan-contaminated soil by using bioaugmentation treatment of fungal inoculant *Aspergillus niger*. *Turkish Journal of Biology, 36*(5), 561–567.

Bhardwaj, D., Ansari, M. W., Sahoo, R. K., & Tuteja, N. (2014). Biofertilizers function as key player in sustainable agriculture by improving soil fertility, plant tolerance and crop productivity. *Microbial Cell Factories, 13*(1), 66.

Bhattacharyya, P. N., & Jha, D. K. (2012). Plant growth-promoting rhizobacteria (PGPR): Emergence in agriculture. *World Journal of Microbiology and Biotechnology, 28*, 1327–1350.

Bisht, S., Pandey, P., Bhargava, B., Sharma, S., Kumar, V., & Sharma, K. (2015). Bioremediation of polyaromatic hydrocarbons (PAHs) using rhizosphere technology. *Brazilian Journal of Microbiology, 46*(1), 7–21.

Boiero, L., Perrig, D., Masciarelli, O., Penna, C., Cassan, F., & Luna, V. (2007). Phytohormone production by three strains of *Bradyrhizobium japonicum* and possible physiological and technological implications. *Applied Microbiology and Biotechnology, 74*(4), 874–880.

Boiteau, R. M., Mende, D. R., Hawco, N. J., McIlvin, M. R., Fitzsimmons, J. N., Saito, M. A., Sedwick, P. N., DeLong, E. F., & Repeta, D. J. (2016). Siderophore-based microbial adaptations to iron scarcity across the eastern Pacific Ocean. *Proceedings of the National Academy of Science, 113*(50), 14237–14242.

Bömke, C., & Tudzynski, B. (2009). Diversity, regulation, and evolution of the gibberellin biosynthetic pathway in fungi compared to plants and bacteria. *Phytochemistry, 70*(15), 1876–1893.

Bottini, R., Cassán, F., & Piccoli, P. (2004). Gibberellin production by bacteria and its involvement in plant growth promotion and yield increase. *Applied Microbiology and Biotechnology, 65*, 497–503.

Boukerma, L., Benchabane, M., Charif, A., & Khélifi, L. (2017). Activity of plant growth promoting rhizobacteria (PGPRs) in the biocontrol of tomato Fusarium wilt. *Plant Protection Science, 53*, 78–84.

Bradford, M. A., Davies, C. A., Frey, S. D., Maddox, T. R., Melillo, J. M., Mohan, J. E., Reynolds, J. F., Treseder, K. K., & Wallenstein, M. D. (2008). Thermal adaptation of soil microbial respiration to elevated temperature. *Ecology Letters, 11*, 1316–1327.

Bravo, A., Likitvivatanavong, S., Gill, S., & Soberon, M. (2011). *Bacillus thuringiensis*: A story of a successful bio-insecticide. *Insect Biochemistry and Molecular Biology, 41*, 423–431.

Burd, G. I., Dixon, D. G., & Glick, B. R. (2000). Plant growth-promoting bacteria that decrease heavy metal toxicity in plants. *Canadian Journal of Microbiology, 46*(3), 237–245.

Campbell, R. (1985). *Plant microbiology* (p. 191). Baltimore: Edward Amold.

Cassán, F., Perrig, D., Sgroy, V., Masciarelli, O., Penna, C., & Luna, V. (2009). *Azospirillum brasilense* Az39 and *Bradyrhizobium japonicum* E109, inoculated singly or in combination, promote seed germination and early seedling growth in corn (*Zea mays* L.) and soybean (*Glycine max* L.). *European Journal of Soil Biology, 45*, 28–35.

Cassán, F., Vanderleyden, J., & Spaepen, S. (2014). Physiological and agronomical aspects of phytohormone production by model plant-growth-promoting rhizobacteria (PGPR) belonging to the genus *Azospirillum*. *Journal of Plant Growth Regulation, 33*, 440–459.

Cawoy, H., Bettiol, W., Fickers, P., & Ongena, M. (2011). *Bacillus* based biological control of plant diseases. In M. Stoytcheva (Ed.), *Pesticides in the modern world-pesticides use and management* (pp. 273–302). Rijeka: InTech.

Chakraborty, U., Chakraborty, B. N., Chakraborty, A. P., & Dey, P. L. (2013). Water stress amelioration and plant growth promotion in wheat plants by osmotic stress tolerant bacteria. *World Journal of Microbiology and Biotechnology, 29*(5), 789–803.

Choudhary, D. K., & Johri, B. N. (2008). Interactions of *Bacillus* spp. and plants -with special reference to induced systemic resistance (ISR). *Microbiological Research, 164*, 493–513.

Chung, H., Park, M., Madhaiyan, M., Seshadri, S., Song, J., Cho, H., & Sa, T. (2005). Isolation and characterization of phosphate solubilizing bacteria from the rhizosphere of crop plants of Korea. *Soil Biology and Biochemistry, 37*, 1970–1974.

Compant, S., Duffy, B., Nowak, J., Clément, C., & Barka, E. A. (2005). Use of plant growth-promoting bacteria for biocontrol of plant diseases: Principles, mechanisms of action, and future prospects. *Applied and Environmental Microbiology, 71*, 4951–4959.

Corbineau, F., Xia, Q., Bailly, C., & El-Maarouf-Bouteau, H. (2014). Ethylene, a key factor in the regulation of seed dormancy. *Frontiers in Plant Science, 5*, 539.

Coutinho, F. P., Felix, W. P., & Yano-Melo, A. M. (2012). Solubilization of phosphates in vitro by *Aspergillus* spp. and *Penicillium* spp. *Ecological Engineering, 42*, 85–89.
Coutinho, B. G., Licastro, D., Mendonça-Previato, L., Cámara, M., & Venturi, V. (2015). Plant-influenced gene expression in the rice endophyte *Burkholderia kururiensis* M130. *Molecular Plant-Microbe Interactions, 28*, 10–21.
CPL Business Consultants. (2010). *The 2010 worldwide biopesticides market summary* (Vol. 1). Wallingford: CAB International Centre.
Crannell, W. K., Tanaka, Y., & Myrold, D. D. (1994). Calcium and pH interaction on root nodulation of nursery-grown red alder (*Alnus rubra* Bong.) seedlings by *Frankia*. *Soil Biology and Biochemistry, 26*, 607–614.
Da Mota, F. F., Gomes, E. A., & Seldin, L. (2008). Auxin production and detection of the gene coding for the auxin efflux carrier (AEC) protein in *Paenibacillus polymyxa*. *The Journal of Microbiology, 56*, 275–264.
Dardanelli, M. S., Carletti, S. M., Paulucci, N. S., Medeot, D. B., Rodriguez Caceres, E. A., Vita, F. A., Bueno, M., Fumero, M. V., & Garcia, M. B. (2010). Benefits of plant growth-promoting rhizobacteria and rhizobia in agriculture. In D. K. Maheshwari (Ed.), *Plant growth and health promoting bacteria, Microbiology monographs* (Vol. 18, pp. 1–20). Berlin: Springer.
Dary, M., Perez, M. A. C., Palomares, A. J., & Pajuelo, E. (2010). In situ phytostabilisation of heavy metal polluted soils using *Lupinus luteus* inoculated with metal resistant plant-growth promoting rhizobacteria. *Journal of Hazardous Materials, 177*, 323–330.
de Bruijn, I., de Kock, M. J. D., Yang, M., de Waard, P., van Beek, T. A., & Raaijmakers, J. M. (2007). Genome-based discovery, structure prediction and functional analysis of cyclic lipopeptide antibiotics in *Pseudomonas* species. *Molecular Microbiology, 63*, 417–428.
De Souza, J. T. A., Arnould, C., Deulvot, C., Lemanceau, P., Gianinazzi-Pearson, V., & Raaijmakers, J. M. (2003). Effect of 2,4-diacetylphloroglucinol on *Pythium*: Cellular responses and variation in sensitivity among propagules and species. *Phytopathology, 93*, 966–975.
De Souza, R., Ambrosini, A., & Passaglia, L. M. (2015). Plant growth-promoting bacteria as inoculants in agricultural soils. *Genetics and Molecular Biology, 38*(4), 401–419.
Delvasto, P., Ballester, A., Muñoz, J. A., González, F., Blázquez, M. L., Igual, J. M., & Valverde, A. (2009). Mobilization of phosphorus from iron ore by the bacterium *Burkholderia caribensis* FeGL03. *Minerals Engineering, 22*(1), 1–9.
Demissie, S., Muleta, D., & Berecha, G. (2013). Effect of phosphate solubilizing bacteria on seed germination and seedling growth of faba bean (*Vicia fabae* L.). *International Journal of Agricultural Research, 8*, 123–136.
Dhanya, R. P., & Adeline, C. S. (2014). A study on the biocontrol of phytopathogens of *Vigna radiata* using *Pseudomonas fluorescence* in sustainable agriculture. *International Journal of Current Microbiology and Applied Sciences, 3*(10), 114–120.
Diagne, N., Arumugam, K., Ngom, M., Nambiar-Veetil, M., Franche, C., Narayanan, K. K., & Laplaze, L. (2013). Use of *Frankia* and actinorhizal plants for degraded lands reclamation. *BioMed Research International, 2013*, 948258.
Dimkpa, C. O., Merten, D., Svatos, A., Büchel, G., & Kothe, E. (2009). Siderophores mediate reduced and increased uptake of cadmium by Streptomyces tendae F4 and sunflower (*Helianthus annuus*), respectively. *Journal of Applied Microbiology, 107*, 1687–1696.
Dixon, R., & Kahn, D. (2004). Genetic regulation of biological nitrogen fixation. *Nature Reviews. Microbiology, 2*, 621–631.
Dorjey, S., Dolkar, D., & Sharma, R. (2017). Plant growth promoting rhizobacteria *Pseudomonas*: A review. *International Journal of Current Microbiology and Applied Sciences, 6*(7), 1335–1344.
du Jardin, P. (2015). Plant biostimulants: Definition, concept, main categories and regulation. *Scientia Horticulturae, 196*, 3–14.
Duc, L., Noll, M., Meier, E., Burgmann, H., & Zeyer, J. (2009). High diversity of diazotrophs in the forefield of a receding alpine glacier. *Microbial Ecology, 57*, 179–190.
Dutta, S. (2015). Biopesticides: An ecofriendly approach for pest control. *World Journal of Pharmacy and Pharmaceutical Sciences, 4*(6), 250–265.

Egamberdieva, D. (2012). *Pseudomonas chlororaphis*: A salt-tolerant bacterial inoculant for plant growth stimulation under saline soil conditions. *Acta Physiologiae Plantarum, 34*(2), 751–756.

Eslamyan, L., Alipour, Z. T., Beidokhty, S. R., & Sobhanipour, A. (2013). *Pseudomonas fluorescens* and sulfur application affect rapeseed growth and nutrient uptake in calcareous soil. *International Journal of Agriculture and Crop Sciences, 5*(1), 39–43.

Etesami, H., Alikhani, H. A., & Hosseini, H. M. (2015). Indole-3-acetic acid (IAA) production trait, a useful screening to select endophytic and rhizosphere competent bacteria for rice growth promoting agents. *Methods X, 2*, 72–78.

Etesami, H., Emami, S., & Alikhani, H. A. (2017). Potassium solubilizing bacteria (KSB): Mechanisms, promotion of plant growth, and future prospects – A review. *Journal of Soil Science and Plant Nutrition, 17*(4).

FAO WHO. (2002). *Human vitamin and mineral requirements*. Food and agriculture organization of the United Nations, Bangkok, Thailand. ISBN:1014–9228.

Fatima, Z., Saleemi, M., Zia, M., Sultan, T., Aslam, M., & Riaz-ur-Rehman, C. M. F. (2009). Antifungal activity of plant growth-promoting rhizobacteria isolates against *Rhizoctonia solani* in wheat. *African Journal of Biotechnology, 8*, 219–225.

Figueiredo, M. V. B., Seldin, L., Araujo, F. F., & Mariano, R. L. R. (2011). Plant growth promoting rhizobacteria: Fundamentals and applications. In D. K. Maheshwari (Ed.), *Plant growth and health promoting bacteria* (pp. 21–42). Berlin/Heidelberg: Springer.

Figueroa-López, A. M., Cordero-Ramírez, J. D., Martínez-Álvarez, J. C., López-Meyer, M., Lizárraga-Sánchez, G. J., Félix-Gastélum, R., Castro-Martínez, C., & Maldonado-Mendoza, I. E. (2016). Rhizospheric bacteria of maize with potential for biocontrol of *Fusarium verticillioides*. *Springer Plus, 5*(1), 330.

Franche, C., Bogusz, D., Perotto, S., & Baluska, F. (2011). *Signalling and communication in actinorhizal symbiosis, Signalling and communication in plant symbiosis* (pp. 73–92). Berlin: Springer.

Franco-Correa, M., Quintanaa, A., Duquea, C., Suarez, C., Rodríguez, M. X., & Barea, J. (2010). Evaluation of actinomycete strains for key traits related with plant growth promotion and mycorrhiza helping activities. *Applied Soil Ecology, 45*, 209–217.

Fravel, D. R. (2005). Commercialization and implementation of biocontrol. *Annual Review of Phytopathology, 43*, 337–359.

Gadd, G. M. (2010). Metals, minerals and microbes: Geomicrobiology and bioremediation. *Microbiology, 156*(3), 609–643.

Gahan, J., & Schmalenberger, A. (2014). The role of bacteria and mycorrhiza in plant sulfur supply. *Frontiers in Plant Science, 5*, 723.

Gamalero, E., Lingua, G., Caprì, F. G., Fusconi, A., Berta, G., & Lemanceau, P. (2004). Colonization pattern of primary tomato roots by *Pseudomonas fluorescens* A6RI characterized by dilution plating, flow cytometry, fluorescence, confocal and scanning electron microscopy. *FEMS Microbiology Ecology, 48*, 79–87.

Gamit, D. A., & Tank, S. K. (2014). Effect of siderophore producing microorganism on plant growth of *Cajanus cajan* (Pigeon pea). *International Journal of Pure and Applied Microbiology, 4*(1), 20–27.

Ganeshan, G., & Kumar, M. A. (2006). *Pseudomonas fluorescens*, a potential bacterial antagonist to control plant diseases. *Journal of Plant Interactions, 1*(3), 123–134.

García-Fraile, P., Menéndez, E., & Rivas, R. (2015). Role of bacterial biofertilizers in agriculture and forestry. *AIMS Bioengineering, 2*(3), 183–205.

García-Gutiérrez, L., Zeriouh, H., Romero, D., Cubero, J., Vicente, A., & Pérez-García, A. (2013). The antagonistic strain *Bacillus subtilis* UMAF6639 also confers protection to melon plants against cucurbit powdery mildew by activation of jasmonate-and salicylic acid-dependent defence responses. *Microbial Biotechnology, 6*(3), 264–274.

Glick, B. R. (2004). Bacterial ACC deaminase and the alleviation of plant stress. *Advances in Applied Microbiology, 56*, 291–312.

Glick, B. R. (2010). Using soil bacteria to facilitate phytoremediation. *Biotechnology Advances, 28*, 367–374.

Glick, B. R. (2012). Plant growth-promoting bacteria: Mechanisms and applications. *Scientifica (Cairo), 2012*, 963401.
Glick, B. R. (2014). Bacteria with ACC deaminase can promote plant growth and help to feed the world. *Microbiological Research, 169*(1), 30–39.
Glick, B. R., Karaturovic, D. M., & Newell, P. C. (1995). A novel procedure for rapid isolation of plant growth promoting pseudomonads. *Canadian Journal of Microbiology, 41*, 533–536.
Glick, B. R., Penrose, D. M., & Li, J. (1998). A model for the lowering of plant ethylene concentrations by plant growth-promoting bacteria. *Journal of Theoretical Biology, 190*(1), 63–68.
Glick, B. R., Cheng, Z., Czarny, J., & Duan, J. (2007). Promotion of plant growth by ACC deaminase-containing soil bacteria. *European Journal of Plant Pathology, 119*, 329–339.
Gontia-Mishra, I., Sapre, S., Sharma, A., & Tiwari, S. (2016). Amelioration of drought tolerance in wheat by the interaction of plant growth promoting rhizobacteria. *Plant Biology, 18*(6), 992–1000.
Gopalakrishnan, S., Sathya, A., Vijayabharathi, R., Varshney, R. K., Gowda, C. L., & Krishnamurthy, L. (2015). Plant growth promoting rhizobia: Challenges and opportunities. *3 Biotech, 5*, 355–377.
Goteti, P. K., Leo, D. A. E., Desai, S., & Ahmed, S. M. H. (2013). Prospective zinc solubilising bacteria for enhanced nutrient uptake and growth promotion in Maize (*Zea mays* L.). *International Journal of Microbiology, 2013*, 1–7.
Gouda, S., Kerry, R. G., Das, G., Paramithiotis, S., Shin, S. H., & Patra, J. K. (2018). Revitalization of plant growth promoting rhizobacteria for sustainable development in agriculture. *Microbiological Research, 206*, 131–140.
Gray, E. J., & Smith, D. L. (2005). Intracellular and extracellular PGPR: Commonalities and distinctions in the plant-bacterium signaling processes. *Soil Biology and Biochemistry, 37*, 395–412.
Grayson, M. (2013). Agriculture and drought. *Nature, 501*, S1.
Grayston, S. J., & Germida, J. J. (1991). Sulfur-oxidizing bacteria as plant growth promoting rhizobacteria for canola. *Canadian Journal of Microbiology, 37*, 521–529.
Griffiths, B. S., & Philippot, L. (2012). Insights into the resistance and resilience of the soil microbial community. *FEMS Microbiology Reviews, 37*, 112–129.
Grover, M., Ali, S. Z., Sandhya, V., Rasul, A., & Venkateswarlu, B. (2010). Role of microorganisms in adaptation of agriculture crops to abiotic stresses. *World Journal of Microbiology and Biotechnology, 27*, 1231–1240.
Guo, Q., Dong, W., Li, S., Lu, X., Wang, P., Zhang, X., Wang, Y., & Ma, P. (2013). Fengycin produced by *Bacillus subtilis* NCD-2 plays a major role in biocontrol of cotton seedling damping-off disease. *Microbiological Research, 169*, 533–540.
Gupta, S., & Dikshit, A. K. (2010). Biopesticides: An eco-friendly approach for pest control. *Journal of Biopesticides, 3*(1), 186–188.
Gusain, Y. S., Kamal, R., Mehta, C. M., Singh, U. S., & Sharma, A. K. (2015). Phosphate solubilizing and indole-3-acetic acid producing bacteria from the soil of Garhwal Himalaya aimed to improve the growth of rice. *Journal of Environmental Biology, 36*, 301–307.
Gutierrez-Luna, F. M., Lopez-Bucio, J., Tamirano-Hernandez, J., Valencia-Cantero, E., de la Cruz, H. R., & Ias-Rodriguez, L. (2010). Plant growth-promoting rhizobacteria modulate root-system architecture in *Arabidopsis thaliana* through volatile organic compound emission. *Symbiosis, 51*, 75–83.
Gutiérrez-Mañero, F. J., Ramos-Solano, B., Probanza, A., Mehouachi, J., Tadeo, F. R., & Talon, M. (2001). The plant-growth promoting rhizobacteria *Bacillus pumilus* and *Bacillus licheniformis* produce high amounts of physiologically active gibberellins. *Physiologia Plantarum, 111*, 206–211.
Haas, D., & Défago, G. (2005). Biological control of soil-borne pathogens by fluorescent pseudomonads. *Nature Reviews. Microbiology, 3*, 307–319.

Habib, S. H., Kausar, H., & Saud, H. M. (2016). Plant growth-promoting rhizobacteria enhance salinity stress tolerance in okra through ROS-scavenging enzymes. *BioMed Research International, 2016*, 6284547.

Haghighi, B. J., Alizadeh, O., & Firoozabadi, A. H. (2011). The role of plant growth promoting rhizobacteria (PGPR) in sustainable agriculture. *Advances in Environmental Biology, 5*, 3079–3083.

Haidar, R., Fermaud, M., Calvo-Garrido, C., Roudet, J., & Deschamps, A. (2016). Modes of action for biological control of *Botrytis cinerea* by antagonistic bacteria. *Phytopathologia Mediterranea, 55*, 301–322.

Han, H. S., & Lee, K. D. (2005). Phosphate and potassium solubilizing bacteria effect on mineral uptake, soil availability, and growth of eggplant. *Research Journal of Agriculture and Biological Sciences, 1*, 176–180.

Hansda, A., Kumar, V., Anshumali, A., & Usmani, Z. (2014). Phytoremediation of heavy metals contaminated soil using plant growth promoting rhizobacteria (PGPR): A current perspective. *Recent Research in Science and Technology, 6*(1), 131–134.

Hansda, A., Kumar, V., & Anshumali. (2017). Cu-resistant *Kocuria* sp. CRB15: A potential PGPR isolated from the dry tailing of Rakha copper mine. *3 Biotech, 7*(2), 132.

Hassan, T. U., Bano, A., & Naz, I. (2016). Alleviation of heavy metals toxicity by the application of plant growth promoting rhizobacteria and effects on wheat grown in saline sodic field. *International Journal of Phytoremediation, 19*(6), 522–529.

Hayat, Q., Hayat, S., Irfan, M., & Ahmad, A. (2010). Effect of exogenous salicylic acid under changing environment: A review. *Environmental and Experimental Botany, 68*, 14–25.

He, L. Y., Zhang, Y. F., Ma, H. Y., Su, L. N., Chen, Z. J., Wang, Q. Y., Qian, M., & Sheng, X. F. (2010). Characterization of copper resistant bacteria and assessment of bacterial communities in rhizosphere soils of copper-tolerant plants. *Applied Soil Ecology, 44*, 49–55.

Herridge, D. F., Peoples, M. B., & Boddey, R. M. (2008). Global inputs of biological nitrogen fixation in agricultural systems. *Plant and Soil, 311*, 1–18.

Hill, D. S., Stein, J. I., Torkewitz, N. R., Morse, A. M., Howell, C. R., Pachlatko, J. P., Becker, J. O., & Ligon, J. M. (1994). Cloning of genes involved in the synthesis of pyrrolnitrin from *Pseudomonas fluorescens* and role of pyrrolnitrin synthesis in biological control of plant disease. *Applied and Environmental Microbiology, 60*(1), 78–85.

Hiltner, L. (1904). Über neuere erfahrungen und probleme auf dem gebiete der bodenbakteriologie unter besonderer berücksichtigung der gründüngung und brache. *Arb Dtsch Landwirtsch Ges, 98*, 59–78.

Hoflich, G., & Kuhn, G. (1996). Forderung das Wachstums und der Nährstoffaufnahme bei kurziferen Ö l- und Zwischenfruhten durch inokulierte Rhizospherenmikroorganismen. *Zeischrift für Pflanzenernährung und Bodenkunde, 159*, 575–578.

Hofte, M., Boelens, J., & Verstraete, W. (1992). Survival and root colonization of mutants of plant growth promoting pseudomonads affected in siderophore biosynthesis or regulation of siderophore production. *Journal of Plant Nutrition, 15*, 2253–2262.

Hrynkiewicz, K., & Baum, C. (2011). The potential of rhizosphere microorganisms to promote the plant growth in disturbed soils. In A. Malik & E. Grohmann (Eds.), *Environmental protection strategies for sustainable development* (pp. 35–64). Berlin: Springer.

Huang, X. D., El-Alawi, Y., Gurska, J., Glick, B. R., & Greenberg, B. M. (2005). A multi-process phytoremediation system for decontamination of persistent total petroleum hydrocarbons (TPHs) from soils. *Microchemical Journal, 8*, 139–147.

Hussain, A., & Hasnain, S. (2009). Cytokinin production by some bacteria: Its impact on cell division in cucumber cotyledons. *African Journal of Microbiology Research, 3*, 704–712.

Hussain, A., Arshad, M., Zahir, Z. A., & Asghar, M. (2015). Prospectus of zinc solubilizing bacteria for improving growth and physiology of maize. *Pakistan Journal of Agricultural Sciences, 52*(4), 915–922.

Ilangumaran, G., & Smith, D. L. (2017). Plant growth promoting rhizobacteria in amelioration of salinity stress: A systems biology perspective. *Frontiers in Plant Science, 8*, 1768.

Iqbal, U., Jamil, N., Ali, I., & Hasnain, S. (2010). Effect of zinc-phosphate-solubilizing bacterial isolates on growth of *Vigna radiata*. *Annales de Microbiologie, 60*, 243–248.

Islam, M., Sultana, T., Joe, M. M., Yim, W., Cho, J. C., & Sa, T. (2013). Nitrogen-fixing bacteria with multiple plant growth promoting activities enhances growth of tomato and red pepper. *Journal of Basic Microbiology, 53*(12), 1004–1015.

Islam, S., Akanda, A. M., Prova, A., Islam, M. T., & Hossain, M. M. (2015). Isolation and identification of plant growth promoting rhizobacteria from cucumber rhizosphere and their effect on plant growth promotion and disease suppression. *Frontiers in Microbiology, 6*, 1360.

Jadhav, H. P., & Sayyed, R. J. (2016). Hydrolytic enzymes of rhizospheric microbes in crop protection. *MedCrave Online Journal of Cell Science & Report, 3*(5), 00070.

Jha, Y., Subramanian, R. B., & Patel, S. (2011). Combination of endophytic and rhizospheric plant growth promoting rhizobacteria in *Oryza sativa* shows higher accumulation of osmoprotectant against saline stress. *Acta Physiologiae Plantarum, 33*, 797–802.

Kageyama, K., & Nelson, E. B. (2003). Differential inactivation of seed exudate stimulation of *Pythium ultimum* sporangium germination by *Enterobacter cloacae* influences biological control efficacy on different plant species. *Applied and Environmental Microbiology, 69*(2), 1114–1120.

Kang, S. M., Khan, A. L., You, Y. H., Kim, J. G., Kamran, M., & Lee, I. J. (2014). Gibberellin production by newly isolated strain *Leifsonia soli* SE134 and its potential to promote plant growth. *Journal of Microbiology and Biotechnology, 24*, 106–112.

Karnwal, A. (2017). Isolation and identification of plant growth promoting rhizobacteria from maize (*Zea mays* L.) rhizosphere and their plant growth promoting effect on rice (*Oryza sativa* L.). *Journal of Plant Protection Research, 57*(2), 144–151.

Kaushal, M., & Wani, S. P. (2016). Plant-growth-promoting rhizobacteria: Drought stress alleviators to ameliorate crop production in drylands. *Annales de Microbiologie, 66*(1), 35–42.

Keel, C., Voisard, C., Berling, C. H., Kahir, G., & Defago, G. (1989). Iron sufficiency is a prerequisite for suppression of tobacco black root rot by *Pseudomonas fluorescens* strain CHA0 under gnotobiotic conditions. *Phytopathology, 79*, 584–589.

Kejela, T., Thakkar, V. R., & Thakor, P. (2016). *Bacillus* species (BT42) isolated from *Coffea arabica* L. rhizosphere antagonizes *Colletotrichum gloeosporioides* and *Fusarium oxysporum* and also exhibits multiple plant growth promoting activity. *BMC Microbiology, 16*(1), 277.

Kertesz, M. A., & Mirleau, P. (2004). The role of microbes in plant sulphur supply. *Journal of Experimental Botany, 55*, 1939–1945.

Khalid, A., Tahir, S., Arshad, M., & Zahir, Z. A. (2004). Relative efficiency of rhizobacteria for auxin biosynthesis in rhizosphere and non-rhizosphere soils. *Australian Journal of Soil Research, 42*, 921–926.

Khalid, S., Akhtar, M. J., Mahmood, M. H., & Arshad, M. (2006). Effect of substrate-dependent microbial ethylene production on plant growth. *Microbiology, 75*, 231–236.

Khan, A. G. (2005). Role of soil microbes in the rhizosphere of plants growing on trace metal contaminated soils in phytoremediation. *Journal of Trace Elements in Medicine and Biology, 18*, 355–364.

Khan, N., & Bano, A. (2016). Role of plant growth promoting rhizobacteria and Ag-nano particle in the bioremediation of heavy metals and maize growth under municipal wastewater irrigation. *International Journal of Phytoremediation, 18*(3), 211–221.

Khan, M. H., & Panda, S. K. (2008). Alterations in root lipid peroxidation and antioxidative responses in two rice cultivars under NaCl salinity stress. *Acta Physiologiae Plantarum, 30*, 89–91.

Khan, M. S., Zaidi, A., & Wani, P. A. (2007). Role of phosphate solubilizing microorganisms in sustainable agriculture – A review. *Agronomy for Sustainable Development, 27*, 29–43.

Khan, A. A., Jilani, G., Akhtar, M. S., Naqvi, S. M. S., & Rasheed, M. (2009). Phosphorus solubilizing bacteria: Occurrence, mechanisms and their role in crop production. *Journal of Agricultural and Biological Science, 1*(1), 48–58.

Khare, E., & Arora, N. K. (2010). Effect of indole-3-acetic acid (IAA) produced by *Pseudomonas aeruginosa* in suppression of charcoal rot disease of chickpea. *Current Microbiology, 61*, 64–68.

Khare, E., & Arora, N. K. (2011). Physiochemical and structural characterization of biosurfactant from fluorescent *Pseudomonas* with biocontrol activity against *Macrophomina phaseolina*. *Proceedings of the 2nd Asian PGPR conference* (pp. 104–109), China.

Khare, E., Kumar, S., & Kim, K. (2018). Role of peptaibols and lytic enzymes of Trichoderma cerinum Gurl in biocontrol of Fusarium oxysporum and chickpea wilt. *Environmental Sustainability, 1*(1), 39–47.

Khatibi, R. (2011). Using sulfur oxidizing bacteria and P solubilizing for enhancing phosphorous availability to *Raphanus sativus*. *African Journal of Plant Science, 5*(8), 430–435.

Kishore, G. K., Pande, S., & Podile, A. R. (2005). Biological control of collar rot disease with broad spectrum antifungal bacteria associated with groundnut. *Canadian Journal of Microbiology, 51*, 122–132.

Kloepper, J. W., & Schroth, M. N. (1978). Plant growth promoting rhizobacteria on radish. In Station de pathologie végétale et phyto-bacteriologie (Ed.), *Proceedings of the 4th conference plant pathogenic bacteria* (pp. 879–882). Angers: INRA.

Kong, Z., & Glick, B. R. (2017). The role of bacteria in phytoremediation. *Applied Bioengineering*, 327–353.

Kong, J., Dong, Y., Xu, L., Liu, S., & Bai, X. (2014). Effects of foliar application of salicylic acid and nitric oxide in alleviating iron deficiency induced chlorosis of *Arachis hypogaea* L. *Botanical Studies, 55*, 9.

Korir, H., Mungai, N. W., Thuita, M., Hamba, Y., & Masso, C. (2017). Co-inoculation effect of rhizobia and plant growth promoting rhizobacteria on common bean growth in a low phosphorus soil. *Frontiers in Plant Science, 8*, 141.

Kotasthane, A. S., Agrawal, T., Zaidi, N. W., & Singh, U. S. (2017). Identification of siderophore producing and cynogenic fluorescent *Pseudomonas* and a simple confrontation assay to identify potential bio-control agent for collar rot of chickpea. *3 Biotech, 7*, 137.

Krapp, A., Berthomé, R., Orsel, M., Mercey-Boutet, S., Yu, A., Castaings, L., Elftieh, S., Major, H., Renou, J. P., & Daniel-Vedele, F. (2011). Arabidopsis roots and shoots show distinct temporal adaptation patterns toward nitrogen starvation. *Plant Physiology, 157*, 1255–1282.

Kravchenko, L. V., Azarova, T. S., Makarova, N. M., & Tikhonovich, I. A. (2004). The effect of tryptophan present in plant root exudates on the phytostimulating activity of rhizobacteria. *Microbiology, 73*(2), 156–158.

Kucera, B., Cohn, M. A., & Leubner-Metzger, G. (2005). Plant hormone interactions during seed dormancy release and germination. *Seed Science Research, 15*, 281–307.

Kudoyarova, G. R., Arkhipova, T. N., & Melent'ev, A. I. (2015). Role of bacterial phytohormones in plant growth regulation and their development. In D. K. Maheshwari (Ed.), *Bacterial metabolites in sustainable agroecosystem* (pp. 69–86). Cham: Springer.

Kumar, S., & Singh, A. (2015). Biopesticides: Present status and the future prospects. *J Fertil Pestic, 6*(2), 100–129.

Kumar, A., Prakash, A., & Johri, B. N. (2011). *Bacillus as* PGPR in Crop Ecosystem. In D. K. Maheshwari (Ed.), *Bacteria in agrobiology: Crop ecosystems* (pp. 37–59). Berlin/Heidelberg: Springer.

Kumari, M. E. R., Gopal, A. V., & Lakshmipathy, R. (2018). Effect of stress tolerant plant growth promoting rhizobacteria on growth of blackgram under stress condition. *International Journal of Current Microbiology and Applied Sciences, 7*(1), 1479–1487.

Kundan, R., Pant, G., Jadon, N., & Agrawal, P. K. (2015). Plant growth promoting rhizobacteria: Mechanism and current prospective. *Journal of Fertilizers & Pesticides, 6*, 155.

Ladau, J., Shi, Y., Jing, X., He, J. S., Chen, L., Lin, X., Fierer, N., Gilbert, J. A., Pollard, K. S., & Chu, H. (2017). *Climate change will lead to pronounced shifts in the diversity of soil microbial communities. bioRxiv*, 180174.

Laranjo, M., Alexandre, A., & Oliveira, S. (2014). Legume growth-promoting rhizobia: An overview on the *Mesorhizobium* genus. *Microbiological Research, 169*(1), 2–17.
Lareen, A., Burton, F., & Schäfer, P. (2016). Plant root-microbe communication in shaping root microbiomes. *Plant Molecular Biology, 90*, 575–587.
Lata, S. A. K., & Tilak, K. V. B. R. (2002). Biofertilizers to augment soil fertility and crop production. In K. R. Krishna (Ed.), *Soil fertility and crop production* (pp. 279–312). Madison: Science Publishers.
Leigh, G. J. (2002). *Nitrogen fixation at the millennium* (p. 470). Amsterdam: Elsevier Science.
Li, F. C., Li, S., Yang, Y. Z., & Cheng, L. J. (2006). Advances in the study of weathering products of primary silicate minerals, exemplified by mica and feldspar. *Acta Petrologica et Mineralogica, 25*, 440–448.
Li, Y., Liu, X., Hao, T., & Chen, S. (2017). Colonization and maize growth promotion induced by phosphate solubilizing bacterial isolates. *International Journal of Molecular Sciences, 18*(7), 1253.
Liu, P., Luoa, L., & Long, C. (2013). Characterization of competition for nutrients in the biocontrol of *Penicillium italicum* by *Kloeckera apiculata*. *Biological Control, 67*, 157–162.
Liu, H., Carvalhais, L. C., Crawford, M., Singh, E., Dennis, P. G., Pieterse, C. M. J., & Schenk, P. M. (2017). Inner plant values: Diversity, colonization and benefits from endophytic bacteria. *Frontiers in Microbiology, 8*, 1–17.
Lobell, D. B., & Gourdji, S. M. (2012). The influence of climate change on global crop productivity. *Plant Physiology, 160*(4), 1686–1697.
Loganathan, M., Garg, R., Venkataravanappa, V., Saha, S., & Rai, A. B. (2014). Plant growth promoting rhizobacteria (PGPR) induces resistance against *Fusarium* wilt and improves lycopene content and texture in tomato. *African Journal of Microbiology Research, 8*(11), 1105–1111.
Long, X. X., Yang, X. E., & Ni, W. Z. (2002). Current status, and prospective on phytoremediation of heavy metal polluted soils. *Journal of Applied Ecology, 13*, 757–762.
Long, S. P., Zhu, X. G., Naidu, S. L., & Ort, D. R. (2006). Can improvement in photosynthesis increase crop yields? *Plant, Cell & Environment, 29*, 315–330.
Loper, J. E., & Buyer, J. S. (1991). Siderophores in microbial interactions of plant surfaces. *Molecular Plant-Microbe Interactions, 4*, 5–13.
Lucas, J. A., Ramos-Solano, B., Montes, F., Ojeda, J., Megias, M., & Gutierrez Mañero, F. J. (2009). Use of two PGPR strains in the integrated management of blast disease in rice (*Oryza sativa* L.) in Southern Spain. *Field Crops Research, 114*, 404–410.
Lucas-Garcia, J., Probanza, A., Ramos, B., Barriuso, J., & Gutierrez-Mañero, F. (2004). Effects of inoculation with plant growth rhizobacteria (PGPRs) and *Sinorhizobium fredii* on biological nitrogen fixation, nodulation and growth of *Glycine max* cv. Osumi. *Plant and Soil, 267*, 143–153.
Lucy, M., Reed, E., & Glick, B. R. (2004). Applications of free living plant growth promoting rhizobacteria. *Antonie Van Leeuwenhoek, 86*, 1–25.
Lugtenberg, B. J. J. (2015). Introduction to plant-microbe interactions. In B. Lugtenberg (Ed.), *Principles of plant-microbe interactions: Microbes for sustainable agriculture* (pp. 1–2). Cham: Springer.
Lugtenberg, B. J. J., & Dekkers, L. C. (1999). What makes *Pseudomonas* bacteria rhizosphere competent? *Environmental Microbiology, 1*, 9–13.
Lugtenberg, B., & Kamilova, F. (2009). Plant-growth-promoting rhizobacteria. *Annual Review of Microbiology, 63*, 541–556.
Lugtenberg, B. J. J., Dekkers, L., & Bloemberg, G. V. (2001). Molecular determinants of rhizosphere colonization by *Pseudomonas*. *Annual Review of Phytopathology, 39*, 461–490.
Lugtenberg, B. J. J., Chin-A-Woeng, T. F. C., & Bloemberg, G. V. (2002). Microbe-plant interactions: Principles and mechanisms. *Antonie Van Leeuwenhoek, 81*, 373–383.
Lukkani, N. J., & Reddy, E. S. (2014). Evaluation of plant growth promoting attributes and biocontrol potential of native fluorescent *Pseudomonas* spp. against *Aspergillus niger* causing

collar rot of ground nut. *International Journal of Plant, Animal and Environmental Sciences, 4*(4), 256–262.

Ma, Y., Prasad, M. N. V., Rajkumar, M., & Freitas, H. (2011). Plant growth promoting rhizobacteria and endophytes accelerate phytoremediation of metalliferous soils. *Biotechnology Advances, 29*, 248–258.

Ma, Y., Oliveira, R. S., Wu, L., Luo, Y., Rajkumar, M., Rocha, I., & Freitas, H. (2015). Inoculation with metal-mobilizing plant-growth-promoting rhizobacterium *Bacillus* sp. SC2b and its role in rhizoremediation. *Journal of Toxicology and Environmental Health, 78*, 931–944.

Ma, Y., Oliveira, R. S., Freitas, H., & Zhang, C. (2016). Biochemical and molecular mechanisms of plant-microbe-metal interactions: Relevance for phytoremediation. *Frontiers in Plant Science, 7*, 918.

Mahanty, T., Bhattacharjee, S., Goswami, M., Bhattacharyya, P., Das, B., Ghosh, A., & Tribedi, P. (2016). Biofertilizers: A potential approach for sustainable agriculture development. *Environmental Science and Pollution Research International, 24*, 3315–3335.

Maheshwari, D. K., Kumar, S., Maheshwari, N. K., Patel, D., & Saraf, M. (2012a). Nutrient availability and management in the rhizosphere by microorganisms. In D. K. Maheshwari (Ed.), *Bacteria in agrobiology: Stress management* (pp. 301–325). Berlin: Springer.

Maheshwari, D. K., Dubey, R. C., Aeron, A., Kumar, B., Kumar, S., Tewari, S., & Arora, N. (2012b). Integrated approach for disease management and growth enhancement of *Sesamum indicum* L. utilizing *Azotobacter chroococcum* TRA2 and chemical fertilizer. *World Journal of Microbiology and Biotechnology, 28*(10), 3015–3024.

Maheshwari, D. K., Dheeman, S., & Agarwal, M. (2015). Phytohormone-producing PGPR for sustainable agriculture. In D. K. Maheshwari (Ed.), *Bacterial metabolites in sustainable agroecosystem* (pp. 159–182). Cham: Springer.

Mahmood, A., Turgay, O. C., Farooq, M., & Hayat, R. (2016). Seed biopriming with plant growth promoting rhizobacteria: A review. *FEMS Microbiology Ecology, 92*(8), 1–14.

Majeed, A., Abbasi, M. K., Hameed, S., Imran, A., & Rahim, N. (2015). Isolation and characterization of plant growth-promoting rhizobacteria from wheat rhizosphere and their effect on plant growth promotion. *Frontiers in Microbiology, 6*, 198.

Maksimov, I. V., Abizgil'dina, R. R., & Pusenkova, L. I. (2011). Plant growth promoting rhizobacteria as alternative to chemical crop protectors from pathogens (Review). *Applied Biochemistry and Microbiology, 47*, 333–345.

Malusá, E., & Vassilev, N. (2014). A contribution to set a legal framework for biofertilisers. *Applied Microbiology and Biotechnology, 98*, 6599–6607.

Mamta, R. P., Pathania, V., Gulati, A., Singhd, B., Bhanwra, R. K., & Tewari, R. (2010). Stimulatory effect of phosphate-solubilizing bacteria on plant growth, stevioside and rebaudioside-A contents of *Stevia rebaudiana Bertoni*. *Applied Soil Ecology, 46*, 222–229.

Marketsandmarkets. (2014). *Biopesticides market by active ingredient, by types, by application, by formulation, by crop type and by geography*. Pune: Marketsandmarkets Available at: MarketResearch.com.

Marketsandmarkets.com. (2016). Biofertilizers market by type (nitrogen-fixing, phosphate-solubilizing, potash-mobilizing), microorganism (*Rhizobium, Azotobacter, Azospirillum*, Cyanobacteria, P-Solubilizer), mode of application, crop type, form, and region – Global forecast to 2022. Available at: www.marketsandmarkets.com.

Marschner, H. (2012). *Marschner's mineral nutrition of higher plants* (3rd ed., p. 672). London: Academic.

Martinez-Viveros, O., Jorquera, M. A., Crowley, D. E., Gajardo, G. M. L. M., & Mora, M. L. (2010). Mechanisms and practical considerations involved in plant growth promotion by rhizobacteria. *Journal of Soil Science and Plant Nutrition, 10*(3), 293–319.

Meena, H., Ahmed, M. A., & Prakash, P. (2015). Amelioration of heat stress in wheat, *Triticum aestivum* by PGPR (*Pseudomonas aeruginosa* strain 2CpS1). *Bioscience Biotechnology Research Communications, 8*(2), 171–174.

Merzaeva, O. V., & Shirokikh, I. G. (2006). Colonization of plant rhizosphere by actinomycetes of different genera. *Microbiology, 75*, 226–230.

Meyer, S. L. F., Everts, K. L., Gardener, B. M., Masler, E. P., Abdelnabby, H. M. E., & Skantar, A. M. (2016). Assessment of DAPG-producing *Pseudomonas fluorescens* for Management of *Meloidogyne incognita* and *Fusarium oxysporum* on Watermelon. *Journal of Nematology, 48*(1), 43–53.

Meziane, H., Vander, S. I., van Loon, L. C., Höfte, M., & Bakker, P. A. H. M. (2005). Determinants of *P. putida* WCS 358 involved in induced systemic resistance in plants. *Molecular Plant Pathology, 6*, 177–185.

Mia, M. A. B., & Shamsuddin, Z. H. (2010). *Rhizobium* as a crop enhancer and biofertilizer for increased cereal production. *African Journal of Biotechnology, 9*(37), 6001–6009.

Miethke, M., & Marahiel, M. A. (2007). Siderophore-based iron acquisition and pathogen control. *Microbiology and Molecular Biology Reviews, 71*(3), 413–451.

Mishra, S., & Arora, N. K. (2011). Evaluation of rhizospheric *Pseudomonas* and *Bacillus* as biocontrol tool for *Xanthomonas campestris* pv *campestris*. *World Journal of Microbiology and Biotechnology, 28*(2), 693–702.

Mishra, S., & Arora, N. K. (2012). Management of black rot in cabbage by rhizospheric *Pseudomonas* sp. and analysis of 2, 4-diacetylphloroglucinol by qRT-PCR. *Biological Control, 61*, 29–32.

Mishra, J., & Arora, N. K. (2018). Secondary metabolites of fluorescent pseudomonads in biocontrol of phytopathogens for sustainable agriculture. *Applied Soil Ecology, 125*, 35–45.

Mishra, P., & Das, D. (2014). Rejuvenation of biofertilizer for sustainable agriculture and economic development. *Consilience: J Sustain Develop, 11*(1), 41–61.

Mishra, P. K., Bisht, S. C., Bisht, J. K., & Bhatt, J. C. (2012). Cold tolerant PGPRs as bioinoculants for stress management. In D. K. Maheshwari (Ed.), *Bacteria in agrobiology: Stress management* (pp. 95–118). Berlin/Heidelberg: Springer.

Mishra, J., Tewari, S., Singh, S., & Arora, N. K. (2015). Biopesticides: Where we stand? In N. K. Arora (Ed.), *Plant microbes symbiosis: Applied facets* (pp. 37–75). New Delhi: Springer.

Mishra, J., Singh, R., & Arora, N. K. (2017a). Alleviation of heavy metal stress in plants and remediation of soil by rhizosphere microorganisms. *Frontiers in Microbiology, 8*, 1706.

Mishra, J., Singh, R., & Arora, N. K. (2017b). Plant growth-promoting microbes: Diverse roles in agriculture and environmental sustainability. In V. Kumar, M. Kumar, S. Sharma, & R. Prasad (Eds.), *Probiotics and plant health* (pp. 71–111). Singapore: Springer.

Mitter, E. K., de Freitas, J. R., & Germida, J. J. (2017). Bacterial root microbiome of plants growing in oil sands reclamation covers. *Frontiers in Microbiology, 8*, 70.

Mohite, B. (2013). Isolation and characterization of indole acetic acid (IAA) producing bacteria from rhizospheric soil and its effect on plant growth. *Journal of Soil Science and Plant Nutrition, 13*(3), 638–649.

Moustaine, M., Elkahkahi, R., Benbouazza, A., Benkirane, R., & Achbani, E. (2017). Effect of plant growth promoting rhizobacterial (PGPR) inoculation on growth in tomato (*Solanum lycopersicum* L.) and characterization for direct PGP abilities in Morocco. *International Journal of Agriculture Environment & Biotechnology, 2*, 590–595.

Mus, F., Crook, M. B., Garcia, K., Costas, A. G., Geddes, B. A., Kouri, E. D., Paramasivan, P., Ryu, M. H., Oldroyd, G. E., Poole, P. S., & Udvardi, M. K. (2016). Symbiotic nitrogen fixation and the challenges to its extension to nonlegumes. *Applied and Environmental Microbiology, 82*(13), 3698–3710.

Nadeem, S. M., Zahir, Z. A., Naveed, M., Asghar, H. N., & Arshad, M. (2010). Rhizobacteria capable of producing ACC deaminase may mitigate salt stress in wheat. *Soil Science Society of America Journal, 74*, 533–542.

Nadeem, S. M., Ahmad, M., Zahir, Z. A., Javaid, A., & Ashraf, M. (2014). The role of mycorrhizae and plant growth promoting rhizobacteria (PGPR) in improving crop productivity under stressful environments. *Biotechnology Advances, 32*, 429–448.

Nahas, E. (1996). Factors determining rock phosphate solubilization by microorganisms isolated from soil. *World Journal of Microbiology and Biotechnology, 12*(6), 567–572.

Nakkeeran, S., Dilantha Fernando, W. G., & Siddiqui, A. (2005). Plant growth promoting rhizobacteria. In Z. A. Siddiqui (Ed.), *PGPR: Biocontrol and biofertilization* (pp. 257–296). Dordrecht: Springer.

Nandi, M., Selin, C., Brassinga, A. K. C., Belmonte, M. F., Fernando, W. G. D., Loewen, P. C., & de Kievit, T. R. (2015). Pyrrolnitrin and hydrogen cyanide production by *Pseudomonas chlororaphis* strain PA23 exhibits nematicidal and repellent activity against *Caenorhabditis elegans*. *PLoS One, 10*, e0123184.

Narula, N., Deubel, A., Gans, W., Behl, R. K., & Merbach, W. (2006). Paranodules and colonization of wheat roots by phytohormone producing bacteria in soil. *Plant, Soil and Environment, 52*(3), 119–129.

Naz, I., Ahmad, H., Khokhar, S. N., Khan, K., & Shah, A. H. (2016). Impact of zinc solubilizing bacteria on zinc contents of wheat. *American-Eurasian Journal of Agricultural & Environmental Sciences, 16*(3), 449–454.

Nazir, Q., Akhtar, M. J., Imran, M., Arshad, M., Hussain, A., Mahmood, S., & Hussain, S. (2017). Simultaneous use of plant growth promoting rhizobacterium and nitrogenous fertilizers may help in promoting growth, yield, and nutritional quality of okra. *Journal of Plant Nutrition, 40*(9), 1339–1350.

Nie, M., Bell, C., Wallenstein, M. D., & Pendall, E. (2015). Increased plant productivity and decreased microbial respiratory C loss by plant growth-promoting rhizobacteria under elevated CO_2. *Scientific Reports, 5*, 9212.

Nobbe, F., & Hiltner, L. (1896). *Inoculation of the soil for cultivating leguminous plants*. U.S. Patent 570 813.

Normile, D. (2008). Reinventing rice to feed the world. *Science, 321*, 330–333.

Noumavo, P. A., Agbodjato, N. A., Baba-Moussa, F., Adjanohoun, A., & Baba-Moussa, L. (2016). Plant growth promoting rhizobacteria: Beneficial effects for healthy and sustainable agriculture. *African Journal of Biotechnology, 15*, 1452–1463.

Okon, G., Okon, E., & Glory, I. (2014). Effect of bioremediation on early seedling growth of *Amaranthus hybridus* L. grown on palm oil mill effluent polluted soil. *International Journal of Biological Research, 2*(2), 84–86.

Omidvari, M., Sharifi, R. A., Ahmadzadeh, M., & Dahaji, P. A. (2010). Role of fluorescent pseudomonads siderophore to increase bean growth factors. *The Journal of Agricultural Science, 2*(3), 242–247.

Ongena, M., & Jacques, P. (2008). Bacillus lipopeptides: Versatile weapons for plant disease biocontrol. *Trends in Microbiology, 16*(3), 115–125.

Ortíz-Castro, R., Contreras-Cornejo, H. A., Macías-Rodríguez, L., & López-Bucio, J. (2009). The role of microbial signals in plant growth and development. *Plant Signaling & Behavior, 4*(8), 701–712.

Oteino, N., Lally, R. D., Kiwanuka, S., Lloyd, A., Ryan, D., Germaine, K. J., & Dowling, D. N. (2015). Plant growth promotion induced by phosphate solubilizing endophytic *Pseudomonas* isolates. *Frontiers in Microbiology, 6*, 745.

Owen, D., Williams, A. P., Griffith, G. W., & Withers, P. J. A. (2015). Use of commercial bio-inoculants to increase agricultural production through improved phosphorus acquisition. *Applied Soil Ecology, 86*, 41–54.

Pal, K. K., & Gardener, B. M. (2006). Biological control of plant pathogens. *Plant Health Instructor, 2*, 1117–1142.

Pandya, N. D., & Desai, P. V. (2014). Screening and characterization of GA3 producing *Pseudomonas monteilii* and its impact on plant growth promotion. *International Journal of Current Microbiology and Applied Sciences, 3*(5), 110–115.

Parmar, P., & Sindhu, S. S. (2013). Potassium solubilization by rhizosphere bacteria: Influence of nutritional and environmental conditions. *Journal of Microbiology Research, 3*(1), 25–31.

Parnell, J. J., Berka, R., Young, H. A., Sturino, J. M., Kang, Y., Barnhart, D. M., & Dileo, M. V. (2016). From the lab to the farm: An industrial perspective of plant beneficial microorganisms. *Frontiers in Plant Science, 7*, 1110.

Patten, C., & Glick, B. (2002). Role of Pseudomonas putida indole acetic acid in development of the host plant root system. *Applied and Environmental Microbiology, 68*, 3795–3801.

Paul, D., & Lade, H. (2014). Plant-growth-promoting rhizobacteria to improve crop growth in saline soils: A review. *Agronomy for Sustainable Development, 34*, 737–752.

Peix, A., Mateos, P. F., Rodríguez-Barrueco, C., Martínez-Molina, E., & Velázquez, E. (2001). Growth promotion of common bean (*Phaseolus vulgaris* L.) by a strain of *Burkholderia cepacia* under growth chamber conditions. *Soil Biology and Biochemistry, 33*, 1927–1935.

Pendall, E., Mosier, A. R., & Morgan, J. A. (2004). Rhizodeposition stimulated by elevated CO_2 in a semiarid grassland. *The New Phytologist, 162*, 447–458.

Penuelas, J., Asensio, D., Tholl, D., Wenke, K., Rosenkranz, M., Piechulla, B., & Schnitzler, J. P. (2014). Biogenic volatile emissions from the soil. *Plant, Cell & Environment, 37*, 1866–1891.

Perez-Fernández, M., & Alexander, V. (2017). Enhanced plant performance in Cicer arietinum L. due to the addition of a combination of plant growth-promoting bacteria. *Agriculture, 7*(5), 40.

Pérez-Montaño, F., Jiménez-Guerrero, I., Contreras Sánchez-Matamoros, R., López-Baena, F. J., Ollero, F. J., Rodríguez-Carvajal, M. A., Bellogín, R. A., & Espuny, M. R. (2013). Rice and bean AHL-mimic quorum-sensing signals specifically interfere with the capacity to form biofilms by plant-associated bacteria. *Microbiological Research, 164*, 749–760.

Pérez-Montaño, F., Cynthia, A., Bellogín, R. A., Del Cerro, P., Espuny, M. R., Jiménez-Guerrero, I., López-Baena, F. J., Ollero, F. J., & Cubo, T. (2014). Plant growth promotion in cereal and leguminous agricultural important plants: From microorganism capacities to crop production. *Microbiological Research, 169*(5), 325–336.

Piccoli, P., Lucangeli, C. D., Schneider, G., & Bottini, R. (1997). Hydrolysis of gibberellin A20-glucoside and gibberellin A20-glucosyl ester by *Azospirillum lipoferum* cultured in a nitrogen-free biotin based chemically-defined medium. *Plant Growth Regulation, 23*, 179–182.

Pieterse, C. M., Zamioudis, C., Berendsen, R. L., Weller, D. M., Van Wees, S. C., & Bakker, P. A. (2014). Induced systemic resistance by beneficial microbes. *Annual Review of Phytopathology, 52*, 347–375.

Ping, L., & Boland, W. (2004). Signals from the underground: Bacterial volatiles promote growth in *Arabidopsis*. *Trends in Plant Science, 9*, 263–266.

Pinter, I. F., Salomon, M. V., Berli, F., Bottini, R., & Piccoli, P. (2017). Characterization of the As (III) tolerance conferred by plant growth promoting rhizobacteria to *in vitro*-grown grapevine. *Applied Soil Ecology, 109*, 60–68.

Pires, C., Franco, A. R., Pereira, S. I. A., Henriques, I., Correia, A., Magan, N., & Castro, P. M. L. (2017). Metal(loid)-contaminated soils as a source of culturable heterotrophic aerobic bacteria for remediation applications. *Geomicrobiology J*, 1–9.

Podile, A. R., & Kishore, G. K. (2006). Plant growth-promoting rhizobacteria. In S. S. Gnanamanickam (Ed.), *Plant-associated bacteria* (pp. 195–230). Dordrecht: Springer.

Postgate, J. R. (1998). *Nitrogen fixation* (p. 252). Cambridge: Cambridge University Press.

PR Newswire. (2017). *Global markets for biopesticides*. Available at: https://www.prnewswire.com/news-releases/global-markets-forbiopesticides300385145

Prajapati, K., & Modi, H. A. (2016). Growth promoting effect of potassium solubilizing *Enterobacter hormaechei* (KSB-8) on Cucumber (*Cucumis sativus*) under hydroponic conditions. *International Journal of Advanced Research in Biological Sciences, 3*(5), 168–173.

Press, C. M., Wilson, M., Tuzun, S., & Kloepper, J. W. (1997). SA produced by *S. marcescens* 90–166 is not the primary determinant of ISR in cucumber/tobacco. *Molecular Plant-Microbe Interactions, 10*, 761–768.

Prithiviraj, B., Zhou, X., Souleimanov, A., Kahn, W. M., & Smith, D. L. (2003). A host-specific bacteria-to-plant signal molecule (Nod factor) enhances germination and early growth of diverse crop plants. *Planta, 21*, 437–445.

Qaisrani, M. M., Mirza, M. S., Zaheer, A., & Malik, K. A. (2014). Isolation and identification by 16s rRNA sequence analysis of *Achromobacter*, *Azospirillum* and *Rhodococcus* strains from the rhizosphere of maize and screening for the beneficial effect on plant growth. *Pakistan Journal of Agricultural Sciences, 51*, 91–99.

Qurashi, A. W., & Sabri, A. N. (2012). Bacterial exopolysaccharide and biofilm formation stimulate chickpea growth and soil aggregation under salt stress. *Brazilian Journal of Microbiology, 43*, 1183–1191.

Raaijmakers, J. M., & Weller, D. M. (2001). Exploiting genotype diversity of 2, 4-diacetylphloroglucinol producing *Pseudomonas* spp.: Characterization of superior root-colonizing *P. fluorescens* strain Q8r1-96. *Applied and Environmental Microbiology, 67*, 2545–2554.

Raaijmakers, J. M., de Bruijn, I., Nybroe, O., & Ongena, M. (2010). Natural functions of lipopeptides from *Bacillus* and *Pseudomonas*: More than surfactants and antibiotics. *FEMS Microbiology Reviews, 34*, 1037–1062.

Rabie, G. H., & Almadini, A. M. (2005). Role of bioinoculants in development of salt-tolerance of *Vicia faba* plants under salinity stress. *African Journal of Biotechnology, 4*, 210–222.

Rabosto, X., Garrau, M., Paz, A., Boido, E., Dellacassa, E., & Carrau, F. (2006). Grapes and vineyard soils as source of microorganisms for biological control of *Botrytis cinerea*. *American Journal of Enology and Viticulture, 57*, 332–338.

Raghavendra, A. S., Gonugunta, V. K., Christmann, A., & Grill, E. (2010). ABA perception and signalling. *Trends in Plant Science, 15*(7), 395–401.

Rais, A., Jabeen, Z., Shair, F., Hafeez, F. Y., & Hassan, M. N. (2017). *Bacillus* spp., a bio-control agent enhances the activity of antioxidant defense enzymes in rice against *Pyricularia oryzae*. *PLoS One, 12*(11), e0187412.

Rajkumar, M., Ae, N., Prasad, M. N. V., & Freitas, H. (2010). Potential of siderophore-producing bacteria for improving heavy metal phytoextraction. *Trends in Biotechnology, 28*, 142–149.

Rajwar, A., Sahgal, M., & Johri, B. N. (2013). Legume-rhizobia symbiosis and interactions in agroecosystems. In N. K. Arora (Ed.), *Plant microbe symbiosis-fundamentals and advances* (pp. 233–265). New Delhi: Springer.

Ramadan, E. M., AbdelHafez, A. A., Hassan, E. A., & Saber, F. M. (2016). Plant growth promoting rhizobacteria and their potential for biocontrol of phytopathogens. *African Journal of Microbiology Research, 10*(15), 486–504.

Ramos-Solano, B., Barriuso-Maicas, J., de la Iglesia, M. T. P., Domenech, J., & Gutiérrez Mañero, F. J. (2008). Systemic disease protection elicited by plant growth promoting rhizobacteria strains: Relationship between metabolic responses, systemic disease protection, and biotic elicitors. *Phytopathology, 98*, 451–457.

Ramyasmruthi, S., Pallavi, O., Pallavi, S., Tilak, K., & Srividya, S. (2012). Chitinolytic and secondary metabolite producing *Pseudomonas fluorescens* isolated from Solanaceae rhizosphere effective against broad spectrum fungal phytopathogens. *Asian Journal of Plant Science & Research, 2*, 16–24.

Rana, A., Kabi, S. R., Verma, S., Adak, A., Pal, M., Shivay, Y. S., Prasanna, R., & Nain, L. (2015). Prospecting plant growth promoting bacteria and cyanobacteria as options for enrichment of macro and micronutrients in grains in rice wheat cropping sequence. *Cogent Food & Agriculture, 1*, 10373–10379.

Rangel, W. M., Thijs, S., Janssen, J., Oliveira Longatti, S. M., Bonaldi, D. S., Ribeiro, P. R., Jambon, I., Eevers, N., Weyens, N., Vangronsveld, J., & Moreira, F. M. S. (2017). Native rhizobia from Zn mining soil promote the growth of *Leucaena leucocephala* on contaminated soil. *International Journal of Phytoremediation, 19*, 142–156.

Rao, D. L. N., Mohanty, S. R., Acharya, C., & Atoliya, N. (2018). *Rhizobial taxonomy-current status* (Newsletter No. 3, pp. 1–4). Indo-UK Nitrogen Fixation Centre (IUNFC).

Reddy, M. S., Ilao, R. I., Faylon, P. S., Dar, W. D., Batchelor, W. D., Sayyed, R., Sudini, H., Vijay Krishna Kumar, K., Armanda, A., & Gopalkrishnan, S. (Eds.). (2014). *Recent advances in biofertilizers and biofungicides (PGPR) for sustainable agriculture* (p. 510). Newcastle upon Tyne: Cambridge Scholars Publishing.

Reetha, S., Bhuvaneswari, G., Thamizhiniyan, P., & Mycin, T. R. (2014). Isolation of indole acetic acid (IAA) producing rhizobacteria of *Pseudomonas fluorescens* and *Bacillus subtilis* and enhance growth of onion (*Allim cepa.* L). *International Journal of Current Microbiology and Applied Sciences, 3*(2), 568–574.

Reitz, M., Oger, P., Meyer, A., Niehaus, K., Farrand, S. K., Hallmann, J., & Sikora, R. A. (2002). Importance of the O-antigen, core-region and lipid A of rhizobial LPS for the induction of SR in potato to *Globodera pallida. Nematology, 4*, 73–79.

Research Nester. (2018). *Biofertilizers market: global demand analysis & opportunity outlook 2023*. Available at: https://www.researchnester.com/reports/biofertilizers-market-global-demand-analysis-opportunity-outlook-2023/193.

Review, M. (2008). Enrichment of cereal grains with zinc: Agronomic or genetic biofortification? *Plant and Soil, 302*(1), 1–17.

Riadh, K., Wided, M., Hans-Werner, K., & Chedly, A. (2010). Responses of halophytes to environmental stresses with special emphasis to salinity. *Advances in Botanical Research, 53*, 117–145.

Ribaudo, C. M., Krumpholz, E. M., Cassán, F. D., Bottini, R., Cantore, M. L., & Cura, J. A. (2006). *Azospirillum* sp. promotes root hair development in tomato plants through a mechanism that involves ethylene. *Journal of Plant Growth Regulation, 25*, 175–185.

Roberson, E. B., & Firestone, M. K. (1992). Relationship between desiccation and exopolysaccharide production in soil *Pseudomonas* sp. *Applied and Environmental Microbiology, 58*, 1284–1291.

Rodriguez, H., Gonzalez, T., Goire, I., & Bashan, Y. (2004). Gluconic acid production and phosphate solubilization by the plant growth-promoting bacterium *Azospirillum* spp. *Naturwissenschaften, 91*(11), 552–555.

Rodríguez, H., Fraga, R., Gonzalez, T., & Bashan, Y. (2006). Genetics of phosphate solubilization and its potential applications for improving plant growth-promoting bacteria. *Plant and Soil, 287*, 15–21.

Rogers, J. R., Bennett, P. C., & Choi, W. J. (1998). Feldspars as a source of nutrients for microorganisms. *American Mineralogist, 83*, 1532–1540.

Romeh, A. A., & Hendawi, M. Y. (2014). Bioremediation of certain organophosphorus pesticides by two biofertilizers, *Paenibacillus (Bacillus) polymyxa* (Prazmowski) and *Azospirillum lipoferum* (*Beijerinck*). *Journal of Agricultural Science and Technology, 16*(2), 265–276.

Roper, M. M., & Gupta, V. S. R. (2016). Enhancing non-symbiotic N_2 fixation in agriculture. *Open Agriculture Journal, 10*, 7–27.

Rosenblueth, M., & Martínez-Romero, E. (2006). Bacterial endophytes and their interactions with hosts. *Molecular Plant-Microbe Interactions, 19*, 827–837.

Ryan, R. P., Germaine, K., Franks, A., Ryan, D. J., & Dowling, D. N. (2008). Bacterial endophytes: Recent developments and applications. *FEMS Microbiology Letters, 278*(1), 1–9.

Ryu, C. M., Farag, M. A., Hu, C. H., Reddy, M. S., Wei, H. X., Paré, P. W., & Kloepper, J. W. (2003). Bacterial volatiles promote growth in Arabidopsis. *Proceedings of the National Academy of Sciences, 100*(8), 4927–4932.

Saha, M., Sarkar, S., Sarkar, B., Sharma, B. K., Bhattacharjee, S., & Tribedi, P. (2016). Microbial siderophores and their potential applications: A review. *Environmental Science and Pollution Research, 23*, 3984–3999.

Saharan, B. S., & Nehra, V. (2011). Plant growth promoting rhizobacteria: A critical review. *Life Sci Med Res, 21*, 1–30.

Sahgal, M., & Johri, B. N. (2003). The changing face of rhizobial systematic. *Current Science, 84*(1), 43–48.

Sahoo, R. K., Bhardwaj, D., & Tuteja, N. (2013). Biofertilizers: A sustainable eco-friendly agricultural approach to crop improvement. In N. Tuteja & S. S. Gill (Eds.), *Plant acclimation to environmental stress* (pp. 403–432). New York: Springer.

Saikia, J., Sarma, R. K., Dhandia, R., Yadav, A., Bharali, R., Gupta, V. K., & Saikia, R. (2018). Alleviation of drought stress in pulse crops with ACC deaminase producing rhizobacteria isolated from acidic soil of Northeast India. *Scientific Reports, 8*, 35–60.

Salam, M. A., & Noguchi, T. (2005). Impact of human activities on carbon dioxide (CO2) emissions: A statistical analysis. *Environmentalist, 25*, 19.

Salamone, I. E. G., Hynes, R. K., & Nelson, L. M. (2001). Cytokinin production by plant growth promoting rhizobacteria and selected mutants. *Canadian Journal of Microbiology, 47*, 404–411.

Salamone, I. E. G., Hynes, R. K., & Nelson, L. M. (2005). Role of cytokinins in plant growth promotion by rhizosphere bacteria. In Z. A. Siddiqui (Ed.), *PGPR: Biocontrol and biofertilization* (pp. 173–195). Dordrecht: Springer.

Saleem, M., Arshad, M., Hussain, S., & Bhatti, A. S. (2007). Perspective of plant growth promoting rhizobacteria (PGPR) containing ACC deaminase in stress agriculture. *Journal of Industrial Microbiology & Biotechnology, 34*, 635–648.

Salimpour, S., Khavazi, K., Nadian, H., Besharati, H., & Miransari, M. (2010). Enhancing phosphorus availability to canola (*Brassica napus* L.) using P solubilizing and sulfur oxidizing bacteria. *Australian Journal of Crop Science, 4*, 330–334.

Salisbury, F. B. (1994). The role of plant hormones. In R. E. Wilkinson (Ed.), *Plant-environment interactions* (pp. 39–81). New York: USA.

Sanchis, V., & Bourguet, D. (2008). *Bacillus thuringiensis*: Applications in agriculture and insect resistance management: A review. *Agronomy for Sustainable Development, 28*(1), 11–20.

Sandhya, V., Ali, S. Z., Grover, M., Reddy, G., & Venkateswarlu, B. (2009). Alleviation of drought stress effects in sunflower seedlings by the exopolysaccharides producing *Pseudomonas putida* strain GAP-P45. *Biology and Fertility of Soils, 46*, 17–26.

Santi, C., Bogusz, D., & Franche, C. (2013). Biological nitrogen fixation in non-legume plants. *Annals of Botany, 111*(5), 743–767.

Santoyo, G., Moreno-Hagelsieb, G., del Carmen Orozco-Mosqueda, M., & Glick, B. R. (2016). Plant growth-promoting bacterial endophytes. *Microbiological Research, 183*, 92–99.

Saraf, M., Pandya, U., & Thakkar, A. (2014). Role of allelochemicals in plant growth promoting rhizobacteria for biocontrol of phytopathogens. *Microbiological Research, 169*(1), 18–29.

Sarr, P. S., Fujimoto, S., & Yamakawa, T. (2015). Nodulation, nitrogen fixation and growth of rhizobia-inoculated cowpea (*Vigna unguiculata* L. Walp) in relation with external nitrogen and light intensity. *International Journal of Plant Biology & Research, 3*(1), 1025.

Sathya, A., Vijayabharathi, R., Srinivas, V., & Gopalakrishnan, S. (2016). Plant growth-promoting actinobacteria on chickpea seed mineral density: An upcoming complementary tool for sustainable biofortification strategy. *3 Biotech, 6*(2), 1–6.

Sathya, A., Vijayabharathi, R., & Gopalakrishnan, S. (2017). Plant growth-promoting actinobacteria: A new strategy for enhancing sustainable production and protection of grain legumes. *3 Biotech, 7*, 102.

Sayyed, R. Z., Badgujar, M. D., Sonawane, H. M., Mhaske, M. M., & Chincholkar, S. B. (2005). Production of microbial iron chelators (siderophores) by fluorescent Pseudomonads. *Indian Journal of Biotechnology, 4*(4), 484–490.

Sayyed, R. Z., Patil, A. S., Gangurde, N. S., Bhamare, H. M., SA, J., & Fulpagare, U. G. (2008). Siderophore producing *A. faecalis:* A potent biofungicide for the control of ground phytopathogens. *Research Journal of Biotechnology*, 411–413.

Schallmey, M., Singh, A., & Ward, O. P. (2004). Developments in the use of *Bacillus* species for industrial production. *Canadian Journal of Microbiology, 50*, 1–17.

Schoebitz, M., Osman, J., & Ciampi, L. (2013). Effect of immobilized *Serratia* sp. by spray-drying technology on plant growth and phosphate uptake. *Chilean Journal of Agricultural & Animal Sciences, 29*, 111–119.

Schulz, B., & Boyle, C. (2006). What are endophytes? In B. J. E. Schulz, C. J. C. Boyle, & T. N. Sieber (Eds.), *Microbial root endophytes* (pp. 1–13). Berlin: Springer.

Schwachtje, J., Karojet, S., Kunz, S., Brouwer, S., & Van Dongen, J. T. (2012). Plant-growth promoting effect of newly isolated rhizobacteria varies between two *Arabidopsis* ecotypes. *Plant Signaling & Behavior, 7*(6), 623–627.

Sellstedt, A., & Richau, K. H. (2013). Aspects of nitrogen-fixing actinobacteria, in particular free-living and symbiotic *Frankia*. *FEMS Microbiology Letters, 342*, 179–186.

Selvakumar, G., Bindu, G. H., Bhatt, R. M., Upreti, K. K., Paul, A. M., Asha, A., Shweta, K., & Sharma, M. (2016). Osmotolerant cytokinin producing microbes enhance tomato growth in deficit irrigation conditions. In *Proceedings of the National Academy of Sciences, India, Section B: Biological Sciences*.

Seshadre, S., Muthukumarasamy, R., Lakshminarasimhan, C., & Ignaacimuthu, S. (2002). Solubilization of inorganic phosphates by *Azospirillum halopraeferans*. *Current Science, 79*(5), 565–567.

Shafi, J., Tian, H., & Ji, M. (2017). *Bacillus* species as versatile weapons for plant pathogens: A review. *Biotechnology and Biotechnological Equipment, 31*, 1–14.

Shahid, M., Hameed, S., Imran, A., Ali, S., & Van Elsas, J. D. (2012). Root colonization and growth promotion of sunflower (*Helianthus annuus* L.) by phosphate solubilizing *Enterobacter* sp. Fs-11. *World Journal of Microbiology and Biotechnology, 28*, 2749–2758.

Shahid, I., Malik, K. A., & Mehnaz, S. (2018). A decade of understanding secondary metabolism in Pseudomonas spp. for sustainable agriculture and pharmaceutical applications. *Environmental Sustainability, 1*(1), 3–17.

Shaikh, S., & Saraf, M. (2017). Zinc biofortification: Strategy to conquer zinc malnutrition through zinc solubilizing PGPR's. *Biomedical Journal of Scientific & Technical Research, 1*(1).

Shakeel, M., Rais, A., Hassan, M. N., & Hafeez, F. Y. (2015). Root associated *Bacillus* sp. improves growth, yield and zinc translocation for basmati rice (*Oryza sativa*) varieties. *Frontiers in Microbiology, 6*, 1286.

Shamseldin, A., Abdelkhalek, A., & Sadowsky, M. J. (2017). Recent changes to the classification of symbiotic, nitrogen-fixing, legume-associating bacteria: A review. *Symbiosis, 71*, 91–109.

Shanware, A. S., Kalkar, S. A., & Trivedi, M. M. (2014). Potassium solubilisers: Occurrence, mechanism and their role as competent biofertilizers. *International Journal of Current Microbiology and Applied Sciences, 3*, 622–629.

Shariatmadari, Z., Riahi, H., Seyed Hashtroudi, M., Ghassempour, A., & Aghashariatmadary, Z. (2013). Plant growth promoting cyanobacteria and their distribution in terrestrial habitats of Iran. *Soil Science & Plant Nutrition, 59*(4), 535–547.

Sharifi, P. (2017). The effect of plant growth promoting rhizobacteria (PGPR), salicylic acid and drought stress on growth indices, the chlorophyll and essential oil of hyssop (*Hyssopus officinalis*). *Biosciences, Biotechnology Research Asia, 14*(3).

Sharma, A., Shankhdar, D., & Shankhdhar, S. C. (2013). Enhancing grain iron content of rice by the application of plant growth promoting rhizobacteria. *Plant, Soil and Environment, 59*, 89–94.

Siddiqui, Z. A. (2006). PGPR: Prospective biocontrol agents of plant pathogens. In Z. A. Siddiqui (Ed.), *PGPR: Biocontrol and biofertilization* (pp. 111–142). Dordrecht: Springer.

Singh, P. P., Shin, Y. C., Park, C. S., & Chung, Y. R. (1999). Biological control of *Fusarium* wilt of cucumber by chitinolytic bacteria. *Phytopathology, 89*, 92–99.

Sivasakthi, S., Usharani, G., & Saranraj, P. (2014). Biocontrol potentiality of plant growth promoting bacteria (PGPR)-*Pseudomonas fluorescens* and *Bacillus subtilis*: A review. *African Journal of Agricultural Research, 9*, 1265–1277.

Someya, N., Tsuchiya, K., Yoshida, T., Noguchi, M. T., Akutsu, K., & Sawada, H. (2007). Co-inoculation of an antibiotic-producing bacterium and a lytic enzyme-producing bacterium for the biocontrol of tomato wilt caused by *Fusarium oxysporum* f. sp. lycopersici. *Biocontrol Science, 12*, 1–6.

Spaepen, S., & Vanderleyden, J. (2011). Auxin and plant-microbe interactions. *Cold Spring Harbor Perspectives in Biology, 3*(4), 1438.

Spaepen, S., Das, F., Luyten, E., Michiels, J., & Vanderleyden, J. (2009). Indole-3-acetic acid-regulated genes in *Rhizobium etli* CNPAF512. *FEMS Microbiology Letters, 291*, 195–200.

Steenhoudt, O., & Vanderleyden, J. (2000). *Azospirillum*, a free-living nitrogen-fixing bacterium closely associated with grasses: Genetics, biochemical and ecological aspects. *FEMS Microbiology Reviews, 24*, 487–506.

Suman, A., Yadav, A. N., & Verma, P. (2016). Endophytic microbes in crops: Diversity and beneficial impact for sustainable agriculture. In D. P. Singh, P. C. Abhilash, & P. Ratna (Eds.), *Microbial inoculants in sustainable agricultural productivity* (pp. 117–143). Dordrecht: Springer.

Sunithakumari, K., Devi, S. N. P., & Vasandha, S. (2016). Zinc solubilizing bacterial isolates from the agricultural fields of Coimbatore, Tamil Nadu, India. *Current Science, 110*, 196–205.

Tahir, H. A. S., Gu, Q., Wu, H., Raza, W., Hanif, A., Wu, L., Colman, M. V., & Gao, X. (2017). Plant growth promotion by volatile organic compounds produced by *Bacillus subtilis* SYST2. *Frontiers in Microbiology, 8*, 171.

Tailor, J. A., & Joshi, B. H. (2014). Harnessing plant growth promoting rhizobacteria beyond nature: A review. *Journal of Plant Nutrition, 37*, 1534–1571.

Tak, A., Gehlot, P., Pathak, R., & Singh, S. K. (2017). Species diversity of rhizobia. In A. Hansen, D. Choudhary, P. Agrawal, & A. Varma (Eds.), *Rhizobium biology and biotechnology. Soil biology* (pp. 215–245). Cham: Springer.

Tang, J. C., Wang, R. G., Niu, X. W., Wang, M., Chu, H. R., & Zhou, Q. X. (2010). Characterisation of the rhizoremediation of petroleum-contaminated soil: Effect of different influencing factors. *Biogeosciences, 7*(12), 3961–3969.

Tewari, S., & Arora, N. K. (2013). Plant growth promoting rhizobacteria for ameliorating abiotic stresses triggered due to climatic variability. *Climate Change Environ Sustain, 1*(2), 95–103.

Tewari, S., & Arora, N. K. (2014). Multifunctional exopolysaccharides from *Pseudomonas aeruginosa* PF23 involved in plant growth stimulation, biocontrol and stress amelioration in sunflower under stress conditions. *Current Microbiology, 69*, 484–494.

Tewari, S., & Arora, N. K. (2015). Plant growth promoting fluorescent pseudomonads enhancing growth of sunflower crop. *International Journal of Science and Technology, 1*(1), 51–53.

Tewari, S., & Arora, N. K. (2016). Soybean production under flooding stress and its mitigation using plant growth-promoting microbes. In M. Miransari (Ed.), *Environmental stresses in soybean production* (pp. 23–40). New York: Academic/Elsevier.

Tewari, S., & Arora, N. K. (2018). Role of salicylic acid from Pseudomonas aeruginosa PF23EPS+ in growth promotion of sunflower in saline soils infested with phytopathogen Macrophomina phaseolina. *Environmental Sustainability, 1*(1), 49–59.

Thakore, Y. (2006). The biopesticide market for global agricultural use. *Industrial Biotechnology, 2*, 192–208.

Tilak, K. V. B. R., Ranganayaki, N., Pal, K. K., De, R., Saxena, A. K., Nautiyal, C. S., Mittal, S., Tripathi, A. K., & Johri, B. N. (2005). Diversity of plant growth and soil health supporting bacteria. *Current Science, 89*, 136–150.

Timmusk, S., Behers, L., Muthoni, J., Muraya, A., & Aronsson, A. C. (2017). Perspectives and challenges of microbial application for crop improvement. *Frontiers in Plant Science, 8*, 49.

Transparency Market Research. (2017). *Biofertilizers market (Nitrogen fixing, phosphate solubilizing and others) for seed treatment and soil treatment applications – Global industry analysis, size, share, growth, trends and forecast, 2013–2019*. Available at: https://www.transparencymarketresearch.com/pressrelease/globalbiofertilizersmarket.htm.

Upadhyay, S. K., Singh, J. S., & Singh, D. P. (2011). Exopolysaccharide-producing plant growth-promoting rhizobacteria under salinity condition. *Pedosphere, 21*, 214–222.

Vacheron, J., Desbrosses, G., Bouffaud, M. L., Touraine, B., Moënne-Loccoz, Y., Muller, D., Legendre, L., Wisniewski-Dyé, F., & Prigent-Combaret, C. (2013). Plant growth-promoting rhizobacteria and root system functioning. *Frontiers in Plant Science, 4*, 356.

Vaid, S., Kumar, B., Sharma, A., Shukla, A., & Srivastava, P. (2014). Effect of Zn solubilizing bacteria on growth promotion and Zn nutrition of rice. *Journal of Soil Science and Plant Nutrition, 14*, 889–910.

Van Loon, L. C. (2007). Plant responses to plant growth-promoting rhizobacteria. *European Journal of Plant Pathology, 119*, 243–354.

Van Loon, L. C., Bakker, P. A., & Pieterse, C. M. J. (1998). Systemic resistance induced by rhizosphere bacteria. *Annual Review of Phytopathology, 36*, 453–483.

Vandenberghe, L. P. S., Garcia, L. M. B., Rodrigues, C., Camara, M. C., Pereira, G. V. M., Oliveira, J., & Soccol, C. R. (2017). Potential applications of plant probiotic microorganisms in agriculture and forestry. *AIMS Microbiology, 3*(3), 629–648.

Vargas, R., Detto, M., Baldocchi, D. D., & Allen, M. F. (2010). Multiscale analysis of temporal variability of soil CO_2 production as influenced by weather and vegetation. *Global Change Biology, 16*(5), 1589–1605.

Vejan, P., Abdullah, R., Khadiran, T., Ismail, S., & Boyce, A. N. (2016). Role of plant growth promoting rhizobacteria in agricultural sustainability –A review. *Molecules, 21*(5), 573.

Verma, V. C., Singh, S. K., & Prakash, S. (2011). Bio-control and plant growth promotion potential of siderophore producing endophytic *Streptomyces* from *Azadirachta indica* A. Juss. *Journal of Basic Microbiology, 51*(5), 550–556.

Verma, P., Yadav, A. N., Kazy, S. K., Saxena, A. K., & Suman, A. (2013). Elucidating the diversity and plant growth promoting attributes of wheat (*Triticum aestivum*) associated acidotolerant bacteria from southern hills zone of India. *National Journal of Life Sciences, 10*(2), 219–226.

Vessey, J. (2003). Plant growth promoting rhizobacteria as biofertilizers. *Plant and Soil, 255*, 571–586.

Vidhyasekaran, P. (2015). Auxin signaling system in plant innate immunity. In P. Vidhyasekaran (Ed.), *Plant hormone signaling systems in plant innate immunity, signaling and communication in plants* (pp. 311–357). Dordrecht: Springer.

Vijayabharathi, R., Sathya, A., & Gopalakrishnan, S. (2016). A renaissance in plant growth-promoting and biocontrol agents by endophytes. In D. P. Singh, H. B. Singh, & R. Prabha (Eds.), *Microbial inoculants in sustainable agricultural productivity* (pp. 37–61). New Delhi: Springer.

Vurukonda, S. S. K. P., Vardharajula, S., Shrivastava, M., & Ali, S. Z. (2016). Enhancement of drought stress tolerance in crops by plant growth promoting rhizobacteria. *Microbiological Research, 184*, 13–24.

Wang, W., Vinocur, B., & Altman, A. (2003). Plant responses to drought, salinity and extreme temperatures: Towards genetic engineering for stress tolerance. *Planta, 218*, 1–14.

Wang, J., Zhou, C., Xiao, X., Xie, Y., Zhu, L., & Ma, Z. (2017). Enhanced iron and selenium uptake in plants by volatile emissions of *Bacillus amyloliquefaciens* (BF06). *Applied Sciences, 7*(1), 85.

Wani, P. A., & Khan, M. S. (2010). *Bacillus* species enhance growth parameters of chickpea (*Cicer arietinum* L.) in chromium stressed soils. *Food and Chemical Toxicology, 48*, 3262–3267.

Wani, P. A., Khan, M. S., & Zaidi, A. (2007). Effect of metal tolerant plant growth promoting *Rhizobium* on the performance of pea grown in metal amended soil. *Archives of Environmental Contamination and Toxicology, 55*, 33–42.

WDR. (2003). *World disaster report: Focus on ethics in aid* (p. 240). Geneva: International Federation of Red Cross and Red Crescent Societies.

Weller, D. M., & Thomashow, L. S. (1994). Current challenges in introducing beneficial microorganisms into the rhizosphere. In F. O'Gara, D. N. Dowling, & B. Boesten (Eds.), *Molecular ecology of rhizosphere microorganisms biotechnology and the release of GMOs* (pp. 1–18). Weinheim: VCH Verlagsgesellschaft.

Weller, D. M., Raaijmakers, J. M., Gardner, B. B. M., & Thomashow, L. S. (2002). Microbial populations responsible for specific soil suppressiveness to plant pathogens. *Annual Review of Phytopathology, 40*, 308–348.

Welsh, A. K., Dawson, J. O., Gottfried, G. J., & Hahn, D. (2009). Diversity of *Frankia* populations in root nodules of geographically isolated Arizona alder trees in central Arizona (United States). *Applied and Environmental Microbiology, 75*(21), 6913–6918.
Wenzel, W. W. (2009). Rhizosphere processes and management in plant-assisted bioremediation (phytoremediation) of soils. *Plant and Soil, 321*, 385–408.
Werner, T., Motyka, V., Laucou, V., Smets, R., Van Onckelen, H., & Schmülling, T. (2003). Cytokinin-deficient transgenic *Arabidopsis* plants show multiple developmental alterations indicating opposite functions of cytokinins in the regulation of shoot and root meristem activity. *Plant Cell, 15*, 2532–2550.
Whipps, J. M. (2001). Microbial interactions and biocontrol in the rhizosphere. *Journal of Experimental Botany, 52*, 487–511.
Wilson, H., Epton, H. A. S., & Sigee, D. C. (1992). Biological control of fire blight of Hawthorn with fluorescent Pseudomonas spp. under protected conditions. *Journal of Phytopathology, 136*, 16–26.
Wu, S. C., Cao, Z. H., Li, Z. G., Cheung, K. C., & Wong, M. H. (2005). Effects of biofertilizer containing N-fixer, P and K solubilizers and AM fungi on maize growth: A greenhouse trial. *Geoderma, 125*, 155–166.
Yadav, A. N., Sachan, S. G., Verma, P., & Saxena, A. K. (2016). Bioprospecting of plant growth promoting psychrotrophic Bacilli from cold desert of north western Indian Himalayas. *Indian Journal of Experimental Biology, 54*(2), 142–150.
Yadav, A. N., Verma, P., Kumar, V., Sachan, S. G., & Saxena, A. K. (2017). Extreme cold environments: A suitable niche for selection of novel psychrotrophic microbes for biotechnological applications. *Advances in Biotechnology and Microbiology, 2*, 1–4.
Yadegari, M., Asadi Rahmani, H., Noormohammadi, G., & Ayneband, A. (2010). Plant growth promoting rhizobacteria increase growth, yield and nitrogen fixation in *Phaseolus vulgaris*. *Journal of Plant Nutrition, 33*(12), 1733–1743.
Yang, J., Kloepper, J. W., & Ryu, C. M. (2009). Rhizosphere bacteria help plants tolerate abiotic stress. *Trends in Plant Science, 14*(1), 1–4.
Yu, X., Ai, C., Xin, L., & Zhou, G. (2011). The siderophore-producing bacterium, *Bacillus subtilis* CAS15, has a biocontrol effect on *Fusarium* wilt and promotes the growth of pepper. *European Journal of Soil Biology, 47*, 138–145.
Zahir, Z. A., Arshad, M., & Frankenberger, W. T. (2004). Plant growth promoting rhizobacteria: Applications and perspectives in agriculture. *Advances in Agronomy, 81*, 97–168.
Zaidi, A., Khan, M. S., Ahemad, M., & Oves, M. (2009). Plant growth promotion by phosphate solubilizing bacteria. *Acta Microbiologica et Immunologica Hungarica, 56*, 263–284.
Zhan, J., & Sun, Q. (2012). Diversity of free-living nitrogen-fixing microorganisms in the rhizosphere and non-rhizosphere of pioneer plants growing on wastelands of copper mine tailings. *Microbiological Research, 167*(3), 157–165.
Zhang, S., Moyne, A. L., Reddy, M. S., & Kloepper, J. W. (2002). The role of salicylic acid in induced systemic resistance elicited by plant growth promoting rhizobacteria against blue mould of tobacco. *Biological Control, 25*, 288–296.
Zhuang, X., Chen, J., Shim, H., & Bai, Z. (2007). New advances in plant growth-promoting rhizobacteria for bioremediation. *Environment International, 33*(3), 406–413.

Chapter 7
Plausible Role of Plant Growth-Promoting Rhizobacteria in Future Climatic Scenario

R. Z. Sayyed, N. Ilyas, B. Tabassum, A. Hashem, E. F. Abd_Allah, and H. P. Jadhav

Contents

Abstract Increasing agro-productivity for feeding growing world population under present climatic scenario requires optimizing the use of resources and adopting the sustainable agriculture methods. This can be achieved by using plant-beneficial bacteria. Target of achieving sustainable agriculture implies the use of varieties that are resistant to disease and tolerant to stress and having desired nutrition value. This can be effectively achieved through the use of rhizospheric microflora including bacteria, fungi, algae, etc. Among these, plant growth-promoting rhizobacteria (PGPR) have been seen as reliable and most promising bioinoculants for promoting plant growth and controlling phytopathogen without causing environmental

R. Z. Sayyed (✉) · H. P. Jadhav
Department of Microbiology, PSGVP Mandal's Arts, Science & Commerce College, Shahada, Maharashtra, India
e-mail: sayyedrz@gmail.com

N. Ilyas
Botany Department, PMAS University of Arid Agriculture, Rawalpindi, Pakistan

B. Tabassum
Department of Zoology, Government Raza PG College, Rampur, Uttar Pradesh, India

A. Hashem
Faculty of Science, Department of Botany and Microbiology, King Saud University, Riyadh, Saudi Arabia

E. F. Abd_Allah
Department of Plant Production, College of Food and Agriculture Sciences, King Saud University, Riyadh, Saudi Arabia

R. C. Sobti et al. (eds.), *Environmental Biotechnology: For Sustainable Future*,
https://doi.org/10.1007/978-981-10-7284-0_7

deterioration. Application of PGPR as bioinoculants can help in achieving the target of global agricultural productivity to feed the world's booming population, which is expected to become 9 billion by 2050. However, to be useful and effective bioinoculants, PGPR strains should be competent in their habitat, safe to the environment, helpful in plant nutrition and biocontrol, compatible with useful soil rhizobacteria, and tolerant to a variety of stress factors and show broad spectrum activity. In the context of the above scenario, this chapter focusses on the use of PGPR to increase agro-productivity and as one of the vital drivers of the agro-economy. In this review we focus on the modes of action of PGPR and their role in environmental protection and agricultural sustainability under increasing climatic variations.

Keywords PGPR · Plant growth promotion · Biocontrol · Abiotic stress · Mechanisms

Introduction

The term PGPR as introduced by Kloepper and Schroth (1978) is used to define bacteria that are found in the rhizosphere and help in plant growth, nutrition, and disease control and thereby crop productivity. Since root exudates are sinks of nutrients for microbial growth, microflora of rhizosphere is relatively different from that of its surrounding (Burdman et al. 2000; Sayyed et al. 2004). Close zone of rhizosphere is a better source of nutrients for microbes as compared to the bulk soil. Population density of PGPR in close vicinity of rhizosphere is generally 10–100 times larger than in the bulk soil (Weller and Thomashow 1993). Bacteria are the most abundant species present in the rhizosphere followed by actinomycetes, fungi, algae, and protozoa that colonize the rhizosphere (Kaymak et al. 2010). PGPR can live in symbiotic association or as free-living microbes (Gray and Smith 2005).

Mechanisms of Action of PGPR

PGPR promote plant growth through direct and indirect mechanisms (Fig. 7.1).

Direct Mechanisms

The direct mechanisms of action of PGPR include supply of nutrients, phytohormones, and stimulation of root growth. Besides symbiotic bacteria, certain bacteria are able to establish mutualistic association with their host plant, while few bacteria can directly integrate their physiology with the plant, causing the formation of specialized structures.

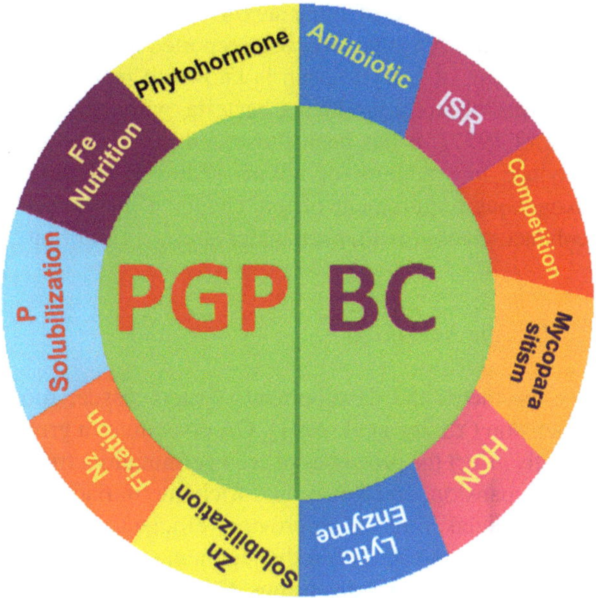

Fig. 7.1 Mechanisms of plant growth promotion and biocontrol by PGPR

Production of Phytohormones

Phytohormones are the plant growth hormones or phytostimulants that influence plant growth. They include auxins (indole-3-acetic acid (IAA), gibberellic acid (GA), cytokinins, and ethylene. These chemical molecules are recognized over the years as four major plant hormones needed for growth and development of plant. PGPR species belonging to the genera *Pseudomonas*, *Bradyrhizobium*, *Rhizobium*, *Enterobacter*, *Alcaligenes*, *Azospirillum*, *Acetobacter*, *Klebsiella*, and *Xanthomonas* and also species of *Bacillus pumilus*, *Bacillus licheniformis*, *Paenibacillus polymyxa*, *Phosphobacteria* sp., *Gluconacetobacter* sp., *Aspergillus* sp., and *Penicillium niger* possess the ability to produce phytohormones (Idris et al. 2007; Karnawal et al. 2009; Shobha and Kumudini 2012).

Auxins

Auxin is a molecule that directly or indirectly regulates most plant processes. Irrespective of endogenous supply of auxins, plants still depend largely on external supply for their optimum growth. This external supply of auxins is made by PGPR (Khalid et al. 2004; Patten and Glick 2002). Auxins trigger various cellular functions like differentiation of vascular tissues, root growth, cell division, stem elongation, and shoot growth in response to the stimuli (Glick 1995). Efficient production of IAA by PGPR depends on the type of species and strain, culture conditions, developmental stage, and availability of nutrient in the rhizosphere (Ashrafuzzaman et al. 2009). Role of other auxins, viz., indole-3-butyric acid (IBA) and phenyl acetic acid (PAA) that have been reported in plants, is yet to be understood. Increase in the level of L-tryptophan is known to increase the biochemical and metabolic activities of auxin-producing bacteria, with a parallel increase in the length of root (Bartel 1997). It has been illustrated that in the absence of L-tryptophan, PGPR produce low amount of auxins (Zahir et al. 2010). Hence, understanding the exogenous requirement of this chemical messenger to bring to the peak so as to enhance plant growth, even under stressed conditions, is crucial.

Gibberellic Acid (GA)

GA promotes the development of stem tissue, root elongation, and lateral root extension (Yaxley et al. 2001). GA constitutes a group of tetracyclic diterpenes that greatly affect the processes of seed germination, fruit development, stem elongation, leaf expansion, and flower and trichome initiation (Yamaguchi 2008). Because of their crucial role in improving efficient photosynthetic processes in plants, gibberellins and strains producing them are of prime importance particularly during environmental stress conditions. GA is thus an important plant growth bioregulator that can increase the stress tolerance in a number of crop plants. Kang et al. (2009) have reported plant growth promotion by GA-producing PGPR. The external supply of GA is useful in amendment of polluted soil and in crop yield (Iqbal et al. 2011). GA-producing PGPR have shown to improve the grain yield in wheat (Radi et al. 2006; Iqbal et al. 2011), barley (Vettakkorumakankav et al. 1999), and tomato by reducing stomatal resistance and enhancing water use efficiency (Maggio et al. 2010). Gibberellin is involved in regulating plant morphology (Van Loon 2007); thus GA functions as a stress tolerance-inducing hormone.

Cytokinins

Cytokinin is involved in cell division, biogenesis of chloroplast, vascular differentiation, nutrient mobilization, leaf senescence, root proliferation, root elongation, shoot differentiation, and apical dominance (Davies and Hartung 2004; Aloni et al. 2006). This molecule can be acquired endogenously and exogenously by either plant or PGPR. Plant increases uptake of endogenous cytokinin via the promotion of biosynthesis (Pospíšilová 2003). Cytokinin also regulates plant adaptation under salt stress conditions (Hadiarto and Tran 2011). Cytokinin helps in regulating the level of other phytohormones; they inhibit formation of abscisic acid (Pospíšilová 2003). Under water scarcity conditions, the plant cytokinin content is drastically reduced resulting in increase in ABA concentration. Mansour et al. (1994) have reported the production of cytokinins by various *Streptomyces* strains. Though this is vital for plant's development, its exact mechanism of action is still not well-elucidated. Cytokinin receptor genes of most plants and organisms are regulated by changes in osmotic conditions and also demonstrate a complex osmotic stress response (Merchan et al. 2007). Inoculation of seedling with cytokinin-producing strains of *B. subtilis* conferred plant resistance against environmental stresses (Merchan et al. 2007).

Ethylene

This is a unique phytohormone with a variety of biological functions. The beneficial role of this biomolecule is best recorded at low concentration, but at high concentration, results show that ethylene is a senescence hormone as it shows inhibitory role in plant growth. To overcome this alarming consequence, an enzyme 1-aminocyclopropane-1-carboxylic acid (ACC) deaminase is needed. The role of this biocatalyst is to degrade the plant ACC which is the direct precursor of ethylene synthesis in plant to α-ketobutyrate and ammonium (Glick et al. 2007). The result of the degradation is the reduction of plant ethylene production through a range of mechanisms, while the PGPR-producing ACC deaminase regulates the ethylene level in plant and prevents the growth inhibition caused by high levels of ethylene (Noumavo et al. 2016). PGPR capable of inducing exogenous production of ethylene via degradation of the endogenous product using enzyme include *Achromobacter*, *Acinetobacter*, *Azospirillum*, *Alcaligenes*, *Agrobacterium*, *Burkholderia*, *Bacillus*, *Pseudomonas*, *Enterobacter*, *Serratia*, *Ralstonia*, and *Rhizobium*. Work has shown that ACC deaminase activity of associated PGPR was vital for *Brassica napus* growth in stress conditions (Dell'Amico et al. 2008). Pierik et al. (2006) found that low concentration of ethylene increases the plant yield, growth performance, and germination properties of *A. thaliana*.

Nitrogen Fixation

Nitrogen is one of the major and important nutrients required by all living organisms for variety of cellular, synthetic, and metabolic functions. Although it is present abundantly (78%) in the atmosphere, its direct availability to plants and animals is restricted due to lack of mechanisms for uptake of elemental form of nitrogen. Nitrogen-fixing PGPR play a significant role in providing nitrogen in the soluble forms to the plants, and from plant the soluble nitrogen reaches to humans via food chain. Several bacteria living freely in soil or in symbiotic association with plants convert (fix) elemental nitrogen into soluble form that can be easily taken up by plants (Sayyed et al. 2012).

Symbiotic Nitrogen Fixers

These are the organisms that live in symbiotic association with leguminous plants. *Rhizobium* and *Frankia* spp. are the best examples of symbiotic nitrogen-fixing PGPR. Besides these *Azorhizobium*, *Bradyrhizobium*, and *Cyanorhizobium* also fix nitrogen in association with leguminous plants. The efficient strains of *Rhizobium* can fix 40–200 kg/ha N in the soil (Sayyed et al. 2012).

Free-Living Nitrogen Fixers

These are the microorganisms that live freely in soil and fix nitrogen. This soluble nitrogen in soil is easily utilized by crop plants. *Azotobacter* sp., *Azospirillum* sp., blue green algae (BGA), *Acetobacter* sp., and *Azolla* sp. are known as best free-living nitrogen-fixing PGPR. They can fix 40–100 kg/ha N in the soil (Sayyed et al. 2012).

Sulfur Solubilization

Although sulfur is present in large quantities (95%) in soil, it is not easily available to crop plants as it is organically bounded in the form of sulfate esters or sulfonates, which are unavailable to the plant and need conversion into inorganic forms via microbial desulfurization. Sulfur-oxidizing PGPR, like *Thiobacillus thiooxidans* and *Thiobacillus novellus*, oxidize insoluble sulfur into soluble form through multienzyme complex responsible for cleaving the S group from the aromatic ring. S-solubilizing PGPR play significant role in sulfur nutrition of crop plants (Gomah et al. 2014).

Potash Solubilization

Potassium (K) is an essential macronutrient for plant growth; it plays important roles in several metabolic processes such as protein synthesis, photosynthesis, and enzyme activation (Raghavendra et al. 2016). Potassium is not just the essential growth nutrient, but it is also an important signaling agent required in mediating a wide range of plant adaptive responses to abiotic and biotic stresses such as drought, salinity, oxidative stress, and apoptosis. Major amount of total K^+ of the soil exists in insoluble form complexed with other elements and organic matter, making it unavailable. In this regard many potassium-solubilizing microorganisms (KSMs) present in soil are capable to solubilize insoluble/unavailable forms of the K into soluble/available form. The main mechanisms of K^+ solubilization include acidolysis, chelation, exchange reactions, and production of organic acid. KSM can increase crop yield by 20–25% (Zahedi 2016).

Phosphate Solubilization

Phosphorus is the second important key element after nitrogen as a mineral nutrient in terms of quantitative plant requirement. It is abundantly available in soil in insoluble forms (Khan et al. 2010) and thus not available for plants. P is generally added in soil as chemical fertilizers, but synthesis and use of chemical P fertilizer pose numerous demerits. In this regard phosphate-solubilizing microorganisms (PSM) have been seen as sustainable means of P nutrition of crops. Several species of bacteria (*Bacillus* and *Pseudomonas*), molds (*Aspergillus* and *Penicillium*), yeasts (*Schizosaccharomyces* and *Pichia*), actinomycetes (*Streptomyces*), cyanobacteria (*Anabaena*, *Calothrix*, *Nostoc* sp.), and vesicular arbuscular mycorrhiza (*Glomus* sp.) are known to solubilize insoluble P into soluble form that is easily absorbed by plants (Saber et al. 2013). PSMs are important components of the P cycle in soil and are vital agents for P solubilization through various mechanisms (Khan et al. 2010) such as (1) the release of P mineral metabolites like organic acid anions, siderophores, protons, hydroxyl ions, and CO_2, (2) the excretion of P-solubilizing phosphatase enzymes (biochemical P mineralization), and (3) the release of P during enzymatic substrate degradation (biological P mineralization). Thus PSMs play a crucial role in P nutrition of crop plants (Sharma et al. 2013).

Iron Nutrition

Iron is an essential element for survival of almost all cell types; it is the fourth most abundant element present in soil, but it exists in insoluble and therefore unavailable form. In order to sequester and solubilize the iron present in soil, PGPR produce a variety of iron-chelating molecules referred to as siderophores (Sayyed et al. 2013). Most important biotechnological exploitation of siderophores is in the rhizosphere region of the plant where they provide iron, serve as first defense against root-invading parasites, and help in removing toxic metals from polluted soil. There are sufficient evidences available regarding iron uptake by plants through microbial siderophores, which convert the insoluble form of iron into soluble form. Siderophore-producing PGPR have been reported to promote growth of various crop plants (Sayyed et al. 2007, 2009).

Zinc Solubilization

Zinc is one of the important micronutrients required in very small quantity (5–100 mg kg^{-1}) for growth and reproduction of plants. Its deficiency affects membrane integrity and synthesis of carbohydrates, IAA, nucleotides, and chlorophyll and makes plant susceptible to heat stress. Zn deficiency in plants is due to its low solubility in soils. Zinc-solubilizing PGPR are potential alternatives to synthetic chemical supplements. Several PGPR including *Pseudomonas* sp. and *Bacillus* sp. are known as Zn solubilizers. Goteti et al. (2013) has reported Zn-solubilizing and plant growth-promoting activity of *Pseudomonas* sp. and *Bacillus* sp. in *Zea mays*. Gontia-Mishra et al. (2017) have also reported Zn-solubilizing and plant growth-promoting ability of *Pseudomonas aeruginosa*, *Ralstonia pickettii*, *Burkholderia cepacia*, and *Klebsiella pneumonia* in rice. *Exiguobacterium aurantiacum* producing array of plant growth-promoting traits have been found effective in Zn solubilization for *Triticum aestivum* (Shaikh and

Saraf 2017). *Pseudomonas fragi*, *Pantoea dispersa*, *Pantoea agglomerans*, *Enterobacter cloacae*, and *Rhizobium* sp. isolated from wheat and sugarcane fields solubilized zinc during pot and field trials and promoted growth of sugarcane and wheat (Boughammoura et al. 2017).

Indirect Mechanisms

PGPR promote plant growth indirectly by inhibiting growth of phytopathogens through the production of antibiotics and induction of systemic resistance and by competing with available nutrients and niches, thus making these unavailable for phytopathogens (Egamberdieva and Lugtenberg 2014).

Antibiotic Production

Antibiotics are low molecular weight compounds produced by PGPR to inhibit growth of phytopathogens. Antibiotic production is one of the most known biocontrol strategies displayed by PGPR. Many antibiotics such as amphisin, 2,4-diacetylphloroglucinol (DAPG) (Vinay et al. 2016; Reshma et al. 2018), oomycin A, phenazine, pyoluteorin, pyrrolnitrin, tensin, tropolone, and cyclic polypeptides (oligomycin A, kanosamine, zwittermicin A, and xanthobaccin) are produced by PGPR (Compant et al. 2005; Loper and Gross 2007). These biochemicals are generally produced by *Pseudomonas*, *Bacillus*, *Streptomyces*, and *Stenotrophomonas* sp. as active chemical agents. Many antibiotics have successfully been utilized to control phytopathogens (Kang et al. 2004; Kaur et al. 2006; Cazorla et al. 2006; Perneel et al. 2008). Increase in crop productivity as a result of biocontrol by PGPR has been reported in pepper root rot (Ezziyyani et al. 2007), leaf blight/seedling blight of rice, *Fusarium* root rot and tomato wilt (Minuto et al. 2006), *S. hygroscopicus* infection, and anthracnose (Prapagdee et al. 2008). Partial list of antibiotic-producing PGPR applied in biocontrol of phytopathogens is given in Table 7.1.

Nutrients and Niche Competition

In order to establish as a dominant species in the soil, PGPR must be able to compete favorably for the available nutrients and space. This is a vital strategy needed for limiting the disease incidence and severity (Kamilova et al. 2005). Rapid and abundant colonization of rhizosphere makes the rhizosphere unavailable for phytopathogens. Aggressive root-colonizing PGPR take charge and control of the metabolic activities in rhizosphere. Besides their inherent property to grow via competition, other properties such as presence of flagellum and lipopolysaccharides,

Table 7.1 List of antibiotic-producing PGPR

PGPR	Antibiotic	Target phytopathogen	Reference
Pseudomonas sp.	2, 4-Diacetylphloroglucinol (2, 4- DAPG)	*Gaeumannomyces graminis* var. *tritici*	de Souza et al. (2003)
P. fluorescens	Phenazine-1-carboxylic acid (PCA)	Pathogenic fungi	Weller (2007)
Bacillus amyloliquefaciens	Lipopeptide and polyketide	Soilborne pathogens	Sacherer et al. (1994)
Pseudomonad sp.	Phenazine	*F. oxysporum* and *Gaeumannomyces graminis*	Chin-A-Woeng et al. (2003)
P. chlororaphis PCL1391	Phena-zine-1-carboxamide	Pathogenic fungi	Hernandez et al. (2004), Haas and Défago (2005)
Bacillus ssp.	Polymyxin, circulin, and colistin	Pathogenic fungi	Maksimov et al. (2011)
P. cepacia	Pyrrolnitrin	*Bipolaris maydis*	Sayyed et al. (2013)
P. fluorescens Pf5		*Sclerotina homoecarpa*	
P. fluorescens2-79	Phenazines	Various sp. of bacteria and fungi	
P. aureofaciens 30-84		*G. graminis tritici*	
P. aureofaciens PGS12		*G. graminis tritici*	
B. subtilis st.	Iturin and fengycin	*Podosphaera fusca*	Romero et al. (2007)
Bacillus, *Streptomyces*, *and Stenotrophomonas* sp.	Oligomycin A, kanosamine, zwittermicin A, and xanthobaccin	Prevents the proliferation of plant pathogens (generally fungi)	Compant et al. (2005a, b)

chemotaxis, and usage of root exudates enhances the survival of PGPR (Lugtenberg and Kamilova 2009). A good illustration can be seen in unavailability of iron to phytopathogenic fungi when chelated by siderophores synthesized by PGPR (Sayyed et al. 2011, 2012, 2013, 2015; Shaikh and Sayyed 2015; Shaikh et al. 2014, 2016). The ability of PGPR to use heterologous siderophores offers added advantage during nutrient competition in soil. This also reduces the metabolic efforts inside the microbial cell (Sayyed and Patel 2011; Sayyed and Chincholkar 2006, 2010). Iron is one of the prerequisite nutrients required for synthesis of ATP and as a metal ion for functioning of various enzymes and protein. In niche competition, a physical occupation of site by PGPR is enhanced through delay tactics, by preventing the colonization of pathogens until the available substrate is exhausted (Heydari and Pessarakli 2010). This feature has been an age-old adaptive property exerted by beneficial soil microorganisms to occupy the root rhizosphere and make nutrients available for their upkeep and unavailable for other microbes.

Induced Systemic Resistance (ISR)

PGPR trigger inducement of defense system in plants that is capable of fighting pathogenic bacteria, fungi, and viruses. This potentially positions the plant as a much stronger and highly adapted species (Van Loon 2007). The gene and gene products involved in such type of biocontrol phenomenon have not been well-documented. Systemic acquired resistance (SAR) is a defense mechanism activated in plant following the primary infection (Ryals et al. 1996; Handelsman and Stabb 1996), while induced systemic resistance (ISR) utilizes organic acids and phytohormones to produce signals for stimulation of the host plant defense response against phytopathogens (Beneduzi et al. 2012; Pieterse et al. 2014; Patel et al. 2016). ISR involves increase in physical and mechanical strength of the cell wall and modulation of physical and biochemical reactions of the cell to environmental stresses (Labuschagne et al. 2010). ISR by PGPR is mediated through the production of salicylic acid, siderophores, lipopolysaccharides, flagella, N-acyl homoserine lactone (AHL) molecules (Van Loon 2007; Shuhegger et al. 2006), and antibiotics. The participating organisms in this form of biocontrol include *Bacillus and Pseudomonas*. On a wider scale, application of PGPR strains has tremendously improved ISR against *Colletotrichum lagenarium*, which causes anthracnose in cucumber, and *Pseudomonas syringae* causing angular leaf spot and bacterial wilt by *Erwinia tracheiphila* in a number of crops (Zehnder et al. 2001).

Lytic Enzymes Production

The production of lytic enzymes such as chitinases, cellulases, lipases, ß-1,3-glucanases, and proteases by rhizobacteria has been suggested to be a vital form of biocontrol (Markovich and Kononova 2003; Jadhav and Sayyed 2016; Jadhav et al. 2017). These hydrolytic enzymes degrade wide range of compounds. In order to hydrolyze the cell wall of phytopathogen, PGPR must secrete multiple hydrolytic enzymes (Whipps 2001) either alone or in combination (Mabood et al. 2014). β-1,3-glucanase produced by *Lysobacter enzymogenes* has been reported to lyse the cell wall of *Pythium* (Palumbo et al. 2005). These hydrolytic enzymes also help the plant in protection from desiccation and from other abiotic stresses (Qurashi and Sabri 2012). Hydrolytic enzyme-producing PGPR have been found effective in controlling blight in pepper caused by *Phytophthora capsici* (Jung et al. 2005), *Fusarium* infection (Hariprasad et al. 2011), and *Pythium ultimum* (Dunne et al. 1997). Chaiharn and Lumyong (2009) reported the antagonistic effect of chitinase, β-1,3-glucanase, protease, and cellulase-producing PGPR. Lytic enzyme-producing *Pseudomonas* sp. is also known to be a good biocontrol agent (Cattelan et al. 1999). Mycoparasites have also been employed in biocontrol activities against *Rosellinia necatrix* and other plant pathogens using chitinases (Harman et al. 2004; Ten Hoopen and Krauss 2006).

Hydrogen Cyanide (HCN)

Hydrogen cyanide is principally produced by *Pseudomonas* sp. (Sayyed and Patel 2011). Methods for quantitative estimation of HCN are now available (Lorck 1948; Castric 1977). Among the volatile compounds, HCN is a well-studied metabolite. Its cyanide ion inhibits most metalloenzymes, especially copper-containing cytochrome c oxidases (Blumer and Haas 2000). Cyanide produced by *Pseudomonas* strains has been used successfully to curb canker of tomato (Lanteigne et al. 2012). It is produced by Gram-negative bacteria as secondary metabolite from glycine under the influence of HCN synthase (Castric 1994). *P. fluorescens* strain CHA0 (Voisard et al. 1989) was used to control tobacco black root rot caused by *Thielaviopsis basicola* (Laville et al. 1998). Due to the aggressive colonizing strength (and HCN production is one of the reasons) of fluorescent pseudomonads, they have effectively been used in the biocontrol of soilborne plant pathogens (Lugtenberg et al. 2001). There are still indications that a good number of rhizobacteria are cyanogenic when provided with glycine in their culture medium.

Use of Hypovirulent Strains

Hypovirulence is a reduced virulence found in few strains of phytopathogens. Application of such hypovirulent strains of phytopathogens has helped in reducing the effect of virulent phytopathogenic strains. *Rhizoctonia solani*, *Gaeumannomyces graminis* var. *tritici*, and *Ophiostoma ulmi* have been used as hypovirulent strains to reduce the severity of plant diseases caused by virulent strains (Sayyed et al. 2013).

Detoxification or Degradation of Virulence Factors

Detoxification of virulence factors of pathogens is another important mechanism of biocontrol. In this mechanism biocontrol strain produces a protein that reversibly binds the toxin leading to irreversible detoxification. Biocontrol strain of *Pseudomonas* sp. is known to detoxify albicidin toxin produced by *Xanthomonas albilineans* (Walsh et al. 2001). Other biocontrol strains, for example, *Burkholderia cepacia* and *Ralstonia solanacearum*, have been reported to hydrolyze a phytotoxin (fusaric acid) produced by various *Fusarium* species.

Mycoparasitism

PGPR with biocontrol potential can be parasites or predators of the pathogens. Mycoparasites, such as *Coniothyrium minitans* and *Sporidesmium sclerotivorum*, have been tested as biocontrol agents, and some of them are efficient in controlling diseases caused by *Sclerotinia* sp. and other sclerotia-forming fungi (Sayyed et al. 2013). *Trichoderma* spp. are being used as biocontrol agents against several plant pathogens. These fungi produce cell wall-degrading enzymes such as beta-1,3-glucanase, chitinase, acid phosphatase, acid proteases, and alginate lyase and toxic volatile metabolites, having significant effects on growth and development of the plant pathogens (Qualhato et al. 2013).

Role of PGPR Under Stress

Soil environmental stresses are the major limiting factors for achieving the sustainability in agricultural production, as they determine the growth of plant. Environmental stresses are of two types – biotic and abiotic. Biotic stress includes phytopathogens and pests such as nematodes, insects, etc., while abiotic stresses include presence of heavy metals in soils, drought, nutrient deficiency, soil salinity, temperature, etc.

Abiotic Stress Tolerance in Plants

In general, PGPR promote plant growth by different ways. Some strains possess more than one mechanism and can help the plant withstand environmental stresses. Under natural, i.e., stress-free, environment most of the plant growth promotion mechanisms used by PGPR are common, but under stress conditions, PGPR may not be able to perform efficiently. However, some PGPR strains tolerate stress conditions and perform their growth-promoting activities even under stressful environmental conditions. The plant growth-promoting ability of PGPR also depends on the interaction with host plant and soil environment. Various mechanisms are adopted by PGPR to mitigate stress-induced adverse effects on plants. Lowering of stress-induced ethylene level, exopolysaccharide (EPS) production, and induction of ISR are some of the examples of such mechanisms (Saravanakumar et al. 2007; Glick et al. 2007; Saharan and Nehra 2011; Upadhyaya et al. 2011). Lowering of ethylene level is one of the major mechanisms elicited by PGPR for promoting plant growth under stress conditions (Mattoo and Suttle 1991; Glick et al. 2007). Under stress conditions the precursor of ethylene, i.e., ACC, is produced in larger quantities (Zapata et al. 2004); as a result more ethylene is produced. ACC regulates root elongation that ultimately affects nutritional as well as physiological functions of plants (Belimov et al. 2002; Alarcon et al. 2012). Therefore, for normal growth of plants, the level of ethylene needs to be kept under control. This is achieved by ACC deaminase produced by PGPR; the enzyme breaks ACC into ammonia and α-ketobutyrate (Glick et al. 2007) resulting in decreased level of ethylene in root vicinity (Whipps 1990; Glick et al. 1998). Increased synthesis and endogenous storage of IAA stimulates ACC synthase that converts S-adenosyl methionine to ACC in plant roots (Patten and Glick 1996). Plant excretes this ACC in their root exudates, which is taken up by the PGPR. This ACC is then converted to ammonia and α-ketobutyrate by ACC deaminase; this in turn lowers the ethylene level, thus offering protection to the plant from toxic concentrations of ethylene. This mechanism suppresses the negative effect of ethylene on the root (Patten and Glick 1996). Many researchers have demonstrated the effectiveness of this mechanism in plant growth promotion (Nadeem et al. 2006a; Cheng et al. 2007; Tank and Saraf 2010; Barnawal et al. 2012; Siddikee et al. 2012; Chen et al. 2013).

Under stress conditions, growth of plant is affected due to imbalance of mineral ions, for instance, saline conditions (high level of Na^+) affect the uptake of other nutrients as well as cause specific ion toxicity (Ashraf 1994). High K^+/Na^+ ratio is one of the prerequisites for salt tolerance and maintenance of osmotic balance in a plant (Hamdia et al. 2004). PGPR produce EPS that chelate Na^+ ions, thus making it unavailable for uptake by plant (Geddie and Sutherland 1993; Khodair et al. 2008; Qurashi and Sabri 2012). The reduced availability of Na^+ results in lowering of its uptake, thereby maintaining high K^+/Na^+ ratio that provides tolerance to plants under high salt concentrations (Ashraf et al. 2004; Han and Lee 2005; Khodair et al. 2008). EPS also protect the plant from drying and desiccation and enable them to continue their growth under water-deficit conditions. *Pseudomonas* species is known to lower stress-induced ethylene level and also decrease the availability of Na^+ by producing EPS. Salinity is one of the major abiotic stresses in crop production particularly in arid and semiarid regions of the world. Studies have revealed that inoculation with PGPR having ability to produce ACC deaminase significantly promoted the plant growth. Inoculation with salt-tolerant PGPR also enhances the uptake of other major nutrients as well as improves the water content of stressed plants (Mayak et al. 2004; Nadeem et al. 2006b, 2007). Inoculation with *Klebsiella oxytoca* (Rs-5) having ACC deaminase resulted in enhanced uptake of ions like N, P, K, and Ca and promoted plant growth by mitigating the negative effects of salt stress (Yue et al. 2007). Inoculation with *Pseudomonas* spp. improved the eggplant growth by lowering the uptake of Na^+ under salinity stress (Fu et al. 2010). Increase in the antioxidant enzyme activities and regulation of mineral uptake may be the two key mechanisms involved in salt stress alleviation. PGPR strains are effective in promoting plant growth under salinity stress, heavy metal stress, drought, and flooding (Glick et al. 2007). Inoculation of *P. fluorescens* protected *Catharanthus roseus* plants from drought stress by increasing the level of ajmalicine (antihypertensive alkaloid) content (Jaleel et al. 2009). PGPR having ACC deaminase also alleviated the adverse effects of drought stress in pea plants (Zahir et al. 2008). EPS-producing PGPR have been successfully used for enhancing drought resistance in sunflower plants (Tewari and Arora 2018). Lowering down of ethylene level also helped in alleviating heavy metal stress (Belimov et al. 2005; Dell'Amico et al. 2008), as PGPR strains can accumulate metals in their cells thereby reducing heavy metal ion concentration. Another important aspect of PGPR is to increase resistance against phytopathogens. Biocontrol potential of PGPR has been demonstrated against large number of phytopathogens even under stress conditions (O'Sullivan and O'Gara 1992; Singh et al. 1999; Ramos Solano et al. 2008; Kotan et al. 2010; Nihorimbere et al. 2011; Bhattacharyya and Jha 2012; Soltani et al. 2012).

PGPR strains are also helpful to overcome the harmful effect of temperature stress, increase the shelf life of flowers, and inhibit parasitic weeds (Bensalim et al.

1998; Nayani et al. 1998; Grichko and Glick 2001; Babalola et al. 2003; Barka et al. 2006). Zahir et al. (2009) reported *P. putida* as better mitigator of salinity stress. *P. fluorescens* and *P. stutzeri* responded better in enhancing growth of canola and tomato plants, respectively, under stressed conditions (Jalili et al. 2009; Tank and Saraf 2010). Such variation in plant growth promotion is due to variable ACC deaminase activity, IAA production, root colonization, and P solubilization abilities (Gamalero et al. 2009). PGPR strains have been found to perform well with other microbial consortia. Figueiredo et al. (2008) claimed good synergistic effect of co-inoculation of *Paenibacillus polymyxa* and *Rhizobium tropici* on growth, nitrogen content, and nodulation in *Phaseolus vulgaris* L. (common bean) under drought stress in a greenhouse experiment. For commercialization of this formulation, this performance was cross-checked under natural field conditions where similar synergistic effect was observed (Heidari et al. 2011). These PGPR were effective under drought stress as well as salinity stress. The growth and yield of groundnut were significantly higher under salt stress conditions when inoculated with PGPR strains. However, the performance of strains was variable. Similar results have been obtained and reported when maize seed was inoculated with rhizobacteria containing ACC deaminase.

Thus PGPR strains are effective bioinoculants for enhancing plant growth under a wide range of stresses such as drought, flooding, salinity, heavy metals, and pathogen attack. This plant growth-promoting effect is attributed to the ability of PGPR of lowering down the ethylene concentration and production of EPS or through ISR. Few studies have been conducted in the field, and further work needs to be done under natural conditions.

Conclusion

PGPR promote plant growth through biofertilization or pathogen suppression, by imparting disease resistance to plants and suppression of phytopathogens in the soil. However, only those strains which possess useful properties like stress tolerance, efficient root colonization, and rhizosphere competence can be potential candidates for bioinoculants to be used in agriculture for sustainable development. PGPR that promote plant growth by efficient nutrient uptake, stress tolerance, biocontrol of plant pathogens, and induction of SAR and ISR have been seen useful in increasing agricultural production while maintaining environmental sustainability. Therefore, the use of PGPR as multifunctional bioinoculants can help in achieving sustainable global agricultural productivity for feeding the booming world's population.

References

Alarcon, M. V., Lloret, P. G., Iglesias, D. J., Talon, M., & Salguero, J. (2012). Comparison of growth responses to auxin 1-naphthaleneacetic acid and the ethylene precursor 1-aminocyclopropane-1-carboxilic acid in maize seedling root. *Acta Biologica Cracoviensia Series Botanica, 54*, 16–23.

Aloni, R., Aloni, E., Langhans, M., & Ullrich, C. I. (2006). Role of cytokinin and auxin in shaping root architecture: Regulating vascular differentiation, lateral root initiation, root apical dominance and root gravitropism. *Annals of Botany, 97*, 883–893.

Ashraf, M. (1994). Organic substances responsible for salt tolerance in *Eruca sativa. Biologia Plantarum, 36*, 255–259.

Ashraf, M., Hasnain, S., Berge, O., & Mahmood, T. (2004). Inoculating wheat seedlings with exopolysaccharide-producing bacteria restricts sodium uptake and stimulates plant growth under salt stress. *Biology and Fertility of Soils, 40*, 157–162.

Ashrafuzzaman, M., Hossen, F. A., Ismail, M. R., Hoque, A., Islam, M. Z., Shahidullah, S. M., & Meon, S. (2009). Efficiency of plant growth-promoting rhizobacteria (PGPR) for the enhancement of rice growth. *African Journal of Biotechnology, 8*(7), 1247–1252.

Babalola, O. O., Osir, E. O., Sanni, A. I., Odhiambo, G. D., & Bulimo, W. D. (2003). Amplification of 1-amino-cyclopropane-1-carboxylic (ACC) deaminase from plant growth promoting rhizobacteria in striga-infested soil. *African Journal of Biotechnology, 2*, 157–160.

Barka, E. A., Nowak, J., & Clément, C. (2006). Enhancement of chilling resistance of inoculated grapevine plantlets with a plant growth-promoting rhizobacterium, *Burkholderia phytofirmans* strain PsJN. *Applied and Environmental Microbiology, 72*, 7246–7252.

Barnawal, D., Bharti, N., Maji, D., Chanotiya, C. S., & Kalra, A. (2012). 1-Aminocyclopropane-1-carboxylic acid (ACC) deaminase-containing rhizobacteria protect *Ocimum sanctum* plants during waterlogging stress via reduced ethylene generation. *Plant Physiology and Biochemistry, 58*, 227–235.

Bartel, B. (1997). Auxin biosynthesis. *Annual Review of Plant Biology, 48*, 51–66.

Belimov, A. A., Safronova, V. I., & Mimura, T. (2002). Response of spring rape (*Brassica napus* var. *oleifera* L.) to inoculation with plant growth promoting rhizobacteria containing 1-aminocyclopropane-1-carboxylate deaminase depends on nutrient status of the plant. *Canadian Journal of Microbiology, 48*, 189–199.

Belimov, A. A., Hontzeas, N., Safronova, V. I., Demchinskaya, S. V., Piluzza, G., Bullitta, S., & Glick, B. R. (2005). Cadmium-tolerant plant growth-promoting bacteria associated with the roots of Indian mustard (*Brassica juncea L.* Czern.). *Soil Biology and Biochemistry, 37*, 241–250.

Beneduzi, A., Ambrosini, A., & Passaglia, L. M. (2012). Plant growth-promoting rhizobacteria (PGPR): Their potential as antagonists and biocontrol agents. *Genetics and Molecular Biology, 35*, 1044–1051.

Bensalim, S., Nowak, J., & Asiedu, S. K. (1998). A plant growth promoting rhizobacterium and temperature effects on performance of 18 clones of potato. *American Journal of Potato Research, 75*, 145–152.

Bhattacharyya, P. N., & Jha, D. K. (2012). Plant growth-promoting rhizobacteria (PGPR): Emergence in agriculture. *World Journal of Microbiology and Biotechnology, 28*, 1327–1350.

Blumer, C., & Haas, D. (2000). Mechanism, regulation, and ecological role of bacterial cyanide biosynthesis. *Archives of Microbiology, 173*, 170–177.

Boughammoura, S., Chemek, M., Mimouna, S. B., Banni, M., & Messaoudi, I. (2017). Involvement of Zn depletion in cd-induced toxicity on prenatal bone formation in rat. *Biology of Trace Elements Research in Press, 180*(1), 70–80. https://doi.org/10.1007/s12011-017-0981-7.

Burdman, S., Jurkevitch, E., & Okon, Y. (2000). Recent advances in the use of plant growth promoting rhizobacteria (PGPR) in agriculture. In *Microbial interactions in agriculture and forestry* (Vol. 2, pp. 229–250).

Castric, P. A. (1977). Glycine metabolism by *Pseudomonas aeruginosa:* Hydrogen cyanide biosynthesis. *Journal of Bactriology, 130*, 826–831.
Castric, P. (1994). Influence of oxygen on the *Pseudomonas aeruginosa* hydrogen cyanide synthase. *Current Microbiology, 29*, 19–21.
Cattelan, A. M., Aversa, S. M. L., Zanchetta, M., Meneghetti, F., De Rossi, A., & Chieco-Bianchi, L. (1999). Regression of AIDS-related Kaposi's sarcoma following antiretroviral therapy with protease inhibitors: Biological correlates of clinical outcome. *European Journal of Cancer, 35*, 1809–1815.
Cazorla, F. M., Duckett, S. B., Bergström, E. T., Noreen, S., Odijk, R., Lugtenberg, B. J., Thomas-Oates, J. E., & Bloemberg, G. V. (2006). Biocontrol of avocado dematophora root rot by antagonistic *Pseudomonas fluorescens PCL1606* correlates with the production of 2-hexyl 5-propyl resorcinol. *Molecular Plant-Microbe Interactions, 19*, 418–428.
Chaiharn, M., & Lumyong, S. (2009). Phosphate solubilization potential and stress tolerance of rhizobacteria from rice soil in northern Thailand. *World Journal of Microbiology and Biotechnology, 25*, 305–314.
Chen, Y., Yan, F., Chai, Y., Liu, H., Kolter, R., Losick, R., & Guo, J. H. (2013). Biocontrol of tomato wilt disease by *Bacillus subtilis* isolates from natural environments depends on conserved genes mediating biofilm formation. *Environmental Microbiology, 15*, 848–864.
Cheng, Z., Park, E., & Glick, B. R. (2007). 1-Aminocyclopropane-1-carboxylate (ACC) deaminase from *Pseudomonas putida UW4* facilitates the growth of canola in the presence of salt. *Canadian Journal of Microbiology, 53*(7), 912–918.
Chin-A-Woeng, T. F., Bloemberg, G. V., & Lugtenberg, B. J. (2003). Phenazines and their role in biocontrol by *Pseudomonas* bacteria. *The New Phytologist, 157*, 503–523.
Compant, S., Duffy, B., Nowak, J., Clément, C., & Barka, E. A. (2005a). Use of plant growth-promoting bacteria for biocontrol of plant diseases: Principles, mechanisms of action, and future prospects. *Applied and Environmental Microbiology, 71*, 4951–4959.
Compant, S., Reiter, B., Sessitsch, A., Nowak, J., Clément, C., & Barka, E. A. (2005b). Endophytic colonization of *Vitis vinifera L.* by plant growth-promoting bacterium *Burkholderia* sp. strain PsJN. *Applied and Environmental Microbiology, 71*, 1685–1693.
Davies, W. J., & Hartung, W. (2004). Has extrapolation from biochemistry to crop functioning worked to sustain plant production under water scarcity. In *Proceeding of the fourth International crop Science Congress* (Vol. 26).
de Souza, J. T., Weller, D. M., & Raaijmakers, J. M. (2003). Frequency, diversity and activity of 2, 4-diacetylphloroglucinol producing fluorescent *Pseudomonas* spp. in Dutch take-all decline soils. *Phytopathology, 93*, 54–63.
Dell'Amico, E., Cavalca, L., & Andreoni, V. (2008). Improvement of Brassica napus growth under cadmium stress by cadmium-resistant rhizobacteria. *Soil Biology and Biochemistry, 40*, 74–84.
Dunne, C., Crowley, J. J., Moënne-Loccoz, Y., Dowling, D. N., & O'Gara, F. (1997). Biological control of Pythium ultimum by *Stenotrophomonas maltophilia* W81 is mediated by an extracellular proteolytic activity. *Microbiology, 143*, 3921–3931.
Egamberdieva D, Lugtenberg B (2014) Use of plant growth-promoting rhizobacteria to alleviate salinity stress in plants. In *Use of microbes for the alleviation of soil stresses* (Vol. 1, pp. 73–96). New York: Springer.
Ezziyyani, M., Requena, M. E., Egea-Gilabert, C., & Candela, M. E. (2007). Biological control of phytophthora root rot of pepper using *Trichoderma harzianum* and *Streptomyces rochei* in combination. *Journal of Phytopathology, 155*, 342–349.
Figueiredo, M. V. B., Martinez, C. R., Burity, H. A., & Chanway, C. P. (2008). Plant growth-promoting rhizobacteria for improving nodulation and nitrogen fixation in the common bean (*Phaseolus vulgaris L.). World Journal of Microbiology and Biotechnology, 24*, 1187–1193.
Fu, Q., Liu, C., Ding, N., Lin, Y., & Guo, B. (2010). Ameliorative effects of inoculation with the plant growth-promoting rhizobacterium *Pseudomonas* sp. DW1 on growth of eggplant (*Solanum melongena L.*) seedlings under salt stress. *Agricultural Water Management, 97*, 1994–2000.

Gamalero, E., Lingua, G., Berta, G., & Glick, B. R. (2009). Beneficial role of plant growth promoting bacteria and arbuscular mycorrhizal fungi on plant responses to heavy metal stress. *Canadian Journal of Microbiology, 55*, 501–514.

Geddie, J. L., & Sutherland, I. W. (1993). Uptake of metals by bacterial polysaccharides. *Journal of Applied Microbiology, 74*, 467–472.

Glick, B. R. (1995). The enhancement of plant growth by free-living bacteria. *Canadian Journal of Microbiology, 41*, 109–117.

Glick, B. R., Penrose, D. M., & Li, J. (1998). A model for the lowering of plant ethylene concentrations by plant growth-promoting bacteria. *Journal of Theoretical Biology, 190*, 63–68.

Glick, B. R., Cheng, Z., Czarny, J., & Duan, J. (2007). Promotion of plant growth by ACC deaminase-producing soil bacteria. *European Journal of Plant Pathology, 119*, 329–339.

Gomah, H. H., Mahmoud, S. M., El-Rewainy, H. M., & Abdrabou, M. R. (2014). Soil solarization and inoculation with Sulphur oxidizing bacteria and their effects on some soil properties. *Journal of Microbial Biochemistry and Technology, S3*, 2.

Gontia-Mishra, I., Sapre, S., & Tiwari, S. (2017). Zinc solubilizing bacteria from the rhizosphere of rice as prospective modulator of zinc biofortification in rice. *Rhizosphere, 3*, 185–190.

Goteti, P. K., Emmanuel, L. D. A., Desai, S., & Shaik, M. H. A. (2013). Prospective zinc solubilising bacteria for enhanced nutrient uptake and growth promotion in maize (Zea mays L.). *International Journal of Microbiology*. https://doi.org/10.1155/2013/869697.

Gray, E. J., & Smith, D. L. (2005). Intracellular and extracellular PGPR: Commonalities and distinctions in the plant–bacterium signaling processes. *Soil Biology and Biochemistry, 37*, 395–412.

Grichko, V. P., & Glick, B. R. (2001). Amelioration of flooding stress by ACC deaminase-containing plant growth-promoting bacteria. *Plant Physiology and Biochemistry, 39*, 11–17.

Haas, D., & Défago, G. (2005). Biological control of soil-borne pathogens by fluorescent pseudomonads. *Nature Reviews. Microbiology, 3*, 307–319.

Hadiarto, T., & Tran, L. S. P. (2011). Progress studies of drought-responsive genes in rice. *Plant Cell Reports, 30*, 297–310.

Hamdia, M. A. E. S., Shaddad, M. A. K., & Doaa, M. M. (2004). Mechanisms of salt tolerance and interactive effects of *Azospirillum brasilense* inoculation on maize cultivars grown under salt stress conditions. *Plant Growth Regulation, 44*, 165–174.

Han, H. S., & Lee, K. D. (2005). Phosphate and potassium solubilizing bacteria effect on mineral uptake, soil availability and growth of eggplant. *Research Journal of Agriculture and Biological Sciences, 1*, 176–180.

Handelsman, J., & Stabb, E. V. (1996). Biocontrol of soil-borne plant pathogens. *The Plant Cell, 8*, 1855–1869.

Hariprasad, P., Divakara, S. T., & Niranjana, S. R. (2011). Isolation and characterization of chitinolytic rhizobacteria for the management of *Fusarium* wilt in tomato. *Crop Protection, 30*, 1606–1612.

Harman, G. E., Howell, C. R., Viterbo, A., Chet, I., & Lorito, M. (2004). *Trichoderma* species-opportunistic, avirulent plant symbionts. *Nature Reviews Microbiology, 2*, 43–56.

Heidari, M., Mousavinik, S. M., & Golpayegani, A. (2011). Plant growth promoting rhizobacteria (PGPR) effect on physiological parameters and mineral uptake in basil (*Ocimum basilicum* L.) under water stress. *ARPN Journal of Agricultural Biological Science, 6*, 6–11.

Hernandez, M. E., Kappler, A., & Newman, D. K. (2004). Phenazines and other redox-active antibiotics promote microbial mineral reduction. *Applied and Environmental Microbiology, 70*, 921–928.

Heydari, A., & Pessarakli, M. (2010). A review on biological control of fungal plant pathogens using microbial antagonists. *Journal of Biological Sciences, 10*, 273–290.

Idris, E. E., Iglesias, D. J., Talon, M., & Borriss, R. (2007). Tryptophan-dependent production of indole-3-acetic acid (IAA) affects level of plant growth promotion by *Bacillus amyloliquefaciens FZB42. Molecular Plant-Microbe Interactions, 20*, 619–626.

Iqbal, H. M. N., Asgher, M., & Bhatti, H. N. (2011). Optimization of physical and nutritional factors for synthesis of lignin degrading enzymes by a novel strain of *Trametes versicolor*. *BioResources, 6*, 1273–1287.

Jadhav, H. P., & Sayyed, R. Z. (2016). Hydrolytic enzymes of rhizospheric microbes in crop protection. *MOJ Cell Science and Report, 3*(5), 00070. https://doi.org/10.15406/mojcsr.2016.03.00070.

Jadhav, H. P., Shaikh, S. S., & Sayyed, R. Z. (2017). Role of hydrolytic enzymes of rhizoflora in biocontrol of fungal phytopathogens: An overview. In *Rhizotrophs: Plant growth promotion to bioremediation* (pp. 183–203). Springer.

Jaleel, C. A., Manivannan, P. M., Wahid, A., Farooq, M., Al-Juburi, H. J., Somasundaram, R., & Panneerselvam, R. (2009). Drought stress in plants: A review on morphological characteristics and pigments composition. *International Journal of Agriculture and Biology, 11*, 100–105.

Jalili, F., Khavazi, K., Pazira, E., Nejati, A., Rahmani, H. A., Sadaghiani, H. R., & Miransari, M. (2009). Isolation and characterization of ACC deaminase-producing fluorescent pseudomonads, to alleviate salinity stress on canola (*Brassica napus L.*) growth. *Journal of Plant Physiology, 166*, 667–674.

Jung, W. J., Kuk, J. H., Kim, K. Y., Kim, T. H., & Park, R. D. (2005). Purification and characterization of chitinase from *Paenibacillus illinoisensis KJA-424*. *Journal of Microbiology and Biotechnology, 15*, 274–280.

Kamilova, F., Validov, S., Azarova, T., Mulders, I., & Lugtenberg, B. (2005). Enrichment for enhanced competitive plant root tip colonizers selects for a new class of biocontrol bacteria. *Environmental Microbiology, 7*, 1809–1817.

Kang, J. G., Shin, S. Y., Kim, M. J., Bajpai, V., Maheshwari, D. K., & Kang, S. C. (2004). Isolation and antifungal activities of 2-hydroxymethyl-chroman-4-one produced by *Burkholderia* sp. MSSP. *The Journal of Antibiotics, 57*, 726–731.

Kang, S. M., Joo, G. J., Hamayun, M., Na, C. I., Shin, D. H., Kim, H. Y., Hong, J. K., & Lee, I. J. (2009). Gibberellin production and phosphate solubilization by newly isolated strain of *Acinetobacter calcoaceticus* and its effect on plant growth. *Biotechnology Letters, 31*, 277–281.

Karnawal, A. (2009). Production of indole acetic acid by *fluorescent Pseudomonas* in the presence of L-tryptophan and rice root exudates. *Journal of Plant Pathology*, 61–63.

Kaur, R., Macleod, J., Foley, W., & Nayudu, M. (2006). Gluconic acid: An antifungal agent produced by *Pseudomonas* species in biological control of take-all. *Phytochemistry, 67*, 595–604.

Kaymak, H. C., Guvenc, I., & Gurol, A. (2010). Elemental analysis of different radish (*Raphanus sativus L.*) cultivars by using wavelength-dispersive x-ray fluorescence spectrometry (wdxrf). *Bulgaraian Journal of Agricultural Science, 16*, 769–774.

Khalid, A., Arshad, M., & Zahir, Z. A. (2004). Screening plant growth-promoting rhizobacteria for improving growth and yield of wheat. *Journal of Applied Microbiology, 96*, 473–480.

Khan, M. S., Zaidi, A., Ahemad, M., Oves, M., & Wani, P. A. (2010). Plant growth promotion by phosphate solubilizing fungi-current perspective. *Archieves in Agronomy Soil Science, 56*, 73–98.

Khodair, T. A., Galal, G. F., & El-Tayeb, T. S. (2008). Effect of inoculating wheat seedlings with exopolysaccharide-producing bacteria in saline soil. *Journal of Applied Sciences Research, 4*, 2065–2070.

Kloepper, J. W., & Schroth, M. N. (1978). Plant growth-promoting rhizobacteria on radishes. In *Proceedings of the 4th international conference on plant pathogenic bacteria* (Vol. 2, pp. 879–882).

Kotan, R., Cakir, A., Dadasoglu, F., Aydin, T., Cakmakci, R., Ozer, H., Kordali, S., Mete, E., & Dikbas, N. (2010). Antibacterial activities of essential oils and extracts of turkish achillea, satureja and thymus species against plant pathogenic bacteria. *Journal of the Science of Food and Agriculture, 90*, 145–160.

Labuschagne, N., Pretorius, T., & Idris, A. H. (2010). Plant growth promoting rhizobacteria as biocontrol agents against soil-borne plant diseases. In *Plant growth and health promoting bacteria* (pp. 211–230). Berlin, Heidelberg: Springer.

Lanteigne, C., Gadkar, V. J., Wallon, T., Novinscak, A., & Filion, M. (2012). Production of DAPG and HCN by *Pseudomonas* sp. *LBUM300* contributes to the biological control of bacterial canker of tomato. *Phytopathology, 102*, 967–973.

Laville, J., Blumer, C., Von Schroetter, C., Gaia, V., Défago, G., Keel, C., & Haas, D. (1998). Characterization of the hcn ABC gene cluster encoding hydrogen cyanide synthase and anaerobic regulation by ANR in the strictly aerobic biocontrol agent *Pseudomonas fluorescens CHA0*. *Journal of Bacteriology, 180*, 3187–3196.

Loper, J. E., & Gross, H. (2007). Genomic analysis of antifungal metabolite production by *Pseudomonas fluorescens Pf-5*. *European Journal of Plant Pathology, 119*, 265–278.

Lorck, H. (1948). Production of hydrocyanic acid by bacteria. *Physiologia Plantarum, 1*, 142–146.

Lugtenberg, B., & Kamilova, F. (2009). Plant-growth-promoting rhizobacteria. *Annual Review of Microbiology, 63*, 541–556.

Lugtenberg, B. J., Dekkers, L., & Bloemberg, G. V. (2001). Molecular determinants of rhizosphere colonization by *Pseudomonas*. *Annual Review of Phytology, 39*, 461–490.

Mabood, F., Zhou, X., & Smith, D. L. (2014). Microbial signaling and plant growth promotion. *Canadian Journal of Plant Science, 94*, 1051–1063.

Maggio, A., Barbieri, G., Raimondi, G., & De Pascale, S. (2010). Contrasting effects of GA3 treatments on tomato plants exposed to increasing salinity. *Journal of Plant Growth Regulation, 29*, 63–72.

Maksimov, I. V., Abizgil'Dina, R. R., & Pusenkova, L. I. (2011). Plant growth promoting rhizobacteria as alternative to chemical crop protectors from pathogens. *Applied Biochemistry and Microbiology, 47*(4), 333–345.

Mansour, F., Aldesuquy, H., & Hamedo, H. (1994). Studies on plant growth regulators and enzymes production by some bacteria. *Qatar University Science Journal, 14*, 281–288.

Markovich, N. A., & Kononova, G. L. (2003). Lytic enzymes of *Trichoderma* and their role in plant defense from fungal diseases: A review. *Applied Biochemistry Microbiology, 39*, 341–351.

Mattoo, A., & Suttle, J. C. (1991). *The plant hormone ethylene* (pp. 352–361). Boca Raton: CRC Press.

Mayak, S., Tirosh, T., & Glick, B. R. (2004). Plant growth-promoting bacteria confer resistance in tomato plants to salt stress. *Plant Physiology and Biochemistry, 42*, 565–572.

Merchan, F., de Lorenzo, L., González-Rizzo, S., Niebel, A., Megías, M., Frugier, F., Sousa, C., & Crespi, M. (2007). Analysis of regulatory pathways involved in the reacquisition of root growth after salt stress in *Medicago truncatula*. *The Plant Journal, 51*, 1–17.

Minuto, A., Spadaro, D., Garibaldi, A., & Gullino, M. L. (2006). Control of soil borne pathogens of tomato using a commercial formulation of *Streptomyces griseoviridis* and solarization. *Crop Protection, 25*, 468–475.

Nadeem, S. M., Zahir, Z. A., Naveed, M., Arshad, M., & Shahzad, S. M. (2006a). Variation in growth and ion uptake of maize due to inoculation with plant growth promoting rhizobacteria under salt stress. *Soil & Environment, 25*, 78–84.

Nadeem, S. M., Hussain, I., Naveed, M., Asghar, H. N., Zahir, Z. A., & Arshad, M. (2006b). Performance of plant growth promoting rhizobacteria containing ACC-deaminase activity for improving growth of maize under salt-stressed conditions. *Pakistan Journal of Agricultural Sciences, 43*, 114–121.

Nadeem, S. M., Zahir, Z. A., Naveed, M., & Arshad, M. (2007). Preliminary investigations on inducing salt tolerance in maize through inoculation with rhizobacteria containing ACC deaminase activity. *Canadian Journal of Microbiology, 53*, 1141–1149.

Nayani, S., Mayak, S., & Glick, B. R. (1998). Effect of plant growth-promoting rhizobacteria on senescence of flower petals. *Indian Journal Experimental Biology, 36*, 836–839.

Nihorimbere, V., Ongena, M., Smargiassi, M., & Thonart, P. (2011). Beneficial effect of the rhizosphere microbial community for plant growth and health. *Biotechnology, Agronomy Society and Environment, 15*, 327.

Noumavo, P. A., Agbodjato, N. A., Baba-Moussa, F., Adjanohoun, A., & Baba-Moussa, L. (2016). Plant growth promoting rhizobacteria: Beneficial effects for healthy and sustainable agriculture. *African Journal of Biotechnology, 15*, 1452–1463.
O'sullivan, D. J., & O'Gara, F. (1992). Traits of *fluorescent Pseudomonas* spp. involved in suppression of plant root pathogens. *Microbiological Reviews, 56*, 662–676.
Palumbo, J. D., Yuen, G. Y., Jochum, C. C., Tatum, K., & Kobayashi, D. Y. (2005). Mutagenesis of β-1,3-glucanase genes in *Lysobacter enzymogenes* strain C3 results in reduced biological control activity toward bipolaris leaf spot of tall fescue and pythium damping-off of sugar beet. *Phytopathology, 95*, 701–707.
Patel, S., Sayyed, R., & Saraf, M. (2016). Bacterial determinants and plant defense induction: Their role as bio-control agent in agriculture. In K. Hakeem (Ed.), *Plant soil-microbes* (pp. 187–204). Cham: Springer.
Patten, C. L., & Glick, B. R. (1996). Bacterial biosynthesis of indole-3-acetic acid. *Canadian Journal of Microbiology, 42*, 207–220.
Patten, C. L., & Glick, B. R. (2002). Role of *Pseudomonas putida* indole acetic acid in development of the host plant root system. *Applied and Environmental Microbiology, 68*, 3795–3801.
Perneel, M., D'hondt, L., De Maeyer, K., Adiobo, A., Rabaey, K., & Höfte, M. (2008). Phenazines and biosurfactants interact in the biological control of soil-borne diseases caused by *Pythium* spp. *Environmental Microbiology, 10*, 778–788.
Pierik, R., Tholen, D., Poorter, H., Visser, E. J., & Voesenek, L. A. (2006). The janus face of ethylene: Growth inhibition and stimulation. *Trends in Plant Science, 11*, 176–183.
Pieterse, C. M., Zamioudis, C., Berendsen, R. L., Weller, D. M., Van Wees, S. C., & Bakker, P. A. (2014). Induced systemic resistance by beneficial microbes. *Annual Review of Phytology, 52*, 347–375.
Pospíšilová, J. (2003). Interaction of cytokinins and abscisic acid during regulation of stomatal opening in bean leaves. *Photosynthetica, 41*, 49–56.
Prapagdee, B., Kuekulvong, C., & Mongkolsuk, S. (2008). Antifungal potential of extracellular metabolites produced by *Streptomyces hygroscopicus* against phytopathogenic fungi. *International Journal of Biological Sciences, 4*, 330.
Qualhato, T. F., Lopes, F. A. C., Steindorff, A. S., Brandao, R. S., Jesuino, R. S. A., & Ulhoa, C. J. (2013). Mycoparasitism studies of Trichoderma species against three phytopathogenic fungi: Evaluation of antagonism and hydrolytic enzyme production. *Biotechnology Letters, 35*(9), 1461–1468.
Qurashi, A. W., & Sabri, A. N. (2012). Bacterial exopolysaccharide and biofilm formation stimulate chickpea growth and soil aggregation under salt stress. *Brazilian Journal of Microbiology, 43*, 1183–1191.
Radi, A. E., Acero Sánchez, J. L., Baldrich, E., & O'Sullivan, C. K. (2006). Reagentless, reusable, ultrasensitive electrochemical molecular beacon aptasensor. *Journal of the American Chemical Society, 28*, 117–124.
Raghavendra, M. P., Nayaka, S. C., & Nuthan, B. R. (2016). Role of Rhizosphere microflora in potassium solubilization. In *Potassium solubilizing microorganisms for sustainable agriculture* (pp. 43–59). New Delhi: Springer.
Ramos Solano, B., Barriuso Maicas, J., Pereyra De La Iglesia, M. T., Domenech, J., & Gutiérrez Mañero, F. J. (2008). Systemic disease protection elicited by plant growth promoting rhizobacteria strains: Relationship between metabolic responses, systemic disease protection, and biotic elicitors. *Phytopathology, 98*, 451–457.
Reshma, P., Naik, M. K., Aiyaz, M., Niranjana, S. R., Chennappa, G., Shaikh, S. S., & Sayyed, R. Z. (2018). Induced systemic resistance by 2, 4-diacetylphloroglucinol positive fluorescent pseudomonas strains against rice sheath blight. *Indian Journal of Experimental Biology, 56*, 207–212.
Romero, D., de Vicente, A., Rakotoaly, R. H., Dufour, S. E., Veening, J. W., Arrebola, E., Cazorla, F. M., Kuipers, O. P., Paquot, M., & Pérez-García, A. (2007). The iturin and fengycin families

of lipopeptides are key factors in antagonism of *Bacillus subtilis* toward *Podosphaera fusca*. *Molecular Plant-Microbe Interactions, 20*, 430–440.
Ryals, J. A., Neuenschwander, U. H., Willits, M. G., Molina, A., Steiner, H. Y., & Hunt, M. D. (1996). Systemic acquired resistance. *Plant Cell, 8*, 1809.
Saber, H. F., Torang, A., Mobaleghi, M., Dehpouri, A., & Saber, H. Z. (2013). Study of nitrogen and potash fertilizers on crop yield, soluble and non soluble sugar in stevia plant (Stevia Rebaudiana Bertoni). *New Findings In Agriculture, 7*(2), 127–135.
Sacherer, P., Défago, G., & Haas, D. (1994). Extracellular protease and phospholipase C are controlled by the global regulatory gene gacA in the biocontrol strain *Pseudomonas fluorescens* CHA0. *FEMS Microbiology Letters, 116*, 155–160.
Saharan, B. S., & Nehra, V. (2011). Plant growth promoting rhizobacteria: A critical review. *Life Sciences and Medicine Research, 21*, 1–30.
Saravanakumar, D., Vijayakumar, C., Kumar, N., & Samiyappan, R. (2007). PGPR-induced defense responses in the tea plant against blister blight disease. *Crop Protection, 26*, 556–565.
Sayyed, R. Z., & Chincholkar, S. B. (2006). Purification of siderophores of *Alcaligenes feacalis* on XAD. *Bioresource Technology, 97*, 1026–1029.
Sayyed, R. Z., Patel, D. C., & Patel, P. R. (2007). Plant growth promoting potential of P solubilizing Pseudomonas sp. occurring in acidic soil of Jalgaon. *Asian Journal of Microbiology, Biotechnology and Environment Science, 4*, 925–928.
Sayyed, R. Z., Naphade, B. S., Joshi, S. A., Gangurde, N. S., Bhamare, H. M., & Chincholkar, S. B. (2009). Consortium of a. Feacalis and P. fluorescens promoted the growth of Arachis hypogea (groundnut). *Asian Journal of Microbiology, Biotechnology and Environment Science, 1*, 48–51.
Sayyed, R. Z., & Chincholkar, S. B. (2010). Growth and siderophore production in *A. faecalis* is influenced by metal ions. *Indian Journal of Microbiology, 50*, 179–182.
Sayyed, R. Z., & Patel, P. R. (2011). Soil microbes & environmental health. *International Journal of Biotechnology & Bioscience, 1*, 41–66.
Sayyed, R. Z., Naphade, B. S., & Chincholkar, S. B. (2004). Ecologically competent rhizobacteria for plant growth promotion and disease management. In M. K. Rai, N. J. Chikhale, P. V. Thakare, P. A. Wadegaonkar, & A. P. Ramteke (Eds.), *Recent trends in biotechnology* (pp. 1–16). Jodhpur: Scientific Publisher.
Sayyed, R. Z., Chincholkar, S. B., Meyer, J. M., & Kale, S. P. (2011). Chemical characterization, crossfeeding and uptake studies on hydroxamate siderophore of *Alcaligenes faecalis*. *Indian Journal of Microbiology, 51*, 176–181.
Sayyed, R. Z., Reddy, M. S., Deshmukh, A. M., Pate, A. S., & Gangurde, N. S. (2012). Potential of plant growth promoting rhizobacteria for sustainable agriculture bacteria. In D. K. Maheshwari (Ed.), *Agrobiology: Plant probiotics* (pp. 287–314). Berlin: Springer.
Sayyed, R. Z., Chincholkar, S. B., Reddy, M. S., Gangurde, N. S., & Patel, P. R. (2013). Siderophore producing PGPR for crop nutrition and phytopathogen suppression bacteria. In D. K. Maheshwari (Ed.), *Agrobiology: Disease management* (pp. 449–471). Dordrecht: Springer.
Sayyed, R. Z., Patel, P. R., & Shaikh, S. S. (2015). Plant growth promotion and root colonization by EPS producing *Enterobacter* sp. RZS5 under heavy metal contaminated soil. *Indian Journal of Experimental Biology, 53*, 116–123.
Shaikh, S. S., & Sayyed, R. Z. (2015). Role of plant growth-promoting rhizobacteria and their formulation in biocontrol of plant diseases. In N. K. Arora (Ed.), *Plant microbes symbiosis: Applied facets* (pp. 337–351). New Delhi: Springer.
Shaikh, S. S., Patel, P. R., Patel, S. S., Nikam, S. D., Rane, T. U., & Sayyed, R. Z. (2014). Production of biocontrol traits by banana field *fluorescent Pseudomonads* and comparison with chemical fungicide. *Indian Journal of Experimental Biology, 52*, 917–920.
Shaikh, S. S., Reddy, M. S., & Sayyed, R. Z. (2016). Plant growth promoting rhizobacteria: An eco-friendly approach for sustainable agroecosystem. In K. Hakeem (Ed.), *Plant soil-microbes* (pp. 182–201). Cham: Springer.

Shaikh, S. S., & Saraf, M. S. (2017). Optimization of growth conditions for zinc solubilizing plant growth associated Bacteria and Fungi. *Journal of Advanced Research in Biotechnology, 2*(1), 9.

Sharma, A., Shankhdhar, D., & Shankhdhar, S. C. (2013). Enhancing grain iron content of rice by the application of plant growth promoting rhizobacteria. *Plant Soil Environment, 59*(2), 89–94.

Shobha, G., & Kumudini, B. S. (2012). Antagonistic effect of the newly isolated PGPR *Bacillus* spp. on *Fusarium oxysporum. International Journal of Applied Science and Engineering Research, 1*, 463–474.

Shuhegger, R., Ihring, A., Gantner, S., Bahnweg, G., Knappe, C., Vogg, G., Hutzler, P., Schmid, M., Van Breusegem, F., Eberl, L., Hartmann, A., & Langebartels, C. (2006). Induction of systemic resistance in tomato by N-acyl-L-homoserine lactone-producing rhizosphere bacteria. *Plant, Cell & Environment, 29*, 909–918.

Siddikee, M. A., Chauhan, P. S., & Sa, T. (2012). Regulation of ethylene biosynthesis under salt stress in red pepper *(Capsicum annuum L.)* by 1-aminocyclopropane-1-carboxylic acid (ACC) deaminase-producing halotolerant bacteria. *Journal of Plant Growth Regulation, 31*, 265–272.

Singh, B. B., Mai-Kodomi, Y., & Terao, T. (1999). A simple screening method for drought tolerance in cowpea. *The Indian Journal of Genetics and Plant and Breeding, 59*, 211–220.

Soltani, A., Khodarahmpour, Z., Jafari, A. A., & Nakhjavan, S. (2012). Selection of alfalfa (*Medicago sativa L.*) cultivars for salt stress tolerance using germination indices. *African Journal of Biotechnology, 11*, 7899–7905.

Tank, N., & Saraf, M. (2010). Salinity-resistant plant growth promoting rhizobacteria ameliorates sodium chloride stress on tomato plants. *Journal of Plant Interactions, 5*, 51–58.

Ten Hoopen, G. M., & Krauss, U. (2006). Biology and control of *Rosellinia bunodes, Rosellinia necatrix* and *Rosellinia pepo*: A review. *Crop Protection, 25*, 89–107.

Tewari, S., & Arora, N. K. (2018). Role of salicylic acid from *Pseudomonas aeruginosa* PF23EPS+ in growth promotion of sunflower in saline soils infested with phytopathogen *Macrophomina phaseolina. Environmental Sustainability, 1*(1), 49–59.

Upadhyaya, C. P., Akula, N., Kim, H. S., Jeon, J. H., Ho, O. M., Chun, S. C., Kim, D. H., & Park, S. W. (2011). Biochemical analysis of enhanced tolerance in transgenic potato plants overexpressing d-galacturonic acid reductase gene in response to various abiotic stresses. *Molecular Breeding, 28*, 105–115.

Van Loon, L. C. (2007). Plant responses to plant growth-promoting rhizobacteria. *European Journal of Plant Pathology, 119*, 243–254.

Vettakkorumakankav, N. N., Falk, D., Saxena, P., & Fletcher, R. A. (1999). A crucial role for gibberellins in stress protection of plants. *Plant & Cell Physiology, 40*, 542–548.

Vinay, J. U., Naik, M. K., Rangeshwaran, R., Chennappa, G., Sohel, S. S., & Sayyed, R. Z. (2016). Detection of antimicrobial traits in fluorescent pseudomonads and molecular characterization of an antibiotic pyoluteorin. *3 Biotech, 6*, 1–11.

Voisard, C., Keel, C., Haas, D., & Dèfago, G. (1989). Cyanide production by *Pseudomonas fluorescens* helps suppress black root rot of tobacco under gnotobiotic conditions. *The EMBO Journal, 8*, 351–358.

Walsh, U. F., Morrissey, J. P., & O'Gara, F. (2001). *Pseudomonas* for biocontrol of phytopathogens: From functional genomics to commercial exploitation. *Current Opinion in Biotechnology, 12*, 289–295.

Weller, D. M. (2007). Pseudomonas biocontrol agents of soil borne pathogens: Looking back over 30 years. *Phytopathology, 97*, 250–256.

Weller, D. M., & Thomashow, L. S. (1993). Use of rhizobacteria for biocontrol. *Current Opinion in Biotechnology, 4*, 306–311.

Whipps, J. M. (1990). Carbon utilization. In J. M. Lynch (Ed.), *The rhizosphere* (pp. 59–97). Chichester: Wiley Interscience.

Whipps, J. M. (2001). Microbial interactions and biocontrol in the rhizosphere. *Journal of Experimental Botany, 52*, 487–511.

Yamaguchi, S. (2008). Gibberellin metabolism and its regulation. *Annual Review of Plant Biology, 59*, 225–251.

Yaxley, J. R., Ross, J. J., Sherriff, L. J., & Reid, J. B. (2001). Gibberellin biosynthesis mutations and root development in pea. *Plant Physiology, 125*, 627–633.

Yue, H., Mo, W., Li, C., Zheng, Y., & Li, H. (2007). The salt stress relief and growth promotion effect of Rs-5 on cotton. *Plant and Soil, 297*, 139–145.

Zahedi, H. (2016). Growth-promoting effect of potassium-solubilizing microorganisms on some crop species. In *Potassium solubilizing microorganisms for sustainable agriculture* (pp. 31–42). New Delhi: Springer.

Zahir, Z. A., Munir, A., Asghar, H. N., Shaharoona, B., & Arshad, M. (2008). Effectiveness of rhizobacteria containing ACC deaminase for growth promotion of peas (*Pisum sativum*) under drought conditions. *Journal of Microbiology and Biotechnology, 18*, 958–963.

Zahir, Z. A., Ghani, U., Naveed, M., Nadeem, S. M., & Asghar, H. N. (2009). Comparative effectiveness of *Pseudomonas* and *Serratia* sp. containing ACC-deaminase for improving growth and yield of wheat (*Triticum aestivum L.*) under salt-stressed conditions. *Archives of Microbiology, 191*, 415–424.

Zahir, Z. A., Shah, M. K., Naveed, M., & Akhter, M. J. (2010). Substrate-dependent auxin production by *Rhizobium phaseoli* improves the growth and yield of *Vigna radiata L.* under salt stress conditions. *Journal of Microbiology and Biotechnology, 20*, 1288–1294.

Zapata, P. J., Serrano, M., Pretel, M. T., Amoros, A., & Botella, M. A. (2004). Polyamines and ethylene changes during germination of different plant species under salinity. *Plant Science, 167*, 781–788.

Zehnder, G. W., Murphy, J. F., Sikora, E. J., & Kloepper, J. W. (2001). Application of rhizobacteria for induced resistance. *European Journal of Plant Pathology, 107*, 39–50.

Chapter 8
Plant Growth-Promoting Microbes: Contribution to Stress Management in Plant Hosts

Krishna Sundari Sattiraju, Srishti Kotiyal, Asmita Arora, and Mahima Maheshwari

Contents

Abstract Plants encounter various challenges that impact on growth and development. In the agricultural scenario, any limiting condition can transform into serious economic losses. Conventional methods employed to deal with biotic and abiotic stresses, including chemical methods, plant breeding, genetic engineering and other modern practices, present a variety of practical concerns. For example, transgenic plants can lead to selection pressure on the parasites thus providing a means to develop resistance. Hence a shift towards exploring the potentialities in plant growth-promoting microbes (PGPM) as a part of mainstream agricultural practices is imperative. In this review, we focus on PGPM (inclusive term for plant growth-

K. S. Sattiraju (✉) · A. Arora · M. Maheshwari
Department of Biotechnology, Jaypee Institute of Information Technology, Noida, Uttar Pradesh, India
e-mail: krishna.sundari@jiit.ac.in

S. Kotiyal
Evason lab, Department of Oncological Sciences, Huntsman Cancer Institute, University of Utah, Salt Lake City, UT, USA

R. C. Sobti et al. (eds.), *Environmental Biotechnology: For Sustainable Future*, https://doi.org/10.1007/978-981-10-7284-0_8

promoting rhizobacteria and fungi), which, apart from their plant growth-promoting activities, also play a role in plant diseases control as well as in alleviating the impact of abiotic stresses. A deeper understanding of the mechanisms by which PGPM modify plant stress responses to boost their resistance and the nuances of the PGPM-host interactions would lead to increased acceptance of PGPM in agricultural applications.

Keywords Plant growth-promoting microbes (PGPM) · Biotic stress · Abiotic stress · Biotechnological interventions · ISR · SAR · Genetically modified PGPB

Introduction

The chapter begins with a discussion on the conventional methods used to manage/mitigate stress in plants, either through selective breeding, hybridization, selection or through molecular biotechnology involving recombinant DNA technology and protein engineering, to develop genetically modified plants (Balconi et al. 2012). Certain operational bottlenecks in these strategies prompt the need to explore more competent alternatives. Plant-associated microbes with plant growth-enhancing effect can serve as promising alternatives for stress management in plants. Current chapter presents various mechanisms through which PGPM are capable of modulating plant responses to biotic or abiotic stressors, promoting stress resistance and/or tolerance in plants (Bach et al. 2016). The chapter further explores the possibility, whether the ability of PGPM to play a central role in nutrient recycling can alleviate stress effects in the soil micro-environment (Sarathambal et al. 2014; Santoyo et al. 2016). PGPM can improve soil fertility by participating in nitrogen fixation, phosphate solubilization, sequestering iron and modulating phytohormone levels (cytokinins, gibberellins, indoleacetic acid, ethylene, etc.), and this ability of theirs can prove useful to plants being grown in nutrient-deficient soils or soils where nutrients are present but are in unavailable form and hence inaccessible for rhizosphere functions (Porcel et al. 2014; Pontes et al. 2015). A synergistic action expressed by certain PGPM is discussed where they elicit cross protection, a phenomenon by which common effector molecules can offer protection against seemingly unrelated stressors. The chapter concludes with a note on how PGPM can be improved by genetic modification and how a system like tripartite approach of strengthening 'host-microbe-soil' as a unit and not as individual entities can help is discussed in context with their prospective role in future plant stress management biology.

Conventional Methods to Deal with Biotic and Abiotic Stress

Plant productivity can be improved by mitigating stress effects, which can be achieved by employing strategies like improving plant material through breeding for tolerance/resistance and/or genetic engineering; improving root health by methods like crop rotation, tillage of soil, control of soil-borne disease; good irrigation practices like ensuring optimal water quality and availability; and by protecting plants against airborne diseases (Balconi et al. 2012). These plant stress management approaches can be broadly classified as: agricultural-based practices, plant breeding-/hybridization-based and genetic engineering-based or biotechnological interventions.

Traditional Agricultural Practices

Good farming practices minimize soil disturbance and contribute to the preservation and improvement of overall soil health. Field rotation is another method that enhances crop nutrition and improves soil health by allowing replenishment of its resources. Maintenance of a protective organic cover on soil surface, by using cover crops or crop residues, is another strategy that not only protects the soil surface but also conserves water and nutrients while promoting biological activity in soil. Use of fertilizers and pesticides also enhances crop yield by their role in management of pathogens via nutrient balancing and biological control properties, respectively (Hobbs et al. 2008). In the recent times, India has shifted from healthy traditional practices and has become more and more reliant on modern systems of agriculture that include unstructured irrigation, usage of chemical fertilizers and harmful pesticides in excess. Some irrigation methods used include strip irrigation, check basin method, furrow irrigation and basin irrigation method. Such irrigation methods are not suitable for all types of crops. Apart from this, gallons of water wasted due to over-irrigation or irrigation run-off not only moves into the drain system, but it also carries off top layers of organic soil (humus) and crop chemicals. Furrow irrigation method particularly involves digging furrows. This requires extra labour, and due to the digging, the salts come up to the surface, increasing the overall salt concentration (Silva et al. 2007). Usage of chemical fertilizers is primarily intended to enhance the yield and reduce the attack of pests on the crop. However, once the crop is harvested, these chemical residues are left in the soil and are not readily degradable and hence become harmful to plants, cattle and human health. Additionally, due to their excessive presence, the soil fertility is decreased, and the chemical composition of the soil is also altered. The biggest negative impact of using chemical fertilizers is the groundwater contamination. Nitrates are produced from nitrogen fertilizers which easily seep into the soil and reach the groundwater. Being insoluble in water, they can stay there for decades (Viets and Lunin 2009). Thus, improper agricultural practices and high use of external inputs like fertilizers/ pesticides, over a period of time, can result in soil and environmental degradation.

Majority of agricultural and soil conservation techniques focus on providing nutrients and water to satisfy the basic needs of plants. Such techniques do not have much to do with understanding the soil as a living system that has a dynamic nature. Lack of such integrated understanding has led to decreased levels of soil organic matter and further increased the use of chemical inputs.

Conventional Breeding Techniques

Introducing genetic resistance in plants by selective breeding has the advantage of negligible maintenance cost once cultivars are developed. However, the risk of these cultivars placing a selection pressure on parasite populations to develop resistance cannot be overruled. For example, R (resistance) genes have traditionally been used in conventional resistance breeding programmes as one R gene has the potential to provide complete resistance to one or more than one strain of a specific pathogen, when transferred to a previously susceptible plant of the same species. Unfortunately, co-evolving pathogens can quickly defeat R genes. Moreover, several R genes also lack durability as they can be nullified by even one, loss-of-function mutation, in corresponding Avr (avirulence) gene. Traditional breeding strategies use R genes in a 'one at a time' manner which exerts strong selection pressure for mutation of the relevant Avr gene, thereby increasing vulnerability to the emergent pathogen. Alternatively, multiple R genes (pyramids) can be bred into individual plant lines which would require the pathogen to accumulate mutations in multiple Avr genes to escape detection (Balconi et al. 2012). But this strategy requires multiple cycles of breeding and rigorous selection norms to be able to arrive at desirable cultivar with multiple R genes incorporated stably. This could be very time-consuming and many a times have unpredictable yield impacts. Particularly, such efforts to breed abiotic stress tolerance in plants gave some survival benefits to plants but exhibited their own set of limitations. Also, such strategies do not play a significant role in increasing the yield. Breeding for such traits generally employs a trade-off at the cost of yield potential, hence making it irrelevant in agricultural scenarios.

Conventional Biotechnological Interventions

Exogenous application of various organic compounds and plant hormones has shown increase in growth and yield in certain host plants that do not exhibit an inherent defence mechanism against stress conditions (Spoel and Dong 2008). In tomato plants, it was reported that GA (gibberellic acid) application decreases stomatal resistance and increases crop growth and yield under saline condition. Under stress, metabolic activities can get disturbed due to altered hormonal balance, and exogenous application of growth hormones might be a useful strategy for stress tolerance (Fahad et al. 2015). Exogenous application of SA (salicylic acid) was found to ameliorate the damage caused by cadmium toxicity in maize and barley. It also conferred tolerance to *Cassia tora* plants exposed to aluminium toxicity, augmented drought tolerance in tomato and bean plants and enhanced tolerance to high temperatures in *Agrostis stolonifera* by preventing oxidative damage. It was also found to relieve the damaging effects of low temperatures in rice, wheat, bean and banana and damaging effects of UV-B radiation in Kentucky bluegrass and tall fescue sod grass. Exogenous application of BRs (brassinosteroids) was found to ameliorate adverse effects of salt stress on seed germination, elongation of roots and subsequent growth of rice plants (Fahad et al. 2015). As per research reports of Duque et al.(2013), increased concentration of CKs (cytokinins) in xylem and their exogenous application can decrease stomatal sensitivity to abscisic acid (ABA), which in return can help in obtaining a better yield from plants experiencing mild drought conditions. CK up-regulation can be achieved by reducing expression of a gene encoding cytokinin oxidase, an enzyme which degrades CKs.

Transgenic crop varieties were also successfully used in combating biotic stressors like viruses. Attempts to introduce a gene coding for whole viral protein or part of a viral protein into the host plants by transformation were successful (Boualem et al. 2015). Virus-resistant plants can be obtained by transferring genes from the pathogen itself into the plant (pathogen-derived resistance), by making transgenic plants expressing viral coat proteins (expression of viral genes disrupt viral infection or its symptoms) or by post-transcriptional gene silencing employing viral replicase genes or RNA-dependent RNA polymerase genes. The later method has been reported to confer resistance to potato leaf roll virus, barley yellow dwarf virus, cucumber mosaic virus and wheat streak mosaic virus in potato, oats, tomato and wheat, respectively. It also induced resistance to rice tungro spherical virus in rice. RIPs (ribosome-inactivating proteins) expression in transgenic plants is also used to protect plants against multiple viruses as RIPs inhibit protein synthesis. Depending on the plant species producing them, they exhibit varying toxicity levels against different pathogens. Another common approach is using antibodies directed against the virus coat proteins that can neutralize virus infection by interacting with newly synthesized coat protein and disrupting viral particle formation in the pathogen.

Stabilization of the functional conformation of proteins is a major concern in plant stress metabolism. Biotechnological approaches for improving abiotic stress response in plants include protein engineering approaches. This involves selection of protein mutants which increase protein stability by strategies such as random mutagenesis and high-throughput screening, functional screens or comparing homologous proteins. There has been a strong research focus on understanding the stabilization of hydrophobic core and internal structural elements of proteins. Protein surfaces also influence stability, and surface residues are generally more flexible. The protein surface structures have free movement than the compact core; therefore, mutations in the protein surface largely affect protein stability and enhance protein stability. Further information is required to understand the rules for protein folding stability and dynamics with the aim to improve protein stability and stress tolerance in plants (Ortbauer 2013).

Insect-resistant transgenic crops are widely used; greater than 30 million hectares of land worldwide is planted with crops expressing Bt (*Bacillus thuringiensis*) d-endotoxins. About 140 genes have been characterized for the Bt d-endotoxins affective against lepidopterans, coleopterans and dipterans, and they are also target specific. Hence, they provide safe alternatives to chemical control agents. Apart from Bt Cry genes, other candidate genes, such as protease inhibitors, alpha-amylase inhibitors, vegetative insecticidal proteins from Bt, cholesterol oxidases and toxins from predators such as mites and scorpions, are also used to make insect-resistant transgenics. Studies have shown that transgenic tobacco plants expressing chitinase show increased resistance to lepidopterans. Development of artificial resistance, via introduction of effector genes into the host plant, was reported to provide a viable molecular strategy for expressing nematode resistance. These effector genes can encode enzymatic inhibitors that block physiological processes within the nematodes, degrading enzymes (e.g. collagenases, chitinases), ingestible toxic compounds (cytotoxins), molecule binding compounds (e.g. lectins, monoclonal antibodies), enzymes which interact with nematodes and substances causing breakdown of particular feeding structures (cytotoxins).

Even though genetically modified (GM) plants have been the centre of attraction of plants researchers, there have been different limitations for the same. First, GM plants are not the natural way of cultivation; hence they pose unexplained and at times perceived threat to the environment. Second, the genes inserted in the plant genome add an extra burden to the host plant itself as it has to partition its metabolic energy resources to fuel these non-native functions, and this reallocation of resources might reflect in decreased crop yield. Such alterations add an extra burden to the roots and result in less plant biomass. GM plants may also influence abundance of soil organisms including the rhizobacteria. The chances of cross contamination are also high when one is dealing with GM plants. Even though there are not many significant facts reported, overall root-plant-soil relationships are perceivably disturbed. There has not been any extensive research about GM plants and their influence on this tri-partite (GM plant _ rhizosphere microbes – soil micro-environment) due to ethical concerns, lack of evidence and apprehension from the market (Domingo and Bordonaba 2011). Figure 8.1 is a pictorial representation of all such aspects and challenges faced by host-soil-microbe systems.

Why the Need for Alternatives to Conventional Methods?

All the methods discussed above in section 'Conventional methods to deal with biotic and abiotic stress' have their own strengths and weaknesses. Primarily, the effect on nontarget species, invasiveness, horizontal gene transfer of transgenes and adverse effects on natural soil biota are causes of great concern. Introduction of insecticide resistance can challenge natural ecosystems with unknown impacts on their associated complex network of nontarget organisms (Cramer et al. 2011). As global decline of biodiversity is a major issue, proactive measures are necessary, and consideration of the likely effects of transgenics on plant and insect biodiversity is essential (Downey 2003). Though engineering genes encoding insecticidal proteins into crop plants have several benefits, researchers expressed this technology could disrupt natural biological control by causing side effects of the plant on the fitness and behaviour of pests. Interactions between transgenic plants and beneficial insects were also taken up to assess issues of incompatibility (Schuler et al. 1999). There are two major concerns regarding the use of Bt transgenic crops: the effect on

Fig. 8.1 Genetically engineered plant and its response to various biotic/abiotic stressors

nontarget organisms and the possibility of resistance development in target insects to the Bt protein. Satellite RNA can also be used to increase resistance against cucumber mosaic virus. Attempts using this approach have been made in tomato but are controversial as a single point mutation in the satellite RNA can transform it into a harmful necrogenic form.

To satisfy the increasing global demand for food, a re-evaluation of existing agricultural practices (including the use of chemical fertilizers, herbicides, fungicides and insecticides) is also required. To this end, there is a shift in agricultural practices towards approaches that are sustainable, as well as environmental friendly. One such beneficial approach with no discernible toxic implications happens to be the application of PGPM in mainstream agriculture (Glick 2012).

Plant Growth-Promoting Microbes

In nature, beneficial relationships between plants and microorganisms are present and defined under many types such as mutualism, symbiosis, cohabitation, commensalism, co-metabolism, biofilms, endophytes and so on. They generally occur in the rhizosphere and aid in improving plant growth or help the plant to cope with biotic and abiotic stress (Zamioudis and Pieterse 2012). Among the diverse range of microscopic life forms found in soil, bacteria are by far the most common (i.e. 95%). However, they are generally not homogenously distributed in the soil, i.e. greater concentration of bacteria would be found in the rhizosphere than in the rest of the soil. This suggests that plants can shape their microbiome by root exudates comprising of nutrients such as sugars, amino acids, organic acids and other small molecules, which account for up to a third of the carbon fixed by a plant. Considering the plant perspective, the interaction between soil bacteria and plants may either be beneficial or harmful or neutral. Also, the effect of particular bacteria on a plant will vary with changing conditions. For example, a bacteria facilitating plant growth by providing fixed nitrogen or phosphorus (which are generally present in limited quantities in soils) are unlikely to prove beneficial to plants when significant amounts of chemical fertilizer are added to the soil. Moreover, a particular bacterium can affect different plants disparately (Glick 2012). Plant growth-promoting rhizobacteria (PGPR) have been the most extensively studied plant growth promoting bacteria (PGPB) (Compant et al. 2005). Depending on plant interactions, PGPR can be divided into two groups: symbiotic bacteria (which live inside plants and exchange metabolites with them directly) and free-living rhizobacteria (which live outside plant cells). Typically symbiotic bacteria live in the intercellular spaces of the plant host, but some bacteria form truly mutualistic interactions and penetrate plant cells. Additionally, some of them integrate their physiology with the plant, leading to formation of specialized structures. Rhizobia are one of the most studied mutualistic bacteria, which live symbiotically with leguminous crop plants and fix atmospheric nitrogen for the plant in root structures called nodules. Other examples of mutualistic bacteria include *Frankia*, which forms nodules wherein it fixes nitrogen in actinorhizic plants such as *Alnus* trees. Several PGPR are used worldwide as biofertilizers, contributing to increased crop yields and soil fertility, and hence hold the potential to contribute to sustainable agriculture and forestry (García-Fraile et al. 2015). *Bacillus mycoides* B38V, *Burkholderia cepacia* 89 and *Paenibacillus riograndensis* SBR5 were studied for their plant growth-promoting characteristics. Bach et al. (2016) evaluated the biocontrol potential and rhizosphere competence of two PGPB. The study was with different cultivars of wheat, and PGPB were added to the substrate. They have recorded remarkable antifungal activity upon inoculation with PGPB in addition to improved growth characteristics in host plants. This and many such examples recommend that PGPB could be successfully used as bioinoculants once host-microbe optimization studies are completed. Regardless of the differences between these bacteria, they all utilize similar mechanisms of support (Glick 2012), as explained in the section 'Role of PGPM in dealing with plant stress' of the chapter.

Endophytic fungi have been widely studied in several geographic and climatic zones and are ubiquitously found within plant tissues and have rich species diversity. They play an important role in providing nutrients to host in exchange for photosynthates, adapting them to their environments, defending them from environmental stresses and promoting biodiversity of plant community (Zhou et al. 2014). Certain rhizospheric fungi belonging to the genera *Penicillium*, *Fusarium*, *Trichoderma* and *Phoma* are involved in promotion of plant growth and development. They impact growth and health of plants by direct and/or indirect mechanisms.

They can affect indirectly by strategies such as antibiotics and siderophore production and directly via solubilization of minerals, etc. They also stimulate plant growth partly by the production of secondary metabolites such as IAA (indoleacetic acid), CK, GAs, ET (ethylene) and other plant growth-promoting substances. They can also protect against pathogens by the production of phytohormones and also through production of molecules which affect hormone homeostasis within the plants (Salas-Marina et al. 2011). P (phosphate)-solubilizing and N (nitrogen)-fixing bacteria synergistically interact with AM (arbuscular mycorrhizal) fungi, increasing P and N availability to the plant and promoting its growth and biotic stress resistance (Alizadeh et al. 2013).

Beneficial associations of PGPM can stimulate plant growth through degradation of soil pollutants and production of phytostimulators (Zamioudis and Pieterse 2012). It has been reported that the PGPM strains of *Pseudomonas alcaligenes*, *B. polymyxa* and *Mycobacterium phlei* promote plant growth significantly when inoculated into the nutrient-deficient soils. PGPM can also foster plant's nutrient uptake efficiency under poor soil conditions. In addition, PGPM can induce plant resistance to phytopathogens, insect pests and nematodes. Since nutrients and PGPM all remain in the soil environment, the soil property could also affect their interactions (Sripontan et al. 2014).

Role of PGPM in Dealing with Plant Stress

PGPM can contribute to plant growth by enhancing its resistance to biotic/abiotic stress or by protecting/priming them against the same. However, to establish this plant-PGPM association, mutual recognition and coordination are needed between them. Substantial evidence indicates that initially plants identify beneficial microbes as potential invaders, thereby triggering an immune response. But at later stages of such interactions, mutualists outwit plant defence responses to successfully colonize host roots (Zamioudis and Pieterse 2012).

How Do PGPM Help in Fighting Biotic Stress?

PGPB can confer induced systemic resistance (ISR) in plants, a phenomenon which resembles systemic acquired resistance (SAR) in phenotypic aspects. Plants exhibiting ISR are called primed (Glick 2012), and biopriming of plants with certain PGPB provides systemic resistance against a broad spectrum of plant pathogens and diseases. ISR is not very specific in its targets and can rather control diseases caused by a variety of pathogens (Compant et al. 2005).The PGPM-mediated ISR is important for disease control under conditions where the PGPM and pathogens are spatially separated. The systemic resistance induction process leads to increases in peroxidase (PO) and phenoloxidase (PPO) activities, which are involved in

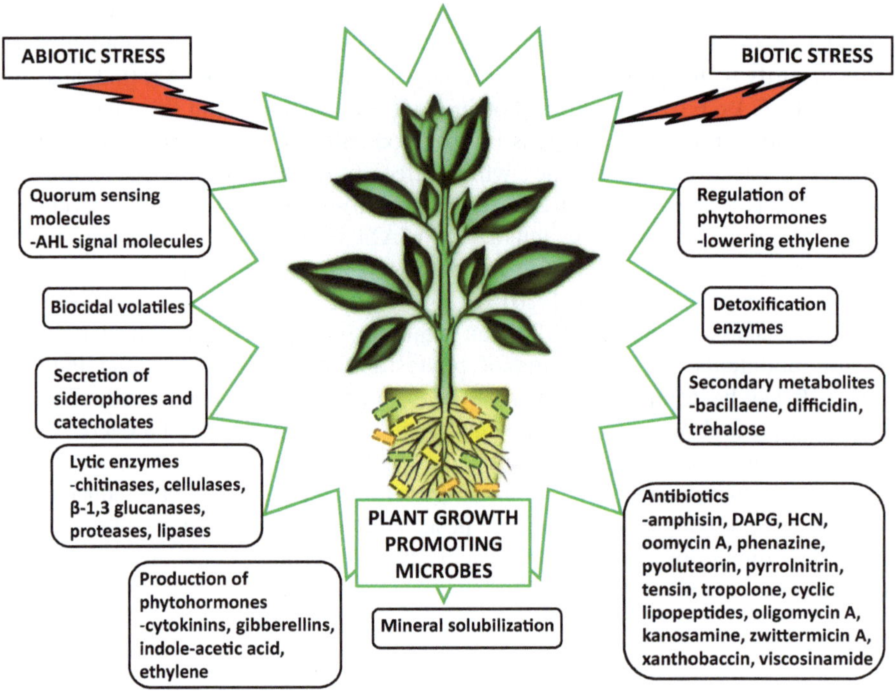

Fig. 8.2 Strategies by which PGPM protect plants against biotic and abiotic stress

catalysing lignin formation and phenyl ammonia lyase (PAL), for the biosynthesis of phytoalexin and phenol (Filippi et al. 2011). Direct interaction between resistance inducing PGPB and pathogen is not required in ISR. Apart from ethylene (ET) and jasmonate, bacterial molecules like O-antigenic side chain of the bacterial outer membrane protein lipopolysaccharide, pyoverdine, flagellar proteins, chitin, β-glucans, cyclic lipopeptide surfactants and salicylic acid (SA), also act as signalling molecules for ISR (Glick 2012). SA can cause an ISR even when present in nanogram amounts (Bloemberg and Lugtenberg 2001). A brief overview of different strategies employed by PGPM to protect plants against biotic and abiotic stressors is given in Fig. 8.2.

The first instances of PGPB-induced ISR were reported in carnation (*Dianthus caryophyllus*) and in cucumber (*Cucumis sativus*) which exhibited reduced susceptibility to wilt caused by *Fusarium* sp. and foliar disease caused by *Colletotrichum orbiculare*, respectively. The combination of host plant and bacterial strain decides manifestation of ISR. Even though most reports of PGPB-induced ISR involve free-living rhizobacterial strains, endophytic bacteria were also found to exhibit ISR activity. *Pseudomonas fluorescens EP1* triggers ISR against red rot caused by *Colletotrichum falcatum* on sugarcane. *Burkholderia phytofirmans PsJN* induces ISR against *Botrytis cinerea* on grapevine and *Verticillium dahliae* on tomato. *Pseudomonas denitrificans* and *Pseudomonas putida* confer resistance against *Ceratocystis fagacearum* in oak, while *P. fluorescens* does the same against

Fusarium oxysporum f. sp. *radicis-lycopersici* in tomato and *Pythium ultimum* and also against *F. oxysporum* f. sp. *pisi* in pea roots. *Bacillus pumilus* SE34 provides ISR against *F. oxysporum* f. sp. *pisi* in pea roots and against *F. oxysporum* f. sp. *vasinfectum* in cotton roots. Both rhizobacteria and bacterial endophytes were observed to have the ability to act as bioprotectants via ISR. Bacterial traits such as flagellation, production of siderophores, lipopolysaccharides, etc. were proposed to trigger ISR. However, no solid evidence exists for an overall ISR signal produced by bacteria. Even though some PGPB trigger an SA-dependent signalling pathway in the rhizosphere, majority of ISR inducing PGPB were shown to utilize a SA-independent pathway involving jasmonate and ET signals. Rather than increasing the production of these hormones, ISR was implicated to increase sensitivity towards them as the former could lead to the activation of a partially different set of defence genes. When PGPB triggers ISR, it was also observed to fortify the strength of the plant cell wall and alter the physiology and metabolic responses of the host to enhance synthesis of plant defence chemicals. After inoculating tomatoes with endophytic *P. fluorescens* WCS417r, the cortical cell walls were found to be thickened upon colonization of epidermal or hypodermal cells. During endophytic colonization by *B. phytofirmans PsJN* in grapevine, accumulation of phenolic compounds and strengthening of the cell walls of exodermis and other cortical cell layers were reported. Plant ISR response also includes formation of structural barriers, like thickening of cell wall papillae by deposition of callose and accumulation of phenolic compounds at the site of attack. Biochemical or physiological changes in plants include accumulation of pathogenesis-related (PR) proteins like PR-1, PR-2, chitinases and some peroxidises. Instead of inducing PR proteins, certain PGPB were known to induce accumulation of phytoalexins, peroxidases, polyphenol oxidase, phenylalanine ammonia lyase and/or chalcone synthase. Production of some of these compounds (e.g. chalcone synthase) in plant defence might be triggered by the same N-acyl homoserine lactones which bacteria also use for intraspecific signalling (Compant et al. 2005).

The list of rhizobacterial *Pseudomonas* species known to induce ISR is rapidly growing as many researchers have worked on the genus. There exists a dependency on plant genotype in generation of ISR as reported in literature. Detailed review of the plant factors involved in the ISR and SAR pathways has shown that induced disease resistance can be increased by simultaneous activation of these two pathways (Bloemberg and Lugtenberg 2001). PGPM were shown to induce ISR in several crops like *Arabidopsis*, cucumber, tomato, potato and so on against fungal, bacteria, nematode and viral pathogens. Studies have reported early and enhanced levels of peroxidase being stimulated in rice plants by seed treatment and seedling root dipping. In the PGPR-treated rice plants inoculated with sheath blight pathogen, *Rhizoctonia solani*, two isoforms of peroxidase were induced. Chilli plants treated with *P. fluorescens* Pf1, when challenged with *Colletotrichum capsici*, reportedly showed higher levels of expression of peroxidases. Similarly, tomato plants treated with PGPR exhibited increased activity of polyphenol peroxidase upon challenging with *F. oxysporum* f. sp. *lycopersici*. *Pseudomonas* strains treated plants have registered higher levels of phenylalanine ammonia lyase as compared to control.

Seedlings dipped in talc-based formulation of *P. fluorescens* were observed to increase the activity of phenylalanine ammonia lyase in finger millet leaves against blast disease. In an experiment, researchers inoculated PGPR strains *P. putida* 89B-27 and *Serratia marcescens* 90-166 with *F. oxysporum* f. sp. *cucumerinum* on two separate halves of roots of cucumber seedlings and have reported induction of systemic resistance against *Fusarium* wilt. They observed delayed development of disease symptoms and reduced number of dead plants. The same PGPR strains were also seen to induce systemic resistance against *P. syringae* pv. *lachrymans* which causes bacterial angular leaf spot in cucumber (Liu et al. 1995). Treatment of maize seeds with *P. fluorescens* witnessed higher activity of peroxidase, polyphenol oxidase and phenylalanine ammonia lyase in the host plant against pathogen *R. solani*. Seeds bactericized with *P. fluorescens* led to the accumulation of higher phenolic compounds and greater activity of polyphenol peroxidase, peroxidase and phenylalanine ammonia lyase, compounds that are known to have a role in multiple defence mechanisms in plants against pathogen (Sivakumar and Sharma 2003). PGPR-induced systemic resistance also controlled diseases caused by nematodes in tomato and bell pepper, and ISR was accredited for a reduction of 42% in nematode penetration (Siddiqui and Shaukat 2002). The experimental set-up consisted of an in vitro split root system, where one half of the split root system was challenged with nematodes, while PGPR strains were applied in the other half of the system (in tomato). A study involving application of PGPR by seed, root and foliar spray treatments separately in different combinations in field revealed that among the different PGPR strains tested, highest activity was by B. subtilis strain GB3, in terms of suppressing bacterial spots and increased activity of defence-related enzymes like peroxidase and phenylalanine ammonia lyase. PGPR which were found effective in greenhouse against bacterial spots also showed sustained ability to induce resistance in tomato under field conditions. Symbiotic association of *Glomus mosseae* with clover plants crop variety Sonja totally prevented infection by *P. ultimum*. Also, disease symptoms induced were systemically reduced even in non-mycorrhizal roots of plants which were grown in split root systems inoculated with AM fungi. Systemic regulation of pathogens induced by AM colonization indicates establishment of ISR. In plants colonized by AM species with biocontrol activities, higher concentrations of ISR-related compounds such as phenolic acids and new isoforms of superoxide dismutases, peroxidises and PR-1 proteins (pathogenesis-related proteins type 1) were detected. Rhizobacteria-mediated ISR in mycorrhizal roots is associated with accumulation of JA (jasmonic acid) which might be related to the systemic pathogen biocontrol. Additionally, local cell wall modifications like callose accumulation were identified around arbuscule-containing cortical cells of tomato roots (Alizadeh et al. 2013).

PGPB are able to colonize and retain their niches in the rhizosphere by production of bacterial allelochemicals, like siderophores, biocidal volatiles, antibiotics and lytic and detoxification enzymes (Compant et al. 2005). Secondary plant metabolites also play a vital role in stress management and plant growth-promoting activities. For example, *Bacillus amyloliquefaciens* FZB42 (a plant-associated bacteria) simultaneously promotes plant growth while producing secondary metabolites like polyketides bacillaene and difficidin which aid in suppression of soil-borne plant pathogens (Chen et al. 2007). Furthermore, colonization of basil plants by *Glomus mosseae* (which protects against *F. oxysporum)* did not increase the concentration of defence-related compounds such as rosmarinic and caffeic acids, phenolics and essential oils, highlighting the role of mechanisms other than the stimulation of systemic and localized plant defence mechanisms in *AM*-mediated biocontrol (Alizadeh et al. 2013).

Over the past two decades, there has been an increase in understanding of antibiosis being employed as biocontrol mechanism by PGPB. Detailed studies of several antibiotics along with their specificity and mode of action have been done, identifying the contribution of compounds like amphisin, DAPG (2,4-diacetylphloroglucinol), HCN (hydrogen cyanide), oomycin A, phenazine, pyoluteorin, pyrrolnitrin, tensin, tropolone and cyclic lipopeptides produced by Pseudomonads. *Bacillus*, *Streptomyces* and *Stenotrophomonas* spp. were also reported to produce oligomycin A, kanosamine, zwittermicin A and xanthobaccin. DAPG is a polyketide compound with broad-spectrum activity against fungi, bacteria and helminths. Phenazines are heterocyclic pigments which contain nitrogen and are synthesized by *Pseudomonas*, *Streptomyces*, *Burkholderia* and *Brevibacterium* species. Pyrrolnitrin is a broad-spectrum antifungal metabolite which can persist actively in the soil for a minimum of 30 days. Pyoluteorin, an aromatic polyketide antibiotic, inhibits oomycetous fungi and has strong activity against *P. ultimum* upon application to seeds, leading to decreased severity of *Pythium* damping off. *P. fluorescens* strain CHAO and its antibiotic over-expressing derivative CHAO/PME 3424 reduce *Meloidogyne incognita* galling in primary growth stages of crops such as tomato and brinjal. There exists a strong negative correlation between rhizobacteria colonization and nematode invasion as reported by Alizadeh et al. (2013). Different PGPR isolates from weedy grass have been used to control rice plant pathogens such as *Pyricularia oryzae*, *R. solani* and *Sarocladium oryzae* (Sarathambal et al. 2014). Many strains of *Pseudomonas* produce AFM (antifungal metabolites) out of which phenazines, pyrrolnitrin, DAPG and pyoluteorin are most common, but new AFMs like viscosinamide and tensin have also been reported. Studies have shown that viscosinamide prevents *P. ultimum* infection in sugar beet. Interestingly, AFM production is also observed in *Pseudomonas* where its biosynthesis happens under complex global regulation and quorum sensing. Global regulators like gacS/gacA genes regulate AFMs (and other extracellular products like protease, HCN) by encoding a two-component regulatory system. GacA has recently been shown to indirectly control the HCN synthase genes (hcnABC) and the protease gene aprA in *P. fluorescens* CHAO via a post-transcriptional mechanism involving a distinct recognition site overlapping the ribosomal binding site. It has recently been established that plants can recognize AHLs (N-acyl homoserine lactones) and their gene expression in roots and shoots can be altered by them. AHLs can also regulate the defence and cell growth responses of plant (Ortiz-Castro et al. 2009). An AHL synthase such as LuxI produces AHL signal molecules, which are believed to be involved in quorum sensing. It was reported that when AHL is at a threshold concentration (depending on density of bacterial cells), it binds to and activates LuxR, a transcriptional regulator. This activated form of the transcriptional regulator further stimulates gene expression (Bloemberg and Lugtenberg 2001).

Moreover, some PGPB-produced antibiotics are being tested for their utility as pharmaceuticals which can be used to tackle the increasing menace of multidrug resistance among human pathogenic bacteria. It is reported that their regulatory cascades involve global regulators GacA/GacS or GrrA/GrrS, the sigma factors RpoD and RpoS and quorum-sensing autoinducers such as AHL derivatives and are under positive autoregulation. Antibiotic synthesis is closely linked to the overall metabolic status of the cell, and the metabolic status is shaped by nutrient availability and other environmental factors. Trace elements specifically zinc and carbon source levels affect the capacity of secondary metabolite producing PGPM by influencing the genetic stability/instability of microbes. It is pertinent that several strains produce pellet of secondary antimicrobial metabolites and that conditions which favour one compound might not favour another. Therefore, the wide variety of biocontrol strains can enable suppression of pathogens under a wide range of environmental conditions. This was illustrated by the reports that the presence of glucose as a carbon source in *P. fluorescens* stimulates CHAO biosynthesis of DAPG and represses pyoluteorin (bacterial aromatic polyketide antibiotic). However, as glucose levels get depleted, pyoluteorin levels increase, and it becomes the more abundant antimicrobial compound produced by this strain which ensures that the antagonist has flexibility when dealing with different or changing environment. Biotic conditions also influence antibiotic biosynthesis. For instance, bacterial metabolites-salicylates and pyoluteorin affect DAPG production by *P. fluorescens*. Additionally, plant growth and development also have an impact on production of antibiotic compounds as biological activity of DAPG producers is induced not by the exudates of young plant roots but by the exudates of older plants. This leads to the creation of selective pressure against other microorganisms in the rhizosphere. Another feature that influences disease-suppressive interaction of plant with a microbial biocontrol agent is the host genotype itself (Compant et al. 2005).

Too much dependence on antibiotic-producing bacteria as biocontrol agents poses the complication of resistance development in phytopathogens against specific antibiotics. To overcome this shortcoming, researchers are utilizing biocontrol strains which synthesize HCN along with one or more antibiotics. This is an effective strategy as even though HCN may not have much biocontrol activity individually, it acts synergistically with bacterially encoded antibiotics (Glick 2012). The cyanide ion is exhaled as HCN and is metabolized further into other compounds. The mechanism through which HCN exerts its biocidal action is by inhibiting electron transport and disrupting the energy supply to the cell leading to the death of the organism. It also disrupts functioning of enzymes and natural receptors that can reverse its impact and inhibits action of cytochrome oxidase. HCN is reportedly produced by several rhizobacteria. HCN has broad-spectrum antimicrobial activity and is involved in biocontrol of several root diseases by many plant-associated fluorescent pseudomonads (Alizadeh et al. 2013). The ability of *P. fluorescens* strain CHAOs to suppress black root rot of tobacco and take-all of wheat was attributed to the production of HCN, and the same isolate was also shown to inhibit in vitro mycelial growth of *Pythium* and suppression of *F. oxysporum* f. sp. *radicis-lycopersici* in tomato. The cyanide producing strain CHAO was also observed to stimulate the formation of root hair presumably by inducing and altering plant physiological activities. In a particular study, four out of six PGPR strains were recognized for inducing systemic resistance in cucumber against *C. orbiculare* through production of HCN. Fluorescent *Pseudomonas* strain RRS1 isolated from Rajnigandha (tuberose) was studied to be positive for HCN production and also contributed to improved seed germination and root length. It has been reported that low oxygen levels are essential for the activity of ANR, a transcription factor responsible for positively regulating HCN biosynthesis.

Several microorganisms exhibit hyper-parasitic activity and attack pathogens by secretion of cell wall hydrolases. PGPB which can synthesize enzymes capable of lysing a portion of the cell walls of pathogenic fungi, such as chitinases, cellulases, β-1,3 glucanases, proteases and lipases, can exhibit biocontrol activity against wide spectrum of pathogenic fungi including *B. cinerea*, *Sclerotium rolfsii*, *F. oxysporum*, *Phytophthora* spp., *R. solani* and *P. ultimum* (Glick 2012). *Serratia plymuthica* C48 produces chitinase to inhibit spore germination and germ-tube elongation in *Botrytis cinerea*. The same enzyme was also responsible for its antagonistic activity against *S. rolfsii*. Suppression of *F. oxysporum* f. sp. *cucumerinum* by *Paenibacillus* sp. 300 and *Streptomyces* sp. strain 385 was also attributed to their ability to produce chitinase (Compant et al. 2005). Extracellular chitinases and laminarinases from *Pseudomonas stutzeri* could digest and lyse mycelia of *F. solani*. In PGPB *S. plymuthica* IC14 suppression of *Sclerotinia sclerotiorum* and *B. cinerea* was due to synthesis of proteases and other biocontrol traits. The β-1,3-glucanases synthesized by both *Paenibacillus* sp. (strain 300) and *Streptomyces* sp. (strain 385) were seen to lyse fungal cell walls of *F. oxysporum* f. sp. *cucumerinum*. The same enzyme synthesized by *B. cepacia* was found to damage the integrity of *R. solani*, *S. rolfsii* and *P. ultimum* cell walls (Compant et al. 2005). Constitutive and additional isoforms of defence-related enzymes were also reported in mycorrhizal roots (Alizadeh et al. 2013).

Iron is a vital element for growth of all living organisms, but the scarcity of bioavailable iron can foment great competition in soil habitats and on plant surfaces. Bacterial strains which don't possess or employ other means of biocontrol can use their capability of producing siderophores to establish themselves as biocontrol agents. Under iron-limiting conditions, PGPB-produced siderophores can help them to efficiently outcompete pathogens in competitively acquiring ferric ion. In fact siderophores from PGPB limit the proliferation of pathogenic fungi by depriving them of iron, an essential element to carry out many metabolic functions. Some PGPB strains can even derive iron from heterologous siderophores being produced by other microorganisms in their vicinity. Biosynthesis of siderophores is under strict regulation of iron-sensitive Fur proteins; GacS and GacA; sigma factors RpoS, PvdS, FpvI and N-acyl homoserine lactone; and site-specific recombinases. However, there are contradictory opinions on this as some studies do not support the involvement of these global regulators in siderophore production. For instance, GacS or RpoS had no significant effect on the level of siderophores synthesized by *Enterobacter cloacae* CAL2 and UW4. Similarly, non-involvement of RpoS of *P.*

putida strain WCS358, and preferential involvement of GrrA/GrrS over GacS/GacA of *S. plymuthica* strain IC1270 in regulation of siderophore synthesis, indicates evolution of genes in the siderophore-producing bacteria. Furthermore various environmental factors were also evidenced to modulate siderophore synthesis, viz. pH, iron levels, form of iron ions, presence or absence of other trace elements and optimal supply of major nutrients like phosphorus, nitrogen and carbon (Compant et al. 2005). Iron depletion in the rhizosphere by siderophores produced by the PGPB does not affect the growth of plants as most plants can thrive at much lower iron concentrations than most microorganisms. Moreover, several plants can bind, take up and utilize the iron-siderophore complexes generated by these PGPB. Evidence for involvement of bacterial siderophores in biocontrol of fungal pathogens comes from various studies. Certain studies used mutants defective in production of siderophores and found them less effective at protecting plants against fungal pathogens than the wild-type strains. Also, another study found that mutants which overproduce siderophores show greater effectiveness in protecting plants against fungal attacks (Glick 2012). Siderophores from endophytic bacteria were reported to limit the growth of *Streptomyces scabies* and *Xanthomonas campestris* in vitro. Compant et al. (2005) proved with their experiment the tissue type and tissue site-specific activities of siderophores by employing endophytic bacteria isolated from potato tubers. A study by Tiwari and Thrimurthy (2007) utilized 21 isolates of siderophore producing *P. fluorescens* and has concluded that isolates, PFR 1 and PFR 2, were superior over others in increasing shoot and root length of rice cv. Bamleshwari. An in vitro evaluation of the *P. fluorescens* isolates confirmed their antagonistic ability against *Pyricularia grisea* and *R. solani*. Pure culture of *Pseudomonas aeruginosa* were also studied for siderophore production, and their antifungal activity was tested against *Fusarium moniliformae*, *Alternaria solani* and *Helminthosporium halodes*, and it was seen that *P. aeruginosa* inhibits these fungal pathogens by production of antifungal secondary metabolites (Alizadeh et al. 2013). Apart from having high affinity for iron, siderophores may show affinity for other metals too. Excretion of catecholate compounds has been reported in *Azotobacter vinelandii*. These were earlier identified as siderophores, and they bind to metal cofactors of nitrogenase (Mo, V and Fe) enzyme (Pontes et al. 2015).

PGPM also employ detoxification of pathogen virulence factors as a mechanism of biocontrol. For example, *Xanthomonas albilineans* produces an albicidin toxin which is detoxified by certain biocontrol agents. In *Klebsiella oxytoca* and *Alcaligenes denitrificans*, this detoxification mechanism involves the production of a protein which reversibly binds the toxin, while *Pantoea dispersa* produces an esterase which irreversibly detoxifies albicidin. *Fusarium* species produce fusaric acid, a phytotoxin which is hydrolysed by many different microorganisms including *B. cepacia* and *Ralstonia solanacearum* strains. However, most pathogen toxins have a broad-spectrum activity capable of suppressing growth of microbial competitors, or as self-defence against biocontrol agents, they can detoxify the antibiotics they produce. PGPB can quench pathogen quorum-sensing capacity by degrading the autoinducer signals which would effectively block the expression of several virulence genes. A majority of plant bacterial pathogens depend on autoinducer-mediated quorum-sensing to activate their gene cascades coding for virulence factors, and this mechanism holds promise for relieving/curing disease even after the onset of infection. Both free-living rhizobacteria and endophytic bacteria share some biocontrol mechanisms. For instance, both are capable of synthesizing metabolites with antagonistic activity towards plant pathogens (Compant et al. 2005).

Lowering a plant's ET response to pathogens can ameliorate the extent of damage caused to plants by phytopathogens. This can be done by treating plants (generally the roots or seeds) with PGPB containing ACC (1-aminocyclopropane-1-carboxylate) deaminase. This technique has lowered the damage in cucumber, potato, castor bean, tomato, carrot and soybean plants, caused by various phytopathogens including *P. ultimum*, *F. oxysporum*, *Erwinia carotovora*, *Agrobacterium tumefaciens*, *Allorhizobium vitis*, *S. rolfsii* and *R. solani*, in both greenhouse and growth chamber experiments (Glick 2014). Transgenic plants expressing bacterial ACC deaminase are protected from damage caused by various phytopathogens to a significant level. Another study by Gamalero et al. (2010) supported the beneficial nature of PGPR for plant growth even under stress conditions. Interaction between ACC (1-aminocyclopropane-1-carboxylate) deaminase-producing bacterial strains and an arbuscular mycorrhizal fungus (AMF) was studied, and the effect of this interaction on cucumber growth was examined under saline conditions. Seeds of cucumber plant were treated with ACC deaminase-producing strain *P. putida* UW4 ($Acds^+$). Inoculation with *P. putida* was shown to have a positive effect on various parameters of assessment including root length, photosynthetic activity and overall plant growth. Inoculation of peas with *Pseudomonas* spp. containing ACC deaminase witnessed a similar growth-promoting effect and also contributed significantly to decrease the adverse effects of drought stress on growth, grain yield and ripening of pea (*Pisum sativum* L.) (Arshad et al. 2008). Salinity has negative effects on growth, but root colonization by ACC deaminase-producing bacteria or AMF can improve the tolerance of plant for such stressful conditions. This study was not only relevant from an ecological point of view, but also it has tremendous application as well. While PGPR stimulate growth rates in wild-type plants, it inhibits growth in ABA-deficient mutant plants. It has also been shown to induce accumulation of ET in ABA-deficient plants correlating with increased expression of the pathogenesis-related gene Sl-PR1b. Such results suggest that in ABA-deficient mutant plants, over-accumulation of ET corresponds with increased expression of Sl-PR1b indicating that maintenance of normal plant endogenous ABA levels might be essential for promotion of growth caused by *Bacillus megaterium* (Porcel et al. 2014). Goel et al. (2008) has elaborated on various type III effector proteins from *P. syringae* strains that act as virulence factors in the host cells. The virulence factors, Pma M6CΔE and HopAM1, were discussed for their role in enhanced nutrient uptake in plants that are grown under drought stress and adaptation to water availability. In *Arabidopsis*, HopAM1 was shown to induce hypersensitivity to ABA, causing stomatal closure and germination arrest. Although a discussion about all the PGPM currently in use is beyond the scope of this chapter, Table 8.1 provides a brief list of some PGPM which are presently being explored/used to protect plants against biotic stress.

Table 8.1 A brief list of PGPM reported to protect plants from biotic stress

S. no	PGPM	Action against	Host	Mechanism	References
1	*Acetobacter* sp., *Xanthomonas*	*Rhizoctonia solani*	Soybean	Antagonistic	Dalal and Kulkarni (2013)
2	*Aureobacterium saperdae*, *Bacillus pumilus*, *Phyllobacterium rubiacearum*, *Pseudomonas putida* and *Burkholderia solanacearum*	Vascular wilt of cotton caused by *Fusarium oxysporum* f. sp. *vasinfectum*	Cotton	Biological control potential	Chen et al. (1995)
3	*Bacillus subtilis* 281 (B1) strain and *Bacillus pumilus* 293 (B2)	Watermelon mosaic poty virus (WMV)	Pumpkin (*Cucurbita maxima L.*)	Induce systemic resistance	Elbeshehy et al. (2015)
4	*Bacillus* sp., *Burkholderia* sp.	*Sclerotium rolfsii*, *Fusarium oxysporum*, *Rhizoctonia solani*	Soybean	Antagonistic	Dalal and Kulkarni (2013)
5	*Klebsiella* sp., *Xanthomonas*	*Fusarium oxysporum*	Soybean	Antagonistic	
6	*Bacillus* sp.	*Colletotrichum truncatum*, *Macrophomina phaseolina*	Soybean	Antagonistic	
7	*Bacillus simplex* 30N-5	*Fusarium*	Cowpea plant	Antagonism	Schwartz et al. (2013)
8	*Bacillus subtilis*	Cowpea fungal pathogens *Fusarium verticillioides*, *F. equiseti* and *Rhizoctonia solani*	Cowpea plant	Antibiosis by antagonism	Killani et al. (2011)
9	*Bacillus thuringiensis*	Silences *Erwinia carotovora* which causes potato soft rot	In vitro on potato tubers	Antagonism and signal interference. Production of AHL-degrading enzyme (AHL-lactonase)	Dong et al. (2004)
10	*Beauveria bassiana*, *Clonostachys solani*, *Metarhizium anisopliae*, *Trichoderma atroviride*, *T. koningiopsis* and *T. gamsii*	*Delia radicum L.* or cabbage maggot	Cabbage	Rhizosphere competence	Razinger et al. (2014)
11	*Cladosporium* sp. and *Ampelomyces* sp.	*Pseudomonas syringae* pv. tomato DC3000	*Arabidopsis thaliana*	SA- and JA-signalling pathways employed in ISR	Naznin et al. (2014)

12	*Fusarium equiseti* GF183	Biocontrol of *Fusarium* wilt of spinach	Spinach	–	Horinouchi et al. (2010)
	Klebsiella sp., *Xanthomonas*	*Fusarium oxysporum*	Soybean	Antagonistic	Dalal and Kulkarni (2013)
13	*Lysobacter enzymogenes* C3	*Bipolaris sorokiniana*	Tall fescue	ISR	Kilic-Ekici and Yuen (2004)
14	*Paenibacillus* sp. 300 and *Streptomyces* sp. 385	*Fusarium wilt* caused by *Fusarium oxysporum* f. sp. *cucumerinum*	*Cucumber (Cucumis sativus)*	–	Singh et al. (1999)
15	*Pseudomonas* sp. DSMZ 13134	Barley leaf pathogen *Rhynchosporium secalis* and barley root pathogen *Gaeumannomyces graminis*	Barley plant	ISR	Fröhlich et al. (2012)
16	*Pseudomonas* sp. (strain PsJN)	Grey mould caused by *Botrytis cinerea*	Grapevine (*Vitis vinifera L.*)	ISR	Barka et al. (2000)
17	*Pseudomonas*	*Alternaria alternata, Colletotrichum truncatum, Rhizoctonia solani, Sclerotium rolfsii, Fusarium oxysporum*	Soybean	Antagonistic	Dalal and Kulkarni (2013)
18	*Pseudomonas fluorescens* WCS417r	*Bipolaris sorokiniana*	Tall fescue	ISR	Kilic-Ekici and Yuen (2004)
19	*Pseudomonas cepacia*	*Rhizoctonia solani, Sclerotium rolfsii* and *Pythium ultimum*		Chitinolytic activity	Fridlender et al. (1993)
20	*Penicillium* sp. (UOM PGPF 27) and *Pythium* sp. (UOM PGPF 41)	Pearl millet downy mildew disease	Pearl millet	–	Murali et al. (2012)
21	*Phoma* sp. (GS6-1, GS7-4 *and* GU23-3) (non-sporulating fungus)	*Gaeumannomyces graminis* var. *tritici (Ggt)* and *Cochliobolus sativus (Cs). GS6-1, GS7-4 (Phoma* sp.*)*	Barley kernels	Competitive root colonization	Shivanna et al. (1996)

(continued)

Table 8.1 (continued)

S. no	PGPM	Action against	Host	Mechanism	References
22	*Penicillium GP17-2, Phoma GS8-2 and sterilefungus GU23-3, Fusarium GF18-3, Penicillium simplicissimum (isolate GP17-2)*	*Colletotrichum orbiculare*	Cucumber plant	ISR	Koike et al. (2001) and Chandanie et al. (2006)
23	*Penicillium GP17-2, Phoma GS8-2, sterilefungus GU23-3*	Bacterial angular leaf spot	Cucumber plant	IR	
24	*Phoma GS8-2, sterilefungus GU23-3, Fusarium GF19-2*	*Fusarium* wilt	Cucumber plant	IR	
25	*Verticillium albo-atrum* and *Diplodia scrobiculata*	*Diplodia pinea*	Aleppo pine (*Pinus halepensis*)	ISR	Muñoz et al. (2008)
26	*Penicillium* spp. *GP15-1 (Penicillium neoechinulatum or Penicillium viridicatum)*	Damping-off disease by *Rhizoctonia solani*	Cucumber plant	ISR	Hossain et al. (2014)
27	*Bacillus pumilus (PGPR strain SE 34)*	*Fusarium oxysporum* f. sp. *radicis-lycopersici*	Tomato (*Lycopersicon esculentum Mill.*)	–	Benhamou et al. (1998)
28	*Pseudomonas denitrificans 1–15*	*Ceratocystis fagacearum*	Live oaks (*Quercus fusiformis*)	–	Brooks et al. (1994)
29	*Pseudomonas (Ps.) chlororaphis Tx-1*	*Pythium aphanidermatum* and *Pythium dissotocum* and controls root rot	Sweet peppers	–	Chatterton et al. (2004)
30	*Pseudomonas aeruginosa 7NSK2*	*Leaf infection by Botrytis cinerea* and resistance to *tobacco mosaic virus*	Bean and in tobacco	ISR by production and accumulation of salicylic acid	De Meyer and Hofte (1997); Audenaert et al. (1999)
31	*Pseudomonas fluorescens PB27*	Aflatoxigenic fungus *Aspergillus flavus*	–	Antibiosis by producing chitinase	Akocak et al. (2015)

32	*Pseudozyma aphidis*	*Clavibacter michiganensis*	Tomato plant	Secretes extracellular metabolites. Activates pathogenesis-related genes. Triggers SA-independent IR response	Barda et al. (2015)
33	*Pseudomonas* sp. *(pf 12)*	*Basal rot disease* by *Fusarium oxysporum Schlechtend:* fr. f. sp. *cepae*	Onion	Reduction of mycelial growth	Malathi (2015)
34	*Serratia plymuthica*	*Botrytis cinerea,* grey mould and *Sclerotinia sclerotiorum,* white mould diseases of leaves	Cucumber	Antagonism	Kamensky et al. (2003)
35	*Streptomyces* NRRL 30562	*Rhizoctonia solani*, *Aspergillus* sp., *Fusarium oxysporum*, *Botrytis allii and Alternaria helianthi*, *Pythium ultimum*, *Phytophthora infestans*, *Penicillium* sp., *Sclerotinia sclerotiorum*, *Erwinia carotovora* and *Cochliobolus carbonum*	Isolated from stems of *Kennedia nigriscans*	Antibiosis. Wide-spectrum antibiotics named Munumbicins	Castillo et al. (2002)
36	*Serratia marcescens* strain 90–166 and *Bacillus pumilus* strain SE34	Cucumber mosaic virus (CMV)	*Arabidopsis thaliana*	ISR by salicylic acid and NPR1-independent and JA-dependent signalling pathway	Ryu et al. (2004)
37	*Streptomyces* sp. *(CAI-24, CAI-121, CAI-127, KAI-32 and KAI-90*	Fusarium wilt disease	Chickpea		Gopalakrishnan et al. (2015)
38	*Streptomyces* sp. *MT7*	Wood-rotting *Basidiomycetes*	–	Cell wall lytic enzymes: chitinase, β-1,3-glucanase and protease; siderophores	Nagpure et al. (2013)

(continued)

Table 8.1 (continued)

S. no	PGPM	Action against	Host	Mechanism	References
39	*Streptomyces violaceusniger MTCC 3959*	*Basidiomycetes* which causes white rot and brown rot fungi	–	Antagonism, mycolytic enzymes: extracellular chitinase, protease, β-1,3-glucanase	Nagpure et al. (2014)
40	*Trichoderma virens* and *T. atroviride*	*Alternaria solani*, *Botrytis cinerea* and *Pseudomonas syringae pv. tomato* (Pst DC3000)	Tomato plant	Induce both ISR and SAR	Salas-Marina et al. (2015)
41	*Trichoderma hamatum 382*	Bacterial spot of tomato by *Xanthomonas euvesicatoria 110c*	Tomato plant	ISR	Alfano et al. (2007)
42	*Trichoderma asperellum T203*	*Pseudomonas syringae* pv. *Lachrymans*	Cucumber plant	ISR	Shoresh et al. (2005)
43	*Trichoderma*	*Botrytis cinerea*	–	ISR, antagonism, competition for nutrients and space	Vos et al. (2015)
44	*Trichoderma* sp., *T. harzianum* (TH 3)	Basal rot disease caused by *Fusarium oxysporum* f. sp. *cepae*	Onion	–	Malathi (2015)

Disease incidence and severity can be minimized when competition between pathogens and PGPB occurs. For example, non-pathogenic soil microbes can rapidly colonize plant surfaces and use up the available nutrients making it difficult for pathogens to grow. Treatment of plants with *Sphingomonas* sp. has been reported to prevent the bacterial pathogen *P. syringae* pv. tomato from causing disease symptoms (Glick 2012). However, poor rhizosphere competence leading to inconsistent performance hampers application of PGPB in field tests. Rhizosphere competence of biocontrol agents is determined by effective root colonization and ability to survive/proliferate along with growing plant roots over a significant time period, in the presence of native microflora. As rhizosphere competence is of utmost importance for effective biological control, detailed understanding of genetic and environmental regulation of root-microbe interaction can considerably help to improve the efficacy of these biocontrol agents. Root colonization is important in competing for root niches and bacterial determinants. Root surface and its surrounding rhizosphere are significant sink for carbon. Photosynthate allocation to this zone can be up to 40%. PGPB compete for these nutrients and niches in order to protect plants from phytopathogens. They reach root surfaces chemotactically, facilitated by flagella. Root exudates contain chemical attractants like organic acids, amino acids and specific sugars. Some of these exudates are also effective as antimicrobial agents providing an ecological niche advantage to organisms that have the ability to detoxify them via adequate enzymatic machinery. Genetic and environmental factors control the quantity and composition of chemoattractants and antimicrobials exuded by plant roots. Thus, it can be deduced that PGPB competence depends heavily upon their ability to take advantage of a specific environment and/or adapt to changing conditions. For example, sugars, amino acids and organic acids induce chemotaxis in *Azospirillum*. However, different strains vary in their degree of chemotactic response to each of these compounds. Rice exudates were shown to induce stronger chemotactic responses of endophytic bacteria compared to non-PGPB present in the rice rhizosphere, suggesting that PGPB are uniquely equipped to sense chemoattractants. Bacterial LPS (lipopolysaccharides) are also involved in root colonization, but their importance in colonization may be strain dependent as the LPS O-antigenic side chain of *P. fluorescens* WCS374 doesn't play a part in potato root adhesion, while on the other hand, the O-antigen chain of *P. fluorescens* PCL1205 does contribute in tomato root colonization. Moreover, the O-antigenic aspect of LPS was shown to have no contribution in rhizoplane colonization of tomato by the PGPB *P. fluorescens* WCS417r, whereas this bacterial determinant did contribute in its endophytic colonization of roots. Properties like high rate of bacterial growth, ability to synthesize vitamin B1 and exuding NADH dehydrogenases enable PGPB to colonize plants. They also indulge in root colonization via type IV pili which are known for their role in the adhesion of bacteria to eukaryotic cells and are also involved in plant colonization by endophytic bacteria. Efficient root colonization in certain PGPB is linked to the ability of secreting even a site-specific recombinase. Transfer of a site-specific recombinase gene from a rhizosphere-competent *P. fluorescens* into a rhizosphere-incompetent *Pseudomonas* strain reportedly increased its ability to colonize root tips (Compant et al. 2005). In field experiments, inadequate biocontrol is often correlated with poor root colonization. A screen for mutants of the rhizobacterial strain *P. putida* KT2440 was able to identify set of putative surface and membrane proteins that have a role in attachment to corn seeds. Some such proteins were homologs of a calcium-binding protein, a hemolysin and a potential multidrug efflux pump. A study used in vivo expression technology to identify *P. fluorescens* genes that are specifically expressed in the rhizosphere (i.e. rhi genes) and found greater than 20 rhi genes. Out of these, 14 were shown to have significant homology to genes involved in nutrient acquisition, stress response or secretion. Various root colonization genes and traits were also observed in the *Pseudomonas* species with biocontrol properties and suggested for use in improving colonization of wild-type *Pseudomonas* strains. The competitiveness of these strains is also increased by their ability to produce siderophores. Authors observed that ability to utilize organic acids is the nutritional basis of tomato rhizosphere colonization where a defect/inability in the utilization of the raid organic acids has led to decreased competitive colonization of the tomato rhizosphere. On the other hand, a defect in sugar utilization found to have no impact on colonization (Bloemberg and Lugtenberg 2001). AMF have to compete with soil-borne pathogens in order to acquire space and nutrients. Plant pathogens may obstruct mycorrhizal colonization if they are present in very large numbers. To avoid this competitive inhibition and have better biocontrol efficiency, AMF pre-inoculation and pre-host treatment are always favoured. Dual inoculation of AMF with rhizobacteria has a synergistic or additive effect on its control of plant growth suggesting that biocontrol properties would depend on the combination of bacterial/fungal species used, soil's nutritional status and other environmental factors (Alizadeh et al. 2013).

How Do PGPM Help in Fighting Abiotic Stress?

Typically, plant growth involves periods of maximum growth interjected randomly with various levels of no growth or growth inhibition periods triggered by external stress stimuli. Upon addition, PGPM can employ any one or more of several different strategies described below, in order to overcome the growth inhibition caused by environmental stress or biotic and abiotic.

ACC deaminase-containing PGPB can be employed to ameliorate abiotic stresses like temperature extremes, metal toxicity, flooding, drought, hypoxia, salt and organic contamination (Glick 2012). Several PGPR containing ACC deaminase are present in the soil which aid in improvement of plant growth, especially under unfavourable environmental conditions. Abiotic stresses trigger increase of ET in plants which is directly related to the concentration of ACC in plant tissues. These bacteria may improve the survival of seedlings in the first few days post-sowing by decreasing ET levels, which helps in longer root formation. Plants with decreased level of ET react better when facing different environmental stresses such as salinity, drought and metal toxicity (Fahad et al. 2015). Both endogenous and exogenous

ACC deaminase genes have been shown to increase the symbiotic performance of several rhizobial strains. In case of flooding, plant roots are typically subjected to hypoxic or oxygen-limiting conditions which lead to increased production of ACC synthase enzyme and other stress proteins. The stressed plant subsequently synthesizes more ACC in its roots, but the newly made ACC cannot be converted to ET in the roots (ET synthesis requires oxygen), hence ACC has to be transported to the shoots where there is an aerobic environment. There ACC gets converted to ET, and production of ET by flooded plants results in epinasty (wilting), chlorosis, necrosis and reduced growth. Treating plants with ACC deaminase-producing PGPB can protect plants from majority of damage caused by flooding. A study reported that PGP bacteria that are endemic to sites of limited rainfall prove better in protecting plant growth against drought stress as compared to similar bacteria from sites of water abundance. Researchers have proved the efficacy of ACC deaminase-containing PGPB in protecting a wide variety of plants against drought-induced growth inhibition. Lowering ET levels using ACC deaminase-containing PGPB might also afford protection against salinity stress as reported in studies by Glick (2014) and Sarathambal et al. (2014). The presence of organic contaminants in the soil would also result in increase in stress-induced ET synthesis. As a number of ACC deaminase-containing PGPB are able to protect plants from a wide range of abiotic stresses, this technology can be explored to become a technology (marketable product) with commercial use in the field. However, certain inhibitions about the use of bacteria on a large scale in agriculture exist as the biosafety level of such PGPB needs confirmation (Glick 2012, 2014).

PGPB that do not contain ACC deaminase nevertheless protect plants from the harmful effects of abiotic stresses by providing IAA to the plant which directly stimulates plant growth, even if other inhibitory compounds are present. However, bacteria producing both IAA and ACC deaminase can be considered effective in protecting plants against a wide range of stresses. The synergistic effect of IAA and ACC deaminase in plant growth promotion can be explained as follows: PGPB bound to plant roots utilize tryptophan (exuded by plant roots) to convert it into IAA. This IAA produced is secreted out by the bacteria in the rhizosphere and is taken up by plant cells. Once there, it joins the plant's IAA to stimulate an auxin signal transduction pathway, which includes various auxin response factors. As a result, growth and proliferation of plant cells occur. Simultaneously, some of the IAA promotes transcription of the ACC synthase-encoding gene which increases concentration of ACC. As ACC oxidase catalyses ACC into ET, the ET levels also shoot up. Various biotic and abiotic stresses can also increase IAA synthesis or stimulate transcription of ACC synthase gene. In the absence of bacterial ACC deaminase, ET-induced cell growth and proliferation limitation take place due to decreased transcription of auxin response factors. It also limits IAA stimulation of the synthesis of more ET. In the presence of ACC deaminase, lesser ET is formed, and therefore in its presence, transcription of auxin response factors is not inhibited, and IAA stimulates cell growth and proliferation without causing a build-up of ET. Therefore, both in the presence and absence of stress, ACC deaminase can decrease inhibition of plant growth by ET and allows IAA to promote plant growth (Glick 2012).

CKs are compounds with structural resemblance to adenine and are named so because of their ability to promote cytokinesis or cell division in plants. Various plants, some yeast strains and a number of soil bacteria including PGPB produce CKs. Transgenic plants which can overproduce CKs under abiotic stress were found to have withstand the harmful effects of stress. However, there aren't any conclusive studies showing that bacterially produced CKs can protect plants from abiotic stresses. A comparative study concentrating on the activity of CK-producing PGPB as against that of CK minus mutants may help in understanding the role of PGPB in CK of those bacteria is needed (Glick 2012).

Gibberellins are plant hormones regulating growth and development processes like germination of seeds, emergence of seedling, floral induction, growth of roots, stems, leaves, fruits and flowers and senescence. In majority of such processes, GAs play a complementary role with other phytohormones and regulatory factors and integrate the signalling pathways (Bottini et al. 2004). It was reported that drought resistance can be acquired by inhibition of GA synthesis (Waqas et al. 2012). Under drought, synthesis of GA and ABA was found to be reduced in maize seedlings, and it was shown that prior inoculation with *Azospirillum lipoferum* promoted growth of both roots and shoots. It has also been reported to relieve water stress effects on wheat, at least in part by GA synthesis (Bottini et al. 2004). It has also been reported that two endophytic fungi, *Phoma glomerata* LWL2 and *Penicillium* sp. LWL3, alleviate drought and salt stress by secreting phytohormones like gibberellins and IAA along with promoting plant growth in cucumber plants (Waqas et al. 2012).

Trehalose is a non-reducing storage disaccharide and is ubiquitous in nature. Increased levels of trehalose act as protectant against various abiotic stresses like drought, high salt and temperature extremes. Trehalose is a highly stable molecule resistant to both acid and high temperature. It forms a gel phase as cells dehydrate, by replacing water and, thereby, decreases drought and salt damage. Additionally, trehalose also prevents protein degradation and aggregation which occur under extreme temperature stresses. PGPM engineered to overproduce trehalose were observed to make plants tolerant against drought and other stresses. Treatment of bean plants with the genetically engineered symbiotic bacteria *Rhizobium etli* (an overproducer of trehalose) has reportedly led to formation of more nodules, more nitrogen fixation, increase in biomass and increased recovery from drought stress as compared to plants inoculated with wild-type *R. etli*. Similarly, treatment of maize plants with the PGPB *Azospirillum brasilense* (modified for overproduction of trehalose) made the plants more resistant to drought, producing more biomass than plants which were treated with wild-type *A. brasilense*. Even though plants can also be directly engineered to overproduce trehalose, it is easier to use genetically modified PGPB to attain the same end. Also, one engineered bacterial strain can be effective in protecting many different crop plants (Glick 2012).

PGPR can also enhance stress resistance by solubilizing minerals and nutrients in the plant-soil system and enhance release of these nutrients into soil solution (Sarathambal et al. 2014). Nitrogen-fixing bacteria (NFB) can play a vital role in plant establishment and development as they provide the limiting nutrient (nitrogen) to plants. The relationships of symbiotic bacteria (like *Rhizobium* and *Frankia)* with leguminous crops have been studied extensively, and its total coverage is beyond the scope of this chapter. It has been reported that diazotrophs, such as *Acetobacter* and *Herbaspirillum*, might be beneficial for plant growth promotion in nonlegumes (Grandlic 2008). Non-rhizobial N_2-fixing bacteria were shown to exhibit endophytic growth in several grasses. *Pseudomonas* species were the most dominant class of nifH (the nitrogen fixation gene) carrying bacteria in the rhizosphere of perennial grasses of South Australia. The nifH gene is present in many non-*Frankia* actinobacteria like *Agromyces*, *Microbacterium*, *Corynebacterium* and *Micromonospora*. Nitrogen fixation is also reported in *Prosopis* sp. under natural conditions (Sarathambal et al. 2014). Other well-known NFB include *Azospirillum*, *Bradyrhizobium*, *Burkholderia*, *Sinorhizobium*, *Mesorhizobium* and *Azorhizobium* (Grandlic 2008). It has been reasonably hypothesized that optimally nitrogen-nourished plants have survival advantage over nitrogen malnourished ones, when challenged by abiotic stresses (Arora 2004). Certain species of *Rhizobium* have been reported to form effective (N_2-fixing) symbioses with legumes under various stresses (salt, heat, acid stress, heavy metal stress) (Zahran 1999).

Some microorganisms are capable of solubilizing the inorganic P (phosphorus) present in the soil and increase its availability to plants. Plant-associated microbes have been reported to solubilize P in various instances. This group comprises of bacteria, fungi and some actinomycetes, and they solubilize unavailable forms of organic P such as tricalcium, iron, aluminium and rock phosphates into soluble forms by releasing various organic acids like malic, fumaric, succinic, citric, glyoxylic and gluconic acids (Pontes et al. 2015). In conditions wherein phosphate is present in an insoluble form, phosphate-solubilizing bacteria (PSB) like strains belonging to the genus *Cedecea* and *Microbacterium* have been reported to promote the growth of barley plants (Sarathambal et al. 2014). The hyphae of AMF were noted to absorb poor mobility nutrients like P from soil beyond the zones that are depleted by roots particularly when nutrient availability is low. Mycorrhizal fungi increase plant growth by enhanced phosphorus uptake, and some AMF aid in enhancing plant resistance towards salinity stress by either modulating the hormonal balance of the host plant or by enhancing water uptake (Alizadeh et al. 2013). At a particular deserted site with low nutrient and organic matter content, a native mycorrhiza, *Geastrum coronatum*, was shown to be an important microbe in plant establishment. In addition, *G. coronatum* was most effective at heightening nitrogen and P content in plants when used together with *Rhizobium* sp. Another mycorrhizal species, *Glomus intraradices*, showed greater affectivity at promoting plant growth with a completely different strain of *Rhizobium* (Grandlic 2008).

Apart from phosphorous, most soils are deficient in micronutrients, such as: Zn, Fe and Mn with Zn registering as foremost nutrient that is deficient across the world. Hence zinc solubilizing bacteria (ZSB) based bioinoculants are highly beneficial for countries like India in which there is high incidence of zinc deficiency (greater than 70%).These ZSB are capable of solubilizing insoluble zinc compounds/minerals in agar plates as well as in the soil. Potassium-solubilizing bacteria like *Bacillus mucilaginosus* and *Bacillus edaphicus* are used in bioinoculants and are capable of solubilizing potassium rock by production and secretion of organic acids. The bacteria being heterotrophic and aerobic can obtain energy and carbon from existing organic sources. Hence, they can also improve soil structure by contributing to the formation and stabilization of water-stable soil aggregates. A study by Sarathambal et al. (2014) reported that *B. subtilis*, *Azospirillum* sp., *A. brasilense* and *Bacillus* sp. have the ability to solubilize minerals such as P, potassium and Zn. The metal-resistant PGPB *Pseudomonas* sp. and *Pseudomonas jessenii* were shown to aid in metal sequestering (nickel, copper and Zn) Ricinus communis when associated with host as well as phosphate solubilization and IAA production. These aforementioned abilities could be highly useful for promoting plant growth to counter abiotic stress polluted soils in metal-stressed soils (Rajkumar and Helena 2008). Sheng et al. (2008) isolated and characterized endophytic lead (Pb)-resistant bacteria (*P. fluorescens* G10 and *Microbacterium* sp. G16) from rape roots of plants grown in heavy metal-contaminated soils. These two strains were shown to possess various heavy metal and antibiotic resistance characteristics and can increase water-soluble Pb (in solution) in Pb-added soil while simultaneously showing plant growth promotion.

For in-field effectiveness, a PGPB should show persistence and proliferation ability in the environment. Some countries have a common spring temperature of 5–10 °C, and PGPB should be functional in these cool soil temperatures and should be able to survive repeated freeze-thaw cycles common during the winter season in several places. Moreover, as many fungal pathogens are most destructive in cold and temperate climates, cold-tolerant (psychrotrophic) PGPB can be expected to show better biocontrol activity in such climates than mesophilic biocontrol strains. It has been studied that in some psychrophilic and psychrotrophic PGPB, exudation of antifreeze proteins into surrounding area takes place when grown at low temperatures. Bacterial antifreeze proteins would regulate the formation of ice crystals outside the bacterium, protecting cell walls of the bacteria from the lethal piercing damage which might be caused by the formation of large crystals at freezing temperatures. Additionally, some of them were also found to possess ice nucleation activity. Even though there have been various studies aiming at isolation and characterization of bacterial antifreeze proteins, none of these studies focused on the possibility of utilizing this activity to enable PGPB functionality in environments that include cold temperatures (Glick 2012).

Other Strategies of Dealing with Abiotic Stress

Recent researches in the field of PGPM have shown tremendous opportunities that have a positive impact on the growth and health of plants. Also, applying PGPM for the remediation of contaminated soils has opened newer possibilities of research. For instance, combining PGPM with contaminant-degrading bacteria can help to eradicate contaminants present in the soil. Rhizobacteria can also be used to increase the uptake of specific metal pollutants from soil. Exploiting and manipulating genetic engineering technologies in this line can be proven to help bioremediation

(Zhuang et al. 2007). Phytodegradation and rhizoremediation are remediation strategies that utilize rhizospheric bacteria to degrade persistent organic compounds. It is well known that ET is integral for initiation of senescence in flowers. Many cut flowers (e.g. carnations and lilies) are treated with silver thiosulfate (an ET inhibitor) prior to their sale. However, high silver thiosulfate levels can be phytotoxic and cause environmental problems. Using ACC deaminase-containing plant PGPR (with ability to limit ET production) for treating cut flowers can provide an environmentally friendly alternative. In the first instance of using ACC deaminase-containing PGPB for metal phytoremediation, it was reported that a nickel-resistant bacteria could alleviate nickel toxicity in canola plants. Since then, there have been several accounts of metal phytoextraction using PGPB including a wide variety of plants, different metals, soils and bacteria. In many of those studies, bacteria were first selected based on their resistance to the toxic metal(s) and then tested for ACC deaminase activity, IAA and siderophore synthesis. Additionally, some other bacterial traits might be involved in metal phytoremediation. Certain bacteria facilitating phytoremediation were shown to have the ability to solubilize phosphate and were proposed to assist in metal uptake. Another study reported the production of biosurfactants in the bacterial strain aiding in phytoremediation which might be involved in increasing the bioavailability of metals. Strategy of using bacteria in combination with phytoremediation was suggested to be effective in removing and/or degrading organic contaminants from impacted soils, both in the lab and under field conditions (Glick 2014).

Timmusk and Wagner (1999) reported changes in plant gene expression up on inoculation with PGPR. The model plant chosen was *Arabidopsis thaliana*, and PGPR inoculated was *Paenibacillus polymyxa*. Abiotic and biotic stress was introduced where abiotic stress was induction of drought and biotic stress was infection by pathogen *E. carotovora*. The results showed that the plants inoculated with PGPR were more tolerant to stress and showed more resistance than the control plants. In a study by De Souza et al. (2015), *Pseudomonas* sp. FeS53a was isolated from rice roots from an area with a history of iron toxicity, and the bacterial genome was sequenced. Its genome was found to contain genes involved in auxin biosynthesis that can regulate metabolic processes and promote plant growth under abiotic stress conditions. This strain was also observed to encode superoxide dismutase and catalase, the enzymatic antioxidants which remove free radicals and prevent damage to cell membranes and DNA. Bacterioferritin genes that can be involved in iron storage systems and the ferric uptake regulation protein (Fur) were also detected in this isolate. In a recent study, Amaresan et al. (2016) isolated salt-tolerant isolates of *Bacillus* PGPR from Tsunami (India)-affected areas and reported the presence of isolates capable of growing even at 10% NaCl concentration. Out of these isolates, 14 showed phosphate solubilization activity, 13 produced siderophores and five produced IAA, while 16 isolates could produce at least one extracellular enzyme. Some of these isolates also exhibited antagonistic activity against *S. rolfsii*. It was hypothesized that such PGPR could be employed as bioinoculants for enhancing crop growth in Tsunami-stressed soils.

The plant-bacteria associations have been a topic for research since decades. But there is a lot more to explore about the mechanisms employed by PGPB. Currently it is known that the bacteria can be good for plant growth and health, whereas a significant factor to be considered is that the plant can also select their microbiome to have beneficial bacteria preferential to their systems (Marasco et al. 2012). This kind of an approach needs to be further explored to maximize the benefits of PGPM-host associations. Bacterial endophytes have been a topic for research due to their direct mechanism of plant growth-promoting capacity as well as for their indirect mechanism as biocontrol agents. The mechanism of rhizopheric bacteria and endophytes are almost similar; therefore much work has been done in the domain of rhizopheric bacteria, assuming a similar mechanism in endophytes. But the fine line that separates the two is the micro-environment in which they dwell. For instance, for rhizopheric bacteria, the variations in temperature, light, soil type and other abiotic factors play a key role. It is also possible that different plant growth-promoting mechanisms which are unexplored in rhizopheric bacteria might be discovered in bacterial endophytes (Santoyo et al. 2016).

Genetic Modification to Improve PGPM

Identification of the genes involved in plant growth-promoting activity of rhizobacterial strains can be used to improve the performance of biocontrol strains and/or to design novel biocontrol strains by genetic modification. Rhizobacterial strains have been transformed using single genes or even complete operons, under control of regulatory genes or regulated by tac or lac promoters. Introduction of a mini-Tn5 vector including the complete operon for biosynthesis of PCA (phenazine-1-carboxylic acid), an antifungal metabolite, has been reported to enhance rhizosphere competence and suppression of fungal diseases by genetically engineered *P. fluorescens*. Similarly, introduction of the phzH gene from *Pseudomonas chlororaphis* PCL1391 was found to enhance the biocontrol ability of *Pseudomonas* strains producing PCA by increasing the production of phenazine-1-carboxamide, additionally. Effective control was seen against tomato foot and root rot. Other studies also reported improved biocontrol and/or plant growth promotion by introduction of such genes like Cry-toxin-encoding cry1Ac7 gene of *B. thuringiensis*, chitinase-encoding chiA gene of *S. marcescens*, and ACC deaminase gene from *E. cloacae* into rhizobacterial strains. Also, it was shown that the transfer of Sss gene of *P. fluorescens* WCS365 could enhance the competitive colonization ability of other *P. fluorescens* strains (Bloemberg and Lugtenberg 2001).

The Sss recombinase genes, *ptsP* and *orfT*, are important in the interaction of *Pseudomonas* spp., with various hosts. The gene *ptsP* encodes a nitrogen-specific EI paralogue called EI^{Ntr} that forms a regulatory PTS phosphoryl transfer chain. This chain is involved in sugar-dependent utilization of certain amino acids and is also linked to metabolism of carbon and nitrogen. Sss gene and orfT were hypothesized to have a contributory role in providing rhizosphere competence and phenotypic variation in fluorescent *Pseudomonas* (Mavrodi et al. 2006). Two genetically modified derivatives of *P. putida* WCS358r carrying the *phz* biosynthetic gene locus of strain *P. fluorescens* 2–79 are known to constitutively produce the antifungal compound PCA, thus imparting improved antifungal activity. Glandorf et al. (2001) suggested that the gene responsible to produce PCA can be introduced into a plant growth-promoting bacterial strain, *P. putida* WCS358r, using the mini-Tn*5* transposon system as a delivery vector. Pectin, a complex plant polysaccharide, when broken down to D-glucuronate and D-galacturonate, serves as a carbon source for bacterial growth and could potentially serve as a nutrient source for efficient root colonization of PGPR. Therefore, the presence of genes that enable D-galacturonate and D-glucuronate utilization could be advantageous for plant growth-promoting activity through efficient root colonization. These genes were present in *B. amyloliquefaciens* subsp. *plantarum* and can be genetically introduced into other species to improve their root colonization ability (Hossain et al. 2015).

Future Prospects and Need for More Intensive Utilization of PGPM

The growing concern and awareness to protect the agriculture, environment and food safety issues has been significant enough to warrant a reduction in the usage of pesticides and other crop chemicals. Biological control through application of PGPB is highly suggested as a potential alternative in crop disease management. Plant productivity and other factors like quality and health have been evidently improved by direct application of PGPM to the soil and through seed inoculation as examined by many researchers and also elaborated in this chapter. A number of studies confirmed that microbial inoculants lead to higher microbial populations in the soil and promote plant growth through improved nutrient acquisition, suppression of plant diseases, increased levels of phytohormones and other growth metabolites and ISR in various crops including cereals (Nelson 2004; Rodriguez et al. 2007; Van Loon 2007; Umashankari and Sekar 2011; Yadav et al. 2011; Basja 2013). Detailed understanding of the mechanisms of plant growth promotion by PGPB has shed light on multiple facets of disease suppression by these biocontrol agents. However, most studies have focused on free-living rhizobacterial strains, especially *Pseudomonas* and *Bacillus*, and much remains to be understood from non-symbiotic endophytes about their unique associations and growth-promoting activity on host plants. Increased understanding of the mechanisms of PGPB action opens up new possibilities for designing strategies for improvement of efficacy of biocontrol agents. Identification of key antimicrobials, like DAPG produced by superior agents, can be exploited for efficient targeted selection of isolates carrying relevant biosynthetic genes. Determining the edaphic parameters that favour disease suppression, production of antibiotics and their activity can be advantageous for identifying target inoculants for soils to support biocontrol. Amending soils or growth substrates with minerals like Zn or priming of inoculants with media amendments during fermentation can also be highly effective as per some scientific recommendations. Similarly, consortium of rhizosphere bacteria can be further augmented by soil aeration, hydrogenation, delivery of molasses, sugars and by apt crop rotations. Identification of the various mechanisms of action, facilitating strain combinations, like bacteria with bacteria or bacteria with fungi, to attack pathogens with a broader arsenal of microbial weapons, etc., can be more beneficial options. Biotechnology can also be applied to further improve strains which have prized qualities (like ease of formulation, stability or exceptional suitability to plant colonization) by creation of transgenic strains that combine multiple means of activity. Continued research on endophytic bacteria holds potential for the development of biocontrol agents which may self-perpetuate by colonizing hosts and getting transferred to progenies, like in the case of associative nitrogen-fixing PGPB on sugarcane. The importance and target-specific approach in plant genetic engineering is unquestionable to further the goals of modern day demands on agriculture systems. And most of the functional advantages that PGPM are offering can be traced back to the genes encoding these functions, and one can either directly engineer the plants with genes from PGPM to overproduce specific metabolites or can easily generate genetically modified PGPB to attain the same end. Also, as many of these PGPM are non-host specific and are predominantly rhizospheric (and some endophytic), the advantage of using PGPM over GM plants would be that one engineered PGPM strain can be effective in protecting many different crop plants. Figure 8.3 is

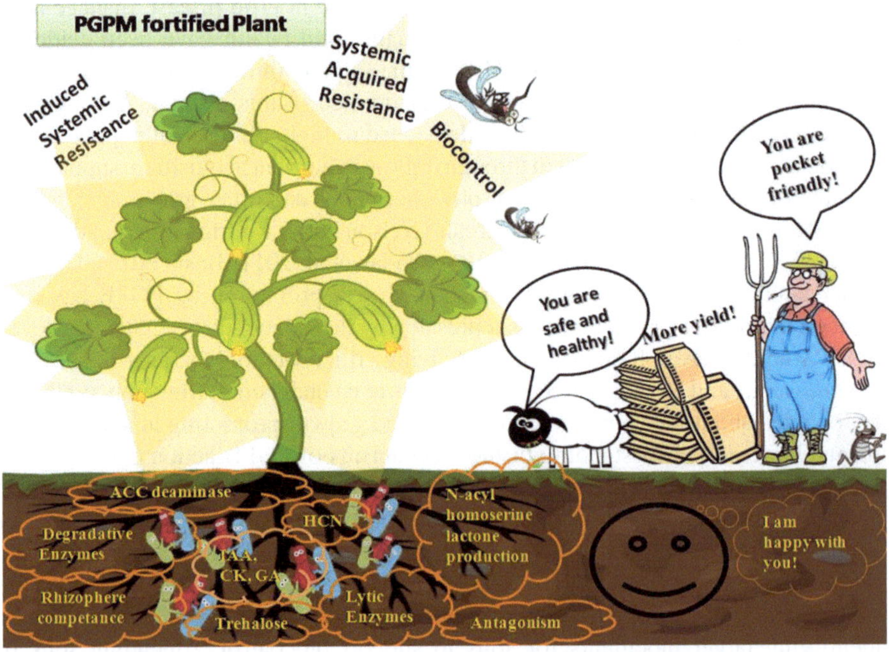

Fig. 8.3 PGPM fortified plant and its response to various biotic/abiotic stressors

a comprehensive pictorial representation of various means by which PGPM can benefit the plants and why they can be a preferred choice over GM plants.

Considering the fact that population growth is far exceeding the availability of food resources, it is imperative that new innovative technologies be developed and implemented to increase crop yield at the face of biotic and abiotic stressors. Plant-associated microorganisms can play a vital role in conferring resistance to several environmental stresses. Use of microbial inoculation for stress alleviation in plants can provide a cost feasible and environmentally sound method in lieu of plant breeding, genetic energy or use of agricultural chemicals. Promoting the use of PGPM, initially in addition to and ultimately instead of agricultural chemicals currently in use, is important to achieve this aim.

References

Akocak, P. B., Churey, J. J., & Worobo, R. W. (2015). Antagonistic effect of chitinolytic *Pseudomonas* and *Bacillus* on growth of fungal hyphae and spores of aflatoxigenic *Aspergillus flavus*. *Food Bioscience, 10*, 48–58.

Alfano, G., Ivey, M. L. L., Cakir, C., Bos, J. I. B., Miller, S. A., Madden, L. V., Kamoun, S., & Hoitink, H. A. J. (2007). Systemic modulation of gene expression in tomato by *Trichoderma hamatum* 382. *Biological Control, 97*, 429–437.

Alizadeh, O., Azarpanah, A., & Ariana, L. (2013). Induction and modulation of resistance in crop plants against disease by bioagent fungi (arbuscular mycorrhiza) and hormonal elicitors and plant growth promoting bacteria. *International Journal of Farming and Allied Sciences, 2*, 982–998.

Amaresan, N., Kumar, K., Madhuri, K., & Usharani, G. K. (2016). Isolation and characterization of salt tolerant plant growth promoting rhizobacteria from plants grown in tsunami affected regions of Andaman and Nicobar Islands. *Geomicrobiology, J36*(20), 942–947.

Arora, R. (2004). *Adaptations and responses of woody plants to environmental stresses* (pp. 1–5). New York: IOS Press.

Arshad, M., Shaharoona, B., & Mahmood, T. (2008). Inoculation with *Pseudomonas* spp. containing ACC-deaminase partially eliminates the effects of drought stress on growth, yield, and ripening of pea (*Pisum sativum* L.). *Pedosphere, 18*(5), 611–620.

Audenaert, K., De Meyer, G., & Höfte, M. (1999). *Pseudomonas aeruginosa* 7NSK2-induced systemic resistance in tobacco depends on in planta salicylic acid accumulation but is not associated with PR1a expression. *European Journal of Plant Pathology, 105*, 513–517.

Bach, E., Seger, G. D. S., Fernandes, G. C., Lisboa, B. B., & Passaglia, L. M. P. (2016). Evaluation of biological control and rhizosphere competence of plant growth promoting bacteria. *Applied Soil Ecology, 99*, 141–149.

Balconi, C., Stevanato, P., Motto, M., & Biancardi, E. (2012). Breeding for biotic stress resistance/tolerance in plants. In M. Ashraf, M. Ozturk, M. S. A. Ahmad, & A. Aksoy (Eds.), *Crop production for agricultural improvement* (pp. 57–114). Springer.

Barda, O., Shalev, O., Alster, S., Buxdorf, K., Gafni, A., & Levy, M. (2015). *Pseudozyma aphidis* induces salicylic-acid-independent resistance to *Clavibacter michiganensis* in tomato plants. *Plant Disease, 99*, 621–626.

Barka, E. A., Belarbi, A., Hachet, C., Nowak, J., & Audran, J. C. (2000). Enhancement of *in vitro* growth and resistance to gray mould of *Vitis vinifera* co-cultured with plant growth-promoting rhizobacteria. *FEMS Microbiology Letters, 186*, 91–95.

Basja, N. (2013). The effect of agricultural practices on resident soil microbial communities: Focus on biocontrol and biofertilization. In B. FJD (Ed.), *Molecular microbial ecology of the rhizosphere* (pp. 687–700). Hoboken: Wiley Inc.

Benhamou, N., Kloepper, J. W., & Tuzun, S. (1998). Induction of resistance against *Fusarium* wilt of tomato by combination of chitosan with an endophytic bacterial strain: Ultra structure and cytochemistry of the host response. *Planta, 204*, 153–168.

Bloemberg, G. V., & Lugtenberg, B. J. J. (2001). Molecular basis of plant growth promotion and biocontrol by rhizobacteria. *Current Opinion in Plant Biology, 4*, 343–350.

Bottini, R., Cassán, F., & Piccoli, P. (2004). Gibberellin production by bacteria and its involvement in plant growth promotion and yield increase. *Applied Microbiology and Biotechnology, 65*, 497–503.

Boualem, A., Dogimont, C., & Bendahmane, A. (2015). The battle for survival between viruses and their host plants. *Current Opinion in Virology, 17*, 32–38.

Brooks, D. S., Gonzalez, C. F., Appel, D. N., & Filer, T. H. (1994). Evaluation of endophytic bacteria as potential biological-control agents for Oak Wilt. *Biological Control, 4*, 373–381.

Castillo, U., Strobel, G., Ford, E., Hess, W., Porter, H., Jensen, J., Albert, H., Robison, R., Condron, M., Teplow, D., Stevens, D., & Yaver, D. (2002). Munumbicins, wide-spectrum antibiotics produced by *Streptomyces* NRRL 30562, endophytic on *Kennedia nigriscans*. *Microbiology, 148*, 2675–2685.

Chandanie, W. A., Kubota, M., & Hyakumachi, M. (2006). Interactions between plant growth promoting fungi and arbuscular mycorrhizal fungus *Glomus mosseae* and induction of systemic resistance to anthracnose disease in cucumber. *Plant and Soil, 286*, 209–217.

Chatterton, S., Sutton, J. C., & Boland, G. J. (2004). Timing *Pseudomonas chlororaphis* applications to control *Pythium aphanidermatum*, *Pythium dissotocum*, and root rot in hydroponic peppers. *Biological Control, 30*, 360–373.

Chen, C., Bauske, E. M., Musson, G., Rodriguezkabana, R., & Kloepper, J. W. (1995). Biological control of *Fusarium* wilt on cotton by use of endophytic bacteria. *Biological Control, 5*, 83–91.

Chen, X. H., Koumoutsi, A., Scholz, R., Eisenreich, A., Schneider, K., Heinemeyer, I., Morgenstern, B., Voss, B., Hess, W. R., Reva, O., & Junge, H. (2007). Comparative analysis of the complete genome sequence of the plant growth–promoting bacterium *Bacillus amyloliquefaciens* FZB42. *Nature Biotechnology, 25*, 1007–1014.

Compant, S., Duffy, B., Nowak, J., Clément, C., & Barka, E. A. (2005). Use of plant growth-promoting bacteria for biocontrol of plant diseases: Principles, mechanisms of action, and future prospects. *Applied and Environmental Microbiology, 71*, 94951–94959.

Cramer, G. R., Urano, K., Delrot, S., Pezzotti, M., & Shinozaki, K. (2011). Effects of abiotic stress on plants: A systems biology perspective. *BMC Plant Biology, 11*, 163–177.

Dalal, J., & Kulkarni, N. (2013). Antagonistic and plant growth promoting potentials of indigenous endophytic bacteria of soybean (*Glycine max* (L) Merril). *Current Research in Microbiology and Biotechnology, 1*, 62–69.

De Meyer, G., & Höfte, M. (1997). Salicylic acid produced by the rhizobacterium *Pseudomonas aeruginosa* 7 NSK2 induces resistance to leaf infection by *Botrytis cinerea* on bean. *Biological Control, 87*, 588–593.

De Souza, R., Sant'Anna, F. H., Ambrosini, A., Tadra-Sfeir, M., Faoro, H., Pedrosa, F. O., Souza, E. M., & Passaglia, L. M. P. (2015). Genome of *Pseudomonas* sp. FeS53a, a putative plant growth- promoting bacterium associated with rice grown in iron-stressed soils. *Genome Announcements, 3*, 1–2.

Domingo, J., & Bordonaba, J. G. (2011). A literature review on the safety assessment of genetically modified plants. *Environment International, 37*, 734–742.

Dong, Y., Zhang, X., Xu, J., & Zhang, L. (2004). Insecticidal *Bacillus thuringiensis* silences *Erwinia carotovora* virulence by a new form of microbial antagonism, signal interference. *Applied and Environmental Microbiology, 70*, 2954–2960.

Downey, R. K. (2003). Ecological, genetic, and social factors affecting environmental assessment of transgenic plants. In B. Bodling (Ed.), *Environmental effects of transgenic plants: The scope and adequacy of regulation* (pp. 17–33). Washington, DC: National Academy Press.

Duque, A. S., de Almeida, A. M., da Silva, A. B., da Silva, J. M., Farinha, A. P., Santos, D., Fevereiro, P., & de Sousa Araújo, S. (2013). Abiotic stress responses in plants: Unravelling the complexity of genes and networks to survive. In K. Vahdati (Ed.), *Abiotic stress-plant responses and applications in agriculture* (pp. 3–23). Rijeka: InTech.

Elbeshehy, E. K. F., Youssef, S. A., & Elazzazy, A. M. (2015). Resistance induction in pumpkin *Cucurbita maxima L.* against watermelon mosaic potyvirus by plant growth-promoting rhizobacteria. *Biocontrol Science and Technology, 25*, 525–542.

Fahad, S., Hussain, S., Bano, A., Saud, S., Hassan, S., Shan, D., Khan, F. A., Khan, F., Chen, Y., Wu, C., Tabassum, M. A., Chun, M. X., Afzal, M., Jan, A., Jan, M. T., & Huang, J. (2015). Potential role of phytohormones and plant growth-promoting rhizobacteria in abiotic stresses: Consequences for changing environment. *Environmental Science and Pollution Research International, 22*, 4907–4921.

Filippi, M. C. C., Silva, G. B., Silva-Lobo, V. L., Cortes, M. V. C. B., Moraes, A. J. G., & Prabhu, A. S. (2011). Leaf blast (*Magnaporthe oryzae*) suppression and growth promotion by rhizobacteria on aerobic rice in Brazil. *Biological Control, 58*, 160–166.

Fridlender, M., Inbar, J., & Chet, I. (1993). Biological control of soilborne plant pathogens by a β-1,3 glucanase-producing *Pseudomonas cepacia. Soil Biology and Biochemistry, 25*, 1211–1221.

Fröhlich, A., Buddrus-Schiemann, K., Durner, J., Hartmann, A., & von Rad, U. (2012). Response of barley to root colonization by *Pseudomonas* sp. DSMZ 13134 under laboratory, greenhouse, and field conditions. *Journal of Plant Interactions, 7*, 1–9.

Gamalero, E., Berta, G., Massa, N., Glick, B. R., & Lingua, G. (2010). Interactions between *Pseudomonas putida* UW4 and *Gigaspora rosea* BEG9 and their consequences for the growth of cucumber under salt-stress conditions. *Journal of Applied Microbiology, 108*, 236–245.

García-Fraile, P., Menéndez, E., & Rivas, R. (2015). Role of bacterial biofertilizers in agriculture and forestry. *AIMS Journal, 2*, 183–205.

Glandorf, C. M., Verheggen, P., Jansen, T., Jorritsma, J. W., Smit, E., Leeflang, P., Wernars, K., Thomashow, L. S., Laureijs, E., Thomas-Oates, J. E., Bakker, P., & Loon, L. C. V. (2001). Effect of genetically modified *pseudomonas putida* WCS358R on the fungal rhizosphere microflora of field-grown wheat. *Applied and Environmental Microbiology, 67*(8), 3371–3378.

Glick, B. R. (2012). *Plant growth-promoting bacteria: Mechanisms and applications*. Scientifica, Article ID 963401, 15 pages.

Glick, B. R. (2014). Bacteria with ACC deaminase can promote plant growth and help to feed the world. *Microbiological Research, 169*, 30–39.

Goel, A. K., Lundberg, D., Torres, M. A., Matthews, R., Tomiyama, C. A., Farmer, L., Dangl, J. L., & Grant, S. R. (2008). The *Pseudomonas syringae* type III effector HopAM1 enhances virulence on water-stressed plants. *Molecular Plant-Microbe Interactions, 21*, 361–370.

Gopalakrishnan, S., Srinivas, V., Alekhya, G., Prakash, B., Kudapa, H., & Varshney, R. K. (2015). Evaluation of *Streptomyces* sp. obtained from herbal vermicompost for broad spectrum of plant growth-promoting activities in chickpea. *Organic Agriculture, 5*, 123–133.

Grandlic CJ (2008) *Plant growth-promoting bacteria suitable for the phytostabilization of mine tailings*. Dissertation, The University of Arizona.

Hobbs, P. R., Sayre, K., & Gupta, R. (2008). The role of conservation agriculture in sustainable agriculture. *Philosophical Transactions of the Royal Society, B363*, 543–555.

Horinouchi, H., Muslim, A., & Hyakumachi, M. (2010). Short communication biocontrol of *Fusarium* wilt of spinach by the plant growth promoting fungus *Fusarium equiseti* gf183. *Journal of Plant Pathology, 92*, 249–254.

Hossain, M. M., Sultana, F., Miyazawa, M., & Hyakumachi, M. (2014). The plant growth-promoting fungus *Penicillium* spp. GP15-1 enhances growth and confers protection against damping-off and anthracnose in the cucumber. *Journal of Oleo Science, 63*, 391–400.

Hossain, M. J., Ran, C., Liu, K., Ryu, C. M., Ivey, C. R., Williams, M. A., Hassan, M. K., Choi, S. K., Jeong, H., Newman, M., Kloepper, J. W., & Liles, M. R. (2015). Deciphering the conserved genetic loci implicated in plant disease control through comparative genomics of *Bacillus amyloliquefaciens* subsp. *plantarum*. *Frontiers in Plant Science, 631*(6), 1–14.

Kamensky, M., Ovadis, M., Chet, I., & Chernin, L. (2003). Soil-borne strain IC14 of *Serratia plymuthica* with multiple mechanisms of antifungal activity provides biocontrol of *Botrytis cinerea* and *Sclerotinia sclerotiorum* diseases. *Soil Biology and Biochemistry, 35*, 323–331.

Kilic-Ekici, O., & Yuen, G. Y. (2004). Comparison of strains of *Lysobacter enzymogenes* and PGPR for induction of resistance against *Bipolaris sorokiniana* in tall fescue. *BiolControl, 30*, 446–455.

Killani, A. S., Abaidoo, R. C., Akintokun, A. K., & Abiala, M. A. (2011). Antagonistic effect of indigenous *bacillus subtilis* on root–/soil-borne fungal pathogens of cowpea. *Research, 3*, 11–18.

Koike, N., Hyakumachi, M., Kageyama, K., Tsuyumu, S., & Doke, N. (2001). Induction of systemic resistance in cucumber against several diseases by plant growth-promoting fungi: Lignification and superoxide generation. *European Journal of Plant Pathology, 107*, 523–533.

Liu, L., Kloepper, J. W., & Tuzun, S. (1995). Induction of systemic resistance in cucumber against *Fusarium* wilt by plant growth promoting rhizobacteria. *Phytopathology, 85*, 695–698.

Malathi, S. (2015). Biological control of onion basal rot caused by *Fusarium oxysporum* f. sp. *cepae*. *Asian Journal of Biological Sciences, 10*, 21–26.

Marasco, R., Rolli, E., Ettoumi, B., Vigani, G., Mapelli, F., Borin, S., Abou-Hadid, A. F., El-Behairy, U. A., Sorlini, C., Cherif, A., Zocchi, G., & Daffonchio, D. (2012). A drought resistance-promoting microbiome is selected by root system under desert farming. *PLoS One, 7*, 1–14.

Mavrodi, O. V., Mavrodi, D. V., Weller, D. M., Linda, S., & Thomashow, L. S. (2006). Role of ptsP, orfT, and sss recombinase genes in root colonization by *Pseudomonas fluorescens* Q8r1-96. *Applied and Environmental Microbiology, 72*(11), 7111–7122.

Muñoz, Z., Moret, A., & Garcés, S. (2008). The use of *Verticillium dahliae* and *Diplodia scrobiculata* to induce resistance in *Pinus halepensis* against *Diplodia pinea* infection. *European Journal of Plant Pathology, 120*, 331–337.

Murali, M., Amruthesh, K., Sudisha, J., & SNaH, S. (2012). Screening for plant growth promoting fungi and their ability for growth promotion and induction of resistance in pearl millet against downy mildew disease. *Journal of Phytology, 4*, 30–36.

Nagpure, A., Choudhary, B., Kumar, S., & Gupta, R. K. (2013). Isolation and characterization of chitinolytic *Streptomyces* sp. MT7 and its antagonism towards wood-rotting fungi. *Annales de Microbiologie, 64*, 531–541.

Nagpure, A., Choudhary, B., & Gupta, R. K. (2014). Mycolytic enzymes produced by *Streptomyces violaceusniger* and their role in antagonism towards wood-rotting fungi. *Journal of Basic Microbiology, 54*, 397–407.

Naznin, H. A., Kiyohara, D., Kimura, M., Miyazawa, M., Shimizu, M., & Hyakumachi, M. (2014). Systemic resistance induced by volatile organic compounds emitted by plant growth-promoting fungi in *Arabidopsis thaliana*. *PLoS One, 9*, e86882.

Nelson, L. M. (2004). Plant growth promoting rhizobacteria (PGPR): Prospects for new inoculants. *Crop Management, 3*, 1–7.

Ortbauer, M. (2013). Abiotic stress adaptation: Protein folding stability and dynamics. In V. Kourosh (Ed.), *Abiotic stress – plant responses and applications in agriculture*. Rijeka: InTech. https://doi.org/10.5772/53129.

Ortíz-Castro, R., Contreras-Cornejo, H. A., Macías-Rodríguez, L., & López-Bucio, J. (2009). The role of microbial signals in plant growth and development. *Plant Signaling & Behavior, 4*(7), 1–12.

Pontes, A. P., de Souza, R., Granada, C. E., & Passaglia, L. M. P. (2015). Screening of plant growth promoting bacteria associated with barley plants (*Hordeum vulgare* L.) cultivated in South Brazil. *Biota Neotropica, 15*, e20140105.

Porcel, R., Zamarreño, A. M., García-Mina, J. M., & Aroca, R. (2014). Involvement of plant endogenous ABA in *Bacillus megaterium* PGPR activity in tomato plants. *BMC Plant Biology, 14*, 36.

Rajkumar, M., & Helena, F. (2008). Influence of metal resistant-plant growth-promoting bacteria on the growth of *Ricinus communis* in soil contaminated with heavy metals. *Chemosphere, 71*, 834–842.

Razinger, J., Lutz, M., Schroers, H. J., Urek, G., & Grunder, J. (2014). Evaluation of insect associated and plant growth promoting fungi in the control of cabbage root flies. *Journal of Economic Entomology, 107*, 1348–1354.

Rodriguez, H., Fraga, R., Gonzalez, T., & Bashan, Y. (2007). Genetics of phosphate solubilization and its potential applications for improving plant growth-promoting rhizobacteria. *Developments in Plant and Soil Sciences, 102*, 15–21.

Ryu, C. M., Murphy, J. F., Mysore, K. S., & Kloepper, J. W. (2004). Plant growth-promoting rhizobacteria systemically protect *Arabidopsis thaliana* against cucumber mosaic virus by a salicylic acid and NPR1-independent and jasmonic acid-dependent signalling pathway. *The Plant Journal, 39*, 381–392.

Salas-Marina, M. A., Silva-Flores, M. A., Cervantes-Badillo, M. G., Rosales-Saavedra, M. T., Islas-Osuna, M. A., & Casas-Flores, S. (2011). The plant growth-promoting fungus *Aspergillus ustus* promotes growth and induces resistance against different lifestyle pathogens in *Arabidopsis thaliana*. *Journal of Microbiology and Biotechnology, 21*, 686–696.

Salas-Marina, M. A., Isordia-Jasso, M. I., Islas-Osuna, M. A., Delgado-Sánchez, P., Jiménez-Bremont, J. F., Rodríguez-Kessler, M., Rosales-Saavedra, M. T., Herrera-Estrella, A., & Casas-Flores, S. (2015). The Epl1 and Sm1 proteins from *Trichoderma atroviride* and *Trichoderma virens* differentially modulate systemic disease resistance against different life style pathogens in *Solanum lycopersicum*. *Frontiers in Plant Science, 6*(77), 1–13.

Santoyo, G., Moreno-Hagelsieb, G., del Carmen Orozco-Mosqueda, M., & Glick, B. (2016). Plant growth-promoting bacterial endophytes. *Microbiological Research, 183*, 92–99.

Sarathambal, C., Ilamurugu, K., Priya, L. S., & Barman, K. K. (2014). A review on weeds as source of novel plant growth promoting microbes for crop improvement. *Journal of Applied and Natural Sciences, 6*, 880–886.

Schuler, T. H., Poppy, G. M., Kerry, B. R., & Denholm, I. (1999). Potential side effects of insect-resistant transgenic plants on arthropod natural enemies. *Trends in Biotechnology, 17*, 210–216.

Schwartz, A. R., Ortiz, I., Maymon, M., Herbold, C. W., Fujishige, N. A., Vijanderan, J. A., Villella, W., Hanamoto, K., Diener, A., Sanders, E. R., DeMason, D. A., & Hirsch, A. M. (2013). *Bacillus simplex*-A little known PGPB with anti-fungal activity alters pea-legume root architecture and nodule morphology when co-inoculated with *Rhizobium leguminosarum* bv. *viciae*. *Agronomy, 3*, 595–620.

Sheng, X. F., Xia, J. J., Jiang, C. Y., He, L. Y., & Qian, M. (2008). Characterization of heavy metal-resistant endophytic bacteria from rape (*Brassica napus*) roots and their potential in promoting the growth and lead accumulation of rape. *Environmental Pollution, 156*, 1164–1170.

Shivanna, M. B., Meera, M. S., & Hyakumachi, M. (1996). Role of root colonization ability of plant growth promoting fungi in the suppression of take-all and common root rot of wheat. *Crop Protection, 15*, 497–504.

Shoresh, M., Yedidia, I., & Chet, I. (2005). Involvement of jasmonic acid/ethylene signalling pathway in the systemic resistance induced in cucumber by *Trichoderma asperellum* T203. *Phytopathology, 95*, 76–84.

Siddiqui, I. A., & Shaukat, S. S. (2002). Rhizobacteria-mediated induction of systemic resistance in tomato against *Meloidogyne javanica*. *Journal of Phytopathology, 150*, 469–472.

Silva, D. C. S., Weatherhead, E. K., Knox, J. W., & Rodriguez-Diaz, J. A. (2007). Predicting the impacts of climate change- a case study of paddy irrigation water requirements in Sri Lanka. *Agricultural Water Management, 93*, 19–29.

Singh, P. P., Shin, Y. C., Park, C. S., & Chung, Y. R. (1999). Biological control of *Fusarium* wilt of cucumber by chitinolytic bacteria. *Phytopathology, 89*, 92–99.

Sivakumar, G., & Sharma, R. C. (2003). Induced biochemical changes due to seed bacterization by *Pseudomonas fluorescens* in maize plants. *Indian Phytopathology, 56*, 134–137.
Spoel, S., & Dong, X. (2008). Making sense of hormone crosstalk during plant immune responses. *Cell Host & Microbe, 3*, 348–351.
Sripontan, Y., Hung, M., Young, C., & Hwang, S. (2014). Effects of soil type and plant growth promoting microorganism on cabbage and *Spodoptera litura* performance. *Journal of Agriculture and Forestry, 63*, 153–161.
Timmusk, S., & Wagner, E. (1999). The plant growth promoting rhizobacterium *Paenibacillus polymyxa* induces changes in *Arabidopsis thaliana* gene expression: A possible connection between biotic and abiotic stress responses. *Phytopathology, 12*, 951–959.
Tiwari, P. K., & Thrimurthy, V. S. (2007). Isolation and characterization of the *Pseudomonas fluorescens* from rhizosphere of different crops. *Journal of Mycology and Plant Pathology, 37*, 231–234.
Umashankari, J., & Sekar, C. (2011). Comparative evaluation of different bio-formulations of PGPR cells on the enhancement of induced systemic resistance (ISR) in rice *P. oryzae* pathosystem under upland condition. *Current Botany, 2*, 12–17.
Van Loon, L. C. (2007). Plant responses to plant growth promoting rhizobacteria. *European Journal of Plant Pathology, 119*, 243–254.
Viets, F. G., & Lunin, J. (2009). The environmental impact of fertilizers. *Critical Reviews in Environmental Control, 5*, 423–453.
Vos, C. M. F., De Cremer, K., Cammue, B. P. A., & De Coninck, B. (2015). The toolbox of *Trichoderma* spp. in the biocontrol of *Botrytis cinerea* disease. *Molecular Plant Pathology, 16*, 400–412.
Waqas, M., Khan, A. L., Kamran, M., Hamayun, M., Kang, S. M., Kim, Y. H., & Lee, I. J. (2012). Endophytic fungi produce gibberellins and indoleacetic acid and promotes host-plant growth during stress. *Molecules, 17*, 10754–10773.
Yadav, J., Verma, J. P., & Tiwari, K. N. (2011). Plant growth promoting activities of fungi and their effect on chickpea plant growth. *Asian Journal of Biological Sciences, 4*, 291–299.
Zahran, H. H. (1999). Rhizobium-legume symbiosis and nitrogen fixation under severe conditions and in an arid climate. *Microbiology and Molecular Biology Reviews, 63*, 968–989.
Zamioudis, C., & Pieterse, C. M. (2012). Modulation of host immunity by beneficial microbes. *Molecular Plant-Microbe Interactions, 25*, 139–150.
Zhou, Z., Zhang, C., Zhou, W., Li, W., Chu, L., Yan, J., & Li, H. (2014). Diversity and plant growth-promoting ability of endophytic fungi from the five flower plant species collected from Yunnan, Southwest China. *Journal of Plant Interactions, 9*, 585–591.
Zhuang, X., Chen, J., Shim, H., & Bai, Z. (2007). New advances in plant growth-promoting rhizobacteria for bioremediation. *Environment International, 33*, 406–413.

Chapter 9
Chemistry, Therapeutic Attributes, and Biological Activities of *Dillenia indica* Linn

Ashok K. Singh and Sudipta Saha

Contents

Abstract *Dillenia indica* Linn. (Dilleniaceae) is generally known as elephant apple and locally known as outenga. The vernacular names are chalta, chulta, bhavya, karambel, ouu, and ramphal. This evergreen deciduous tree is markedly disseminated in the seasonal tropics of many Asian countries, in India from Himalaya to south India. The different parts of this plant have been prevalently investigated for

A. K. Singh · S. Saha (✉)
Department of Pharmaceutical Sciences, Babasaheb Bhimrao Ambedkar University, Lucknow, Uttar Pradesh, India
e-mail: sudiptapharm@gmail.com

R. C. Sobti et al. (eds.), *Environmental Biotechnology: For Sustainable Future*,
https://doi.org/10.1007/978-981-10-7284-0_9

the plethora of biological activities including anticancer, antidiabetic, antihyperlipidemic, antileukemic, antioxidant, antimutagenic, antimicrobial, antinociceptive, antidiarrheal, and hairweaving activities. Differently prepared extracts of this plant have been reported mainly to contain a wide range of flavonoids, triterpenoids (lupene-type), phytosteroids, phenolics, alcohols, and ketones and an anthraquinone. Several phytochemical investigations revealed substantial presence of various types of active constituents including β-sitosterol, stigmasterol, betulin, betulinic acid, kaempferol, myricetin, quercetin, dillenetin and rhamnetin. Among these the major chemical constituents are betulin and betulinic acid (lupene-type triterpenoids) that show a wide spectrum of pharmacological activities like anti-HIV, anticancer, antimalarial, anti-inflammatory, etc. The present chapter thus approaches to highlight on phytochemistry, traditional and therapeutic uses, and biological activities of *Dillenia indica*.

Keywords *Dillenia indica* · Therapeutic attributes · Phytochemistry · Therapeutic uses · Pharmacology

Introduction

A vast amount of knowledge and practices on herbal medicinal systems have been transmitted through the ages. For centuries, medicinal plants were the only resources available for the treatment of several diseases which afflicted humanity (Ozdemir and Alpınar 2015). Numerous of these plants are uncommon, endemic, and found only in forest region. There is neither biological data nor satisfactory information that prompted their rarity in the natural surroundings (Kerrigan et al. 2011). Correspondingly, there are many plant species which have been utilized by tribal and folk communities of different forest regions of India; however, their medicinal and also pharmacological esteem is yet obscure as these plants are hardly available. There are many plant species which have been utilized by tribal groups of India; however, their restorative and also pharmacological knowledge is yet obscure as these plants are not easily accessible and studied. Among these, there are few plants belonging to family Dilleniaceae which have not gained much popularity but have interesting medicinal values. The genus *Dillenia* has 60 species; however, only a few of them are reported to have important phytochemicals and thereby enrich their medicinal values. These species are *D. indica*, *D. pentagyna*, *D. suffruticosa*, *D. andamanica*, *D. serrata*, *D. sumatrana*, *D. aurea*, *D. bracteata*, *D. excelsa*, *D. ovata*, *D. papuana*, *D. parviflora*, *D. philippinensis*, *D. pulchella*, *D. reticulata*, *D. scabrella*, *D. eximia,* and *D. triquetra.* Only two plants *D. indica* Linn. and *D. pentagyna* Roxb. are available in India. *D. indica* has been extensively studied and a more commonly employed medicinal plant in different parts of India (Dickison 1979). Several research works have been conducted on the isolation and quantification of the different phytochemicals from various parts of *D. indica*; however, very few phytochemical investigations have been performed from *D. pentagyna.*

D. indica grows in moist and evergreen forests of India (The Wealth of India 1952). This plant is known for its lemon-flavored fruits that are used to prepare jam and jellies and as flavoring agent for curries (Sharma et al. 2001). This plant has been exploited by tribal and folk communities of various regions where the fruits of *D. indica* are eaten raw, but people are not much familiar with its medicinal values (Pradhan and Badola 2008; Dubey et al. 2009; Sharma and Pegu 2011). The leaf, stem bark, fruit, and flower of *D. indica* contain a wide range of flavonoids, triterpenoids (lupene-type), phytosteroids, phenolics, alcohols, and ketones and an anthraquinone. The substantial presence of various types of active constituents including β-sitosterol, stigmasterol, betulin, betulinic acid, kaempferol, myricetin, quercetin, dillenetin and rhamnetin enriches the diversity of therapeutically important phytochemicals in *D. indica*. The major chemical constituents among these are betulin and betulinic acid (lupene-type triterpenoids) that show vast and wide range of medicinal values.

The present review is an attempt to compile the detailed phytochemistry, traditional and therapeutic uses, as well as biological activities of this plant. This review may be helpful to explore further excellent phytochemistry and medicinal potentials of *D. indica* for the preparation of various types of formulations in future.

Chemistry

The significant classes of chemical constituents extracted from *D. indica* are flavonoids and triterpenoids (lupene-type). Other isolated compounds including phytosteroids, diterpene, ionone, phenolics, anthraquinone, alcohols, and ketones also enhance the diversity of phytochemistry in *D. indica*. As per our extensive search, a total of 34 compounds isolated from *D. indica* are included in this review which may lead to further research and noble challenge to discover new chemical constituents from this plant. These compounds are listed in Table 9.1, and their chemical structures are displayed in Fig. 9.1.

Stem bark of *D. indica* contains triterpenoids like lupeol, betunaldehyde, and betulinic acid; flavonoids like kaempferol, dillenetin, rhamnetin, dihydroisorhamnetin, myricetin, naringenin, and quercetin; and 10% tannin (Shah 1978; Khanum et al. 2007; Khare 2007). The ethanol extract of stem bark is enriched with two flavonoids, kaempferol and quercetin, as well as a triterpenoid (Srivastava and Pande 1981). Parvinet al. (2009) acquired methanolic extract of stem after partitioning with n-hexane and isolated four compounds, viz., lupeol, betulinic acid, betunaldehyde, and stigmasterol, using column chromatographic separation.

Leaves of *D. indica* contain betulinic acid, betulin, lupeol, and β-sitosterol (Dan and Dan 1980). The petroleum ether extract of leaves contains betulin, β-sitosterol, cycloartenone, and n-hentriacontanol, whereas chloroform extract has betulinic acid (Mukherjee and Badruddoza 1981). Methanolic extract of leaves after fractionation with n-hexane and chloroform also has compounds like betulinic acid, β-sitosterol, dillenetin, and stigmasterol (Muhit et al. 2010). Phytochemicals have

Table 9.1 Compounds isolated from *D. indica*

Compound's structure no.	Isolated compound	Chemical type	Plant part	References
1	Kaempferol	Flavonol	Pericarp, twig, stem bark	Pavanasasivam and Sultanbawa (1975a)
			Stem	Srivastava and Pande (1981)
2	Myricetin	Flavonol	Stem bark	Banerji et al. (1975)
3	Quercetin	Flavonol	Leaf	Bate-Smith and Harborne (1971) and Kumar et al. (2013)
4	Dillenetin	Flavonol	Pericarp	Pavanasasivam and Sultanbawa (1975b)
			Leaf	Muhit et al. (2010)
5	Rhamnetin	Flavonol	Leaf	Bate-Smith and Harborne (1971)
6	Isorhamnetin	Flavonol	Fruit, twig, stem bark	Pavanasasivam and Sultanbawa (1975a)
7	Kaempferide	Flavonol	Leaf	Bate-Smith and Harborne (1971)
8	Kaempferide 3-*O*-diglucoside	Substituted flavonol	Leaf	Bate-Smith and Harborne (1971)
9	3′,5-Dihydroxy-4′,3-dimethoxy flavone-7-*O*-*β*-D--glucopyranoside	Substituted flavonol	Stem bark	Tiwari and Srivastava (1979)
10	5,7-Dihydroxy-4′-methoxyflavone-3-*O*-*β*-D--glucopyranoside	Substituted flavonol	Stem bark	Tiwari and Srivastava (1979)
11	(+)-Dihydroxykaempferol	Dihydroflavonol	Twig	Pavanasasivam and Sultanbawa (1975a)
12	(+)-3′-Methoxy-dihydroquercetin	Dihydroflavonol	Stem bark	Pavanasasivam and Sultanbawa (1975a)
13	(+)-Dihydroisorhamnetin	Dihydroflavonol	Stem bark	Pavanasasivam and Sultanbawa (1975b)
14	Dihydrokaempferide	Dihydroflavonol	Leaf	Bate-Smith and Harborne (1971)
15	Dihydrokaempferide 7-diglucoside	Dihydroflavonol	Leaf	Bate-Smith and Harborne (1971)
16	4,5,7,3′,4'-Pentahydroxy flavan-3-*O*-*β*-D--glucopyranoside	Flavan	Stem bark	Tiwari and Srivastava (1979)
17	Leucocyanidin	Flavan-3-ol	Leaf	Bate-Smith and Harborne (1971)

(continued)

Table 9.1 (continued)

Compound's structure no.	Isolated compound	Chemical type	Plant part	References
18	Naringenin	Flavanone	Stem bark	Pavanasasivam and Sultanbawa (1975a)
			Leaf	Bate-Smith and Harborne (1971)
19	Naringenin 7-diglucoside	Flavanone	Leaf	Bate-Smith and Harborne (1971)
20	3,5,7-Trihydroxy-2-(4-hydroxy-benzyl)-chroman-4-one	Chromane	Leaf	Kaur et al. (2016)
21	Lupeol	Lupene	Stem bark	Parvin et al. (2009) and Banerji et al. (1975)
			Leaf	Dan and Dan (1980)
			Fruit	Sundararamaiah et al. (1976)
22	Betulin	Lupene	Stem bark	Banerji et al. (1975) and Bhattacharjee and Chatterjee (1962)
			Leaf	Dan and Dan (1980)
			Fruit	Sundararamaiah et al. (1976)
23	Betulinaldehyde	Lupene	Stem bark	Parvin et al. (2009) and Banerji et al. (1975)
24	Betulinic acid	Lupene	Stem bark	Parvin et al. (2009), Banerji et al. (1975) and Bhattacharjee and Chatterjee (1962)
			Leaf	Dan and Dan (1980), Mukherjee and Badruddoza (1981), Muhit et al. (2010) and Kumar et al. (2013)
25	3β-Hydroxylupane-13β,28-lactone	Lupene	Stem bark	Banerji et al. (1975)

(continued)

Table 9.1 (continued)

Compound's structure no.	Isolated compound	Chemical type	Plant part	References
26	β-Sitosterol	Phytosteroid	Pericarp, twig, stem bark	Pavanasasivam and Sultanbawa (1975a, b)
			Fruit	Sundararamaiah et al. (1976)
			Leaf	Mukherjee and Badruddoza (1981), Muhit et al. (2010) and Kumar et al. (2013)
27	Stigmasterol	Phytosteroid	Stem bark	Parvin et al. (2009)
			Leaf	Kumar et al. (2013)
28	Stigmasteryl palmitate	Phytosteroid	Leaf	Kumar et al. (2013)
29	Cycloartenone	Phytosteroid	Leaf	Mukherjee and Badruddoza (1981)
30	Gallic acid	Phenolic	Twig	Pavanasasivam and Sultanbawa (1975b)
31	1,8-Dihydroxy-2-methyl-anthraquinone-3-*O*-β-D-glucopyranoside	Anthraquinone	Stem bark	Tiwari and Srivastava (1979)
32	n-Hentriacontanol	Alcohol	Leaf	Mukherjee and Badruddoza (1981)
33	n-Heptacosan-7-one	Ketone	Leaf	Kumar et al. (2013)
34	n-Honatriacontan-18-one	Ketone	Leaf	Kumar et al. (2013)

also been investigated from acid hydrolyzed extracts of dried leaves which demonstrated the presence of kaempferol, whereas fresh leaves were found to contain dihydrokaempferide and naringenin-7-diglucoside which get further oxidized to ten corresponding flavonols (Bate-Smith and Harborne 1971). Kumar et al. (2010) isolated and quantified betulinic acid using validated HPLC method from various fractions such as methanol, ethyl acetate, n-butanol, and water. The highest concentration among them was found in ethyl acetate fraction.

Fruit of *D. indica* contains about 34% of total phenolics in methanolic extract (Abdille et al. 2005), isorhamnetin (Pavanasasivam and Sultanbawa 1975a), lupeol, betulin, β-sitosterol (Sundararamaiah et al. 1976), and polysaccharide like arabinogalactan. Uppalapati and Rao (1980) reported the presence of steroids, saponins, fixed oil, free amino acids, glycosides, tannins, and sugars in the seeds of *D. indica*. These scientific reports collectively revealed that betulin, betulinic acid, and β-sitosterol are present in almost all parts of *D. indica*.

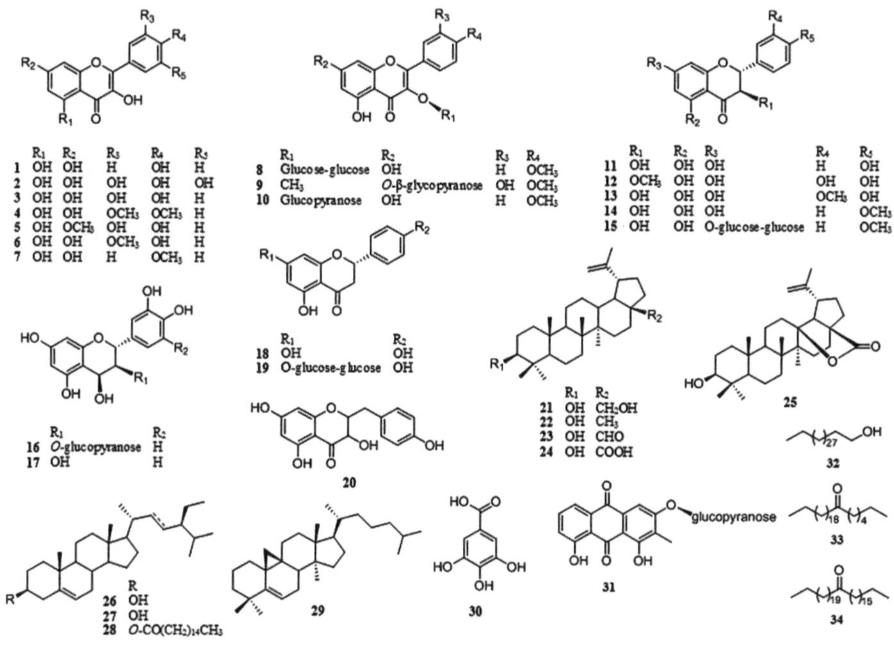

Fig. 9.1 Structure of compounds isolated from *Dillenia indica*

Traditional and Therapeutic Uses

Usually all parts of *D. indica* are traditionally exploited for therapeutic purposes. The jelly-like content inside the fruit of *D. indica* is applied for hair treatment against dandruff and falling hair. The mixed juices of leaves and stem bark are used orally for the prevention of diarrhea and cancer (Yeshwante et al. 2009a, b; Sunil et al. 2011). The leaves and stem bark are also used as laxative and astringent (Sharma et al. 2009). Apart from this, the stem bark is used for the production of charcoal. The fresh and dried materials of various parts of *D. indica* are processed as juice, decoction, poultice, and mucilage for the medical care of diabetes, wounds, diarrhea, cancer, rheumatism, urinary problems, skin diseases, aches, fever, cough, and falling hair. Almost all the known medicinal uses of *D. indica* are enlisted in Table 9.2. Skin diseases including leukoderma, eczema, skin itches, and skin rash can be treated using the leaf, fruit, and stem bark of *D. indica* (Quattrocchi 2012; Boer et al. 2012; Bhat et al. 2014). Leaves of *D. indica* as well as decoction and juice of the fruit and stem bark are exploited in daily practices to attenuate cancerous growth, particularly breast and gastric cancers (Das et al. 2009; Sharma et al. 2001). Furthermore, fruit juice of *D. indica* is supplemented orally to eliminate fever and cough-associated symptoms (Angami et al. 2006; Quattrocchi 2012). The mixed juice of fruit and calyx of *D. indica* is used in daily practices for the treatment of diabetes (Pavani et al. 2012; Ripunjoy 2013). The root of *D. indica* is generally

Table 9.2 Traditional and therapeutic uses of *D. indica*

Plant part	Usable form	Traditional and therapeutic uses	References
Fruit	Juice	Flatulence, boils, fever, and cough as well as to help in semen production	Islam et al. (2014)
		Laxative and abdominal pain	Kirtikar and Basu (2003)
		Fever	Khongsai et al. (2011) and Anisuzzaman et al. (2007)
		Dysentery	Das et al. (2008) and Kalita and Deb (2004)
		Diabetes	Pavani et al. (2012) and Tag et al. (2012)
	Decoction	Jaundice	Rai and Lalramnghinglova (2010) and Sharma et al. (2012)
		Stomach disorders	Lalfakzuala et al. (2007)
	Raw fruit	To enhance appetite and weakness of the body	Poonam and Singh (2009)
		Laxative	Prasad et al. (2008)
		Laxative, carminative, bechic, febrifuge, antispasmodic (abdominal pains)	Khare (2007)
		Constipation and stomachache	Ripunjoy (2013)
		Stomachache	Srivastava et al. (2010)
	Mucilage	Clean hair	Sarmah et al. (2008)
		Skin lice	Quattrocchi (2012)
	Pickle	Cough, fever, and weakness	Angami et al. (2006)
	Dried fruit	Cough	Khongsai et al. (2011)
	Paste	Hydrocele	Quattrocchi (2012)
		Blood dysentery	Quattrocchi (2012)
		Cholera	Hazarika et al. (2016)
	–	Hair loss	Rahman et al. (2011a)
	–	Laxative, carminative, bechic, febrifuge, antispasmodic (abdominal pains)	Khare (2007)
	–	Used in flavoring of curries and production of jam and jelly	Abdille et al. (2005) and Kar and Borthakur (2007)
	–	Cancer	Azam et al. (2016)
Leaf	Paste	Diarrhea of domestic animals	Quattrocchi (2012)
	Decoction	Malaria	Nguyen-Pouplin et al. (2007)
	Juice	Cancer and diarrhea	Sharma et al. (2001)
	–	Cancer	Azam et al. (2016)
	–	Astringent	Khare (2007)

(continued)

Table 9.2 (continued)

Plant part	Usable form	Traditional and therapeutic uses	References
Stem bark	Dry powdered	Urinary diseases	Quattrocchi (2012)
	Juice	Blood cancer	Majumdar et al. (2006)
		Drink for cough, cold, fever, and diarrhea	Quattrocchi (2012)
	Decoction	Diarrhea, dysentery, and cholera	Hazarika et al. (2016)
		Neutralizer for food poisoning	Islam et al. (2014)
	Poultice	Sores caused by septicemial infection	Singh et al. (2002)
	Paste	To apply on blistering boils	Quattrocchi (2012)
	–	Astringent	Khare (2007)
Root	Decoction	To discharge blood in urine	Quattrocchi (2012)
		Biliousness	Quattrocchi (2012)
		Neutralizer for food poisoning	Islam et al. (2014)
	Paste	Abortion	Quattrocchi (2012)
Flower	Fresh flower	Dysentery	Sarmah et al. (2008)
	Decoction	Amoebic dysentery	Purkayastha et al. (2005)
	Extract	Diabetes mellitus	Tarak et al. (2011)
Calyx	Fleshy calyx	Diabetes	Ripunjoy (2013)
		Stomach disorders	Quattrocchi (2012) and Kagyung et al. (2010)

utilized for the purpose of abortion (Quattrocchi 2012). Moreover, the mucilage of *D. indica* fruits is used to treat falling hair, to clean hair, as well as to remove dandruff from hair (Saikia et al. 2006). The sweetish-sour edible fruits of this plant may be consumed directly or juiced with sugar to take as fresh and healthy drink. An interesting event is that the bark of the stem and roots has extensively been used as a food-poisoning neutralizer (Grosvenor et al. 1995a, b; Islam et al. 2014). Altogether, it is concluded that *D. indica* contains the chemical constituents that can treat a broad spectrum of human ailments.

Biological Activities of *D. indica*

Cytotoxicity

Cytotoxic activity of phytochemicals present in different parts of *D. indica* was screened against numerous cancer cell lines. In vitro cytotoxic activity against leukemia, carcinoma, and lymphoma cells was reported using methanol extracts of *D. indica* fruit (Kumar et al. 2010) and leaves (Akter et al. 2014). In particular, the methanol extract of the fruit inhibited the growth of U937, HL60, and K562 cancer cell lines with IC_{50} of 328.80, 297.69, and 275.40 mg/mL, respectively, comparable to standard drugs Ara C and Gleevec (Kumar et al. 2010), whereas the methanol extract of the leaves inhibited the growth of AGS, MCF-7, and MDA-MB-231 cancer cell lines with IC_{50} values of 1.18, 0.34, and 0.54 mg/mL, respectively, as compared to cycloheximide with IC_{50} values of 0.0010, 0.061, and 0.0004 mg/mL, respectively (Akter et al. 2014). On the contrary, the ethanol and aqueous extracts were noticed noncytotoxic toward the cancer cell lines (Nguyen-Pouplin et al. 2007; Armania et al. 2013a, b). Ultimately, a significant correlation was established between the presence of betulinic acid and cytotoxic activity of extracts and fractions of *D. indica*. For example, the ethyl acetate fraction of the methanol extract of *D. indica* calyx containing considerable amounts of betulinic acid exhibited prominent cytotoxicity when compared to the n-butanol fraction (Kumar et al. 2010). Other than betulinic acid, three compounds (lupeol, betulin, and gallic acid) isolated from *D. indica* were also proven to have cytotoxic activity toward cancer cell lines. Finally, the result revealed that lupane-type triterpenoids isolated from *D. indica* (lupeol, betulin, and betulinic acid) displayed considerable cytotoxic properties. Although the cytotoxicity of this plant is very less studied and needs to be performed on variety of cell lines, however, these findings may provide a great opportunity for further development of anticancer compounds from *D. indica*.

Immunomodulatory

The cornerstone of good health is no doubt a strong, well-functioning immune system. Phytochemicals such as flavonoids, terpenoids, glycosides, and phenolic compounds act as a natural defense system not only for the host plant, but also they can serve in immunomodulatory activities in humans (Venkatalakshmi et al. 2016). For example, the aqueous methanolic extract (70%) of *D. indica* fruit enhanced the production of polyclonal immunoglobulin M (IgM) in cultured BALB/c female mice spleen cells at a concentration of 200 mg/mL as compared to lipopolysaccharide (0.1 mg/mL) (Sarker et al. 2012).

Antidiabetic Activity

Indian natives have since long used *D. indica* to treat diabetes. The leaf extract of *D. indica* was found to inhibit enzymatic activity of rat intestinal sucrose and maltase (Jong-Anurakkun et al. 2007). Moreover, the leaf extract of *D. indica* assayed on streptozotocin (STZ)- and alloxan-induced type 1 and type 2 diabetic rats reduced the levels of blood glucose, hypertriglyceridemia, and hypercholesterolemia. In fact, the extract enhanced the production of insulin and high-density lipoprotein cholesterol (HDL-c) (Jong-Anurakkun et al. 2007; Kumar et al. 2011a, b). Additionally, the extract inhibited overproduction of liver function enzymes such as aspartate transaminase (AST), alanin transaminase (ALT), and alkaline phosphatase (ALP) in diabetic rats (Kumar et al. 2011a, b). Histopathological studies showed that the liver, pancreas, and kidney in the treated rats restored to normal conditions after treatment with the extract of *D. indica* (Kumar et al. 2011b). In an experiment performed by Kumar et al. (2011a, b), the ethyl acetate fraction of the methanol extract of the leaves assayed in vivo on STZ-induced type 1 and type 2 diabetic rats showed a reduction in blood glucose, serum cholesterol, and triglyceride levels after 21 days of treatment at the doses of 200 and 400 mg/kg body weight (bw). This fraction also increased the level of HDL-C in the treated rats. Likewise, the defatted methanol extract of the leaves assayed in vivo on type 1 diabetic rats induced by STZ and alloxan decreased the blood glucose, hypertriglyceridemia, and hypercholesterolemia levels as well as increased the production of insulin and HDL-C at the doses of 250 and 500 mg/kg bw during 21 days of treatment (Jong-Anurakkun et al. 2007; Kumar et al. 2011a, b). In another experiment, three compounds (quercetin, β-sitosterol, and stigmasterol) isolated from the ethyl acetate fraction of this extract were found to reduce the blood glucose level of type 2 STZ-nicotinamide-induced diabetic mice which was comparable to standard drug glibenclamide (Kumar et al. 2013). Later, similar activity was performed by the extract of *D. indica* fresh leaves assayed on STZ-induced diabetic rats. Further investigation of active constituent of this extract led to the isolation of 3,5,7-trihydroxy-2-(4-hydroxy-benzyl)-chroman-4-one, which significantly demonstrated antidiabetic activity by reducing blood glucose, cholesterol, and triglycerides levels. The treatment also increased the levels of insulin, HDL-C, as well as body weight when compared with diabetic rats. Histopathological study showed that treatment with the extract and 3,5,7-Trihydroxy-2-(4-hydroxy-benzyl)-chroman-4-one restored the hyperglycemic conditions and kidney structure abnormality owing to oxidative stress in the diabetic rats (Kaur et al. 2016). The alcoholic extract of the fresh leaves assayed in vivo on STZ-induced diabetic rats reduced levels of blood glucose, cholesterol, and triglycerides at doses of 100, 200, and 400 mg/kg bw for 21 days of treatment, comparable to glimepiride (10 mg/kg bw). The extract significantly increased the body weight, serum insulin and HDL-C levels (Kaur et al. 2016). These findings well supported the traditional use of *D. indica* for the treatment of diabetes among Indian natives.

Anti-inflammatory and Antinociceptive Activity

Many of *D. indica* extracts were tested for anti-inflammatory and antinociceptive activities. The alcoholic extract of *D. indica* leaves obliterated carrageenan-induced paw edema at doses of 200 and 400 mg/kg bw with the impact comparable to indomethacin (10 mg/kg bw) (Yeshwante et al. 2009a, b). Moreover, the methanol extract of the leaves and its nonpolar fractions assayed on acetic acid-induced colitis mice at doses of 200 and 800 mg/kg bw restored the colon weight and macroscopic damage in acetic acid-induced colitis in mice. The non-polar fractions also attenuated the production of tumor necrosis factor alpha (TNF-α) and myeloperoxidase released from azurophilic granules of neutrophils (Somani et al. 2014). On the other hand, the glycolic extract and emulsion of *D. indica* fruit were not found suitable to accelerate wound healing process on skin injuries made in rats (Miglianto et al. 2011). The alcoholic extracts of the leaf and stem bark of *D. indica* were examined for antinociceptive activity using the hot plate method, tail immersion test, and writhing model in mice induced by acetic acid and found to exhibit central and peripheral analgesia (Bose et al. 2010; Yeshwante et al. 2011; Alam et al. 2012). In particular, the methanol extract of the leaves was evaluated on acetic acid-induced writhing mice at doses of 250 and 500 mg/kg bw which obliterated the writhing behavior by 48.82% and 55.88% inhibition comparable to that of diclofenac (60% at 25 mg/kg bw) (Bose et al. 2010). Furthermore, the methanol extract of the leaves measured by hot tail, tail immersion, formalin-induced nociception, and acetic acid-induced writhing model on mice at doses of 400 mg/kg bw considerably exhibited central and peripheral analgesia when comparable to those of the standard drugs, pentazocine (15 mg/kg bw) and indomethacin (20 mg/kg bw) (Yeshwante et al. 2011). Later, the methanol extract of the stem bark was evaluated using hot plate method, tail immersion test, and acetic acid-induced writhing model in mice at doses of 200 and 400 mg/kg bw that demonstrated dose-dependent analgesic activity comparable to those of the standard drugs nalbuphine (10 mg/kg bw) and diclofenac sodium (10 mg/kg bw) (Alam et al. 2012).

Antioxidant Activity

Flavonoids, terpenoids, tannins, and phenolics in *D. indica* plants are major compounds responsible for primary antioxidants or free radical scavenging effects (Polterai 1997). A compound 3,5,7-trihydroxy-2-(4-hydroxy-benzyl)-chroman-4-one isolated from *D. indica* displayed remarkable antioxidant property when assayed on 2,2-diphenyl-1-picryl-hydrazyl-hydrate (DPPH), hydrogen peroxide, and superoxide radicals as well as ferric ion. This compound also elevated the production of antioxidant enzymes (superoxide dismutase and glutathione) in streptozocin (STZ)-induced diabetic rats. Treatment with this compound remarkably reduced the level of lipid peroxidation marker, i.e., thiobarbituric acid-reactive substances (TBARS) in diabetic rats (Kaur et al. 2016). In an experiment, methanol, ethyl acetate, and aqueous extracts of the fruit reduced molybdenum (IV) to molybdenum (V) with the capacity of 1904.80, 1067.00, and 594.60 mmol/g of extract, respectively, when compared to ascorbic acid. The extracts also exhibited DPPH free radical scavenging activity over the concentration range of 25–100 mg/mL with the activity in order of methanol extract > ethyl acetate extract > water extract. The extracts when evaluated on β-carotene bleaching produced antioxidant activity with the capacity of 80.20, 55.50, and 45.50%, respectively, at 100 mg/mL when compared to BHA (97.50%) (Abdille et al. 2005). In another experiment, the aqueous acetone extract of the stem bark reduced molybdenum (IV) to molybdenum (V) with capacity of 3.12 mmol/g of extract at 50 mg/mL as compared to ascorbic acid and exhibited DPPH and superoxide radical scavenging activity causing 90.90%

and 31.73% inhibition at 25 and 50 mg/mL which were found to be comparable of BHA (91.00%) and gallic acid (47.73%), respectively. The extract also assayed using hydroxyl radical-induced deoxyribose damage which showed radical scavenging activity with percentage inhibition of 53.90–74.66% at 100–500 mg/mL (Deepa and Jena 2011). Furthermore, the ethanol extract of the leaves assayed in vitro by DPPH, hydroxyl, and hydrogen peroxide radicals displayed antioxidant activity with IC_{50} of 34.80, 64.40, and 51.00 mg/mL, respectively, as compared to ascorbic acid with IC_{50} of 24.00, 48.00, and 34.40 mg/mL, respectively. The extract also measured using ferric-reducing antioxidant power (FRAP) assay caused the reduction of ferric ion as compared with ascorbic acid (Shendge et al. 2011). Later, similar extract also evaluated in vivo on doxorubicin-induced rats that restored the levels of GSH and cardiac malondialdehyde (MDA) at doses of 250 and 500 mg/kg bw (Shendge and Gadge 2012). In another experiment on Swiss albino mice, the methanol extract of the stem bark reduced the production of ROS in kidney cells with IC_{50} of 34.72 mg/mL as compared to trolox (IC_{50} 8.66 mg/mL) (Alam et al. 2012). Further, Singh et al. (2012) also demonstrated that methanol, acetone, and water extracts of the stem bark measured by DPPH radical produced antioxidant activity with IC_{50} values of 188.08, 177.42, and 163.68 mg/mL of fresh mass, respectively. It was also investigated that the methanol extract of the leaves and its nonpolar fractions reduced the level of MDA and, however, enhanced the levels of the antioxidant enzymes including catalase (CAT), superoxide dismutase (SOD), and glutathione (GSH) in acetic acid-induced colitis at doses of 200 and 800 mg/kg (Somani et al. 2014). Interestingly, a compound proanthocyanidins isolated from *D. indica* fresh fruit produced significant antioxidant activity measured by FRAP and oxygen radical absorbance capacity (ORAC) assays with values of 2.32×10^3 mmol Fe(II)/g and 1.06×10^4mmol trolox equivalent/g (Fu et al. 2015). Lastly, the alcoholic extract of the fresh leaves obtained from sequential extraction with petroleum ether, chloroform, alcohol, and aqueous alcohol (40%) exhibited free radical scavenging activity toward DPPH, hydrogen peroxide, and superoxide radicals with IC_{50} of 2.98, 228.69, and 75.09 mg/mL, respectively, and ferric-reducing antioxidant power with EC_{50} of 111 mg/mL. In the similar experiment, it was also investigated that the extract enhanced the production of antioxidant enzymes (SOD and GSH) in STZ-induced diabetic rats at doses of 100, 200, and 400 mg/kg bw after 21 days of treatment when compared to glimepiride at 10 mg/kg bw (Kaur et al. 2016).

Antimicrobial Activity

D. indica was investigated for antibacterial, antifungal, and antiviral activities. The extracts and fractions of *D. indica* were documented to show growth inhibition against Gram-positive and Gram-negative bacteria. However, in comparison to bacteria, they attenuated the fungi including *Aspergillus fumigatus*, *Aspergillus niger*, *Penicillium* sp., *Candida albicans*, *Candida krusei*, *Rhizopus oryzae*, *Saccharomyces cerevisiae*, and *Trichoderma viride* (Nick et al. 1995a, b; Wiart et al. 2004; Haque et al. 2008; Apu et al. 2010; Smitha et al. 2012). Betulinic acid isolated from

D. indica has proven antimicrobial activity (Nick et al. 1994, 1995a, b; Ragasa et al. 2009). Meanwhile, a few nonpolar fractions (chloroform, carbon tetrachloride, and hexane) of the methanol extract of the leaves weakened the growth of *Escherichia coli, Bacillus cereus, Bacillus subtilis, Bacillus megaterium, Staphylococcus aureus, Sarcina lutea, Pseudomonas aeruginosa, Vibrio mimicus, Vibrio parahemolyticus, Salmonella typhi, Salmonella paratyphi, Shigella boydii*, and *Shigella dysenteriae* and fungal inhibition of *A. niger, C. albicans*, and *S. cerevisiae* with inhibition zones ranging from 6 to 8 mm at 400 mg/disc when compared with kanamycin (30 mg/disc; 30–40 mm) (Apu et al. 2010). In another experiment, hexane, dichloromethane, and ethyl acetate fractions of the methanol extract of the stem bark inhibited the growth of *E. coli, B. cereus, B. subtilis, S. aureus, S. lutea, P. aeruginosa, V. mimicus, V. parahemolyticus, S. paratyphi, S. typhi*, and *S. dysenteriae* with minimum inhibitory concentration (MIC) ranging from 0.31 to 20.00 mg/mL as compared to those of kanamycin (30 mg/disc; 22–30 mm) and amoxicillin (10 mg/disc; 14–22 mm). These fractions also attenuated the growth of *A. niger, C. albicans*, and *S. cerevisiae* with the inhibitory zones ranging from 7 to 13 mm as compared to ketoconazole (50 mg/disc; 19–23 mm) (Alam et al. 2011). Jaiswal et al. (2014) concluded that the aqueous acetone (70%) extract of the fruit and stem bark inhibited the growth of food-borne pathogens (*B. cereus, S. aureus, Yersinia enterocolitica*, and *E. coli*) with minimum inhibitory concentration ranging from 1.25 to 10.00 mg/mL. These findings suggest that *D. indica* has a powerful potential as antimicrobial agent, which supports it as traditional therapeutic remedy against the diseases caused by microbial infection, like diarrhea, dysentery, septicemia, and skin diseases.

Antidiarrheal Activity

In an experiment, castor oil-induced mice were used to check the antidiarrheal activity of *D. indica.* The aqueous and methanolic extracts of the leaves prolonged the onset and reduced the total number of feces after 2 h of treatment at doses of 200 and 400 mg/kg bw (Yeshwante et al. 2009a, b). The polar extracts of the leaf and fruit caused the prolongation of onset and curtailment in defecation frequency in the treated mice. The assay of extracts using charcoal meal revealed that it had a capacity to reduce the motility of the gastrointestinal tract (Yeshwante et al. 2009a, b; Bose et al. 2010; Rahman et al. 2011a, b). Further, the methanol extract of the leaves significantly diminished the frequency of defecation and number of total stool count at a dose of 500 mg/kg bw as compared to loperamide (25 mg/kg bw; 77.22%) (Bose et al. 2010). Ethanol extracts of the fruit and leaves minimized the total number of wet feces and also decreased the motility of gastrointestinal tract in castor oil-induced diarrheal mice at doses of 200 and 400 mg/kg bw when compared with loperamide (5 mg/kg bw) (Rahman et al. 2011b). The antidiarrheal action of this plant is suggested due to inhibition of inflammatory mediators by flavonoids and tannins (Yeshwante et al. 2009a, b).

Antiprotozoal Activity

Antiprotozoal activities of *D. indica* against malaria and leishmaniasis have been tested. Cyclohexane fraction of ethanolic extract of *D. indica* leaves was reported to inhibit the growth of parasite *Plasmodium falciparum*. In this experiment, cyclohexane fraction of the 80% ethanol extract of the leaves exhibited 53% inhibition against parasite *P. falciparum* at a concentration of 10 mg/mL as compared to chloroquione (Nguyen-Pouplin et al. 2007).

Antimutagenic Activity

Mutations are the cause of innate metabolic defects in cellular mechanism which trigger initiation and progression of several human diseases including cancer. The antimutagenic and protective effect has been ascribed to many classes of phytochemicals preferably flavanoids and phenolic compounds (Aqil et al. 2008). Meanwhile, Jaiswal et al. (2014) found that 70% aqueous acetone extract of the stem bark of *D. indica* demonstrated antimutagenic activity against sodium azide-induced mutation in *Salmonella* tester strain (TA-1531).

Anticholinergic Activity

Anticholinergics generally inhibit the action of acetylcholine from binding to its receptor sites on certain nerve cells and block parasympathetic nerve impulses. Bhadra et al. (2014) found that the standardized methanol extract of the fruit of *D. indica* inhibited acetylcholinesterase (AchE) and butyrylcholinesterase (BchE) activity with IC_{50} values of 67.26 and 122.39 mg/mL, respectively.

Protoscolicidal Activity

The scolicidal agents are generally employed in surgical manipulation of the hydatid cysts in hydatid diseases (García et al. 1997). Chowdhury et al. (2013) investigated the protoscolicidal activity in the stem bark of *D. indica* where an assay of the methanolic extract of the stem of *D. indica* was performed on earthworm *Pheretima posthuma*, and the result demonstrated the paralysis (194–136 min) and death (237–176 min) of the worms at 10–25 mg/mL comparable to albendazole at 10 mg/mL.

Hemolytic Activity

Hemolytic activity of any compounds is a measure of general cytotoxicity toward normal healthy cells (Da Silva et al. 2004). This activity was investigated by Jaiswal et al. (2014) in *D. indica* plant where the aqueous acetone (70%) extract of the fruit and stem bark of *D. indica* was assayed using rat whole blood which exhibited low inhibition against erythrocytes.

Hair Treatment Activity

On the basis of the previous information that keratin is the main component and mechanical strength of the hair, an experiment was performed where the aqueous extract of the mucilaginous *D. indica* seed sap protected human hair from loss of keratin after treatment with 10 mg hair/1 mL for 12 h. The physical structure of hair and keratin degradation were further affirmed by Fourier transform infrared spezctroscopy (FTIR), thermogravimetric analysis (TGA), and scanning electron microscopy (SEM) analysis, and a good hair weaving activity was noticed (Saikia 2013).

Enzyme Inhibitory Activity

Inhibition is one major mechanism for physiological enzyme regulation. Further, enzyme inhibition has a number of medicinal significances, and a large number of drugs generally act by the mechanism involving enzyme inhibition. In this view Jong-Anurakkun et al. (2007) performed an experiment to assure whether *D. indica* has an enzyme inhibitory action. As a result, the 50% aqueous methanolic extract of the leaves attenuated the intestinal sucrose and maltase activity with percentage inhibition of 40% and 56%, respectively.

Toxicology of *D. indica*

Several types of extracts of *D. indica* were tested for their toxicity action against *Artemia salina*. Despite the toxicity, some extracts of *D. indica* were found nontoxic (Kumar et al. 2010) and exhibited ameliorative (Shendge and Gadge 2012) and hepatoprotective (Padhya et al. 2008; Himakar et al. 2010) actions when assayed using in vivo models. The ethanolic extract of the leaves restored the level of myocardial enzymes such as alanine transaminase (ALT), aspartate transaminase (AST), lactate dehydrogenase (LDH), and creatine phosphokinase (CK) on myocardium of doxorubicin-induced rats at doses of 250 and 500 mg/kg bw (Shendge and Gadge 2012).

However, the ethanolic extract of the leaves reduced the levels of serum AST, ALT, ALP, bilirubin, and lipid peroxidation in the the liver induced by carbon tetrachloride at a dose of 300 mg/kg bw (Padhya et al. 2008). The nonpolar hexane extract of the seeds decreased the levels of serum enzymes, bilirubin, urea, creatinine, and lipid peroxidation but increased the levels of SOD,CAT, glutathione reductase (GR), glutathione peroxidase (GPx), and glutathione S-transferase (GST) in CCl_4-induced rats at doses of 250 and 500 mg/kg bw (Himakar et al. 2010). Besides these, *D. indica* has been used as food poisoning neutralizer (Islam et al. 2014; Grosvenor et al. 1995a, b) which also evidenced that *D. indica* is safe and not toxic. The methanolic extract of the leaves was nontoxic and did not produce any mortality in mice during 24 h of treatment with 100–1500 mg/kg bw of extract by intraperitoneal administration (Kumar et al. 2010). Apart from this, water, chloroform, carbon tetrachloride, and n-hexane fractions of methanol extract of the leaves demonstrated lethality on brine shrimp as compared to vincristine sulfate (Apu et al. 2010). However, the methanol, ethyl acetate, dichloromethane, and n-hexane extracts of the stem bark exhibited inhibition on brine shrimp with less lethality when compared to the leaf extracts (Parvin et al. 2009; Alam et al. 2011; Chowdhury et al. 2013).

D. indica in Drug Formulation and Drug Delivery

The mucilage obtained from seeds of *D. indica* fruit contains a natural mucoadhesive hydrophilic polymer, which is generally used in the formulation for drug delivery (Sharma et al. 2009; Bal et al. 2012a, b). The mucoadhesive and viscous properties of mucilage of *D. indica* fruit were used as better substitute of synthetic polymers, i.e., Carbopol 934 (Kuotsu and Bandyopadhyay 2007; Bal et al. 2012b) and hydroxylpropyl methyl cellulose. Novel mucoadhesive buccal tablets of oxytocin and formulation of nasal gels were prepared from *D. indica* (Kuotsu and Bandyopadhyay 2007; Metia and Bandyopadhyay 2008). Bal et al. (2012a, b) developed a mucoadhesive carvedilol microcapsule using the mucilage obtained from seeds of *D. indica* for encapsulation purpose. This microcapsule was found to be free flowing and usually spherical in shape. It was concluded that this mucilage was effective for sustained drug release, which can be employed to reduce the hypertension for a period of 24 h. Further, Sharma et al. (2013) conducted an experiment where they encapsulated pantoprazole sodium and metformin hydrochloride, respectively, using this mucilage. Good swelling properties and mucoadhesivity was found to perform at the intestinal pH. This finding suggested that the seed mucilage of *D. indica* has good potency for the purpose of drug encapsulation (Sharma et al. 2010). Nanoparticles are being extensively used as drug carrier for the treatment of diseases nowadays. Singh et al. (2013) prepared colloidal silver nanoparticles (SNP) using the aqueous extract of *D. indica* fruits as reducing agent and as a better substitute of sodium borohydride. This extract reduced $AgNO_3$ to silver nanoparticles with stability more than 6 days, which suggested that *D. indica* could be employed as natural reducing agent for silver nanoparticles loaded formulation.

Future Prospects

Despite a broad range of medicinal properties of *D. indica*, very few investigations regarding chemical constituents and pharmacological aspects have been carried out. There is little evidence over the quantification of different active phytoconstituents responsible for important pharmacological activities. It is evident from the available literature that *D. indica* possesses adequate therapeutic potential and need to be explored further for chemical and pharmacological investigations. Current knowledge of *D. indica* shows great lacunae that need more biological investigations to be done on its plant extracts. Hence, further studies are highly required to explore the potential of its plant extracts against various diseases and search for molecular mechanisms underlying their action. Future studies are also required to evaluate the adverse effects, safety profile, and different biological activities of extracts as well as particular chemical constituents from *D. indica* in order to facilitate their clinical applications as modern medicines for human health.

Conclusion

Herbal medicines are the most extensively used therapeutics worldwide. To promote their proper use and to establish their potential as sources for new drugs, it is necessary to study medicinal plants having folklore reputation in a better and intensified way. The extensive literature survey as well as research reports revealed that *D. indica* is highly regarded to have good potential in the herbal medicine. Betulin and betulinic acid are the major constituents found to be present in almost all parts of this plant which can cure various human ailments and diseases. As the raw fruits are eaten by tribal communities, the juice of *D. indica* may be taken as energy drink due to their good nutritional value. It has previously been confirmed that *D. indica* have curing properties in wound healing, diabetes, cuts and burns, abdominal pains, and many more, but scientific evidence of these reports is yet not much developed. Various pharmacological investigations have been done using different plant parts such as leaves which have various activities like antioxidant, cytotoxic, antimicrobial, antidiarrheal, and anxiolytic, seeds which are hepatoprotective and antimicrobial, and fruits which are antileukemic. Despite a few toxicity reports, *D. indica* and most of its extracts were found to be non-toxic and exhibited ameliorative and hepatoprotective activities in several in vivo studies. The ethnopharmacological use of *D. indica* as food poisoning neutralizer also indicates its better safety profile and non-toxic nature. The current status of *D. indica* demands some biotechnological investigations including protein and gene expression for target identification and exploration of molecular mechanisms underlying the action at molecular levels.

Acknowledgments The authors are grateful to the Vice Chancellor of Babasaheb Bhimrao Ambedkar University, Lucknow, India, for the support.

References

Abdille, M. H., Sigh, R. P., Jayaprakasha, G. K., & Jena, B. S. (2005). Antioxidant activity of the extracts from *Dillenia indica* fruits. *Food Chemistry, 90*, 891–896.

Akter, R., Uddin, S. J., Grice, I. D., & Tiralongo, E. (2014). Cytotoxic activity screening of Bangladeshi medicinal plant extracts. *Journal of Natural Medicines, 68*, 246–252.

Alam, M. D., Chowdhury, N. S., Mazumder, M. E. H., & Haque, M. E. (2011). Antimicrobial and toxicity study of different fractions of *Dillenia indica* Linn. bark extract. *International Journal of Pharmaceutical Sciences and Research, 2*, 860–866.

Alam, B. M., Rahman, M. S., Hasan, M., Khan, M. M., Nahar, K., & Sultana, S. (2012). Antinociceptive and antioxidant activities of the *Dillenia indica* bark. *International Journal of Pharmacology, 8*, 243–253.

Angami, A., Gajurel, P. R., Rethy, P., Singh, B., & Kalita, S. K. (2006). Status and potential of wild edible plants of Arunachal Pradesh. *The Indian Journal of Traditional Knowledge, 5*, 541–550.

Anisuzzaman, M., Rahman, A. H. M. M., Harun-Or-Rashid, M., Naderuzzaman, A. T. M., & Islam, A. K. M. R. (2007). An ethnobotanical study of Madhupur, Tangail. *Journal of Applied Sciences Research, 3*, 519–530.

Apu, A. S., Muhit, M. A., Tareq, S. M., Pathan, A. H., Jamaluddin, A. T. M., & Ahmed, M. (2010). Antimicrobial activity and brine shrimp lethality bioassay of the leaves extract of *Dillenia indica* Linn. *Journal of Young Pharmacists, 2*(1), 50–53.

Aqil, F., Zahin, M., & Ahmad, I. (2008). Antimutagenic activity of methanolic extracts of four ayurvedic medicinal plants. *Indian Journal of Experimental Biology, 46*(9), 668–672.

Armania, N., Yazan, L. S., Ismail, I. S., Foo, J. B., Tor, Y. S., Ishak, N., Ismail, N., & Ismail, M. (2013a). *Dillenia suffruticosa* extract inhibits proliferation of human breast cancer cell lines (MCF-7 and MDA-MB-231) via induction of G2/M arrest and apoptosis. *Molecules, 18*, 13320–13339.

Armania, N., Yazan, L. S., Musa, S. N., Ismail, I. S., Foo, J. B., Chan, K. W., Noreen, H., Hisyam, A. H., Zulfahmi, S., & Ismail, M. (2013b). *Dillenia suffruticosa* exhibited antioxidant and cytotoxic activity through induction of apoptosis and G2/M cell cycle arrest. *Journal of Ethnopharmacology, 146*, 525–535.

Azam, M. N. K., Rahman, M. M., Biswas, S., & Ahmed, M. N. (2016). Appraisals of Bangladeshi medicinal plants used by folk medicine practitioners in the prevention and management of malignant neoplastic diseases. *International Scholarly Research Notices*. https://doi.org/10.1155/2016/7832120.

Bal, T., Murthy, P. N., & Pandey, A. (2012a). Evaluation of mucoadhesive carvedilol microcapsules prepared using orifice gelatin technique. *Journal of Pharmacy Research, 5*(1), 519–525.

Bal, T., Murthy, P. N., & Sengupta, S. (2012b). Isolation and analytical studies of mucilage obtained from the seeds of *Dillenia indica* (family Dilleniaceae) by use of various analytical techniques. *Asian Journal of Pharmaceutical and Clinical Research, 5*(3), 65–68.

Banerji, N., Majumder, P., & Dutta, N. L. (1975). New pentacyclic triterpene lactone from *Dillenia indica*. *Phytochemistry, 14*, 1447–1448.

Bate-Smith, E. C., & Harborne, J. B. (1971). Differences in flavonoid content between fresh and herbarium leaf tissue in Dillenia. *Phytochemistry, 10*, 1055–1058.

Bhat, P., Hegde, G. R., Hegde, G., & Mulgund, G. S. (2014). Ethnomedicinal plants to cure skin diseases- an account of the traditional knowledge in the coastal parts of Central Western Ghats, Karnataka, India. *The Journal of Ethnopharmacology, 151*, 493–502.

Bhattacharjee, S. R., & Chatterjee, A. (1962). Betulinic acid and betulin, the triterpenoid constituents of *Dillenia indica*. *Journal of the Indian Chemical Society, 39*, 276–284.

Boer, H. J., Lamxay, V., & Bjork, L. (2012). Comparing medicinal plant knowledge using similarity indices, a case of the Brou, Saek and Kry in Lao PDR. *Journal of Ethnopharmacology, 141*, 481–500.

Bose, U., Gunasekaran, K., Bala, V., & Rahman, A. A. (2010). Evaluation of phytochemical and pharmacological properties of *Dillenia indica* Linn. Leaves. *Journal of Pharmacology and Toxicology, 10*(5), 222–228.

Chowdhury, M. I., Dewan, S. M. R., Ahamed, S. K., Moghal, M. M. R., & Ahmed, J. (2013). Biological activities of *Dillenia indica* L. bark growing in Bangladesh. *Journal of Biological & Scientific Opinion, 1*(2), 45–49.

Da Silva, E., Shahgaldian, P., & Coleman, A. W. (2004). Haemolytic properties of some water-soluble *para*-sulphonato-calix-(*n*)-arenes. *International Journal of Pharmaceutics, 273*, 57–62.

Dan, S., & Dan, S. S. (1980). Triterpenoids of Indian Dilleniaceae. *Journal of the Chemical Society, 57*, 760.

Das, A. K., Dutta, B. K., & Sharma, G. D. (2008). Medicinal plants used by different tribes of Cachar district Assam. *The Indian Journal of Traditional Knowledge, 7*(3), 446–454.

Das, S., Khan, M. L., Rabha, A., & Bhattacharjya, D. K. (2009). Ethnomedicinal plants of Manas National Park, Assam, Northeast India. *The Indian Journal of Traditional Knowledge, 8*, 514–517.

Deepa, N., & Jena, B. S. (2011). Antioxidant fraction from bark of *Dillenia indica*. *International Journal of Food Properties, 14*, 1152–1159.

Dickison, W. C. (1979). A note on the wood anatomy of Dillenia(Dilleniaceae). *IAWA Bulletin, 2–3*, 57–60.

Dubey, P. C., Sikarwar, R. L. S., Khanna, K. K., & Tiwari, A. P. (2009). Ethnobotany of *Dillenia pentagyna* Roxb. in Vindhya region of Madhya Pradesh, India. *Natural Product Radiance, 8*(5), 546–548.

Fu, C., Yang, D., Peh, W. Y. E., Lai, S., Feng, X., & Yang, H. (2015). Structure and antioxidant activities of proanthocyanidins from elephant apple (*Dillenia indica* Linn.). *Journal of Food Science, 80*(10), 2191–2199.

García, J. I. L., Alonso, E., Gonzalez-Uriarte, J., & Romano, D. R. (1997). Evaluation of scolicidal agents in an experimental hydatid disease model. *European Surgical Research, 29*(3), 202–208.

Grosvenor, P. W., Gothard, P. K., McWilliam, N. C., Supriono, A., & Gray, D. O. (1995a). Medicinal plants from Riau Province, Sumatra, Indonesia. Part 1 uses. *The Journal of Ethnopharmacology, 45*, 75–95.

Grosvenor, P. W., Supriono, A., & Gray, D. O. (1995b). Medicinal plants from Riau Province, Sumatra, Indonesia. Part 2, antibacterial and antifungal activity. *Journal of Ethnopharmacology, 45*, 97–111.

Haque, M. E., Islam, M. N., Hossain, M., Mohamad, A. U., Karim, M. F., & Rahman, M. A. (2008). Antimicrobial and cytotoxic activities of *Dillenia pentagyna*. Dhaka Univ. *Journal of Pharmaceutical Sciences, 7*(1), 103–105.

Hazarika, T. K., Marak, S., Mandal, D., Upadhyaya, K., Nautiyal, B. P., & Shukla, A. C. (2016). Underutilized and unexploited fruits of Indo-Burma hot spot, Meghalaya, north-east India: Ethno-medicinal evaluation, socio-economic importance and conservation strategies. *Genetic Resources and Crop Evolution, 63*(2), 289–304.

Himakar, R. K., Tharanath, V., Nagi, R. K. B., Sharma, P. V. G. K., & Reddy, O. V. S. (2010). Studies on hepatoprotective effect of hexane extract of *Dillenia indica* against CCl4 induced toxicity and its safety evaluation in Wistar albino rats. *RJPBCS, 1*(3), 441–450.

Islam, M. K., Saha, S., Mahmud, I., Mohamad, K., Awang, K., Uddin, S. J., Rahman, M. M., & Shilpi, J. A. (2014). An ethnobotanical study of medicinal plants used by tribal and native people of Madhupur forest area Bangladesh. *The Journal of Ethnopharmacology, 151*, 921–930.

Jaiswal, S., Mansa, N., Prasad, M. S. P., Jena, B. S., & Negi, P. S. (2014). Antibacterial and antimutagenic activities of *Dillenia indica* extracts. *Food Bioscience, 5*, 47–53.

Jong-Anurakkun, N., Bhandari, M. R., & Kawabata, J. (2007). α-Glucosidase inhibitors from Devil tree (*Alstonia scholaris*). *Food Chemistry, 103*, 1319–1323.

Kagyung, R., Gajurel, P. R., Rethy, P., & Singh, B. (2010). Ethnomedicinal plants used for gastro-intestinal diseases by Adi tribes of Dehang-Debang biosphere reserve in Arunachal Pradesh. *The Indian Journal of Traditional Knowledge, 9*(3), 496–501.

Kalita, D., & Deb, B. (2004). Some folk medicines used by the Sonowal Kacharis tribe of the Brahmaputra valley Assam. *Natural Product Radiance, 3*(4), 240–246.

Kar, A., & Borthakur, S. K. (2007). Wild vegetables sold in local markets of Karbi Anglong Assam. *The Indian Journal of Traditional Knowledge, 6*(1), 169–172.

Kaur, N., Kishore, L., & Singh, R. (2016). Antidiabetic effect of new chromane isolated from *Dillenia indica* L. leaves in streptozotocin induced diabetic rats. *Journal of Functional Foods, 22*, 547–555.

Kerrigan, R. A., Craven, L. A., & Dunlop, C. R. (2011). Dilleniaceae. In P. S. Short & I. D. Cowie (Eds.), *Flora of the Darwin region* (pp. 1–6). Palmerston: Northern Territory Government.

Khanum, A., Khan, I., & Ali, A. (2007). Ethnomedicine and human welfare. *Ukaaz Publications, 4*, 52.

Khare, C. P. (2007). *Indian medicinal plants* (Vol. 214). Berlin: Springer.

Khongsai, M., Saikia, S. P., & Kayang, H. (2011). Ethnomedicinal plants used by different tribes of Arunachal Pradesh. *The Indian Journal of Traditional Knowledge, 10*(3), 541–546.

Kirtikar, K. R., & Basu, B. D. (2003). Indian medicinal plants. *Oriental Enterprizes, Dehradun, 1*, 75–77.

Kumar, D., Mallick, S., Vedasiromoni, J. R., & Pal, B. C. (2010). Anti-leukemic activity of *Dillenia indica* L. fruit extract and quantification of betulinic acid by HPLC. *Phytomedicine, 17*, 431–435.

Kumar, S., Kumar, V., & Prakash, O. (2011a). Antidiabetic and hypolipidemic activities of *Dillenia indica* extract in diabetic rats. *Chinese Journal of Integrative Medicine, 9*(5), 570–574.

Kumar, S., Kumar, V., & Prakash, O. (2011b). Antidiabetic and antihyperlipidemic effects of *Dillenia indica* (L.) leaves extract. *Brazilian Journal of Pharmaceutical Sciences, 47*(2), 373–378.

Kumar, S., Kumar, V., & Prakash, O. (2013). Enzymes inhibition and antidiabetic effect of isolated constituents from *Dillenia indica*. *BioMed Research International, 2013*, 1–7.

Kuotsu, K., & Bandyopadhyay, A. K. (2007). Development of oxytocin nasal gel using natural mucoadhesive agent obtained from the fruits of *Delliniaindica* L. *Science Asia, 33*, 57–60.

Lalfakzuala, R., Lalramnghinglova, H., & Kayang, H. (2007). Ethnobotanical usages of plants in western Mizoram. *The Indian Journal of Traditional Knowledge, 6*, 486–493.

Majumdar, K., Saha, R., Datta, B. K., & Bhakta, T. (2006). Medicinal plants prescribed by different tribal and non-tribal medicine men of Tripura state. *The Indian Journal of Traditional Knowledge, 5*(4), 559–562.

Metia, P. K., & Bandyopadhyay, A. K. (2008). In vitro and in vivo evaluation of a novel mucoadhesive buccal oxytocin table prepared with *Dillenia indica* fruit mucilage. *Pharmazie, 63*, 270–274.

Miglianto, K. F., Chiosini, M. A., Mendonça, F. A. S., Esquisatto, M. A. M., Salgado, H. R., & Santos, G. M. T. (2011). Effect of glycolic extract of *Dillenia indica* L. combined with microcurrent stimulation on experimental lesions in Wistar rats. *Wounds, 23*(5), 111–120.

Muhit, A. M., Tareq, S. M., Apu, A. S., Basak, D., & Islam, M. S. (2010). Isolation and identification of compounds from the leaf extract of *Dillenia indica* Linn. *Bangladesh Pharmaceutical Journal, 13*(1), 49–53.

Mukherjee, K. S., & Badruddoza, S. (1981). Chemical constituents of *Dillenia indica* Linn. and Vitex negundo Linn. *Journal of the Indian Chemical Society, 58*, 97–98.

Nguyen-Pouplin, J., Tran, H., Phan, T. A., Dolecek, C., Farrar, J., Tran, T. H., Caron, P., Bodo, B., & Grellier, P. (2007). Antimalarial and cytotoxic activities of ethnopharmacologically selected medicinal plants from South Vietnam. *Journal of Ethnopharmacology, 109*, 417–427.

Nick, A., Wright, A. D., Sticher, O., & Rali, T. (1994). Antibacterial triterpenoid acids from *Dillenia papuana*. *Journal of Natural Products, 57*, 1245–1250.

Nick, A., Rali, T., & Sticher, O. (1995a). Biological screening of traditional medicinal plants from Papua New Guinea. *Journal of Ethnopharmacology, 49*, 147–156.

Nick, A., Wright, A. D., Rali, T., & Sticher, O. (1995b). Antibacterial triterpenoids from *Dillenia papuana* and their structure-activity relationships. *Phytochemistry, 40*, 1691–1695.

Ozdemir, E., & Alpınar, K. (2015). An ethnobotanical survey of medicinal plants in western part of central Taurus Mountains: Aladaglar (Nigde-Turkey). *Journal of Ethnopharmacology, 166*, 53–65.

Padhya, I. P., Choudhary, N. S., Padhy, S. K., & Das, S. (2008). Effect of *Dillenia indica* leaves against carbon tetrachloride induced hepatotoxicity. *Journal of Pharmacy and Chemistry, 2*, 190–193.

Parvin, M. N., Rahman, M. S., Islam, M. S., & Rashid, M. A. (2009). Chemical and biological investigations of *Dillenia indica* Linn. *Bangladesh Journal of Pharmacology, 4*, 122–125.

Pavanasasivam, G., & Sultanbawa, M. U. S. (1975a). Flavonoids of some Dilleniaceae species. *Phytochemistry, 14*, 1127–1128.

Pavanasasivam, G., & Sultanbawa, M. U. S. (1975b). Chemical investigation of Ceylonese plants. XII. (+)-3,4′,5,7-tetrahydroxy-3′-methoxyflavanone ((+)-dihydroisorhamnetin) and 3,5,7-trihydroxy-3′,4′-dimethoxyflavone (dillenetin), two new natural products from *Dillenia indica*. *Journal of the Chemical Society, 6*, 612–613.

Pavani, M., Rao, M. S., Nath, M. M., & Rao, C. A. (2012). Ethnobotanical explorations on antidiabetic plants used by tribal inhabitants of Seshachalam forest of Andhra Pradesh, India. *Indian Journal of Fundamental and Applied Life Sciences, 2*(3), 100–105.

Polterai, O. (1997). Antioxidants and free-radical scavengers of natural origin. *Current Organic Chemistry, 1*, 415–440.

Poonam, K., & Singh, G. S. (2009). Ethnobotanical study of medicinal plants used by the Taungya community in Terai Arc landscape, India. *The Journal of Ethnopharmacology, 123*, 167–176.

Pradhan, B. K., & Badola, H. K. (2008). Ethnomedicinal plant use by Lepcha tribe of Dzongu valley, bordering Khangchendzonga Biosphere Reserve, in North Sikkim, India. *Journal of Ethnobiology and Ethnomedicine, 4*, 22.

Prasad, P. R. C., Reddy, C. S., Raza, S. H., & Dutt, C. B. S. (2008). Folklore medicinal plants of North Andaman Islands, India. *Fitoterapia, 79*, 458–464.

Purkayastha, J., Nath, S. C., & Islam, M. (2005). Ethnobotany of medicinal plants from Dibru-Saikhowa biosphere reserve of Northeast India. *Fitoterapia, 76*, 121–127.

Quattrocchi, U. F. L. S. (2012). *CRC world dictionary of medicinal and poisonous plants* (pp. 1407–1408). New York: CRC Press.

Ragasa, C. Y., Alimboyoguen, A. B., & Shen, C.-C. (2009). Antimicrobial triterpenes from *Dillenia philippinensis*. *The Philippine Scientist, 46*, 78–87.

Rahman, M. D., Rahman, M., Islam, M. M., & Reza, M. S. (2011a). The importance of forests to protect medicinal plants, a case study of Khadimnagar National Park, Bangladesh. *IJBSESM, 7*, 283–294.

Rahman, M. S., Shams-Ud-Doha, K. M., & Rahman, R. (2011b). Antidiarrhoeal activity of the leaf and fruit extracts of *Dillenia indica*. *International Journal of Biosciences, 1*(6), 39–46.

Rai, P. L., & Lalramnghinglova, H. (2010). Ethnomedicinal plant resources of Mizoram, India, implication of traditional knowledge in health care system. *Ethnobotanical Leaflets, 14*, 274–305.

Ripunjoy, S. (2013). Indigenous knowledge on the utilization of medicinal plants by the Sonowal Kachari Tribe of Dibrugarh district in Assam, North-East India. *International Research Journal of Biological Sciences, 2*(4), 44–50.

Saikia, J. P. (2013). Hair waving natural product: *Dillenia indica* seed sap. *Colloids and Surfaces. B, Biointerfaces, 102*, 905–907.

Saikia, A. P., Ryakala, V. K., Sharma, P., Goswami, P., & Bora, U. (2006). Ethnobotany of medicinal plants used by Assamese people for various skin ailments and cosmetics. *Journal of Ethnopharmacology, 106*, 149–157.

Sarker, M. M. R., Nimmi, I., & Kawsar, M. H. (2012). Preliminary screening of six popular fruits of Bangladesh for in vitro IgM production and proliferation of splenocytes. *Bangladesh Pharmaceutical Journal, 15*(1), 31–37.

Sarmah, R., Adhikari, D., Majumdar, M., & Arunachalan, A. (2008). Traditional medicobotany of Chakma community residing in the Northwestern periphery of Namdapha National Park in Arunachal Pradesh. *The Indian Journal of Traditional Knowledge, 7*(4), 587–593.

Shah, G. L. (1978). *Dillenia indica* and *Dillenia pentagyna. Flora of Gujarat, 49*, 214.

Sharma, U. M., & Pegu, S. (2011). Ethnobotany of religious and supernatural beliefs of the Mishing tribes of Assam with special reference to the 'DoburUie'. *Journal of Ethnobiology and Ethnomedicine, 7*, 16.

Sharma, H. K., Chhangte, L., & Dolui, A. K. (2001). Traditional medicinal plants in Mizoram, India. *Fitoterapia, 72*, 146–161.

Sharma, H. K., Sarangi, B., & Pradhan, S. P. (2009). Preparation and in vitro evaluation of mucoadhesive microbeads containing timolol maleate using mucoadhesive substances of *Dillenia indica* L. *Archives of Pharmaceutical Science and Research, 1*(2), 181–188.

Sharma, H. K., Pradhan, S. P., & Sarangi, B. (2010). Preparation and in vitro evaluation of enteric controlled release pantoprazole loaded microbeads using natural mucoadhesive substance from *Dillenia indica* L. *International Journal of PharmTech Research, 2*(1), 542–551.

Sharma, J., Gairola, S., Gaur, R. D., & Painuli, R. M. (2012). The treatment of jaundice with medicinal plants in indigenous communities of the Sub-Himalayan region of Uttarakhand, India. *Journal of Ethnopharmacology, 143*, 262–291.

Sharma, H. K., Lahkar, S., & Nath, L. K. (2013). Formulation and in vitro evaluation of metformin hydrochloride loaded microspheres prepared with polysaccharide extracted from natural sources. *Acta Pharmaceutica, 63*, 209–222.

Shendge, P., & Gadge, M. (2012). Protective effect of *Dillenia indica* L. on doxorubicin induced cardiotoxicity in rats. *International Journal of Drug Discovery and Medical Research, 1*(2), 9–13.

Shendge, P., Patil, L., & Kadam, V. (2011). In vitro evaluation of antioxidant activity of *Dillenia indica* Linn. leaf extract. *International Journal of Pharmaceutical Sciences and Research, 2*(7), 1814–1818.

Singh, A. K., Raghubanshi, A. S., & Singh, J. S. (2002). Medical ethnobotany of the tribals of Sonaghati of Sonbhadra district, Uttar Pradesh, India. *The Journal of Ethnopharmacology, 81*, 31–41.

Singh, S., Saikia, J. P., & Buragohain, A. K. (2013). A novel "green" synthesis of colloidal silver nanoparticles (SNP) using *Dillenia indica* fruit extract. *Colloids and Surfaces B: Biointerfaces, 102*, 83–85.

Singh, D. R., Singh, S., Salim, K. M., & Srivastava, R. C. (2012). Estimation of phytochemicals and antioxidant activity of underutilized fruits of Andaman Islands (India). *International Journal of Food Sciences and Nutrition, 63*(4), 446–452.

Smitha, V. P., Ch, M. M., Kandra, P., Sravani, R., & Akondi, R. B. (2012). Screening of antimicrobial and antioxidant potentials of Acacia caesia, *Dillenia pentagyna* and *Buchanania lanzan* from Maredumilli forest of India. *Journal of Pharmacy Research, 5*(3), 1734–1738.

Somani, S. J., Badgujar, L. B., Sutariya, B. K., & Saraf, M. N. (2014). Protective effect of *Dillenia indica* L. on acetic acid induced colitis in mice. *Indian Journal of Experimental Biology, 52*, 876–881.

Srivastava, B. K., & Pande, C. S. (1981). Chemical examination of bark of *Dillenia indica. Acta Cienc Indica, 7*(1–4), 170–174.

Srivastava, R. C., Singh, R. K., Community, A., & Mukherjee, T. K. (2010). Indigenous biodiversity of Apatani plateau: Learning on biocultural knowledge of Apatani tribe of Arunachal Pradesh for sustainable livelihoods. *The Indian Journal of Traditional Knowledge, 9*(3), 432–442.

Sundararamaiah, T., Ramraj, S. K., Rao, K. L., & Vimalabai, M. V. (1976). Isolation of the lupeol group of triterpenes from *Dillenia indica* Linn. and *Diospyros perigrina. Journal of the Indian Chemical Society, 53*, 638.

Sunil, K., Vipin, K., & OM, P. (2011). Free radicals scavenging effect of *Dillenia indica* leaves. *Asian Journal of Pharmaceutical and Biological Research, 1*, 169–173.

Tag, H., Kalita, P., Dwivedi, P., Das, A. K., & Namsa, N. D. (2012). Herbal medicines used in the treatment of diabetes mellitus in Arunachal Himalaya, northeast, India. *Journal of Ethnopharmacology, 141*, 786–795.

Tarak, D., Namsa, N. D., Tangjang, S., Arya, S. C., Rajbonshi, B., Samal, P. K., & Mandal, M. (2011). An inventory of the ethnobotanicals used as anti-diabetic by a rural community of Dhemaji district of Assam, Northeast India. *Journal of Ethnopharmacology, 138*, 345–350.

The wealth of India, Raw Materials. (1952). *The wealth of India, raw materials* (Vol. 3, pp. 64–65). New Delhi: CSIR.

Tiwari, K. P., & Srivastava, S. S. D. (1979). Pigments from the stem bark of *Dillenia indica*. *Planta Medica, 35*, 188–190.

Uppalapati, S. L., & Rao, J. T. (1980). Protein analysis of the seeds of *Dillenia indica* Linn. *Journal of the Institution of Chemists, 52*, 111–112.

Venkatalakshmi, P., Vadivel, V., & Brindha, P. (2016). Role of phytochemicals as immunomodulatory agents: A review. *International Journal of Green Pharmacy, 10*(1), 1–18.

Wiart, C., Mogana, S., Khalifah, S., Mahan, M., Ismail, S., Buckle, M., Narayana, A. K., & Sulaiman, M. (2004). Antimicrobial screening of plants used for traditional medicine in the state of Perak, Peninsular Malaysia. *Fitoterapia, 75*, 68–73.

Yeshwante, S. B., Juvekar, A. R., Nagmoti, D. M., Wankhede, S. S., Shah, A. S., Pimprikar, R. B., & Saindane, D. S. (2009a). Anti-inflammatory activity of methanolic extracts of *Dillenia indica* L. leaves. *Journal of Young Pharmacists, 1*, 63–66.

Yeshwante, S. B., Juvekar, A. R., Pimprikar, R. B., Kakade, R. T., Tabrej, M., Kale, M. K., & Firke, S. D. (2009b). Anti-diarrheal activity of methanolic and aqueous extracts of *Dillenia indica* L. *Research Journal of Pharmacology Pharmacodynamics, 1*(3), 140–142.

Yeshwante, S. B., Juvekar, A. R., Nagmoti, D. M., & Wankhede, S. (2011). In vivo analgesic activity of methanolic extract of *Dillenia indica* (L) leaves. *Pharmacology, 3*, 1084–1096.

Part III
Aquatics and Wastewater Treatment

Chapter 10
Implication of Algal Microbiology for Wastewater Treatment and Bioenergy Production

Vinayak V. Pathak, Shamshad Ahmad, and Richa Kothari

Contents

Abstract The Indian power sector has developed huge infrastructure for power generation, transmission and distribution of energy since the major utilization of fossil fuel has been an obstacle to achieve energy sustainability. Moreover, inadequate efficiency to treat the municipal and industrial wastewater is also generating serious environmental hazards. Thus concurrent management roadmap

V. V. Pathak
Department of Chemistry, Manav Rachna University, Faridabad, Haryana, India

S. Ahmad
Department of Environmental Science, School for Environmental Science, Babasaheb Bhimrao Ambedkar University (A Central University), Lucknow, Uttar Pradesh, India

R. Kothari (✉)
Department of Environmental Sciences, Central University of Jammu, Jammu and Kashmir, India

Babasaheb Bhimrao Ambedkar University, Lucknow, Uttar Pradesh, India
e-mail: kotharirricha21@gmail.com

R. C. Sobti et al. (eds.), *Environmental Biotechnology: For Sustainable Future*,
https://doi.org/10.1007/978-981-10-7284-0_10

is required to tackle these challenges. In this context, algae have enormous efficiency to deal the challenges related to environment as well as energy sustainability. Algae have unique potential to grow under variety of environmental conditions due to flexible metabolic pathways. Wastewater generation and its adequate treatment are two of the major challenges to achieve the environmental sustainability that can be resolved through integration of algal cultivation in wastewater treatment systems. In this article, a holistic view on implication of algal microbiology for wastewater remediation and bioenergy production is provided. The procedures, associated challenges, and advancement in line of algal biofuel production explored by various authors are studied to project better ways for algal biofuel production process. A variety of pollutants as nutrients and their impact on lipid production in algal biomass are considered as the main advantages of this holistic approach. Advancements and challenges in algal harvesting and conversion processes for biodiesel production are also reviewed.

Keywords Bioenergy · Microalgae · Biodiesel · Algal harvesting · Environmental hazard

Introduction

Industrial revolution, rapid urbanization, and the transportation sector have generated several major environmental issues such as energy crisis, greenhouse gas (GHG) emissions, global warming, water crisis, and subsequent climate challenges. Holistic strategies are required to manage these issues simultaneously. India being one of the most populous countries (18% of the global population) uses only 6% of the world's primary energy. However, India's consumption for primary energy has almost doubled since 2000 and is expected to grow enormously in the coming years. Despite the unsustainability of fossil fuels and their contribution in GHG emissions, three quarters of energy demand in India is still met by these fuels. In the power sector, coal remains as a backbone, accounting for 70% generation (Indian Energy Outlook 2015). Moreover, poor energy transmission and distribution are also challenges for Indian power sector. Thus to ensure the energy security and its homogenous distribution, renewable energy sources are now being adopted in Indian power sector. In 2012, the world relied on renewable sources for around 13.2% of its total primary energy supply. In 2013, renewables accounted for almost 22% of global electricity generation, and the IEA Medium-Term Renewable Energy Report, 2015, foresees share reaching at least 26% increase in 2020. In India, renewable energy accounts for 23% of total energy consumption, whereas the share of biomass in total renewable-based generation is only 12.8% (REN-21 2014).

The other major environmental issue is associated with treatment of urban and industrial wastewater. The adverse quality of water due to anthropogenic influences caused a major issue of concern across the globe. Wastewater comprises liquid waste discharged from industrial, domestic, and commercial activities and agricultural sectors. Wastewater comprises of a long range of contaminants such as organic and mineral, which exhibit toxic effects on the microorganisms and resist the degradation and mineralization processes. Wastewater contains major contaminants such as nitrogen (inorganic or organic) in the form of ammonia, nitrite, nitrate, phosphorus, proteins, drugs, thiosulfate, and thiocyanates; pathogens such as bacteria, viruses, and parasitic worms; inorganic particles such as sand, grit, metal particles, and ceramics; gases such as hydrogen sulfide and carbon dioxide; and aerosols such as paint, hair colorant, and emulsified oils. Commonly, alum was used as a conventional method to remove solids together for easy removal of solid particle. Disinfection is also a part of this step, which is carried out through chlorination to remove microorganisms, parasites, and harmful larva from the treated wastewater (USEPA 1999). Industrial discharge from various industries has been continuously adding lot of wastewater containing a high level of nutrients, heavy metals, and hazardous substances to the agricultural fields, river streams, etc. (Baccella et al. 2000). These industrial effluents not only increase the nutrient level in aquatic bodies but also cause the toxicity to aquatic organisms, which finally affect the availability of potable water (Kisku et al. 2000).

The integration of wastewater treatment and energy production provides advantages over the conventional practices of wastewater treatment. The Ministry of New and Renewable Energy (MNRE) has classified waste to energy as a renewable source and provided subsidies and incentives to encourage the projects (http://mnre.gov.in/schemes/offgrid/waste-to-energy/).

Conventional Wastewater Treatment

Conventional wastewater treatment systems are found as energy intensive processes and inadequate to achieve discharge standards in effluent prescribed by various agencies (Caputo and Pelagagge 2001). It is associated with a series of physical, chemical, and biological processes to minimize the organic content and nutrient load of wastewater. Conventional treatment plants involve primary and secondary processes followed by disinfection for disposal of wastewater. The primary treatment process is based on physical unit operations such as screening, grit removal, and sedimentation, which are used to remove solid wastewater constituents and part of suspended solids from wastewater. Normally, the first step is the screening in which trash and large solids are removed or reduced in size (Bourgeous et al. 2001). These solids are collected on screen and scraped off for subsequent disposal. Solid content in wastewater is present in the form of dissolved, suspended, and insoluble solids. Collectively these are 0.1% of the total volume of wastewater. Separation of solid content reduces the suspended solid and organic load. It also maintains the homogeneity of waste stream which helps in the treatment process.

On the other hand, secondary wastewater treatment is mainly driven by biological methods such as activated sludge process (ASP) (Sonune and Ghate 2004). This process leads to the oxidation of organic compounds for their breakdown in simpler compounds. ASP efficiently removes the dissolved organic matter and converts the remaining colloidal material to a biological sludge. It has been found that efficiency of activated sludge process to reduce sludge through predation of *Aeolosoma hemprichi* was found to be about 39–65% (Muga and Mihelcic 2008). The ASP is also found beneficial for removal of total phosphorus. The effluent after this step is transferred into secondary clarifier, where biological solids or solids are settled through influence of gravity. The effluent discharge from this step is further introduced in anaerobic chamber for further reduction of organic content and nutrient from effluent. Fixed film systems involve the passage of raw wastewater onto a filter medium in which bacteria can attach, build up and accumulate in biomass which is subsequently removed (Naidoo and Olaniran 2013).

Advance Wastewater Treatment

Advance wastewater treatment incorporates the technologies used to remove nutrients, harmful substances (heavy metals), and oxygen demand and improve wastewater quality for reuse. Tertiary treatment can be defined as any treatment process in which one additional unit operation is added with conventional treatment process. It may be as simple to add filter to remove suspended solids or to add complex system to remove organic or nutrient load and harmful substances. Tertiary treatment of wastewater is generally employed to remove the nutrient load and pathogens from the wastewater. Generally, nutrients (nitrate and phosphate) remaining in effluent after secondary treatment are the major drivers of eutrophication and elevated oxygen demand (Schaar et al. 2010). Biological wastewater treatment has become popular in recent years, whereas aerobic treatment process has traditionally been employed to reduce organic load and nutrient from wastewater. But simultaneous reduction of most of the contaminants is not accomplished properly without coupling of aerobic treatment with anaerobic treatment. Microorganisms have capability to treat complex wastewater (Okoh et al. 2010). Wastewater itself contains numerous microorganisms such as bacteria (Table 10.1), protozoans, fungi, algae, helminths, etc. (Naidoo and Olaniran 2013). Algae breakdown the organic material and obtain energy and nourishment for their cell growth and reproduction. Among the different advanced wastewater treatment options, bioprocessing of wastewater using algal biomass in particular is appearing to be the cheapest due to its non-pathogenic nature and capability to grow in industrial, municipal, and agricultural effluents in fresh as well as seawater (Chinnasamy et al. 2010). In the next section, the wastewater treatment by using algal biomass is discussed due to its great degree of adaptability to the adverse environmental conditions.

Table 10.1 Assimilation of various compounds by algal biomass

	Nutrient recovery	Endocrine disruptors	Heavy metals	Oil/grease	[a](PAHs) and (PCBs)	Carbon Dioxide
Sources	Wastewater from municipal and industrial discharge, agricultural runoff, industrial exhaust, anaerobic digestion of waste	Pharmaceuticals, plasticizers, hormones, pesticides, poly-aromatic hydrocarbons	Industrial/municipal wastewater	Dairy effluent spills, mining activity	Oil/coal industry, diesel gas engine, incinerators, asphalt, production coke stove	Emissions from power plants, biomass combustion, etc.
Potential effects	Excess nitrogen leads to methemoglobinemia, and excess P leads to kidney damage in humans Eutrophication of lake due to uncontrolled algal growth	Neurological disorders, birth defects, reproductive health problems	Bioaccumulation in food chain, health impacts leads to organ dysfunctions	Lethal to aquatic wildlife, bioaccumulation issues	Carcinogenic, mutagenic and teratogenic	Leads to the phenomenon of climate change, global warming
Changes in microalgae	Enhanced biomass accumulation Change in biomass composition	Enhanced growth in cyanobacteria <100 mg/L Photosynthesis completely inhibited in marine algae	Sulfur accumulation, limits uses of algal biomass	Higher biomass production with long growth phase	Bioaccumulation and biotransformation of PAHs, PCBs accumulate in lipids	Carbon sequestration by alga lead to its high biomass productivity

[a]Polychlorinated biphenyls (PCB); poly-aromatic hydrocarbons (PAH)

Wastewater Treatment Using Algal Microbiology: Potential Approach

Algae are effective biological mediators that own potential to convert solar energy and carbon dioxide into valuable products such as biodiesel, bioethanol, and bio-butanol. Despite having higher photosynthetic efficiency (> 20%) than higher plants, global algal production is limited to only 9000 tons per year^{-1}. On the other hand, cost of algal biomass production (20–200$) is too high to use for commercial-level biofuel production (Wang et al. 2016). Conventional mode algal cultivation in freshwater is also found unsustainable in long term in view of blue water footprint. Algal cultivation is practiced for a long time in aerobic ponds or maturation ponds, which can be used in a more cost-effective manner in conjunction with wastewater treatment (Green et al. 1995). Researchers observed that algal cultivation using wastewater not only reduces the cost of algal biomass production but also provides additional advantage of simultaneous wastewater treatment. Algae have been demonstrated for the treatment of wastewater from municipal, industrial (sugar mill effluent, pulp and paper effluent, petroleum wastewater, coal-fired contaminated wastewater, textile industry wastewater, tannery industry wastewater, pharmaceutical industry, and electroplating industry), and agricultural sectors (dairy wastewater, fish farm wastewater, piggery wastewater) (Ruiz-Marin et al. 2010; Mata et al. 2010; Kong et al. 2010; Pathak et al. 2015). Industrial wastewater containing high chemical oxygen demand (COD) and nutrient load has been coupled with preliminary anaerobic treatment to avoid inhibition of algal growth.

Algae have found to assimilate a number of organic and inorganic compounds (Table 10.2); in this context a number of algal strains have been found to reduce the concentrations of nitrogen and phosphorus (2.2 mgL^{-1} and 0.15 mg L^{-1}). A number of algal strains such as *Botryococcus braunii*, *Chlorella vulgaris*, *Scenedesmus obliquus*, and *Chlorella pyrenoidosa* were found to have the potential for treatment of sewage wastewater, urban wastewater, olive oil industry wastewater, and dye wastewater, respectively (Hodaifa et al. 2008; Pathak et al. 2015). Algae have been found to reduce the organic load from industrial wastewater by 19–67% depending on the strength of wastewater (Munoz et al. 2009). Besides the removal of organic and nutrient load, algal strains such as *C. pyrenoidosa*, *C. vulgaris*, and *Oscillatoria tenuis* were found to have potential for removal of more than 30 azo compounds (Mona et al. 2011; Kousha et al. 2012; Khatee et al. 2013; El-Kassas and Mohamed 2014). It has been reported that algal growth is negatively influenced by higher dye concentration because it reduces the electron transport and metabolic rate of alga (Hamendez-Zamora et al. 2014). It was also observed that mixotrophic algal cultivation in heterotrophic mode removed biological oxygen demand (BOD) and COD by 70% and 65%, respectively (Cheng et al. 2009; Liang et al. 2009), on the part of pollutant removal.

Table 10.2 Different ways for energy generation through wastewater

Energy in wastewater	Description	Application
Bio-hydrogen/hydrogen	Bio-hydrogen can be produced by electrolysis of wastewater using conventional/nonconventional energy sources	Transportation and electricity generation
Bioelectricity	Chemical energy of wastewater is converted in to electricity through microorganisms in microbial fuel cell application (MFC)	Bioelectricity and table water
Biosolids	Biosolids present in wastewater can be incinerated to produce electricity	Electricity generation
Algae	Wastewater can be used as nutritive medium for algal cultivation, which can be used for various energy products	Produced algae can be used as a source for biofuel application
Heat	Heated water from house and industry is spent to the drain; energy can be recovered as thermal energy	Heat recovery for heating applications
Anaerobic digestion	Sludge can be digested by anaerobic microorganism and produces methane gas	Methane is used for cooking, lighting, and electricity generation

Energy Recovery with Alga-Based Wastewater Treatment: An Integrated Technology

Wastewater treatment can be an alternative option to generate energy through various conversion technologies such as microbial fuel cell, algal cultivation, biohydrogen production, etc. Extraction of energy from wastewater is easy in biological form in comparison to other modes of energy extraction. This approach develops the integrated biological wastewater treatment and simultaneous energy production as shown in Table 10.2. These energy recovery strategies could help to compensate the electricity consumption during the treatment of wastewater and represent effective policy implication to develop sustainable energy planning.

Biogas

Estimated energy recovery from all wastewater treatment plants in the USA for biogas was about 2401 million m^3/day through anaerobic digestion (Stillwell et al. 2011). Champagne and Anderson (2015) used municipal wastewater containing high concentration of fat, oil, and grease as a biogas feedstock. They reported enhanced biogas production (25.14 ± 2.14 L/d) including increased COD solubilization and volatile fatty acid (VFA) concentrations in two-stage thermophilic semicontinuous flow digestion system. Molinuevo-Salces et al. (2010) investigated indirect application of wastewater for generation of biogas feedstock. It was observed that feedstock cultivation in wastewater is a promising strategy to produce renewable energy by coupling wastewater bioremediation and biomass production for lipid (Kothari et al. 2017c).

Bioelectricity

Domestic wastewater was found to produce about 146 mW/m^2 of bioelectricity (Hong and Logan 2004). Significant relation was found between COD of wastewater and maximum power density. In this context, a maximum loading of 3500 mg/L of COD was found to produce a maximum power density of around 37.4 m W/m^2 (Zhu et al. 2011). Similarly, higher extent of bioelectricity was produced from wastewater by using microorganisms (Rahimnejad et al. 2011). Industrial wastewater was found more feasible to produce bioelectricity in comparison to other wastewater. The maximum current intensity and power density were 3.74 mA and 621.13 mW/m^2, respectively, on the anode surface at organic loading rate (OLR) equal to 53.22 kg COD/m^3 d. Thus, microbial fuel cell coupled with wastewater treatment is a possible and economic way to achieve the goal of wastewater treatment and bioenergy production (Mansoorian et al. 2016).

Bio-hydrogen

Wastewater can be used as a substrate to produce feedstock that can be used for biohydrogen production through different routes such as photolysis, indirect biohydrogen production, fermentation, etc. In this context, the cyanobacterium *Anabaena variabilis* was found to produce about 12.5 ml H_2/h (Markov et al. 1997). In photolysis, processed electrons are released from water due to the photochemical oxidation by PS II system containing plastoquinone oxidoreductase enzyme. These electrons are then energized by photon from sunlight and transferred to the hydrocarbon cluster of [Fe]-hydrogenase through the thylakoid electron transport chain in PS I. Plastoquinone is reduced in to plastoquinol through these electrons, which reduces $NADP^+$ to NADPH. Energized electrons are replaced by oxidizing water to form hydrogen ion and oxygen molecule (Kothari et al. 2017b). These electrons are used in photosynthesis process via PS II system. Protons generated through oxidation of water result in proton gradient, which is used to generate ATP in the presence of ATP synthase enzyme. Protons are terminal acceptors of these photosynthetically generated electrons in the algal chloroplast. The whole process results in the simultaneous production of hydrogen and oxygen. Direct photolysis enables the algae and cyanobacteria to split water directly into oxygen and hydrogen

(Florin et al. 2001). Algae are suggested to have potential to harness solar energy by extracting protons and electrons by water splitting reactions (Kothari et al. 2015).

Biodiesel

Algae have attracted a wide attention of researchers to be used as a biofuel feedstock due to its efficiency to grow in minimal freshwater input in comparison to other known energy crops. Algae can grow well on several types of less desirable sources of water including domestic and industrial wastewaters, saline and brackish water, sea water, etc., because of which they are considered as approaches that will not threaten food security of the region. Oleaginous algae from industrial and municipal wastewater were isolated and processed for biodiesel production (Sheehan et al. 1998). Various researchers (Chisti 2007; Converti et al. 2009; Chinnasamy et al. 2010) have investigated the cultivation of algal biomass in saline, brackish, or wastewater for production of biodiesel. Wastewater containing higher concentrations of ammonia was found as potential substrate to support the growth of *C. vulgaris*, which accumulated at approximate concentration of about 23.3 $mgL^{-1}d^{-1}$ lipid at 39 mgL^{-1} NH_3 concentration (He et al. 2013). Wastewater contains a number of bacterial populations, which affect algal growth positively (Ma et al. 2014). Algal biomass cultivated in high raceway algal pond accumulated 70% fatty acid content, which is significantly high to produce biodiesel (Drira et al. 2016). *C. vulgaris* under polluted environment was found to produce about 0.41 g biomass, and 0.15 g L^{-1} of crude oil was obtained (Kalhor et al. 2016).

Algae-Based Biofuel Production Processes

Biofuel production with the algal feedstock involves a number of steps like isolation of algal strain, its screening and then the selection of appropriate strain, an apt selection of culture media for its proper growth, the technological aspects involved in algal cultivation, and then its biomass harvesting, lipid extraction, and transesterification. Among the various upstream processes of algal biofuel production, isolation of algal strain is one of the decisive stages as it ensures the further accessibility of that particular algal strain (Osundeko et al. 2013; Talebi et al. 2015). To succeed in getting lipid-rich algal strain, various efficacious techniques are still under exploration, yet each technique has its own pros and cons (Table 10.3). There exist a number of research institutions as well as collection centers both at national and international levels that make available this highly productive cultured algal biomass. Naming a few is the National Collection of Industrial Microorganisms (NCIM), Pune, India; University of Texas (UTEX), USA; Australian National Algae Culture Collection (ANACC), Australia; Culture Collection of Algae and Protozoa (CCAP), UK; Nippon Institute For Biological Science (NIBS), Japan; Sammlung

Table 10.3 Relative advantages and disadvantages of cultured and indigenous algal strains

Strain type	Advantages	Disadvantages
Cultured strain	Strains are recognized and specific to the demand for high lipid production and value-added compounds	Limited number of species, which survive under specific conditions High cost of development
Indigenous strain	Vast diversity of strains Adapted to the local environmental conditions No charge for procurement	It involves cost and time for isolation and screening practices Optimal culture conditions are difficult to achieve in field conditions

von Algenkulturen der Universität Göttingen (SAG), Germany; etc. Regardless of the fact that these cultured algal strains are readily available, various researchers and industrial stakeholders are still involved in the exploration of different native strains that have the potential of biofuel production and adaptation in local environment (Lynch et al. 2015).

Specifying an algal strain for its biofuel production involves a number of factors like the state of its aquatic environment and augmented conditions of the abiotic factors like pH, temperature, nutrient concentration, and various other physicochemical dynamics that provide the optimal environmental conditions to boost the biomass growth and act as a nutrient medium for algal cultures.

Cultivation of algal biomass is another important step and performed under the photoautotrophic, heterotrophic, mixotrophic, and heteroautotrophic mode. Usually two types of cultivation systems are used for scale up of algal biomass such as open and close systems (Shamshad et al. 2017). The most frequent design of pond used for cultivation includes shallow pond, raceway pond, tank, and circular ponds. Raceway ponds generally contain an oval-shaped shallow pond lined with photovoltaics (PV) cement or clay within an area of 1–200 ha (Anderson et al. 2005). Cultivation of algae in open pond is similar to natural method of growing algae (Chisti 2007). Raceway ponds involve low capital investment and operational cost with effective mixing (via paddle wheel) of nutrients and gaseous exchange. Outdoor cultivation has various drawbacks such as poor light penetration, loss of water due to evaporation, diffusion of carbon dioxide from the atmosphere, requirement of large area, contamination, temperature regulation, poor mass transfer and uneven light intensity, etc. (Harun et al. 2010). Despite of having such drawbacks, open pond cultivation is the most industrially applied technology due its low-cost input in operational process (Pandey et al. 2014). Recently it has been demonstrated that cultivation of *C. pyrenoidosa* in outdoor open raceway pond using domestic wastewater as a nutrient medium under arid climate has great potential for algal-based wastewater treatment (Dahmani et al. 2016).

Close cultivation system provides the facility to grow algae under controlled conditions, which is also referred as photobioreactors (PBRs). One of the important criteria for PBR design is high mass transfer, especially for sequestration of carbon dioxide (Ugwu et al. 2008). Agitation in PBR is provided either mechanically or nonmechanically. Nonmechanical agitation can be seen in the case of airlift, bubble column, tubular reactor, and flat panel operation. PBR can be operated in continuous as well as in batch modes. Comparative to open system, close system is easy to control with regard to environmental parameters and able to produce high biomass productivity (Sierra et al. 2008). Photobioreactor facilitates better control of culture environment such as supply of carbon dioxide, water supply, optimal temperature, efficient exposure to light, density of culture, pH levels, gas supply rate, effective mixing, etc. (Mata et al. 2012).

Integration of open pond and closed system is termed as hybrid culture system; this combination is used to gain greater results. Open ponds are economical, and they have advantage of low operational cost, but they tend to get contaminated by pathogenic microorganisms; in order to get the culture free from contamination, an integration of closed system is better. The initial inoculum size should be large enough to cultivate the desired algal species in open system before its contamination. To avoid the contamination, cleaning and flushing the pond should be part of the algal culture. It involves two-stage cultivation to integrate the distinct growth stages in the photobioreactor and open pond. The first stage in photobioreactor involves controlled condition, whereas the second production stage is in open pond with nutrient stress, which enhances the synthesis of lipid productivity (Victoria et al. 2014).

Though progress has been achieved in photobioreactors for mass cultivation of alga, further development is still required for improvement, especially regarding the cost reduction of bioreactor design. Large outdoor cultivation system requires large land area, which is a critical issue in densely populated countries like India and China. Thus, low-cost closed photobioreactors can greatly influence the autotrophic algal cultivation. It is noted by various researchers that biomass yield is higher in closed PBRs in comparison to open raceway ponds and this is due to controlled nutritional mechanism and single specific cultures in photobioreactors (Kasiri et al. 2015).

Factors Affecting Biomass and Lipid Productivity

Algal biomass can be converted into various bioenergy products depending on the conversion routes and quantity of targeted biochemical constituents. According to research findings, lipid/oil composition of the cell is used as a raw material to produce biodiesel through transesterification process. Biochemical composition of various microalgal strains in freshwater or marine habitats, observed by various investigators under optimized and varying environmental conditions is mentioned in Table 10.4.

Starch and lipid are the main storage carbons in the alga required for the survival under unfavorable conditions. Lipid content in alga increases where there is change of environment and microclimate between optimal and suboptimal conditions. During the nitrogen starvation, algal cells stop dividing and accumulate the storage product in non-nitrogen-limited cell, thus resulting in the double or triple

Table 10.4 Biochemical composition of different algae on dry matter basis

Microalgal species (marine and freshwater)	Lipid content (% dry weight biomass)	Lipid productivity (mg/L/day)	Protein % dry weight	Carbohydrate content % dry matter basis
Ankistrodesmus sp.	24.0–31.0	–	–	–
Botryococcus braunii	25.0–75.0	–	–	–
Chaetoceros muelleri	33.6	21.8	–	–
Chaetoceros calcitrans	14.6–16.4/39.8	17.6	–	–
Chlorella emersonii	25.0–63.0	10.3–50.0	–	–
Chlorella protothecoides	14.6–57.8	1214	–	–
Chlamydomonas reinhardtii	21	–	48	17
Chlorella sorokiniana	19.0–22.0	44.7	–	–
Chlorella vulgaris	5.0–58.0	11.2–40.0	51–58	12–17
Chlorella sp.	10.0–48.0	42.1	–	–
Chlorella	18.0–57.0	18.7	–	–
Chlorococcum sp.	19.3	53.7	–	–
Crypthecodinium cohnii	20.0–51.1	–	–	–
Dunaliella salina	6.0–25.0	116.0	57	32
Dunaliella primolecta	23.1	–	–	–
Dunaliella tertiolecta	16.7–71.0	–	–	–
Dunaliella sp.	17.5–67.0	33.5	–	–
Ellipsoidion sp.	27.4	47.3	–	–
Euglena gracilis	14.0–20.0	–	39–61	14–18
Haematococcus pluvialis	25.0	–	–	–
Isochrysis galbana	7.0–40.0	–	–	–
Isochrysis sp.	7.1–33	37.8	–	–
Monodus subterraneus	16.0	30.4	–	–
Monallanthus salina	20.0–22.0	–	–	–
Nannochloris sp.	20.0–56.0	60.9–76.5	–	–
Neochloris oleoabundans	29.0–65.0	90.0–134.0	3	2
Nitzschia sp.	16.0–47.0	–	–	–
Nostoc sp.	18.3	–	56.2	25.5
Nodularia sp.	24.5	–	58.2	17.3
Oscillatoria sp.	11.6	–	64.6	23.9
Oocystis pusilla	10.5	–	–	–
Pavlova salina	30.9	49.4	–	–
Pavlova lutheri	35.5	40.2	–	–
Phaeodactylum tricornutum	18.0–57.0	44.8	2	2
Prymnesium parvum	22–38	–	28–45	25–33
Scenedesmus dimorphus	16–40	–	8–18	21–52
Scenedesmus sp.	19.6–21.1	40.8–53.9	–	–
Skeletonema sp.	13.3–31.8	27.3	–	–

Source: Mata et al. (2010, 2012)

lipid content by dry weight. Microalgal lipids include neutral lipid, polar lipids, wax esters, sterols, and hydrocarbons; prenyl derivatives such as tocopherols, carotenoids, terpenes, and quinines; and pyrrole derivatives such as chlorophyll. In general, lipid in algae can be grouped in two classes, viz., storage lipids (nonpolar lipid) and structural lipids (polar lipid). Triacylglyceride (TAG) is the main form of storage lipid, which is made of mostly saturated fatty acids and some unsaturated fatty acids.

Photosynthesis provides base for lipid synthesis by using carbon as a source, and through Calvin cycle, sugars are produced which are converted into sucrose and starch. Lipid is synthesized through glycolysis pathway using the stored carbon sources (Levine et al. 2010). Light, CO_2, temperature, etc. affect the photosynthesis and also affect the lipid synthesis in alga (Pandey et al. 2014). Most of the algae which have high lipid content are inefficient due to their slow growth rate, so by varying the medium composition and external factors, lipid productivity of alga and growth can be simultaneously increased (Fakhry and El Maghraby 2015). There are mainly two ways to increase the lipid productivity in a microalgal cell, viz., (i) increasing the rate of biomass of accumulation and (ii) increasing the proportion of biomass that contain useful lipid.

Light and Temperature

Light and temperature are essential external factors, which affect the growth of autotrophic algae as well as its photosynthetic activity. Chl "a" and Chl "b" are considered as major light-harvesting pigments. Temperature affects the algal metabolic process and thus causes changes in cellular composition, alternation of CO_2 uptake, etc. Each species of algae grow well under the optimal conditions of light intensity as well as accumulate lipid, which is considered as precursor for biodiesel production. *Chlorella minutissima* was observed for its growth at a temperature of 35 °C and irradiance of 30–550 $\mu molm^{-2}\ s^{-1}$. Minimum irradiance was found to sustain net growth rate. About 81.8% increment in growth rate was observed as the temperature increased from 10 °C to 30 °C. Similarly, *Chlorella pyrenoidosa* was found to accumulate maximum biomass and lipid at 30 °C. Photosynthetic activity directly depends on the light availability. Lipid content and its composition directly depend on the quantity and quality of light, and highest fatty acid content was found at light above the saturating intensity (Wagenen et al. 2012). Impact of exposure time of light on *Dunaliella tertiolecta* was observed, and a photoperiod of 24 h was found to produce highest biomass (Tang et al. 2011). Temperature as a strong influencing parameter was found to affect lipid productivity in *Nannochloropsis oculata* and *C. vulgaris*. Lipid content of algae was strongly influenced by temperature within a range of 20–25 °C, and increase in lipid content from 7.90 to 14.92% in *N. oculata* was observed. However, Converti et al. (2009) observed decrease in lipid content (14.71–5.91%) in *C. vulgaris*.

pH and CO_2

Variation in pH has significant impact on algal growth. It not only affects the growth but also distribution of carbon dioxide and carbon availability and alters the availability of trace metal and nutrients. A high pH directly affects the physiological process, and it not only affects the freshwater algae but also the marine algae, which have strong carbonated buffering system. Increase in pH from 7.5 to 10 was found to affect the structural lipids in diatoms and reduce the total lipid content (Spilling et al. 2013).

CO_2 is a very important factor for photosynthesis of algal cells; 1–5% by volume of CO_2 concentration can result in the maximum growth. CO_2 supply through glycerol and sodium acetate induced the lipid accumulation in *C. vulgaris* and produces 80% higher biomass than the control (Estévez-Landazábal et al. 2015). Ponnuswamy et al. (2014) studied the lipid productivity of *Chlorella* sp. under sodium bicarbonate concentration ranging from 50 to 60 mgL^{-1}. They achieved a maximum lipid content of 20% at carbon dioxide concentration of 40%. Lin and Wu (2015) observed the lipid productivity of *Chlorella* sp. under varying carbon sources. They achieved maximum lipid content of 35.5% under mixotrophic cultivation mode with sucrose as a carbon source.

Nutrient Starvation

Nutrients play significant role in the growth of algae as well as lipid production and fatty acid composition. Under stress conditions, cell division declines, and active biosynthesis of fatty acid begins in some algae. During slow growth of algal cell, when there is no requirement to form new membrane, the deposited fatty acid converted into the TAG as a result of protective mechanism. It has been observed that lipid productivity in *C. vulgaris* was 77.8 $mgL^{-1}D^{-1}$ in nutrient medium supplemented with no nitrogen and 2 mgL^{-1} of phosphorus (Mujtaba et al. 2012). Nitrogen plays significant role in the lipid metabolism and starvation of nitrogen induced higher lipid productivity in algal cell (Kumar et al. 2012). In another observation, *C. vulgaris* was found to accumulate about 40.7% lipid under the nitrogen deprivation condition.

Thus, physical factors (light, temperature) and composition of nutrient medium affect the biomass productivity of algae. Lipid stimulation through varying nutrient concentrations is found significant within a certain range; however, overall biomass yield is found decreased. Thus, an optimized method to obtain high lipid production with high biomass productivity is still under research and development.

Harvesting of Algal Biomass

Collection of algal biomass is another bottleneck for industrial biofuel production process. This process is considered as one of the main steps of downstream processing to achieve the solid content from >1.0% to up to 20% of solid, and it costs about 20–30% of the total cost in biofuel production (Mata et al. 2010). For the downstream processing of algal biomass for biodiesel production, various processes like centrifugation, filtration, and flocculation have been introduced by the researchers (Mata et al. 2010; Zhang et al. 2016). Centrifugation of the algal biomass is exorbitant and requires an excessive amount of energy during the process. A number of researchers have noticed that centrifugation of dilute culture or small size of algae must be managed with large volumes. Hence, it becomes quite difficult to harvest the algal biomass for biodiesel production through the process of centrifugation (Girma et al. 2003; Milledge and Heaven 2013; Zhang et al. 2016). In the case of sedimentation and filtration, colloidal stability of algae limits the separation efficiency. Thus, flocculation of algal biomass is suggested as more feasible way to harvest the algal biomass (Mata et al. 2010).

Chemical-mediated harvesting of algal biomass such as calcium hydroxide ($Ca(OH)_2$) is also found efficient (99.5%) for high extent of precipitation (Zhang et al. 2016). pH acts as an influencing factor in chemical-mediated coagulation process, and a pH range between 10.8 and 11.8 was found critical in coagulation process. Another innovative way for chemical-mediated algal biomass harvesting was introduced by Salama et al. (2015) by employing acid mine drainage (AMD) for coagulation of microalgal biomass. They achieved 89% flocculation efficiency for *S. obliquus* and 93% for *C. vulgaris*. The AMD was observed as a cost-effective method, which might further allow reuse of flocculated medium for algal cultivation. Chemical-mediated algal harvesting process is effective but too expensive to use in large-scale operations. Alternations in culture medium such as pH and interrupting supply of carbon dioxide can cause flocculation of algal biomass by its own, which is termed as "auto-flocculation." A change in pH from 8.5 to 12.0 was found more efficient for harvesting of *Chlorococcum* sp. R-AP13 (Ummalyma et al. 2016). Bioflocculation of algal biomass using bioflocculant is another innovative approach for harvesting of algal biomass (Kothari et al. 2017a). Researchers have successfully investigated the potential of chitosan as a bioflocculant for *Spirulina*, *Oscillatoria*, *Chlorella*, *Synechocystis*, *Tetraselmis chui*, *Thalassiosira pseudonana*, and *Isochrysis* sp. (Divakaran and Pillai 2002; Heasman et al. 2000). Oh et al. (2001) also studied microbial flocculants (AM49) for the harvesting of *C. vulgaris*. This flocculant was found to be better than other commonly used flocculants. Recovery of more than 83% solids when operating in the pH range 5–11 was recorded; this is higher than that when using aluminum sulfate (72%) or the cationic polymer polyacrylamide (78%).The first evidence on bioflocculation of algal biomass using *Shinella albus* (bacteria) for harvesting of *C. vulgaris* was studied by Yi et al. (2016). According to Kothari et al. (2017a, b, c), bioflocculant was stable with thermal and pH change, and its efficiency in association of Ca^{2+} was highest for harvesting of algal biomass. According to literature, harvesting and dewatering processes are performed in two categories. First category involves the algal culture, while second category involves the conversion of algal biomasses into macroscopic masses by the process of agglomeration to facilitate the dewatering process. The former one is of low cost compared to the second category due to the input of chemicals.

Extraction of Lipid from Algae

Organic solvents are used in chemical processes for recovery of lipid from microalgal cells involving hexane soxhlet extraction (Soxhlet 1879) and mixed methanol-chloroform (2:1 V/V) (Bligh and Dyer 1959). In solvent extraction process, the solvents (hexane, chloroform, and methanol) being toxic in nature pose harm to the environment, and it is also a high-energy intensive process. Organic solvents have potential to penetrate the cell membrane and dissolve the biological matrix of the cell (Pathak et al. 2016). Organic solvents such as methanol and isopropyl alcohol have potential to break this hydrogen bond linkage and extract neutral lipids from the cell. The solvent lipid complex is transported in solvent medium through the diffusion process by concentration gradient. Thus, a suitable solvent should possess characteristics like (1) high affinity to oily composition, (2) low boiling point and stability at normal temperature, and (3) non-toxic and biodegradability. The European Food Safety authority has approved dimethyl ether (DME) as a safe organic solvent. DME showed better yield of lipid content (40.1% of the dry weight) when applied on *Microcystis aeruginosa* (Kanda and Li 2011). It was demonstrated that DME method has potential for extraction of lipid directly from wet algal biomass.

Most of the researchers used both polar and nonpolar solvents for complete extraction of lipid chloroform, and methanol mixture with various cell disruption methods (autoclaving, bead beating, microwaves, sonication, and a 10% NaCl solution) was used to assess their relative potential; however, microwave oven method was observed as the most simple, easy, and effective process of lipid extraction from algae with a lipid productivity of 5.7 $mgL^{-1}d^{-1}$. Mixture of hexane/methanol/acetone was used as organic solvent to extract the lipid from the activated sludge, and 27.43 ± 0.98% of lipid content was obtained with three-time extraction (Dufreche et al. 2007). The addition of isopropanol as a cosolvent in the hexane solvent system showed improvement in the total lipid yield by more than 300% (Halim et al. 2012).

Supercritical fluid (CO_2) extraction (SFE) method is also explored by researchers (Mendes et al. 2006; Cho et al. 2011) to extract the lipid from algal biomass; however, most of them achieved lower lipid yield in comparison to the extraction performed by mixture of chloroform, methanol and water (Bligh and Dyer method), acetone, and ethanol. Hence, modified supercritical CO_2 with cosolvent is also studied for recovery of lipid and carotenoid from microalgal species of *Nannochloropsis oculata* (Cho et al. 2011). It has been observed that addition ratio of ethanol and supercritical fluid was an important factor to increase the extraction efficiency for lipid.

Transesterification

Transesterification refers to the chemical way to convert the fatty acid into fatty acid methyl esters (FAME), which are termed as biodiesel. This process reduces the viscosity of FAME (Knothe et al. 1997). Initially transesterification process was carried out by employing the base or acid catalysis, but now researchers have explored other potential ways such as direct methanolysis, enzymatic transesterification, and microwave-assisted transesterification (Ehimen et al. 2010; Lee et al. 2012). Among conventional transesterification processes, base-catalyzed transesterification of algal lipids is considered as more frequent conversion process in comparison to acid-catalyzed transesterification process and mainly used in industrial-level biodiesel production (Meher et al. 2006; Demirbas 2000). Acid-catalyzed transesterification has relatively lower conversion efficiency; however, it is feasible and less affected by presence of free fatty acid (FFA) (Helwani et al. 2009). In contrast to homogenous catalyst, heterogeneous catalyst can be separated by simple filtration and save water during purification of biodiesel. Alkaline catalyst is used at an atmospheric pressure with temperature ranges 60–70 °C (Srivastava and Prasad 2000). The main challenge with base-catalyzed reaction is the formation of soap at high free fatty acid content; hence prior removal of water and free fatty acid is essential for this reaction (Demirbas 2000).

On the other hand, conventional transesterification processes are energy intensive and require longer time to complete the reaction (Mohan et al. 2015). Advancement in transesterification process has introduced the direct methanolysis process, which possesses potential to eliminate drawbacks associated with conventional process. Various researchers (Patil et al. 2009; Sathish and Sins 2012; Ehimen et al. 2010) have proven the potential of direct methanolysis for biodiesel production using wet algal biomass (90% water). Direct methanolysis also reduces the amount of solvent and reaction time and improves the biodiesel yield. The feasibility of the microwave-assisted transesterification process was found efficient and easier than the chemical-mediated transesterification using algal oil obtained from *Botryococcus* sp., *C. vulgaris*, and *Scenedesmus* sp. (Lee et al. 2012). It was found that microwave-enhanced methanolysis process, improved the extraction of algae significantly with a high efficiency, reduced extraction and transesterification time, and increased yield. Despite the advantages of transesterification, most of the industries are still using base-catalyzed transesterification, which is due to various technology barriers such as lack of operational viability and high cost of operation and recovery option of catalyst. Therefore, studies are required to prove the sustainability of the advanced method of transesterification process so that it could be implemented at industrial-level.

Future Prospects

Algal-based wastewater treatment offers cost-effective and efficient treatment technology as microalgae biomass is the only renewable resource that possesses high uptake capacity of pollutants and generates oxygen that triggers the aerobic oxidation of organic compounds. Despite the fact that algal cultivation for biofuel industry alone involves huge expenditure, integration of the wastewater from different sources, to be used as nutrient source for algal cultivation can be very effective. This integrated aspect can be more sustainable in function than the conventional process. Wastewater with lower nutrient concentration can be a limiting factor for microalgal growth such as secondary-/tertiary-treated wastewater, municipal wastewater, etc.; supplementation of nutrient to promote the algal growth can be used as an alternative method; however, it adds cost to microalgal cultivation. Waste material (whey permeate, cattle slurry, agricultural waste, food waste) rich in nutrient can be used as alternative to the chemical nutrient to support the algal growth in low nutrient wastewater. However, these substrates require pretreatment to release the nutrient as well as possible contaminants. Emerging organic pollutants (EOCs) such as personal care products, detergents, flame retardant, and surfactants are another challenge for algal-based wastewater treatment system as very little literature is available on removing of EOCs from wastewater.

Selection of suitable strain with higher tolerance to varying culture conditions is another important factor in wastewater-based microalgal growth. Algal tolerance toward the wastewater toxicity can be increased by genetic engineering. In this context, Halder (2014) proposed that transgenic microalgal cells with higher metal binding capacity could be produced by overexpressing the gene related to metal binding proteins. The unknown consequences of released genetic engineered microorganism in the environment are still a challenge for developing genetic engineered microalgae for wastewater treatment.

Recently, Elrayies (2018) explored the prospects of microalgae for greener future buildings, which is quite modest, and the technology is still in initial phase. Integration of photo-bioreactor (PBR) in the building's facades can be used to control the thermal heating of the building. PBR facades prevent the building from direct heating of sun and create sun shading for the building's interior and keeping it cooler during intense sunny days. In addition to the shading effect, PBR filters the thermal load and protects the building.

Conclusion

It is indicated that species, substrate, reactor type, nutrient concentration, temperature, pH, and light are the main influencing factors for biodiesel production from algae. The biodiesel production from algae cultivated on industrial wastewater is a cost-effective technology in biofuel sector. The biggest challenge over the next few years in algal biodiesel production will be to reduce the production cost and selection or development of suitable strains with high stress tolerance capacity and high lipid productivity. Advancement in the downstream process steps is urgently required to develop a significant methodology to explore maximum yield of algal biofuel. It can be concluded that many challenges are in the way of processing and harvesting to commercialize this integrated approach, and it demands proper research to fill the lacunae in the way of algal biodiesel production, a part of green fuel economy.

References

Anderson, J. L., Peterson, R. C., & Swainson, I. P. (2005). Combined neutron powder and X-ray single-crystal diffraction refinement of the atomic structure and hydrogen bonding of goslarite ($ZnSO_4.7H_2O$). *Mineralogical Magazine, 69*, 259–271.

Baccella, S., Cerichelli, G., Chiarini, M., Ercole, C., Fantauzzi, E., Lepidi, A., Toro, L., & Veglio, F. (2000). Biological treatment of alkaline industrial waste waters. *Process Biochemistry, 35*, 595–602.

Bligh, E. G., & Dyer, W. J. (1959). A rapid method of total lipid extraction and purification. *Canadian Journal of Biochemistry and Physiology, 37*(8), 911–917.

Bourgeous, K. N., Darby, J. L., & Tchobanoglous, G. (2001). Ultrafiltration of wastewater effects of particles, mode of operation, and backwash effectiveness. *Water Research, 35*(1), 77–90.

Caputo, A. C., & Pelagagge, P. M. (2001). Waste-to-energy plant for paper industry sludges disposal: Technical-economic study. *Journal of Hazardous Materials, 81*(3), 265–283.

Champagne, P., & Anderson, B. C. (2015). Enhanced biogas production from anaerobic co-digestion of municipal wastewater treatment sludge and fat, oil and grease (FOG) by a modified two-stage thermophilic digester system with selected thermo-chemical pre-treatment. *Renewable Energy, 83*, 474–482.

Cheng, Y., Lu, Y., Gao, C., & Wu, Q. (2009). Alga-based biodiesel production and optimization using sugar cane as the feedstock. *Energy & Fuels, 23*, 4166–4173.

Chinnasamy, S., Bhatnagar, A., Claxton, R., & Das, K. C. (2010). Biomass and bioenergy production potential of algae consortium in open and closed bioreactors using untreated carpet industry effluent as growth medium. *Bioresource Technology, 101*, 6751–6760.

Chisti, Y. (2007). Biodiesel from algae. *Biotechnology Advances, 25*, 294–306.

Cho, Y. C., Cheng, J. H., Hsu, S. L., Hong, S. E., Lee, T. M., & Chang, C. M. J. (2011). Supercritical carbon dioxide anti-solvent precipitation of anti-oxidative zeaxanthin highly recovered by elution chromatography from *Nannochloropsis oculata*. *Separation and Purification Technology, 78*(3), 274–280.

Converti, A., Casazza, A. A., Ortiz, E. Y., Perego, P., & Del Borghi, M. (2009). Effect of temperature and nitrogen concentration on the growth and lipid content of *Nannochloropsis oculata* and *Chlorella vulgaris* for biodiesel production. *Chemical Engineering and Processing, 48*, 1146–1151.

Dahmani, S., Zerrouki, D., Ramanna, L., Rawat, I., & Bux, F. (2016). Cultivation of *Chlorella pyrenoidosa* in outdoor open raceway pond using domestic wastewater as medium in arid desert region. *Bioresource Technology, 219*, 749–752.

Demirbas, A. (2000). Progress and recent trends in biodiesel fuels. *Energy Conversion and Management, 50*, 14–34.

Divakaran, R., & Pillai, V. N. S. (2002). Flocculation of algae using chitosan. *Journal of Applied Phycology, 14*, 419–422.

Drira, N., Piras, A., Rosa, A., Porcedda, S., & Dhaouadi, H. (2016). Algae from domestic wastewater facility's high rate algal pond: Lipids extraction, characterization and biodiesel production. *Bioresource Technology, 206*, 239–244.

Dufreche, S. R., Hernandez, T., French, D., Sparks, M., & Zappi, E. (2007). Extraction of lipids from municipal wastewater plant microorganisms for production of biodiesel. *Journal of the American Oil Chemists' Society, 84*, 181–187.

Ehimen, E. A., Sun, Z. F., & Carrington, C. G. (2010). Variables affecting the in situ transesterification of algae lipids. *Fuel, 89*(3), 677–684.

El-Kassas, H. Y., & Mohamed, L. A. (2014). Bioremediation of textile waste effluent by *Chlorella vulgaris*. *Egyptian Journel of Aquatic Research, 40*(3), 301–208.

Elrayies, G. M. (2018). Microalgae: Prospects for greener future buildings. *Renewable and Sustainable Energy Reviews, 81*, 1175–1191.

Estévez-Landazábal, L. L., Barajas-Solano, A. F., Barajas-Ferreira, C., & Kafarov, V. (2015). Improvement of lipid productivity on *Chlorella vulgaris* using waste glycerol and sodium acetate. *Ciencia Tecnologíay Futuro, 5*(2), 113–126.

Fakhry, E. M., & El Maghraby, D. M. (2015). Lipid accumulation in response to nitrogen limitation and variation of temperature in *Nannochloropsis salina*. *Botanical Studies, 56*, 6.

Florin, L., Tsokoglou, A., & Happe, T. A. (2001). Novel type of Fe-hydrogenase in the green alga Scenedesmus obliquus is linked to the photosynthetical electron transport chain. *The Journal of Biological Chemistry, 276*, 6125–6132.

Girma, E., Belarbi, E. H., Fernandez, G. A., Medina, A. R., & Chisti, Y. (2003). Recovery of microalgal biomass and metabolites: Process options and economics. *Biotech Advances, 20*, 491–515.

Green, B., Lundquist, T., & Oswald, W. J. (1995). Energetics of advanced integrated wastewater pond systems. *Water Science and Technology, 31*(12), 9–20.

Halder, S. (2014). Bioremediation of heavy metals through freshwater microalgae: A review. *Scholars Academic Journal of Biosciences, 2*, 825–830.

Halim, R., Danquah, M. K., & Webley, P. A. (2012). Extraction of oil from algae for biodiesel production: A review. *Biotechnology Advances, 30*, 709–732.

Hamendez-Zamora, M., Perales-Vela, H. V., Flores-Ortiz, C. M., & Canizares-Villanueva, R. O. (2014). Physiological and biochemical responses of *Chlorella vulgaris* to Congo red. *Ecotoxicology and Environmental Safety, 129*, 189–198.

Harun, R., Davidson, M., Doyle, M., Gopiraj, R., Danquah, M., & Forde, G. (2010). Techno-economic analysis of an integrated algae photobioreactor, biodiesel and biogas production facility. *Biomass and Bioenergy, 35*(1), 741–747.

He, P. J., Mao, B., Shen, C. M., Shao, L. M., Lee, D. J., & Chang, J. S. (2013). Cultivation of Chlorella vulgaris on wastewater containing high levels of ammonia for biodiesel production. *Bioresource Technology, 129*, 177–181.

Heasman, M., Diemar, J., O'Connor, W., Sushames, T., & Foulkes, L. (2000). Development of extended shelf-life algae concentrate diets harvested by centrifugation for bivalve molluscs—A summary. *Aquaculture Research, 31*, 637–659.

Helwani, Z., Othman, M. R., Aziz, N., Fernando, W. J. N., & Kim, J. (2009). Technologies for production of biodiesel focusing on green catalytic techniques: A review. *Fuel Processing Technology, 90*, 1502–1514.

Hodaifa, G., Martinez, M. E., & Sanchez, S. (2008). Use of industrial wastewater from olive oil extraction for biomass production of *Scenedesmus obliquus*. *Bioresource Technology, 99*, 1111–1117.

Hong, L., & Logan, B. (2004). Electricity generation using air cathode single chamber microbial fuel cell in the presence and absence of proton exchange membrane. *Environmental Science & Technology, 38*, 4040–4046.

http://mnre.gov.in/scheme/offgrid/wasteto energy posted 16.10.2014.

Indian Energy Outlook. (2015). *International Energy Agency*. Available at: https://www.iea.org/publications/freepublications/publication/IndiaEnergyOutlook_WEO2015.pdf

Kalhor, A. X., Dabbagh, A., Mohammadi, N., Ehsan, A., Bahrami, A., & Movafeghi, A. (2016). Biodiesel production in crude oil contaminated environment using *Chlorella vulgaris*. *Bioresource Technology, 222*, 190–194.

Kanda, H., & Li, P. (2011). *Simple extraction method of green crude from natural blue green algae by di methyl ether: Extraction efficiency on several species compared to the Bligh –Dyer's method*. Sweden: World Renewable Energy Congress.

Kasiri, S., Ulrich, A., & Prasad, V. (2015). Optimization of CO_2 fixation by Chlorella kessleri cultivated in closed raceway photobioreactor. *Bioresource Technology, 194*, 144–155.

Khatee, A. R., Vafaei, F., & Jannatkhah. (2013). Biosorption of three textile dyes from contaminated water by filamentous green alga *Spirogyra* sp. kinetic isotherm and thermodynamic studies. *International Biodeterioration and Biodegradation, 83*, 33–40.

Kisku, G. C., Barman, S. C., & Bhargava, S. K. (2000). Contamination of soil and plants potentially toxic elements irrigated with mixed industrial effluent and impact in the environment. *Water, Air, and Soil Pollution, 120*, 121–137.

Knothe, G., Bagby, M. O., & Ryan, T. A. (1997). Cetane numbers of fatty compounds: Influence of compound structure and of various potential cetane improvers. *SAE Technical Paper, 971681*, 127–132.

Kong, Q., Li, L., Martinez, B., Chen, P., & Ruan, R. (2010). Culture of algae *Chlamydomonas reinhardtii* in wastewater for biomass feedstock production. *Applied Biochemistry and Biotechnology, 160*, 9–18.

Kothari, R., Pathak, V. V., Chopra, A. K., Ahmad, S., Allen, T., & Yadav, B. C. (2015). Developments in bioenergy and sustainable agriculture sectors for climate change mitigation in Indian context: A state of art. *Climate Change and Environment Sustainability, 3*(2), 93–103.

Kothari, R., Kumar, V., Pathak, V. V., Ahmad, S., Aoyi, O., & Tyagi, V. V. (2017a). A critical review on factors influencing fermentative hydrogen production. *Frontier Bioscience (Landmark Edition), 22*, 1195.

Kothari, R., Pathak, V. V., Pandey, A., Ahmad, S., Srivastava, C., & Tyagi, V. V. (2017b). A novel method to harvest *Chlorella* sp. via low cost bioflocculant: Influence of temperature with kinetic an thermodynamic functions. *Bioresource Technology, 225*, 84–89.

Kothari, R., Pandey, A., Ahmad, S., Kumar, A., Pathak, V. V., & Tyagi, V. V. (2017c). Microalgal cultivation for value-added products: A critical enviro-economical assessment. *3 Biotech, 7*(4), 243.

Kousha, A., Daneshvar, E., Esmaeli, A. R., Jokar, M., & Khatee, A. R. (2012). Optimization of Aci blue 25 removal from aqueous solutions by raw esterified and protonated *Janiaadhaerens* biomass. *International Journal of Biodeterioration & Biodegradation, 69*, 97–105.

Kumar, P. R., Sameera, K., Mahalakshmi, G., Akbersha, M. A. A., & Thajuddin, N. (2012). Influence of nutrient deprivations on lipid accumulation in a dominant indigenous microalga, *Chlorella* sp., BUM11008: Evaluation for biodiesel production. *Biomass and Bioenergy, 37*, 60–66.

Lee, A. K., Lewis, D. M., & Ashman, P. J. (2012). Disruption of microalgal cells for the extraction of lipids for biofuels: Processes and specific energy requirements. *Biomass and Bioenergy, 46*, 89–101.

Levine, R. B., Pinnarat, T., & Savage, P. E. (2010). Biodiesel production from wet algal biomass through in-situ lipid hydrolysis and supercritical transesterification. *Energy & Fuels, 24*, 5235–5243.

Liang, Y., Sarkany, N., & Cui, Y. (2009). Biomass and lipid productivities of *Chlorella vulgaris* under autotrophic, heterotrophic and mixotrophic growth conditions. *Biotechnology Letters, 31*, 1043–1049.

Lin, T. S., & Wu, J. Y. (2015). Effect of carbon sources on growth and lipid accumulation of newly isolated algae cultured under mixotrophic condition. *Bioresource Technology, 184*, 100–107.

Lynch, F., Santana-Sanchez, A., Jamsa, M., Sivonen, K., Aro, E. M., & Allahverdiyeva, Y. (2015). Screening native isolates of cyanobacteria and a green alga for integrated wastewater treatment, biomass accumulation and neutral lipid production. *Algal Research, 11*, 411–420.

Ma, L. P., Li, B., & Zhang, T. (2014). Abundant rifampin resistance genes and significant correlations of antibiotic resistance genes and plasmids in various environments revealed by metagenomic analysis. *Applied Microbiology and Biotechnology, 98*, 5195–5204.

Mansoorian, H. J., Amir, H. M., Ahmad, J. J., & Narges, K. (2016). Evaluation of dairy industry wastewater treatment and simultaneous bioelectricity generation in a catalyst-less and mediator-less membrane microbial fuel cell. *Journal of Saudi Chemical Society, 20*, 88–100.

Markov, S. A., Thomas, A. D., Bazin, M. J., & Hall, D. O. (1997). Photoproduction of hydrogen by cyanobacteria under partial vacuum in batch culture or in a photobioreactor. *International Journal of Hydrogen Energy, 22*, 521.

Mata, T. M., Martins, A. A., & Caetano, N. S. (2010). Algae for biodiesel production and other applications: A review. *Renewable and Sustainable Energy Reviews, 14*, 217–232.

Mata, T. M., Martins, A. A., & Caetane, N. S. (2012). Algae processing for biodiesel production. In *Advances in biodiesel production* (pp. 204–231).

Meher, L. C., Dharmagadda, V. S. S., & Naik, S. N. (2006). Optimization of alkali catalyzedtrans-esterification of *Pongamiapinnata* oil for production of biodiesel. *Bioresource Technology, 97*, 1392.

Mendes Rui, L., Reis Alberto, D., & Palvara Antonio, F. (2006). Supercritical CO_2 extraction of γ-linolenic acid and other lipids from *Arthrospira (Spirulina) maxima*: Comparison with organic solvent extraction. *Food Chemistry, 99*(1), 57–63.

Milledge, J. J., & Heaven, S. (2013). A review on the harvesting of algae for biofuel production. *Review in Environmental Science and Biotechnology, 12*(2), 165–178.

Mohan, V. S., Rohit, M. V., Chandra, R., & Goud, K. K. (2015) Algal biorefinary. In J. Shibu & B. Thallada (Eds.), *Advanced biorefineries for sustainable production and distribution* (pp. 199–216).

Molinuevo-Salces, B., García-González, M. C., González-Fernández, C., Cuetos, M. J., Morán, A., & Gómez, X. (2010). Anaerobic co-digestion of livestock wastes with vegetable processing wastes: A statistical analysis. *Bioresource Technology, 101*(24), 9479–9485.

Mona, S., Kaushik, A., & Kaushik, C. P. (2011). Waste biomass of *Nostoclinckia* as adsorbent of crystal violet dye: Optimization based on statistical model. *International Journal of Biodeterioration & Biodegradation, 65*(3), 513–521.

Muga, H. E., & Mihelcic, J. R. (2008). Sustainability of wastewater treatment technologies. *Journal of Environmental Management, 88*, 437–447.

Mujtaba, G., Choi, W., Lee, C. G., & Lee, K. (2012). Lipid production by *Chlorella vulgaris* after a shift from nutrient-rich to nitrogen starvation conditions. *Bioresource Technology, 123*, 279–283.

Munoz, R., Köllner, C., & Guieysse, B. (2009). Biofilm photobioreactors for the treatment of industrial wastewaters. *Journal of Hazardous Materials, 161*, 29–34.

Naidoo, S., & Olaniran, A. O. (2013). Treated wastewater effluent as a source of microbial pollution of surface water resources. *International Journal of Environmental Research and Public Health, 11*(1), 249–270.

Oh, H. M., Lee, S. J., Park, M. H., Kim, H. S., Kim, H. C., Yoon, J. H., Kwon, G. S., & Yoon, B. D. (2001). Harvesting of Chlorella vulgaris using a bioflocculant from *Paenibacillus* sp. AM49. *Biotechnology Letters, 23*, 1229–1234.

Okoh, A. I., Sibanda, T., & Gusha, S. S. (2010). Inadequately treated wastewater as a source of human enteric viruses in the environment. *International Journal of Environmental Research and Public Health, 7*(6), 2620–2637.

Osundeko, O., Davies, H., & Pittman, J. K. (2013). Oxidative stress tolerant algae strains are highly efficient for biofuel feedstock production on wastewater. *Biomass and Bioenergy, 56*, 284–294

Pandey, A., Lee, D. J., Chisti, Y., & Scccol, C. (2014). *Biofuels from algae* (p. 348). San Diego: Elsevier.

Pathak, V. V., Kothari, R., Chopra, A. K., & Singh, D. P. (2015). Experimental and kinetic studies for Phycoremediation and dye removal by *Chlorella pyrenoidosa* from textile wastewater. *Journal of Environmental Management, 2015*, 1–8.

Pathak, V. V., Kothari, R., Chopra, A. K., Ahmad, S., Pandey, A. K., & Rahim, N. A. (2016). *Effect of solvent extraction methods on oil yield and its parametric feasibility with C. Pyrenoidosa* (pp. 87–86).

Patil, V., Tran, K. Q., & Giselrød, H. R. (2009). Towards sustainable production of biofuels from algae. *International Journal of Molecular Sciences, 9*(7), 1188–1195.

Ponnuswamy, I., Soundararajan, M., Shabudeen, S., & Shoba, U. S. (2014). Resolution of lipid content from algal growth in carbon sequestration studies. *International Journal of Advance Science and Technology, 67*, 23–32.

Rahimnejad, M., Ghoreyshi, A. A., Najafpour, G., & Jafary, T. (2011). Power generation from organic substrate in batch and continuous flow microbial fuel cell operations. *Applied Energy, 88*(11), 3999–4004.

REN 21 Renewables. (2014). *Global status report*. Available at: http://ren21.net/

Ruiz-Marin, A., Mendoza-Espinosa, L. G., & Stephenson, T. (2010). Growth and nutrient removal in free and immobilized green algae in batch and semi-continuous cultures treating real wastewater. *Bioresource Technology, 101*, 58–64.

Salama, E. S., Kim, J. R., Ji, M. K., Cho, D. W., Abou-Shanab, R. A., & Kabra, A. N. (2015). Application of acid mine drainage for coagulation/flocculation of microalgal biomass. *Bioresource Technology, 186*, 232–237.

Sathish, A., & Sins, R. C. (2012). Biodiesel from mixed culture algae via a wet lipid extraction procedure. *Bioresource Technology, 118*, 643–647.

Schaar, H., Clara, M., Gans, O., & Kreuzinger, N. (2010). Micropollutant removal during biological wastewater treatment and a subsequent ozonation step. *Environmental Pollution, 158*(5), 1399–1404.

Shamshad, A., Pandey, A., Kothari, R., Pathak, V. V., & Tyagi, V. V. (2017) Closed photobioreactors: Construction material and influencing parameters at the commercial scale. In: Y.F. Tsang (Ed.), *Photobioreactors: Advancements, applications and research* (pp. 149–161). NOVA Publication. ISBN: 978-1-53612-354-8.

Sheehan, J., Dunahay, T., Benemann, J., Roessler, P. (1998). A look back at the U.S. Department of Energy's Aquatic Species Program—biodiesel from algae. National Renewable Energy Laboratory, Golden, CO. Report NREL/TP-580–24190.

Sierra, E., Acien, F. G., Fernandez, J. M., Garcia, J. L., Gonzalez, C., & Molina, E. (2008). Characterization of flat plate photobioreactor for the production of algae. *Chemical Engineering Journal, 138*, 136–147.

Sonune, A., & Ghate, R. (2004). Developments in wastewater treatment methods. *Desalination, 167*, 55–63.

Soxhlet, F. (1879). Die gewichtsanalytischebestimmung des milchfettes. *Dinglers' Polytechisches Journal, 232*, 461–465.

Spilling, K., Ásа, B., Dagmar, E., Heiko, R., & Halldór, G. S. (2013). The effect of high pH on structural lipids in diatoms. *Journal of Applied Phycology, 25*(5), 1435–1439.

Srivastava, A., & Prasad, R. (2000). Triglycerides-based diesel fuels. *Renewable and Sustainable Energy Reviews, 4*, 111–133.

Stillwell, A. S., King, C. W., Webber, M. E., Duncan, I. J., & Hardberger, A. (2011). The energy water nexus in Texas. *Ecology and Society, 16*(1), 2.

Talebi, A. F., Mohtashami, S. K., Tabatabaei, M., Tohidfar, M., Bagheri, A., Zeinalabedini, M., Mirzaei, H. H., Mirzajanzadeh, M., Shafaroudi, S. M., & Bakhtiari, S. (2015). Fatty acid profiling: A selective criterion for screening algae strains for biodiesel production. *Algal Research, 2*(3), 258–267.

Tang, H., Abunasser, N., Garcia, M. E. D., Chen, M., Ng, K. S., & Salley, S. O. (2011). Potential of algae oil from Dunaliella tertiolecta as a feedstock for biodiesel. *Applied Energy, 88*(10), 3324–3330.

Ugwu, C. U., Aoyagi, H., & Uchiyama, H. (2008). Photobioreactor for mass cultivation of algae. *Bioresource Technology, 99*(10), 4021–4028.

Ummalyma, S. B., Anil, K. M., Pandey, A., & Sukumaran, R. K. (2016). Harvesting of microalgal biomass: Efficient method for flocculation through pH modulation. *Bioresource Technology, 213*, 216–221.

USEPA. (1999). *Guidelines for Carcinogen risk assessment review draft*. NCEA-F-0644.

Victoria, O., Adesanya, E. C., Stuart, A. S., & Alison, G. S. (2014). Life cycle assessment on microalgal biodiesel production using a hybrid cultivation system. *Bioresource Technology, 163*, 343–355.

Wagenen, V. J., Miller, T. W., Hobbs, S., Hook, P., Crowe, B., & Huesemann, M. (2012). Effects of light and temperature on fatty acid production in Nannochloropsis Salina. *Energies, 5*, 731.

Wang, Y., Ho, S. H., Cheng, C. L., Guo, W. Q., Nagarajan, D., Ren, N. Q., Lee, D. J., & Chang, J. S. (2016). Perspectives on the feasibility of using microalgae for industrial wastewater treatment. *Bioresource Technology, 222*, 485–497.

Yi, L., Yanting, X., Lei, L., Xiaobing, J., Kun, Z., Tianling, Z., & Wang, H. (2016). First evidence of bioflocculant from Shinella albus with flocculation activity on harvesting of Chlorella vulgaris biomass. *Bioresource Technology, 218*, 807–815.

Zhang, X., Wang, L., Sommerfeld, M., & Hu, Q. (2016). Harvesting microalgal biomass using magnesium coagulation-dissolved air flotation. *Biomass and Bioenergy, 93*, 43–49.

Zhu, F., Wang, W., & ZhangX, T. G. (2011). Electricity generation in a membrane-less microbial fuel cell with down-flow feeding onto the cathode. *Bioresource Technology, 102*(15), 7324–7328.

Chapter 11
Efficiency of Constructed Wetland Microcosms (CWMs) for the Treatment of Domestic Wastewater Using Aquatic Macrophytes

Saroj Kumar and Venkatesh Dutta

Contents

Abstract Constructed wetland microcosms (CWMs) are engineered wastewater treatment systems that are designed to treat wastewater from small communities, involving aquatic plants, a variety of substrate materials, soils and their associated microbial fauna. CWMs are considered as promising ecological technology that requires low or no energy input, low operational cost and provides more benefits and better alternative to conventional wastewater treatment systems. In CWMs dissolved oxygen (DO), pH and temperature are controlled to achieve the desirable treatment efficiency. Several other components such as plant, substrate, water depth, hydraulic loading rates (HLRs) and hydraulic retention time (HRT) are also critical to establishing viable CWMs for the better performance. The literature on CWMs suggests excellent nutrient removal performances which are achieved with low and

S. Kumar · V. Dutta (✉)
Department of Environmental Science, School of Environmental Sciences, Babasaheb Bhimrao Ambedkar University, Lucknow, Uttar Pradesh, India
e-mail: dvenks@gmail.com

R. C. Sobti et al. (eds.), *Environmental Biotechnology: For Sustainable Future*,
https://doi.org/10.1007/978-981-10-7284-0_11

stable effluent concentrations. Further, the choice of appropriate macrophyte species having high uptake of pollutants and high pollutant tolerance and choice of substrate materials are critical for treatment performance. CWMs can be differentiated based on existing native vegetation type (such as floating leaved macrophytes, free-floating macrophytes, emergent macrophytes and submerged macrophytes, in which emergent macrophytes are common) and, hydrology (surface flow constructed wetlands (SFCWs), subsurface flow constructed wetlands (SSFCWs) and hybrid systems). The focus of this paper is to review the state of the art in improving the overall efficiency of CWMs for wastewater treatment. The paper documents both the design and operation of CWMs which are critically dependent on environmental, operational and hydraulic factors. It further outlines key challenges and future prospects for their wider replication.

Keywords Constructed wetland microcosms · Hydraulic loading rates · Hydraulic retention time · Macrophytes · Treatment efficiency

Introduction

Rapid urbanization due to enormous population growth and changing living standards, intensification of agricultural activities and over-exploitation of freshwater ecosystems have caused both global and regional water scarcities (Wang et al. 2017). There has been an emergent need of moving towards new and alternative technologies for improving the quality of water in both developed and developing countries. The treatment of wastewater containing high proportion of nutrients and organic matter (OM) or refuse water from communities has been a great challenge and sometimes hard to achieve in conventional treatment processes (Wojciechowska et al. 2017). Therefore, wastewater treatment technologies such as constructed wetlands (CWs) have emerged as an innovative, economical and sustainable way of protecting and rehabilitating freshwater ecosystems in developing countries (Vymazal 2011). Designer CWs have emerged as novel engineered systems that have primarily been developed and implemented in Europe and the USA. These systems are now routinely used in subtropical and tropical regions in countries like India and Brazil (Machado et al. 2017). CWs with better control systems have been also widely implemented in Central and Eastern Europe having higher proportion of inhabitants living in small rural settlements (Istenic et al. 2015). In China, CWs have been used for ecological engineering since more than 20 years (Zhang et al. 2012). In India, CWs are used as decentralized wastewater systems for smaller communities as well as for small drains outfalling in large rivers (Rai et al. 2013). The use of this technology has grown more progressively in recent decades because of their low and easy operational and maintenance cost, reliable efficiency and environmental friendliness, relying fully on natural and continuous ongoing processes compared with other conventional treatment technologies (Zhang et al. 2014). Natural wetlands provide us a wide range of ecosystem services, such as CO_2 uptake

Table 11.1 Treatment of different types of wastewater in CWs using emergent macrophytes

S. no.	Type of wastewater	Vegetation	References
1	Domestic wastewater	*H. psittacorum, P. australis ,P. karka, T. latifolia, T. angustifolia, A. halimus, J. acutus, S. perennis*	Bohórquez et al. (2017), Butterworth et al. (2016), Fountoulakis et al. (2017)
1	Industrial wastewater	*T. latifolia, T. domingensis, S. cyperinus, P. australis, J. articulates*	Khan et al. (2009)
2	Sewage	*T. latifolia, S. acutus, S. validus, P. australis, P. karka*	Ladu et al. (2012), Mulling et al. (2013)
3	Agriculture runoff	*P. karka, T. angustifolia, S. mucronatus*	Sim et al. (2011)
4	Runoff + sewage	*P. australis, T. orientalis, C. malaccensis*	Wang et al. (2011)
5	Pesticides in runoff	*P. australis, T. latifolia*	Elsaesser et al. (2011)
6	Eutrophic water	*T. angustifolia*	Li et al. (2008)

and release as a regulating service (Yamochi et al. 2017), while CWs are promising technology for the treatment of various types of wastewater such as domestic sewage, industrial drainage, storm water runoff, animal wastewaters, agricultural runoff, leachates and polluted river water (Wu et al. 2015a; Maine et al. 2017; Li et al. 2017) (Table 11.1). In a CW, community composition and species richness both represent a diverse effect on nutrient removal. Higher plant species richness habitually results in increased primary production (Naylor et al. 2003) that reduces effluent nutrient concentration due to increased plant uptake (Wang et al. 2013; Han et al. 2016; Zhao et al. 2016a, b).

Constructed Wetland Microcosms (CWMs)

CWs are engineered systems that are used to forecast the behaviour of natural wetlands under more controlled conditions (Zhang et al. 2014). A CWM unit has different kind of filter material (substrates), planted with different macrophytes (Fig. 11.1). Wastewater passes through the basin and flows over the surface to meet with substrate and is discharged out from the CWM unit through a discharge point (Sudarsan et al. 2015).

A CWM unit has the following main components (Sudarsan et al. 2015):

a. Basin
b. Substrate
c. Vegetation
d. Inlet system
e. Outlet system

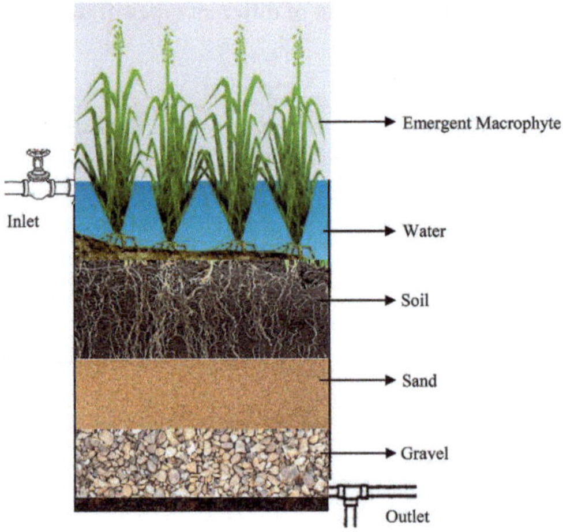

Fig. 11.1 CWM unit planted with emergent macrophyte. (Shao et al. 2014)

On the basis of hydrology, CWs are characterized mainly into surface flow constructed wetlands (SFCWs) or free water surface (FWS), subsurface flow constructed wetlands (SSFCWs) and hybrid systems (or mixed systems) (Wu et al. 2015a). SSFCWs may further be categorized into horizontal flow constructed wetlands (HFCWs) and vertical flow constructed wetlands (VFCWs) on the basis of effluent flow. According to the macrophytic growth, they are further categorized into emergent, free floating, submerged and floating leaved. Most widely used CWs are subsurface flow, and nowadays hybrid system has gained great attention in comparison with others because of their high treatment efficiency. They are designed to acquire benefit of the natural wetlands under controlled environment. Gaining a better knowledge of the mechanisms linked with CWs, various designs and operational mode are available to achieve greater efficiency of domestic sewage treatment, e.g. single-stage modification (Kumari and Tripathi 2014), multistaged in series (Melián et al. 2010) and/or combination with other treatment technologies (Singh et al. 2009). Accordingly, a number of researchers have published review articles related to the use of CWs for wastewater treatment (Haynes 2015; Liu et al. 2015; Vymazal and Březinová 2015; Wu et al. 2015a). Nevertheless, there are relatively few studies on the present knowledge aimed at on-site treatment of wastewater. Still, there is an uncertainty about the selection of the suitable type of CWs which is more appropriate for domestic wastewater treatment in decentralized system. Most of the research on the use of macrophytes in CWs has been done in temperate regions; while there are much more untested macrophytes in tropical regions. Tropical conditions can lead to considerable uptake of wastewater nutrients by macrophytes (Zhang et al. 2014). The roots of the macrophytes provide substrate for microbial growth and transfer oxygen and dissolved organic matter from leaves and aerial parts to the rhizosphere (Meng et al. 2014).

More recent research on CWs has primarily focused on water purification (Ávila et al. 2014), selection of appropriate plant and configuration (Wang and Sample 2013), choice of substrates (Ge et al. 2015), hydraulic loading rates (HLR) (Mexicano et al. 2013) and hydraulic retention time (HRT) (Dzakpasu et al. 2015). Some studies have also found how physical properties of the substrates, such as substrate depth and size, influence pollutant removal.

Mechanism Involved in CWs for the Domestic Wastewater Treatment

Components Involved

The main components involved in CWs are wetland vegetation, media material (which are either natural, industrial by-product or artificially prepared material) and microbial communities. Together, these systems utilize a combination of biological, chemical and physical processes to remove most of the contaminants from wastewater.

Wetland Vegetation

In CWs, a number of wetland plants have been employed having several properties required for the treatment process. Most often used macrophytes in CWs systems are broadly categorized into free-floating plants, submerged plants, floating-leaved plants and emergent plants, typically grown in water or soil media. Even though more than 150 macrophytic plant species have been reported that are used in CWs globally, only a few of these are very frequently used (Saeed and Sun 2012; Vymazal 2013a). Highly dense macrophytes provide more substrates to the biofilms for microbial action to enhance treatment (Badhe et al. 2014; Zheng et al. 2015, 2016; Wang et al. 2016; Wu et al. 2016).

Floating-Leaved Macrophytes

These are rooted in submerged sediments with the water depth of 0.5–3.0 m and have slightly aerial or floating leaves; examples include *Nymphaea odorata* and *Nuphar lutea*.

Free-Floating Macrophytes

They freely float on surface water. These plants are able to remove nitrogen (N) and phosphorus (P) by means of increased plant biomass and by denitrification, and they also remove suspended solids; main examples are *Eichhornia crassipes* (Pontederiaceae), *Nymphaea tetragona* (Nymphaeaceae), *Trapa bispinosa* (Lythraceae), *Marsilea quadrifolia* (Marsileaceae), *Salvinia natans* (Salviniaceae), *Azolla* spp. (Salviniaceae) and *Lemna minor* (Arecaceae).

Emergent Macrophytes

Emergent plants are generally observed on water-saturated or submerged soil and are able to grow in water depth of 0.5 m or more. Commonly used emergent macrophytes are *Phragmites* spp. (Poaceae), *Typha* spp. (Typhaceae), *Canna indica* (Cannaceae), *Scirpus* spp. (Cyperaceae), *Iris* spp. (Iridaceae), *Juncus* spp. (Juncaceae) and *Acorus calamus* (Acoraceae). They transfer oxygen from roots to rhizosphere, which gives rise to degradation of pollutants aerobically.

Use of emergent macrophytes in CWs greatly reduces surface speed, enhances sedimentation and makes the available substrate for periphyton breeding to support pollutant degradation. The most often used macrophyte species are *Typha*, *Scirpus*, *Phragmites* and *Juncus* (Vymazal 2013b).

Submerged Macrophytes

These have their tissues submerged in water, grow healthy in oxygenated water and are principally used for polishing wastewater after secondary treatment. Examples include *Hydrilla verticillata* (Hydrocharitaceae), *Ceratophyllum demersum* (Ceratophyllaceae), *Vallisneria natans* (Hydrocharitaceae), *Potamogeton crispus* (Potamogetonaceae) and *Myriophyllum spicatum* (Haloragaceae).

From the above-mentioned macrophytes, emergent macrophytes are the key species in CWs for wastewater treatments because of their high treatment efficiency (Vymazal 2013b); amongst them *Phragmites australis* is the most frequent species in Asia and Europe (Vymazal 2011).

Media material (Substrates)

Substrates are selected on the basis of hydraulic permeability and the ability to absorb pollutants. Poor hydraulic conductivity may cause clogging of systems, decrease the efficiency of the system, lower the adsorption and also affect the performance of CWs for long-term applications (Wang et al. 2010). Previous studies

Table 11.2 Common media substrates used in CWs systems

S. no.	Substrate type	Source
1	Artificial material	
	Compost	Saeed and Sun (2012)
	Activated carbon	Ren et al. (2007)
	Lightweight aggregates	Saeed and Sun (2012)
	Calcium silicate hydrate Basic oxygen furnace slag (BOFS)	Li et al. (2011) Barca et al. (2014)
2	Industrial by-products	
	Fly ash	Xu et al. (2006)
	Coal cinder	Ren et al. (2007)
	Slag	Cui et al. (2010)
	Alum sludge	Babatunde et al. (2010)
	Oil palm shell	Chong et al. (2013)
	Hollow brick crumbs	Ren et al. (2007)
3	Natural material	
	Sand	Saeed and Sun (2013)
	Gravel	Calheiros et al. (2008)
	Clay	Calheiros et al. (2008)
	Calcite	Ann et al. (1999)
	Limestone	Tao and Wang (2009)
	Zeolite	Bruch et al. (2011)

Wu et al. (2015a)

for the choice of wetland substrate media especially for phosphorus removal from wastewater explain that the substrates mainly include natural materials, artificial media and industrial by-products (Table 11.2) (Yan and Xu 2014). From these studies it is proved that most of the natural substrates are less efficient for long-term phosphorus removal; in contrast, industrial and artificial products with high hydraulic conductivity have high phosphorus sorption capacity. Several other studies also provided some knowledge on substrate choice in order to maximize the removal efficiency of nitrogen and organics. Substrates such as alum sludge, compost, peat, rice husk and marble are the best choices (Babatunde et al. 2010).

Microorganisms

The well-known microbial population in CWs is present in the form of the biofilms associated with plant's roots or attached with the surface of the filter media (Faulwetter et al. 2009). The structure of microbial community in various layers of planted soil in wetlands system for the treatment of domestic wastewater was given by Truu et al. (2005). They observed that the depth is a crucial factor affecting the microbial community composition and microbial activity (Truu et al. 2009) in

CWMs (Iasur-Kruh et al. 2010). Various studies have experimented microbial populations in full-scale CWs and laboratory-scale units under controlled conditions (Zhang et al. 2010; Dong and Reddy 2010). However, there is a short of information on the changes of the microbial communities and diversity in long-term operations for the domestic wastewater treatment (Adrados et al. 2014). It is represented by several studies that the below- and above-ground parts of the macrophytic plants increase the diversity of microorganisms which make available large surface area for the growth of well-defined biofilms, responsible for nearly all of the microbial processes taking place in the wetlands (Chen et al. 2014; Button et al. 2015). Excessive nutrients such as N and P (eutrophication) (Giaramida et al. 2013) and the presence of other toxic substances affect biofilms and their structure (Calheiros et al. 2009) in the wetland system. In CWs different wetland plants, rhizospheric zones are able to provide unique add-on sites for certain microbial populations and mediate the environment by the release of oxygen and root exudates which can control the function and development of certain microbial communities (Lv et al. 2017; Zhang et al. 2016).

Treatment Efficiency of CWMs

Recent research in CWs for domestic wastewater treatment using halophytes shows that they have great potential to build up salts in their tissues (Fountoulakis et al. 2017). The design parameters and operational phase must be chosen according to the environmental conditions of the site and the effluent quality needed after treatment (Bohórquez et al. 2017). HLR and HRT both are significant design parameters for determining the treatment efficiency of a CW; removal efficiencies decreased with increasing HLR and decreasing HRT (Abou-Elela et al. 2017).To date nearly all of the developing countries have warm tropical and subtropical climates throughout the year, and it is commonly known that CWs are more feasible in tropical regions compared to temperate regions (Zhang et al. 2015). In tropical regions, wetlands are exposed to higher temperatures and direct sunlight throughout the year and show higher year-round plant productivity and a simultaneous decrease in the time needed for microbial biodegradation. A warm climate is favourable for plant growth and microbiological activity, which have positive effects on treatment performances (Zhang et al. 2014). In CWs, the core mechanisms associated with contaminant removal are microbial activities. However macrophytes also play a central role in contaminant removal from wastewater. They utilize nutrients and add them into plant tissue and consequently increase plant biomass (Zhang et al. 2007; Mthembu et al. 2013).

Different types of wastewater such as industrial, agricultural, landfill leachate and storm water runoff are hard to be treated in a single-stage system. Recently hybrid systems of different configurations were built together for the treatment of combined sewer overflow (Ávila et al. 2013) or refinery effluent (Vymazal 2005; Wallace and Kadlec 2005; Elfanssi et al. 2017). It is reported that CWs with different designs and planted with different macrophytes obtain high percentage reduction of organic load, total phosphorus and ammonium ions, at short detention times in small communities (Kadlec and Wallace 2008).

Removal of Organics

In CWs organic matter degradation involves both aerobic and anaerobic microbes (Table 11.3). Removal efficiency of organic pollutants which are present in wastewater used in CWs is dependent on influent strength (Saeed and Sun 2012; Wu et al. 2015b). The aerobic heterotrophic bacteria have comparatively faster metabolic rate than autotrophs to oxidize organics that make use of oxygen as the final electron acceptor and release carbon dioxide, ammonia and other stable chemical compounds (Garcia et al. 2010). The intensity of organic matter biodegradation in CWs is also dependent on the biodegradability of the organic matters; such characteristics are best represented by the biological oxygen demand (BOD) and chemical oxygen demand (COD) ratio of the wastewater. Usually, the ratio of BOD and COD for untreated domestic wastewater ranges from 0.3 to 0.8. A BOD and COD ratio of 0.5

Table 11.3 Mechanisms of wastewater treatment by using CWs

S. no.	Wastewater components	Removal mechanisms
1	Suspended solids	Sedimentation
		Filtration
2	Soluble organics	Aerobic microbial degradation (biotransformation)
		Anaerobic microbial degradation
3	Nitrogen	Ammonification and microbial nitrification
		Denitrification (conversion of NO_3 to N_2)
		Plant uptake (accumulation into plants parts)
		Matrix sorption (sorption through the substrates)
		Ammonia volatilization (vaporization)
4	Phosphorus	Matrix sorption
		Plant uptake
5	Metals	Adsorption and cation exchange
		Complexation (formation of coordination compounds)
		Plant uptake
		Precipitation (formation of insoluble compound)
		Microbial oxidation/reduction
6	Pathogens	Sedimentation
		Filtration
		Natural die-off
		Predation
		UV irradiation
		Excretion of antibiotics from macrophytes

Cooper et al. (1997), Mthembu et al. (2013)

or more indicates that the organics are simply degraded, while the ratio below 0.3 shows that the available organics which are present are difficult to degrade by the microorganism (Saeed and Sun 2012). Organic matter degradation is enhanced with sufficient and efficient oxygen supply (Vymazal and Kröpfelová 2009; Ong et al. 2010). Therefore, by increasing the airflow rate, COD removal efficiencies were progressively enhanced because of additional oxygen supply and the highest efficiency to be found at the aeration rate of 2.0 L $min^{-(1)}$ (Saeed and Sun 2012).

Earlier researches pointed out that intermittent aeration strategy greatly enhances the removal efficiency in CWs (Jiang et al. 2017). Removal efficiencies in intermittently aerated CWs with biochar or without biochar were better than non-aerated CWs with or without biochar that means a significant improvement was achieved in organic matter removal through artificial aeration (Headley et al. 2013), while removal efficiency of COD was greater than other CW treatments such as bioaugmentation (Zhao et al. 2016a, b), polyvinyl alcohol immobilized nitrifier, (Wang et al. 2016) and earthworm eco-filters (Zhao et al. 2014).

Removal of Nitrogen

Discharge of nitrogen in excessive is able to cause serious environmental consequences, like eutrophication, which deteriorates water quality and downgrades the aquatic ecosystems (Li et al. 2014; Fan et al. 2016). Nitrogen in wastewater is present mainly in two forms, organic and inorganic (Stefanakis et al. 2014), and removal mechanisms include ammonification (conversion of organic nitrogen to ammonia), nitrification (conversion of ammonia to nitrite and then nitrite to nitrate), denitrification (conversion of nitrate to N_2 gas), nitrate usually used as electron sink and to end with dinitrogen gas (Drizo et al. 1997; Elfanssi et al. 2017), plant uptake (nitrogen taken by plants as nutrients in the form of mainly nitrates and ammonia) and adsorption (mostly ammonia adsorbed on the media material) (Table 11.4) (Tsihrintzis 2017).

In CWs, both nitrification and denitrification are extensively accepted pathways for biological nitrogen removal (Fig. 11.2). The process requires both aerobic and anaerobic environments, while nitrification can convert nitrogen into various forms but cannot achieve its removal from the wastewater (Fan et al. 2013). A continuous aeration strategy has been developed and adopted to attain complete nitrification (Ong et al. 2010; Wu et al. 2015b). However intermittent aeration mode is known to be a more cost-effective strategy because it has more nitrifying and other viable bacteria in comparison with non-aerated CWs (Foladori et al. 2013; Fan et al. 2013). They greatly increased total nitrogen (TN) removal efficiency by creating favourable conditions (alternate aerobic and anaerobic conditions). It is reported that the removal efficiency of TN in CWs can be altered by using different designing models, by controlling environmental conditions (e.g., temperature, pH and dissolved oxygen, etc.) and by different operational factors (e.g., C/N ratio, HLRs, HRT, etc.) (Saeed and Sun 2012). Recently, a lot of investigations were carried out by the

Table 11.4 Significance of novel and classical routes for the removal of nitrogen in wastewater

Mode	Route	Significance
Microbiological	Partial Nitrification–denitrification	NO_2 removal over NO_3, reducing TN content of wastewater
Microbiological	Anammox	Anaerobic NH_4^+ oxidation into N_2
Microbiological	Canon	Completely autotrophic NO_2 removal over NO_3
Microbiological	Ammonification	Transforms N_2 in wastewater, e.g. from organic nitrogen to NH_4–N
Microbiological	Nitrification	Changes NH_4–N to NO_2–N and NO_3–N. Net quantity of TN remains constant
Microbiological	Denitrification	Reduces NO_3–N to N_2 gas. The process also reduces TN when combined with nitrification
Microbiological	Dissimilatory nitrate reduction	Reduces NO_2–N and NO_3–N to NH_4–N. As such, the quantity of TN remains constant
Microbiological	Biomass assimilation	Adsorbs NH_4–N, thereby reducing nitrogen content of wastewater
Biological	Plant uptake	Remove nitrogen from wastewater by accumulation into plants parts
Physico-chemical	Volatilization	Converts NH_4^+ to NH_3 gas, followed by gaseous strip, and then eliminates N_2 from wastewater
Physico-chemical	Adsorption	Adsorbs NH_4–N from wastewater, reducing TN content. However, aerobic environment can nitrify the adsorbed NH_4–N, followed by desorption

Saeed and Sun (2012)

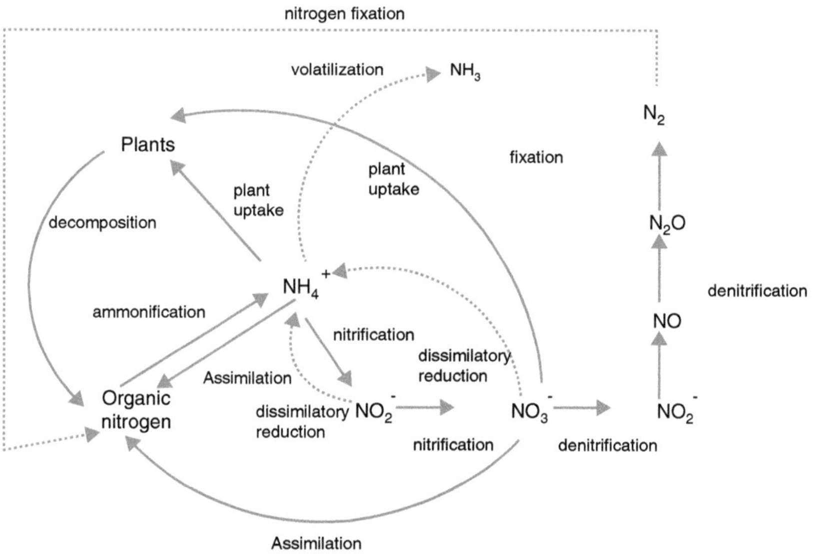

Fig. 11.2 Classical nitrogen removal routes in CWs. (Source: Saeed and Sun 2012)

researchers to study the role of C/N ratio on nitrogen removal in treatment of wastewater (Zhao et al. 2010). From the study done previously, it is stated that the TN removal efficiency was found to be higher at C/N ratio of 2.5–5. In addition, another study (Fan et al. 2013) shows that the high removal rate of TN (82%) was found in aerated SSFCWs with C/N ratio of 10. Later (Zhu et al. 2014) it was reported that the highest removal efficiency of TN was at a C/N ratio of 5, and the removal efficiency rose with an increase of C/N ratio. Nevertheless, the best possible C/N ratios to attain maximum nitrogen removal in SFCWs still remain uncertain especially for purifying the effluent of sewage treatment plant. Actually, the higher removal efficiencies for TN are always coupled with higher C/N ratios. In CWs, degradation of organic matter consumes more DO which threatens the activity of nitrifying microorganisms (Zhu et al. 2014).

Removal of Total Phosphates (TP)

Anthropogenic activities such as agricultural practices and rapid urbanization have altered the biogeochemical cycling of phosphorus (Bouwman et al. 2013; Penuelas et al. 2013; Geng et al. 2017). In wastewater, phosphorus can be found in organic or inorganic forms; orthophosphates ($PO4^{3-}$) is the common form. In CWs phosphate removal is done primarily by adsorption, precipitation and immobilization by microbes (Seo et al. 2005) and high removal efficiency is achieved when there is more oxygen exposure to the rhizosphere through the vascular bundle transformation (Wu et al. 2015c). Dissolved phosphorus is taken by macrophytes or adsorbed onto the substrate media and precipitated, predominantly when Al, Fe, Ca or Mg cations are present at high proportion. Some specialized media materials such as zeolite, bauxite, dolomite, limestone, etc. are probably used to enhance phosphorus adsorption (Stefanakis and Tsihrintzis 2012; Stefanakis et al. 2014). However, high water depth, subsequent to a low flow velocity, is complimentary to increase the rate of this removal process (Guo et al. 2017). TP removal rates varied according to the seasons, linked to the rising of plant biomass and microbial activity from cold to warm one. A positive correlation was found in between total phosphorus removal and seasonal variation (Zhao et al. 2011). Precipitation and adsorption can easily saturate the adsorption sites during pollutant treatment, thus decreasing the treatment efficiency. Consequently, the selection of filter media with high adsorption capacity is necessary to achieve higher treatment efficiency and for the longevity of a CW. Therefore, ongoing study to develop new filter media with enhanced phosphate adsorption capacity has become a main concern for researchers in the last two decades (Park et al. 2017).

Recently a number of substrate materials have been used in CWs to improve phosphate treatment capacity among which basic oxygen furnace slag (BOFS) (Barca et al. 2014) and electric arc furnace (EAF) slag (Okochi and McMartin 2011) are promising substrates.

Sustainability of CWs

The physical functions and chemical composition of wetlands affect all natural biological processes. The DO, pH and temperature are the most significant factors affecting the performance of CWs (Kadlec and Wallace 2008). The criteria for suitable CW design and sustainable operation include site, substrate selection, wastewater type, plant selection, (based on their role in treatment process as a whole plant or by their tissues) (Table 11.5), HLR, HRT and water depth (Akratos et al. 2009; Kadlec 2009; Wu et al. 2014). Particularly, the factors such as plant, substrate, water depth, HLR, HRT and feeding mood are vital for development of sustainable CW system to achieve maximum treatment performance (Fig. 11.3). Brundtland Commission on Sustainable Development (formally known as the World Commission on Environment and Development (WCED)) defined cost–benefit analysis for the sustainability of any project that aims at improving the quality of the environment. In CWs criteria such as land acquisition, energy consumption, ecological benefits, investment and operation costs must be considered during construction and operation phase. A number of earlier studies point out that CWs have an evident advantage of construction and operation cost savings in comparison with other conventional wastewater treatment plants (WWTPs) (Zhang et al. 2012; Wu et al. 2014).

Table 11.5 Parts of macrophytes and their role in treatment process

S. no.	Plant parts	Role in treatment process
1	Aerial plant tissue	Reduced growth of phytoplanktons by light attenuation Reduced wind velocity and risk of resuspension Influence of microclimate–insulation during winter Store nutrients and add aesthetic values
2	Plant tissue in water	Filtering effect – filter out bulky debris Amplified rate of sedimentation, reduced current velocity and risk of resuspension Enhanced aerobic degradation Nutrient uptake Acts as a filter medium, provides oxygen
3	Roots and rhizomes in the sediment	Stabilizing the sediment surface – less erosion Offers surface for bacterial growth Prevents clogging of the medium Increases degradation by release of oxygen Release of antibiotics Promotes biodegradation

Vymazal (2011)

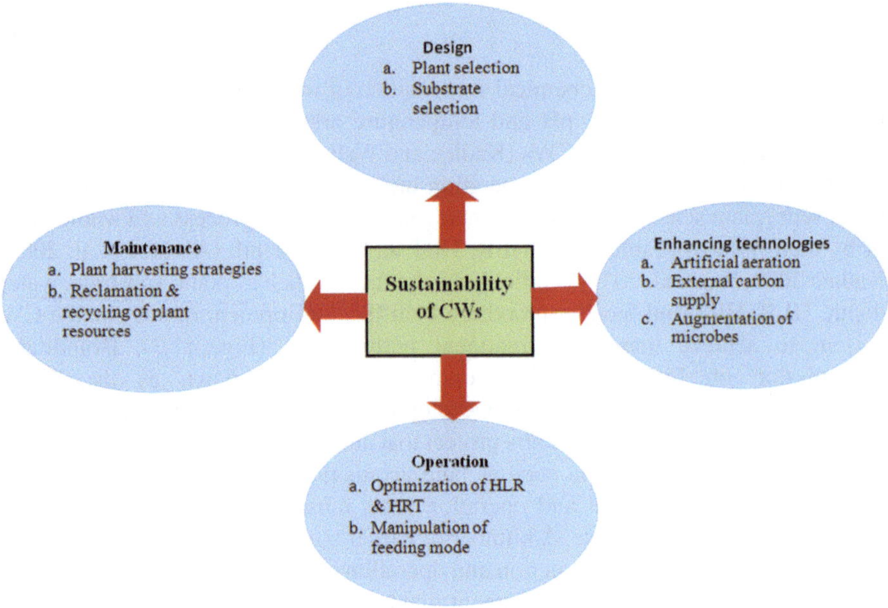

Fig. 11.3 Recent developments and future considerations for improving the sustainability of the CWs. (Wu et al. 2015a)

Future Prospects

The performance of CWs has improved considerably by innovation in the design and mode of operation in recent years. The exceptional treatment efficiency and performance of CWs for treating high strength wastewater containing nutrients can be achieved by appropriate selection of plants and substrates, proper management of the hydraulic loads, mode of operation and pollutant loading rate. These factors can be effectively controlled through innovations in design criteria. Therefore, optimization of these conditions requires extensive research in the future. The challenge is to develop appropriate plant harvest strategies as well as recycling of plant resources because when they die and decay, they could release nutrients and other pollutants into receiving water which may decrease the overall removal performance. There is an emergent need for more research and improvement for traditional CWs to develop new technologies for the enhancement in treatment efficiencies, which are required for sustainable water quality improvement, especially in developing countries. Future research should be devoted to develop artificial aeration, various filter media (non-conventional media materials such as industrial by-products, agricultural wastes, etc.), additional carbon addition, tidal operation, step feeding, microbial augmentation, baffled flow and hybrid CWs.

Conclusion

CWs are considered as an environmental-friendly wastewater treatment technology. CWs have emerged as an alternate, cost-effective solution for treatment of different types of wastewater especially in remote locations of developing countries. The focus of this review has been on the efficiency of CWMs for domestic wastewater treatment. Both the design and operation of a CW are crucial to achieve the sustainable treatment performance which is critically dependent on environmental, operational and hydraulic conditions. There will also be a significant change in removal efficiencies with HLR and HRT, as pollutant removal efficiencies decreased when the HLR is increased and HRT is decreased and removal efficiency increased when HLR is decreased and HRT is increased. However, the removal of plant nutrients (N and P) is highly variable. Still, the choice of appropriate macrophyte species (i.e. supply more oxygen, high uptake of pollutants, and tolerate high pollutant loadings) and substrates are critical for the sustainable wastewater treatment performance.

Acknowledgement The authors are grateful to the Department of Environmental Science, Babasaheb Bhimrao Ambedkar University (a Central University), Lucknow, India, for their continuous support throughout this study. Junior Research Fellowship (JRF) from University Grants Commission, New Delhi to the first author is greatly acknowledged.

References

Abou-Elela, S. I., Elekhnawy, M. A., Khalil, M. T., & Hellal, M. S. (2017). Factors affecting the performance of horizontal flow constructed treatment wetland vegetated with *Cyperus papyrus* for municipal wastewater treatment. *International Journal of Phytoremediation, 19*(11), 1023–1028.

Adrados, B., Sánchez, O., Arias, C. A., Becares, E., Garrido, L., Mas, J., Brix, H., & Morató, J. (2014). Microbial communities from different types of natural wastewater treatment systems: Vertical and horizontal flow constructed wetlands and biofilters. *Water Research, 55*, 304–312.

Akratos, C. S., Papaspyros, J. N., & Tsihrintzis, V. A. (2009). Total nitrogen and ammonia removal prediction in horizontal subsurface flow constructed wetlands: use of artificial neural networks and development of a design equation. *Bioresource Technology, 100*, 586–596.

Ann, Y., Reddy, K. R., & Delfino, J. J. (1999). Influence of chemicals amendments on phosphorus immobilization in soils from a constructed wetland. *Ecological Engineering, 14*, 157–167.

Ávila, C., Salas, J. J., Martín, I., Aragón, C., & García, J. (2013). Integrated treatment of combined sewer wastewater and storm water in a hybrid constructed wetland system in southern Spain and its further reuse. *Ecological Engineering, 50*, 13–20.

Ávila, C., Matamoros, V., Reyes-Contreras, C., Piña, B., Casado, M., Mita, L., Rivetti, C., Barata, C., García, J., & Bayona, J. M. (2014). Attenuation of emerging organic contaminants in a hybrid constructed wetland system under different hydraulic loading rates and their associated toxicological effects in wastewater. *The Science of the Total Environment, 470*, 1272–1280.

Babatunde, A. O., Zhao, Y. Q., & Zhao, X. H. (2010). Alum sludge-based constructed wetland system for enhanced removal of P and OM from wastewater: Concept, design and performance analysis. *Bioresource Technology, 101*(16), 6576–6579.

Badhe, N., Saha, S., Biswas, R., & Nandy, T. (2014). Role of algal biofilm in improving the performance of free surface, up-flow constructed wetland. *Bioresource Technology, 169*, 596–604.

Barca, C., Meyer, D., Liira, M., Drissen, P., Comeau, Y., Andrès, Y., & Chazarenc, F. (2014). Steel slag filters to upgrade phosphorus removal in small wastewater treatment plants: Removal mechanisms and performance. *Ecological Engineering, 68*, 214–222.

Bohórquez, E., Paredes, D., & Arias, C. A. (2017). Vertical flow-constructed wetlands for domestic wastewater treatment under tropical conditions: Effect of different design and operational parameters. *Environmental Technology, 38*, 199–208.

Bouwman, L., Goldewijk, K. K., Van Der Hoek, K. W., Beusen, A. H., Van Vuuren, D. P., Willems, J., Rufino, M. C., & Stehfest, E. (2013). *Exploring global changes in nitrogen and phosphorus cycles in agriculture induced by livestock production over the 1900–2050 periods*. In Proceedings of the National Academy of Sciences (Vol. 110, pp. 20882–20887).

Bruch, I., Fritsche, J., Bänninger, D., Alewell, U., Sendelov, M., Hürlimann, H., Hasselbach, R., & Alewell, C. (2011). Improving the treatment efficiency of constructed wetlands with zeolite-containing filter sands. *Bioresource Technology, 102*, 937–941.

Butterworth, E., Richards, A., Jones, M., Mansi, G., Ranieri, E., Dotro, G., & Jefferson, B. (2016). Performance of four full-scale artificially aerated horizontal flow constructed wetlands for domestic wastewater treatment. *Water, 8*, 365.

Button, M., Nivala, J., Weber, K. P., Aubron, T., & Müller, R. A. (2015). Microbial community metabolic function in subsurface flow constructed wetlands of different designs. *Ecological Engineering, 80*, 162–171.

Calheiros, C. S., Rangel, A. O., & Castro, P. M. (2008). Evaluation of different substrates to support the growth of *Typha latifolia* in constructed wetlands treating tannery wastewater over long-term operation. *Bioresource Technology, 99*, 6866–6877.

Calheiros, C. S., Rangel, A. O., & Castro, P. M. (2009). Treatment of industrial wastewater with two-stage constructed wetlands planted with *Typha latifolia* and *Phragmites australis*. *Bioresource Technology, 100*, 3205–3213.

Chen, Y., Wen, Y., Zhou, Q., & Vymazal, J. (2014). Effects of plant biomass on nitrogen transformation in subsurface-batch constructed wetlands: A stable isotope and mass balance assessment. *Water Research, 63*, 158–167.

Chong, H. L. H., Chia, P. S., & Ahmad, M. N. (2013). The adsorption of heavy metal by Bornean oil palm shell and its potential application as constructed wetland media. *Bioresource Technology, 130*, 181–186.

Cooper, P. F., Job, G. D., Green, M. B., & Shutes, R. B. E. (1997). Reed beds and constructed wetlands for wastewater treatment. *European Water Pollution Control, 6*, 49.

Cui, L., Ouyang, Y., Lou, Q., Yang, F., Chen, Y., Zhu, W., & Luo, S. (2010). Removal of nutrients from wastewater with *Canna indica* L. Under different vertical-flow constructed wetland conditions. *Ecological Engineering, 36*, 1083–1088.

Dong, X., & Reddy, G. B. (2010). Soil bacterial communities in constructed wetlands treated with swine wastewater using PCR-DGGE technique. *Bioresource Technology, 101*, 1175–1182.

Drizo, A. F. C. A., Frost, C. A., Smith, K. A., & Grace, J. (1997). Phosphate and ammonium removal by constructed wetlands with horizontal subsurface flow, using shale as a substrate. *Water Science and Technology, 35*, 95–102.

Dzakpasu, M., Scholz, M., McCarthy, V., & Jordan, S. N. (2015). Assessment of long-term phosphorus retention in an integrated constructed wetland treating domestic wastewater. *Environmental Science and Pollution Research, 22*, 305–313.

Elfanssi, S., Ouazzani, N., Latrach, L., Hejjaj, A., & Mandi, L. (2017). Phytoremediation of domestic wastewater using a hybrid constructed wetlands in mountainous rural area. *International Journal of Phytoremediation, 20*(1), 75–87.

Elsaesser, D., Blankenberg, A. G. B., Geist, A., Mæhlum, T., & Schulz, R. (2011). Assessing the influence of vegetation on reduction of pesticide concentration in experimental surface flow constructed wetlands: Application of the toxic units approach. *Ecological Engineering, 37*, 955–962.

Fan, J., Wang, W., Zhang, B., Guo, Y., Ngo, H. H., Guo, W., Zhang, J., & Wu, H. (2013). Nitrogen removal in intermittently aerated vertical flow constructed wetlands: impact of influent COD/N ratios. *Bioresource Technology, 143*, 461–466.

Fan, J., Zhang, J., Guo, W., Liang, S., & Wu, H. (2016). Enhanced long-term organics and nitrogen removal and associated microbial community in intermittently aerated subsurface flow constructed wetlands. *Bioresource Technology, 214*, 871–875.

Faulwetter, J. L., Gagnon, V., Sundberg, C., Chazarenc, F., Burr, M. D., Brisson, J., Camper, A. K., & Stein, O. R. (2009). Microbial processes influencing performance of treatment wetlands: A review. *Ecological Engineering, 35*, 987–1004.

Foladori, P., Ruaben, J., & Ortigara, A. R. (2013). Recirculation or artificial aeration in vertical flow constructed wetlands: A comparative study for treating high load wastewater. *Bioresource Technology, 149*, 398–405.

Fountoulakis, M. S., Sabathianakis, G., Kritsotakis, I., Kabourakis, E. M., & Manios, T. (2017). Halophytes as vertical-flow constructed wetland vegetation for domestic wastewater treatment. *The Science of the Total Environment, 583*, 432–439.

Garcia, J., Rousseau, D. P., Morato, J., Lesage, E. L. S., Matamoros, V., & Bayona, J. M. (2010). Contaminant removal processes in subsurface-flow constructed wetlands: A review. *Critical Reviews in Environmental Science and Technology, 40*, 561–661.

Ge, Y., Wang, X., Zheng, Y., Dzakpasu, M., Zhao, Y., & Xiong, J. (2015). Functions of slags and gravels as substrates in large-scale demonstration constructed wetland systems for polluted river water treatment. *Environmental Science and Pollution Research, 22*, 12982–12991.

Geng, Y., Han, W., Yu, C., Jiang, Q., Wu, J., Chang, J., & Ge, Y. (2017). Effect of plant diversity on phosphorus removal in hydroponic microcosms simulating floating constructed wetlands. *Ecological Engineering, 107*, 110–119.

Giaramida, L., Manage, P. M., Edwards, C., Singh, B. K., & Lawton, L. A. (2013). Bacterial communities' response to microcystins exposure and nutrient availability: Linking degradation capacity to community structure. *International Biodeterioration and Biodegradation, 84*, 111–117.

Guo, C., Cui, Y., Dong, B., Luo, Y., Liu, F., Zhao, S., & Wu, H. (2017). Test study of the optimal design for hydraulic performance and treatment performance of free water surface flow constructed wetland. *Bioresource Technology, 238*, 461–471.

Han, W., Chang, J., Fan, X., Du, Y., Chang, S. X., Zhang, C., & Ge, Y. (2016). Plant species diversity impacts nitrogen removal and nitrous oxide emissions as much as carbon addition in constructed wetland microcosms. *Ecological Engineering, 93*, 144–151.

Haynes, R. J. (2015). Use of industrial wastes as media in constructed wetlands and filter beds—prospects for removal of phosphate and metals from wastewater streams. *Critical Reviews in Environmental Science and Technology, 45*, 1041–1103.

Headley, T., Nivala, J., Kassa, K., Olsson, L., Wallace, S., Brix, H., van Afferden, M., & Müller, R. (2013). *Escherichia coli* removal and internal dynamics in subsurface flow eco-technologies: Effects of design and plants. *Ecological Engineering, 61*, 564–574.

Iasur-Kruh, L., Hadar, Y., Milstein, D., Gasith, A., & Minz, D. (2010). Microbial population and activity in wetland microcosms constructed for improving treated municipal wastewater. *Microbial Ecology, 59*, 700–709.

Istenic, D., Bodík, I., & Bulc, T. (2015). Status of decentralised wastewater treatment systems and barriers for implementation of nature-based systems in central and eastern Europe. *Environmental Science and Pollution Research, 22*, 12879–12884.

Jiang, Y., Sun, Y., Pan, J., Qi, S., Chen, Q., & Tong, D. (2017). Nitrogen removal and N_2O emission in subsurface wastewater infiltration systems with/without intermittent aeration under different organic loading rates. *Bioresource Technology, 244*, 8–14.

Kadlec, R. H. (2009). Comparison of free water and horizontal subsurface flow treatment wetlands. *Ecological Engineering, 35*, 159–174.

Kadlec, R. H., & Wallace, S. (2008). *Treatment wetlands*. Boca Raton: CRC Press.

Khan, S., Ahmad, I., Shah, M. T., Rehman, S., & Khaliq, A. (2009). Use of constructed wetland for the removal of heavy metals from industrial wastewater. *Journal of Environmental Management, 90*, 3451–3457.

Kumari, M., & Tripathi, B. D. (2014). Effect of aeration and mixed culture of *Eichhornia crassipes* and *Salvinia natans* on removal of wastewater pollutants. *Ecological Engineering, 62*, 48–53.

Ladu, J. L. C., Loboka, M. K., & Lukaw, Y. S. (2012). Integrated constructed wetland for nitrogen elimination from domestic sewage: The case study of Soba rural area in Khartoum South, Sudan. *Natural Science, 10*, 30–36.

Li, L., Li, Y., Biswas, D. K., Nian, Y., & Jiang, G. (2008). Potential of constructed wetlands in treating the eutrophic water: evidence from Taihu Lake of China. *Bioresource Technology, 99*, 1656–1663.

Li, C. J., Wan, M. H., Dong, Y., Men, Z. Y., Lin, Y., Wu, D. Y., & Kong, H. N. (2011). Treating surface water with low nutrients concentration by mixed substrates constructed wetlands. *Journal of Environment Science and Health Part A, 46*, 771–776.

Li, F., Lu, L., Zheng, X., Ngo, H. H., Liang, S., Guo, W., & Zhang, X. (2014). Enhanced nitrogen removal in constructed wetlands: effects of dissolved oxygen and step-feeding. *Bioresource Technology, 169*, 395–402.

Li, M., Wu, H., Zhang, J., Ngo, H. H., Guo, W., & Kong, Q. (2017). Nitrogen removal and nitrous oxide emission in surface flow constructed wetlands for treating sewage treatment plant effluent: Effect of C/N ratios. *Bioresource Technology, 240*, 157–164.

Liu, R., Zhao, Y., Doherty, L., Hu, Y., & Hao, X. (2015). A review of incorporation of constructed wetland with other treatment processes. *Chemical Engineering Journal, 279*, 220–230.

Lv, T., Zhang, Y., Carvalho, P. N., Zhang, L., Button, M., Arias, C. A., Weber, K. P., & Brix, H. (2017). Microbial community metabolic function in constructed wetland mesocosms treating the pesticides imazalil and tebuconazole. *Ecological Engineering, 98*, 378–387.

Machado, A. I., Beretta, M., Fragoso, R., & Duarte, E. (2017). Overview of the state of the art of constructed wetlands for decentralized wastewater management in Brazil. *Journal of Environment Management, 187*, 560–570.

Maine, M. A., Hadad, H. R., Sánchez, G. C., Di Luca, G. A., Mufarrege, M. M., Caffaratti, S. E., & Pedro, M. C. (2017). Long-term performance of two free-water surface wetlands for metallurgical effluent treatment. *Ecological Engineering, 98*, 372–377.

Melián, J. H., Rodríguez, A. M., Arana, J., Díaz, O. G., & Henríquez, J. G. (2010). Hybrid constructed wetlands for wastewater treatment and reuse in the Canary Islands. *Ecological Engineering, 36*, 891–899.

Meng, P., Pei, H., Hu, W., Shao, Y., & Li, Z. (2014). How to increase microbial degradation in constructed wetlands: Influencing factors and improvement measures. *Bioresource Technology, 157*, 316–326.

Mexicano, L., Glenn, E. P., Hinojosa-Huerta, O., Garcia-Hernandez, J., Flessa, K., & Hinojosa-Corona, A. (2013). Long-term sustainability of the hydrology and vegetation of Cienega de Santa Clara, an anthropogenic wetland created by disposal of agricultural drain water in the delta of the Colorado River, Mexico. *Ecological Engineering, 59*, 111–120.

Mthembu, M. S., Odinga, C. A., Swalaha, F. M., & Bux, F. (2013). Constructed wetlands: A future alternative wastewater treatment technology. *African Journal of Biotechnology, 12*(29), 4542–4553.

Mulling, B. T., van den Boomen, R. M., van der Geest, H. G., Kappelhof, J. W., & Admiraal, W. (2013). Suspended particle and pathogen peak discharge buffering by a surface-flow constructed wetland. *Water Research, 47*, 1091–1100.

Naylor, S., Brisson, J., Labelle, M. A., Drizo, A., & Comeau, Y. (2003). Treatment of freshwater fish farm effluent using constructed wetlands: The role of plants and substrate. *Water Science and Technology, 48*, 215–222.

Okochi, N. C., & McMartin, D. W. (2011). Laboratory investigations of storm water remediation via slag: Effects of metals on phosphorus removal. *Journal of Hazardous Materials, 187*, 250–257.

Ong, S. A., Uchiyama, K., Inadama, D., Ishida, Y., & Yamagiwa, K. (2010). Performance evaluation of laboratory scale up-flow constructed wetlands with different designs and emergent plants. *Bioresource Technology, 101*, 7239–7244.
Park, J. H., Wang, J. J., Kim, S. H., Cho, J. S., Kang, S. W., Delaune, R. D., & Seo, D. C. (2017). Phosphate removal in constructed wetland with rapid cooled basic oxygen furnace slag. *Chemical Engineering Journal, 327*, 713–724.
Penuelas, J., Poulter, B., Sardans, J., Ciais, P., Van Der Velde, M., Bopp, L., Boucher, O., Godderis, Y., Hinsinger, P., Llusia, J., & Nardin, E. (2013). Human-induced nitrogen–phosphorus imbalances alter natural and managed ecosystems across the globe. *Nature Communications, 4*, 2934.
Rai, U. N., Tripathi, R. D., Singh, N. K., Upadhyay, A. K., Dwivedi, S., Shukla, M. K., Mallick, S., Singh, S. N., & Nautiyal, C. S. (2013). Constructed wetland as an eco-technological tool for pollution treatment for conservation of Ganga River. *Bioresource Technology, 148*, 535–541.
Ren, Y., Zhang, B., Liu, Z., & Wang, J. (2007). Optimization of four kinds of constructed wetlands substrate combination treating domestic sewage. *Wuhan University Journal of Natural Science, 12*, 1136–1142.
Saeed, T., & Sun, G. (2012). A review on nitrogen and organics removal mechanisms in subsurface flow constructed wetlands: Dependency on environmental parameters, operating conditions and supporting media. *Journal of Environmental Management, 112*, 429–448.
Saeed, T., & Sun, G. (2013). A lab-scale study of constructed wetlands with sugarcane bagasse and sand media for the treatment of textile wastewater. *Bioresource Technology, 128*, 438–447.
Seo, D. C., Cho, J. S., Lee, H. J., & Heo, J. S. (2005). Phosphorus retention capacity of filter media for estimating the longevity of constructed wetland. *Water Research, 39*, 2445–2457.
Shao, Y., Pei, H., Hu, W., Chanway, C. P., Meng, P., Ji, Y., & Li, Z. (2014). Bioaugmentation in lab scale constructed wetland microcosms for treating polluted river water and domestic wastewater in northern China. *International Biodeterioration and Biodegradation, 95*, 151–159.
Sim, C. H., Eikaas, H. S., Chan, S. H., & Gan, J. (2011). Nutrient removal and plant biomass of 5 wetland plant species in Singapore. *Water Practice Technology, 6*, 2011053.
Singh, S., Haberl, R., Moog, O., Shrestha, R. R., Shrestha, P., & Shrestha, R. (2009). Performance of an anaerobic baffled reactor and hybrid constructed wetland treating high-strength wastewater in Nepal—A model for DEWATS. *Ecological Engineering, 35*, 654–660.
Stefanakis, A. I., & Tsihrintzis, V. A. (2012). Effects of loading, resting period, temperature, porous media, vegetation and aeration on performance of pilot-scale vertical flow constructed wetlands. *Chemical Engineering Journal, 181*, 416–430.
Stefanakis, A., Akratos, C. S., & Tsihrintzis, V. A. (2014). *Vertical flow constructed wetlands: Eco-engineering systems for wastewater and sludge treatment*. Newnes
Sudarsan, J. S., Roy, R. L., Baskar, G., Deeptha, V. T., & Nithiyanantham, S. (2015). Domestic wastewater treatment performance using constructed wetland. *Sustainable Water Resources Management, 1*, 89–96.
Tao, W., & Wang, J. (2009). Effect of vegetation, limestone and aeration on nitritation, anammox and denitrification in wetland treatment systems. *Ecological Engineering, 35*, 836–842.
Truu, J., Nurk, K., Juhanson, J., & Mander, Ü. (2005). Variation of microbiological parameters within planted soil filter for domestic wastewater treatment. *Journal of Environmental Science and Health, 40*, 1191–1200.
Truu, M., Juhanson, J., & Truu, J. (2009). Microbial biomass, activity and community composition in constructed wetlands. *The Science of the Total Environment, 407*, 3958–3971.
Tsihrintzis, V. A. (2017). The use of vertical flow constructed wetlands in wastewater treatment. *Water Resources Management*, 1–26.
Vymazal, J. (2005). Horizontal sub-surface flow and hybrid constructed wetlands systems for wastewater treatment. *Ecological Engineering, 25*, 478–490.
Vymazal, J. (2011). Plants used in constructed wetlands with horizontal subsurface flow. *Hydrobiologia, 10*, 738–749.
Vymazal, J. (2013a). Emergent plants used in free water surface constructed wetlands: a review. *Ecological Engineering, 61*, 582–592.

Vymazal, J. (2013b). The use of hybrid constructed wetlands for wastewater treatment with special attention to nitrogen removal: a review of a recent development. *Water Research, 47*, 4795–4811.

Vymazal, J., & Březinová, T. (2015). The use of constructed wetlands for removal of pesticides from agricultural runoff and drainage: A review. *Environ Int, 75*, 11–20.

Vymazal, J., & Kröpfelová, L. (2009). Removal of organics in constructed wetlands with horizontal sub-surface flow: A review of the field experience. *The Science of the Total Environment, 407*, 3911–3922.

Wallace, S., & Kadlec, R. (2005). BTEX degradation in a cold-climate wetland system. *Water Science Technology, 51*, 165–171.

Wang, C. Y., & Sample, D. J. (2013). Assessing floating treatment wetlands nutrient removal performance through a first order kinetics model and statistical inference. *Ecological Engineering, 61*, 292–302.

Wang, G., Wang, Y., & Gao, Z. (2010). Use of steel slag as a granular material: Volume expansion prediction and usability criteria. *Journal of Hazardous Materials, 184*, 555–560.

Wang, Y. C., Ko, C. H., Chang, F. C., Chen, P. Y., Liu, T. F., Sheu, Y. S., Shih, T. L., & Teng, C. J. (2011). Bioenergy production potential for aboveground biomass from a subtropical constructed wetland. *Biomass and Bioenergy, 35*, 50–58.

Wang, H., Chen, Z. X., Zhang, X. Y., Zhu, S. X., Ge, Y., Chang, S. X., Zhang, C. B., Huang, C. C., & Chang, J. (2013). Plant species richness increased belowground plant biomass and substrate nitrogen removal in a constructed wetland. *Clean Soil Air Water, 41*, 657–664.

Wang, W., Ding, Y., Wang, Y., Song, X., Ambrose, R. F., Ullman, J. L., Winfrey, B. K., Wang, J., & Gong, J. (2016). Treatment of rich ammonia nitrogen wastewater with polyvinyl alcohol immobilized nitrifier biofortified constructed wetlands. *Ecological Engineering, 94*, 7–11.

Wang, X. J., Zhang, J. Y., Gao, J., Shahid, S., Xia, X. H., Geng, Z., & Tang, L. (2017). The new concept of water resources management in China: ensuring water security in changing environment. *Environment Development Sustainability*, 1–13.

Wojciechowska, E., Gajewska, M., & Ostojski, A. (2017). Reliability of nitrogen removal processes in multi-stage treatment wetlands receiving high-strength wastewater. *Ecological Engineering, 98*, 365–371.

Wu, S., Kuschk, P., Brix, H., Vymazal, J., & Dong, R. (2014). Development of constructed wetlands in performance intensifications for wastewater treatment: A nitrogen and organic matter targeted review. *Water Research, 57*, 40–55.

Wu, H., Zhang, J., Ngo, H. H., Guo, W., Hu, Z., Liang, S., Fan, J., & Liu, H. (2015a). A review on the sustainability of constructed wetlands for wastewater treatment: Design and operation. *Bioresource Technology, 175*, 594–601.

Wu, H., Fan, J., Zhang, J., Ngo, H. H., Guo, W., Hu, Z., & Liang, S. (2015b). Decentralized domestic wastewater treatment using intermittently aerated vertical flow constructed wetlands: Impact of influent strengths. *Bioresource Technology, 176*, 163–168.

Wu, H., Fan, J., Zhang, J., Ngo, H. H., GuoW, L. S., Hu, Z., & Liu, H. (2015c). Strategies and techniques to enhance constructed wetland performance for sustainable wastewater treatment. *Environmental Science and Pollution Research, 22*, 14637–14650.

Wu, H., Lin, L., Zhang, J., Guo, W., Liang, S., & Liu, H. (2016). Purification ability and carbon dioxide flux from surface flow constructed wetlands treating sewage treatment plant effluent. *Bioresource Technology, 219*, 768–772.

Xu, D., Xu, J., Wu, J., & Muhammad, A. (2006). Studies on the phosphorus sorption capacity of substrates used in constructed wetland systems. *Chemosphere, 63*, 344–352.

Yamochi, S., Tanaka, T., Otani, Y., & Endo, T. (2017). Effects of light, temperature and ground water level on the CO2 flux of the sediment in the high water temperature seasons at the artificial north salt marsh of Osaka Nanko bird sanctuary, Japan. *Ecological Engineering, 98*, 330–338.

Yan, Y., & Xu, J. (2014). Improving winter performance of constructed wetlands for wastewater treatment in northern China: A review. *Wetlands, 34*, 243–253.

Zhang, Z., Rengel, Z., & Meney, K. (2007). Nutrient removal from simulated wastewater using Canna indica and *Schoenoplectus validus* in mono-and mixed-culture in wetland microcosms. *Water Air Soil Pollution, 183*, 95–105.

Zhang, C. B., Wang, J., Liu, W. L., Zhu, S. X., Ge, H. L., Chang, S. X., Chang, J., & Ge, Y. (2010). Effects of plant diversity on microbial biomass and community metabolic profiles in a full-scale constructed wetland. *Ecological Engineering, 36*, 62–68.

Zhang, T., Xu, D., He, F., Zhang, Y., & Wu, Z. (2012). Application of constructed wetland for water pollution control in China during 1990-2010. *Ecological Engineering, 47*, 189–197.

Zhang, D. Q., Jinadasa, K. B. S. N., Gersberg, R. M., Liu, Y., Ng, W. J., & Tan, S. K. (2014). Application of constructed wetlands for wastewater treatment in developing countries–a review of recent developments (2000–2013). *Journal of Environmental Management, 141*, 116–131.

Zhang, D. Q., Jinadasa, K. B. S. N., Gersberg, R. M., Liu, Y., Tan, S. K., & Ng, W. J. (2015). Application of constructed wetlands for wastewater treatment in tropical and subtropical regions (2000–2013). *Journal of Environmental Sciences, 30*, 30–46.

Zhang, Y., Carvalho, P. N., Lv, T., Arias, C., Brix, H., & Chen, Z. (2016). Microbial density and diversity in constructed wetland systems and the relation to pollutant removal efficiency. *Water Science & Technology, 73*, 679–686.

Zhao, Y. J., Liu, B., Zhang, W. G., Ouyang, Y., & An, S. Q. (2010). Performance of pilot-scale vertical-flow constructed wetlands in responding to variation in influent C/N ratios of simulated urban sewage. *Bioresource Technology, 101*, 1693–1700.

Zhao, Y. J., Hui, Z., Chao, X., Nie, E., Li, H. J., He, J., & Zheng, Z. (2011). Efficiency of two-stage combinations of subsurface vertical down-flow and up-flow constructed wetland systems for treating variation in influent C/N ratios of domestic wastewater. *Ecological Engineering, 37*, 1546–1554.

Zhao, Y., Zhang, Y., Ge, Z., Hu, C., & Zhang, H. (2014). Effects of influent C/N ratios on wastewater nutrient removal and simultaneous greenhouse gas emission from the combinations of vertical subsurface flow constructed wetlands and earthworm eco-filters for treating synthetic wastewater. *Environmental Science: Processes & Impacts, 16*, 567–575.

Zhao, Z., Chang, J., Han, W., Wang, M., Ma, D., Du, Y., Qu, Z., Chang, S. X., & Ge, Y. (2016a). Effects of plant diversity and sand particle size on methane emission and nitrogen removal in microcosms of constructed wetlands. *Ecological Engineering, 95*, 390–398.

Zhao, X., Yang, J., Bai, S., Ma, F., & Wang, L. (2016b). Microbial population dynamics in response to bioaugmentation in a constructed wetland system under 10.C. *Bioresource Technology, 205*, 166–173.

Zheng, Y., Wang, X. C., Ge, Y., Dzakpasu, M., Zhao, Y., & Xiong, J. (2015). Effects of annual harvesting on plants growth and nutrients removal in surface flow constructed wetlands in north-western China. *Ecological Engineering, 83*, 268–275.

Zheng, Y., Wang, X., Dzakpasu, M., Zhao, Y., Ngo, H. H., Guo, W., Ge, Y., & Xiong, J. (2016). Effects of interspecific competition on the growth of macrophytes and nutrient removal in constructed wetlands: A comparative assessment of free water surface and horizontal subsurface flow systems. *Bioresource Technology, 207*, 134–141.

Zhu, H., Yan, B., Xu, Y., Guan, J., & Liu, S. (2014). Removal of nitrogen and COD in horizontal subsurface flow constructed wetlands under different influent C/N ratios. *Ecological Engineering, 63*, 58–63.

Chapter 12
Modelling Water Temperature's Sensitivity to Atmospheric Warming and River Flow

Shaik Rehana, Francisco Munoz-Arriola, Daniel A. Rico, and Shannon L. Bartelt-Hunt

Contents

Abstract River water bodies serve as prominent water sources for various purposes ranging from drinking water, waste load allocation, irrigation, hydropower generation and ecosystem services. Human activities and natural processes require a balanced water supply and demand, while population growth, land use and climate change are the external forces which try to change the stream and river water quantity and quality. Water temperature is an inherent property of its quality and a controlling factor of health of freshwater environments. It is often considered as a driver of metabolic activity in water bodies, which influence the biological and chemical processes affecting the metabolic responses from organisms to ecosystems. The present work aims to explore sources of predictability of river water temperature (RWT) as a keen driver of hydrological and ecological processes at multiple scales. Increasing RWT in response to climate change and local-to-regional anthropogenic

S. Rehana (✉)
Lab for Spatial Informatics, International Institute of Information Technology, Hyderabad, Telangana, India
e-mail: rehana.s@iiit.ac.in

F. Munoz-Arriola
Department of Biological Systems Engineering, University of Nebraska-Lincoln, Lincoln, USA

D. A. Rico
Department of Electrical and Computer Engineering, University of Nebraska-Lincoln, Lincoln, USA

S. L. Bartelt-Hunt
Department of Civil Engineering, University of Nebraska-Lincoln, Lincoln, USA

R. C. Sobti et al. (eds.), *Environmental Biotechnology: For Sustainable Future*,
https://doi.org/10.1007/978-981-10-7284-0_12

activities result in decreasing dissolved oxygen (DO) levels and anaerobic conditions in the aquatic system, thereby affecting marine life and the consequent availability of food, reproduction and migration. An assessment of integrated RWT and streamflow fluctuations is proposed to evidence biological activity, chemical speciation, oxygen solubility and self-purification capacity of a river system and fluctuations of flows responsive to hydro-climate pulses. The independent and integrated contributions of air temperatures and flow fluctuations to RWT in the Missouri River near Nebraska City, USA, represent the stream responses to global raising temperatures. To quantify the contributions of multiple variations of predictor variables in the air-water interfaces to RWT variability, we use a multiple regression. The performance of the model was tested along Missouri River near Nebraska City, USA, using historical series of daily river water temperature, air temperature and river discharges for the 1947–2014 period. A sensitivity analysis on river water temperature is performed, under air temperature increase of +2 °C, +4 °C and + 6 °C with a decrease of discharge of ±20%. Overall, the increase of RWT for the Missouri River is observed as about 2.76 °C under various air temperature and discharge changes when compared with the observed conditions at mean annual scale. The study results provide a comprehensive analysis of the impacts of river discharge and air temperature changes under climate change over RWT.

Keywords River water temperature · Climate change · River water quality

Introduction

Infrastructure development, such as for water and agricultural resources and energy generation, is human intervention that affects the water quality and quantity of rivers and other water bodies in multiple forms and across scales. Altering the water cycle and the biogeochemical cycles of the elements poses challenges to the water, ranging from drinking water, waste load allocation, irrigation, hydropower generation and ecosystem services. Interdependence between climate and water systems can be evidenced in spatio-temporal changes of air temperature and precipitation and an evolving intensification and increasing frequency of extreme events. Further, the pronounced effect in physical properties of river temperatures and quantitative changes of river flows could alter non-conservative properties of streams leading to changes in metabolic and chemical processes, reaction kinetics and speciation across living systems and their scales. These anomalous fluctuations also overlap with changes of nonpoint source pollutants and consequently raising the complexity of RWT. For example, an increase in RWT decreases the oxygen-holding capacity, causing a decrease in dissolved oxygen (DO) concentration, which directly affects the "self-purification" capacity of the river system. Biological systems and the metabolic rates that drive them are temperature-dependent, and therefore, RWT is considered as an indicator for the biological/biogeochemical component of the water system's metabolism (Mackey and Berrie 1991). Further, fish, insects and other aquatic species generally have their own temperature ranges and metabolic rates, and any change may lead to fluctuations in the population of species. Therefore, reliable prediction of RWT has become the main issue for many environmental applications, hydrology and ecology. RWT has become prominent due to the use of water to cool thermal power plants, construction of reservoirs, withdrawal of city effluents and urbanisation. Also, the accurate prediction of river water temperature has become an important field due to the anthropogenic changes of climate change. In recent years, water temperature trends in major rivers are identified as a complex function of both climate and hydrological changes (Webb and Nobilis 2007; van Vliet et al. 2011; van Vliet et al. 2012; Toffolon and Piccolroaz 2015a, b). Various approaches are developed in the research community to estimate the RWT as a function of hydro-climatic variables integrating the complexity and data input requirements. The most complex approaches are physically based or deterministic models based on the heat advection-dispersion transport equations (Yearsley 2009). Few physical models are based on the equilibrium temperature concept, which considers the net heat transfer processes at the water surface (Edinger et al. 1968; O'Driscoll and DeWalle 2006). Physical approaches require detailed data sets such as stream geometry, meteorological conditions, discharge of the upstream and downstream sections, topography, heat fluxes at the stream and river bed interfaces, land use and vegetative covers (Kim and Chapra 1997; Bogan et al. 2006; Caissie et al. 2007). The data required to implement the physical models is difficult to measure over a reasonable time frame, and limitations persist towards the integration into the estimation procedures. The lack of detailed hydrometeorological and river sediment data sets and the knowledge on physical process involved in the river thermal systems limits the use of phenomenological models to estimate the RWT. Water temperature regression models have been widely applied in the research community as an alternative to predict and forecast river water temperature using air temperature (Neumann et al. 2003; Caissie 2006; Rehana and Mujumdar 2011). Also, regression models became more popular due to their simplicity and limited requirement of hydrometeorological data in the absence of detailed heat fluxes information (Webb and Nobilis 2007; Mohseni and Stefan 1999). Air temperature is the most commonly used predictor variable in river water temperature regression models, as it is the representative variable for the net changes of heat flux at the water surface. The net excahnge of heat energy from atmosphere is zero at equilibrium temperature, which would be equal to the air temperature near the water surface (Dingman 1972). Therefore, most of the studies found that air temperature is an appropriate variable in river water temperature modelling using regression models. There are several studies stressing on the use of air temperature to model RWT, under limited meteorological data availability constraints (Marce and Armengol 2008). Rehana and Mujumdar (2011, 2012) applied a linear regression model relating air and RWT to model the daily RWT for Tungabhadra River, India, in the impact assessment study of climate change. Streamflow is also one of the prominent variables to define the pollutant transport and dilution capacity of a river system. In addition to air temperature, one more readily available variable is streamflow, and several authors have related river water temperature with streamflow and air temperatures (Hockey et al. 1982; Webb 1996; Webb and Nobilis 2007) with a negative relationship between RWT and streamflow. Moreover, streamflow is negatively correlated with water temperature as more water is available for dilution from various sources such as outfalls from a sewage treatment plant and downstream of a thermal plant (Webb et al. 2003; van Vliet et al. 2011).

Raising atmospheric temperatures anticipate an influence on river water temperatures together with modifications in flow regulation, water withdrawals and alterations in available water quantity (Caissie 2006). Few studies focused on the impact of climate change on river water temperatures by using air temperature and river flows using regression models at daily, weekly, and monthly basis (Mohseni et al. 1998; van Vliet et al. 2011; Rehana and Mujumdar 2012). A fundamental sensitivity analysis for predicting thc changes in RWT under the altered conditions of air temperature and river flows will provide a preliminary understanding about the river ecosystem response to changing climate. Further, the sensitivity of river water temperatures under atmospheric warming and changes in river flows will possibly have negative consequences on fresh water ecosystem. A global appraisal indicated that a decrease in river discharge by 20 and 40% would exacerbate water temperature increase by 0.3 °C and 0.8 °C on average, respectively, as in further to a 2–6 °C increase due to rising air temperature (van Vliet et al. 2011). Hence, the present study aims to test the performance of river water temperature regression model, air temperature and river discharge and as predictors of river water temperature. The inherent RWT in response to increasing air temperatures and decrease in river discharges will evidence the sensitivity of watersheds in agricultural landscapes to natural and anthropogenic drivers of change. To address these objectives, a multiple linear regression model (MLRM) with air temperature and discharge as predictor variables are used to model daily RWT along Missouri River near Nebraska City, USA. The performance of the model was tested for the selected river station, and a sensitivity study of river water temperature was assessed under decrease in river discharge and increase in air temperature under the context of climate change.

Data and Methods

In order to test the performance of the river water temperature model, we have selected a station along Missouri River near Nebraska City, USA, characterised by different hydrological conditions. The location of the station is at latitude 40°40′55″ and longitude of 95°50′48″, Otoe County, Nebraska, on the right bank 0.7 mi upstream from the bridge on State Highway 2 in Nebraska City and 562.6 mi upstream from the mouth with drainage area as 410,000 mi^2. The daily river discharge data was collected for the site number USGS 06807000 Missouri River at Nebraska City from US Geological Survey (USGS) for the period of 1929–2017. The daily river water temperature data was obtained from USGS for the period of September 15, 2012, to January 10, 2017. The daily air temperature data was obtained from the Global Historical Climatology Network (GHCN) for the station

USC00251825, named COLUMBUS 3 NE, for the period of 1910–2014. To set up the regression model, the common data availability period of September 15, 2012, to December 31, 2014, is considered for the analysis out of which 70% of the data samples of 593 (September15, 2012, to April 30, 2014) are considered for training the MLRM and the remaining samples of 250 (April 01, 2014, to December 31, 2014) are used for testing the model. The MLRM developed based on the training is given in the following equation:

$$T_w = a + bT_{\text{air}} + cQ \tag{12.1}$$

where T_w is the daily river water temperature in °C, $T_{\text{air}}$is the daily air temperature in °C, Q is the daily discharge in m^3/s, and a, b, c are the parameters estimated based on training the MLRM. The model performance of MLRM was tested based on the Nash-Sutcliffe coefficient (NSC) (Nash and Sutcliffe, 1970) (Eq. (12.2), to show the efficiency of the model fit. The quality of the MLRM is analysed using Root Mean Square Error (RMSE) (Janssen and Heuberger 1995) (Eq. (12.3).

$$\text{NSC} = \frac{\sum_{n}^{i=1}\left(T_{W_{\text{Sim}}} - T_{W_{\text{Obs}}}\right)^2}{\sum_{n}^{i=1}\left(T_{W_{\text{Obs}}} - T_{W_{\text{Obs,Avg}}}\right)^2} \tag{12.2}$$

$$\text{RMSE} = \sqrt{\frac{\sum_{n}^{i=1}\left(T_{W_{\text{Sim}}} - T_{W_{\text{Obs}}}\right)^2}{n}} \tag{12.3}$$

where $T_{W_{\text{Sim}}}$ is the simulated daily river water temperature at time step i in °C; $T_{W_{\text{Obs}}}$ is the observed daily river water temperature at time step i in °C; $T_{W_{\text{Obs,Avg}}}$ is the average daily river water temperature at time step i in °C; n is the number of data pairs in comparison. In order to explore the sensitivity of river water temperature to atmospheric warming and changes in discharges, we applied the trained MLRM for the period of September 15, 2012, to April 30, 2014, with perturbed air temperature and river flow series. The regression parameters obtained in the training were kept same for the sensitivity analysis, under the assumption that river geometry and other physical setting of the river remains the same (Mohseni and Stefan 1999). The historical daily air temperature data and discharge for the period of 2012–2014 is perturbed incrementally with air temperature increase of +2 °C, +4 °C and + 6 °C and changes in river flow as ±20%. The selected increase in air temperature range is based on the global projected surface temperature increase of about 2.6 °C to 4.8 °C under Representative Concentration Pathway (RCP) 8.5 by the end of the twenty-first century (2081–2100) based on Assessment Report 5 (AR5), Intergovernmental Panel on Climate Change (IPCC) (IPCC 2014). Further, the water yield for the Missouri River basin is predicted to decrease by 10–20% during spring and summer months according to Stone et al. (2001) based on Regional Climate Model (RegCM) and a physically based rainfall-runoff model. Moreover, a significant increasing trend for river flows is observed at 5% significance level using Mann-Kendall trend test, which is a nonparametric trend detection method in the climate and hydrological data analysis (Mann 1945; Kendall 1955). The linear trend analysis of the observed data has shown an increasing trend with slope of 0.71 m^3/s per day. Thus, the selected rates of increase in air temperature and decrease in discharge are plausible under climate change perspective.

Results and Discussion

Time series of observed daily air and water temperature for the Missouri River near Nebraska City, NE, USA, for the period 2012–2014 is shown in Fig. 12.1. The daily river water temperature is following the variation of daily air temperature indicating air temperature as the prominent independent variable in the prediction of river water temperature. Further, the relationship between river discharge, air temperature and water temperature are studied using correlation coefficients between them and presented in Table 12.1. A strong positive correlation is observed between air and river water temperatures at daily time scale for the case study. A strong positive correlation between discharge and river water temperature is also observed for the Missouri River at Nebraska City. The observed annual river water temperature is noted as 12.09 °C, which is more than the observed annual air temperature of about 9.05 °C. It should be noted that there can be various anthropogenic heat sources persisting, such as emission of industrial cooling water, downstream of a thermal plant or reservoir, outfalls from a sewage treatment plant, net exchange from ground water temperature, etc. which may cause higher river water temperature at any particular river location.

The predicted river water temperature with MLRM is compared with observed values for training and testing periods in Fig. 12.2. A good agreement between the measured and predicted river water temperature is observed with clustering around

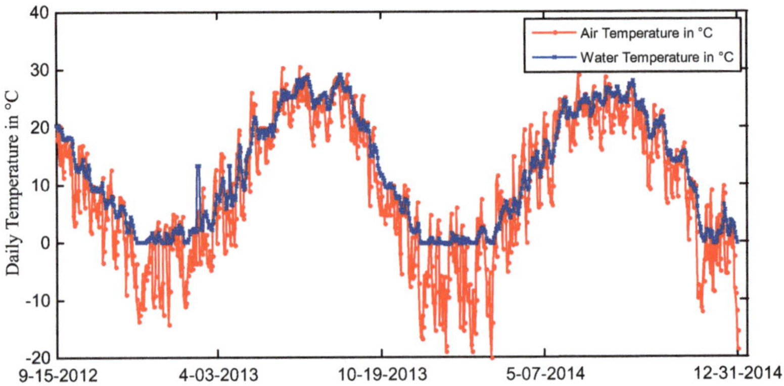

Fig. 12.1 Daily air and water temperatures for the period of 2012–2014 at Missouri River at Nebraska City, NE, USA

Table 12.1 Correlation coefficients between air temperature, water temperature and discharge along Missouri, Nebraska City, NE, USA

	Air temperature	Discharge	River water temperature
Air temperature	1.0	0.52	0.91
Discharge	0.52	1.0	0.55
River water temperature	0.91	0.55	1.0

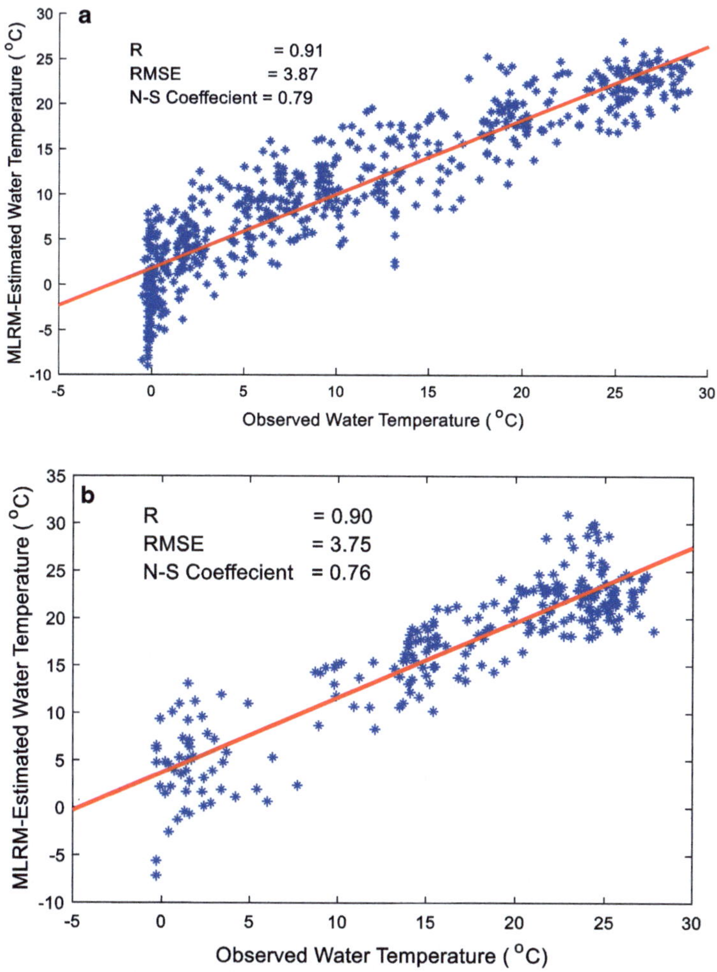

Fig. 12.2 Multiple Linear Regression Model (MLRM) estimated and observed river water temperature for (**a**) training (**b**) testing periods for Missouri River at Nebraska City, NE, USA. The solid line represents for 1:1 line

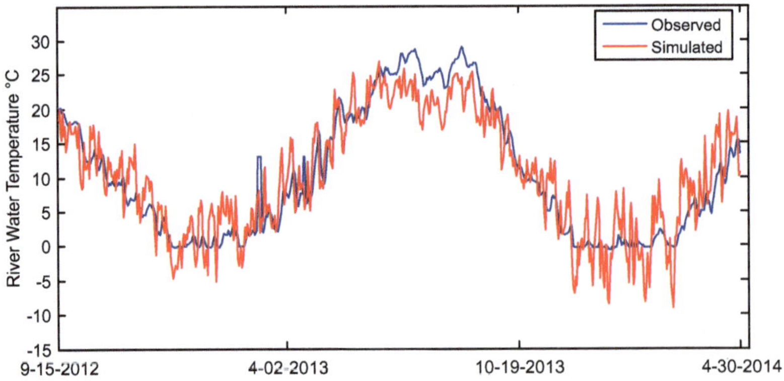

Fig. 12.3 Observed and simulated river water temperature for the training period of September 15, 2012, to April 30, 2014, for Missouri River at Nebraska City, NE, USA

the 1:1 line (Fig. 12.2) for both training and testing periods. The performance of the river water temperature model developed is studied with Nash-Sutcliffe (N-S) coefficient as 0.79 and 0.76 for the training and testing period, respectively, while the Root Mean Square Error (RMSE) for the training and testing of the model is observed as 3.87 and 3.75, respectively. The coefficient of determination (R) is obtained as 0.91 and 0.90 for training and testing of the MLRM, respectively.

Figure 12.3 shows the observed and simulated river water temperature in °C for the training period. The regression equation developed in the training model is given in the Eq. (12.4).

$$T_w = 2.25 + 0.67T_{\text{air}} + 0.004Q \tag{12.4}$$

The regression model showed an underestimation of river water temperature, particularly during the June, July, August, September, and October, when the river water temperatures are the highest. The linear regression model developed is not able to capture the higher river water temperatures. As the air temperature increases, the moisture holding capacity of the atmosphere increases, and the rate of evaporative cooling often increases, and therefore the stream temperature no longer increases linearly with air temperature (Mohseni et al. 1998; Bogan et al. 2003). Therefore, the air and water temperature linear regression models may not be preferable at higher river temperatures, whereas a non-linear logistic regression model (Mohseni et al. 1998) can be more promising.

The daily river water temperature model developed is used for the sensitivity analysis to predict the changes in RWT under the changes in air temperature and river flows. The annual mean river water temperature is observed to be increased at the river station under the increase of air temperatures of +2 °C, +4 °C and + 6 °C, with an annual mean (range) increase of 12.09 (−0.5 to 29) °C, 13.52 (−7.67 to 32.22) °C, 14.85 (−6.34 to 33.54) °C and 16.18 (−5.01 to 34.89) °C, respectively. The annual mean river water temperature increase under different air temperature and discharge changes is given in Table 12.2. The increase in air temperature of

Table 12.2 Mean annual river water temperature increase (°C) under various air temperature increases and changes in discharges for the river station

Historical	+2 °C AT	+4 °C AT	+6 °C AT	+2 °C AT, +20%Q	+2 °C AT, −20%Q	+4 °C AT, +20%Q	+4 °C AT, −20%Q	+6 °C AT, +20%Q	+6 °C AT, −20%Q
12.09	13.52	14.85	16.18	14.29	12.74	15.63	14.07	16.96	15.40

+2 °C, +4 °C and + 6 °C generally increased the RWT, while decrease of 20% in river flow intensified water temperature increase for the river location. The highest increase in water temperature is observed for the increase in air temperature of +6 °C with an increase in river flow of 20% compared with the observed annual average water temperatures. The increase in water temperature only with +6 °C (16.18 °C) and with +6 °C AT and + 20%Q (16.96 °C) are not showing any considerable difference, indicating influence of increase in air temperature will have more pronounced increase in river water temperature for the river location. Overall, the average annual increase of RWT for the Missouri River at Nebraska City, Nebraska, USA, under various air temperature and discharge changes is observed as 14.85 °C with an increase of about 2.76 °C when compared with the observed conditions.

Conclusions

Quantification of river water quality under climate change is prominent to know the possible risk of low water quality. Climate change-inducing water quality projections are essential for developing adaptive responses for the river water quality control models. The linear regression water temperature model developed based on air temperature and discharge as predictor variables is simple and robust to predict the daily river water temperature with good agreement of performance. The river water temperature sensitivity analysis presented in this chapter will provide a preliminary overview of the changes in river water quality under the increase in air temperature and changes in river flows. The results of the present study with RWT changes under air temperature increase, and river flow alterations indicate that the impact of air temperature increases were generally more pronounced compared to discharge. In fact, the impact of river flow changes on RWT were observed as moderate at mean annual basis compared to air temperature increase. Overall, the increase of RWT for the Missouri River at Nebraska City, Nebraska, USA, under various air temperature and discharge changes is observed as about 2.76 °C when compared with the observed conditions at mean annual scale. The study results are preliminary estimates of RWT for the alterations of air temperature and river flows. The selected range of air temperature and discharge are representatives of the local observed trends and global climate change scenarios. A detailed research is further needed to address the impact of climate change and anthropogenic changes such as reservoir operation, thermal effluents and cooling of water discharges.

Future Directions

Quantification of river water temperature has emerged as a vital problem in the river water quality modelling in the context of management and adaptation perspective. The current river water temperature models are based on the average conditions of air and river flows. The climate change sensitivity is more predominant under the extreme scenarios of hydro-climatic variables. Therefore, the future challenges may be perceived towards the modelling strategies to include hydro-climatic extremes in the RWT prediction models. Further, such modelled RWT under extreme scenarios can be integrated with the risk assessment models to study the extreme risk of river water quality under climate change (Rehana and Dhanya 2018). The advancement of hydrological models can be utilised to integrate the effect of meteorological and hydrological conditions for the analysis of alterations of RWT under hydro-climatic extremes of climate change (Ficklin et al. 2012). To this end, while climate change is one dominant factor, in impairing the river water quality, anthropogenic activities will be yet another dominant factor which need to be incorporated in the river water quality control problems.

Acknowledgements The research work presented in the manuscript is a part of Water Advanced Research and Innovation (WARI) Fellowship Program supported by the Department of Science and Technology (DST), Government of India, the University of Nebraska-Lincoln (UNL), the Daugherty Water for Food Institute (DWFI) and the Indo-US Science Technology Forum (IUSSTF).

References

Bogan, T., Mohseni, O., & Stefan, H. G. (2003). Stream temperature-equilibrium temperature relationship. *Water Resources Research, 39*(9), 1245. https://doi.org/10.1029/2003WR002034.

Bogan, T., Othmer, J., Mohseni, O., & Stefan, H. (2006). Estimating extreme stream temperatures by the standard deviate method. *Journal of Hydrology, 317*, 173–189.

Caissie, D. (2006). The thermal regime of rivers: A review. *Freshwater Biology, 51*(8), 1389–1406.

Caissie, D., Satish, M. G., & El-Jabi, N. (2007). Predicting water temperatures using a deterministic model: Application on Miramichi River catchments (New Brunswick, Canada). *Journal of Hydrology, 336*, 303–315.

Dingman, S. L. (1972). Equilibrium temperature of water surface as related to air temperature and solar radiation. *Water Resources Research, 8*(1).

Edinger, J. E., Duttweiler, D. W., & Geyer, J. C. (1968). The response of water temperatures to meteorological conditions. *Water Resources Research, 4*(1), 1137–1143.

Ficklin, D. L., Luo, Y., Stewart, I. T., & Maurer, E. P. (2012). Development and application of a hydroclimatological stream temperature model within the soil and water assessment tool. *Water Resources Research, 48*, W01511. https://doi.org/10.1029/2011WR011256.

Hockey, J. B., Owens, I. F., & Tapper, N. J. (1982). Empirical and theoretical models to isolate the effect of discharge on summer water temperatures in the Hurunui River. *Journal of Hydrology, 21*(1), 1–12.

IPCC. (2014). *Climate Change Synthesis report*. Summary for policymakers. https://www.ipcc.ch/pdf/assessmentreport/ar5/syr/AR5_SYR_FINAL_SPM.pdf.

Janssen, P. H. M., & Heuberger, P. S. C. (1995). Calibration of process oriented models. *Ecological Modelling, 83*(1– 2), 55–66.
Kendall, M. G. (1955). *Rank correlation methods*. New York: Hafner Publishing Co.
Kim, K. S., & Chapra, S. C. (1997). Temperature model for highly transient shallow streams. *Journal of Hydraulic Engineering, 123*, 30–40.
Mackey, A. P., & Berrie, A. D. (1991). The prediction of water temperatures in chalk streams from air temperatures. *Hydrobiologia, 210*, 183–189.
Mann, H. B. N. T. (1945). Against trend. *Econometrica, 13*, 245–259. https://doi.org/10.2307/1907187.
Marce, R., & Armengol, J. (2008). Modelling river water temperature using deterministic, empirical, and hybrid formulations in a Mediterranean stream. *Hydrological Processes, 22*, 3418–3430. https://doi.org/10.1002/hyp.6955.
Mohseni, O., & Stefan, H. G. (1999). Stream temperature air temperature relationship: A physical interpretation. *Journal of Hydrology, 218*(3–4), 128–141.
Mohseni, O., Stefan, H. G., & Erickson, T. R. (1998). A nonlinear regression model for weekly stream temperatures. *Water Resources Research, 34*, 2685–2692.
Nash, J. E., & Sutcliffe, J. V. (1970). River flow forecasting through conceptual models, part 1: A discussion of principles. *Journal of Hydrology, 10*(3), 282–290.
Neumann, D. W., Rajagopalan, B., & Zagona, E. A. (2003). Regression model for daily maximum stream temperature. *Journal of Environmental Engineering, 129*, 667–674.
O'Driscoll, M. A., & DeWalle, D. R. (2006). Stream–air temperature relations to classify stream–ground water interactions in a karst setting, Central Pennsylvania, USA. *Journal of Hydrology, 329*, 140–153.
Rehana, S., & Dhanya, C. T. (2018). Modeling of extreme risk in river water quality under climate change. *Journal of Water and Climate Change.* MS. No: JWC-D-17-00024.
Rehana, S., & Mujumdar, P. P. (2011). River water quality response under hypothetical climate change scenarios in Tungabhadra river, India. *Hydrological Processes, 25*(22), 3373–3386.
Rehana, S., & Mujumdar, P. P. (2012). Climate change induced risk in water quality control problems. *Journal of Hydrology, 444–445*, 63–77.
Stone, M. C., Hotchkiss, R. H., Hubbard, C. M., Fontaine, T. A., Mearns, L. O., & Arnold, J. G. (2001). Impacts of climate change on Missouri River basin water yield. *Journal of the American Water Resource Association, 37*(5), 1119–1129.
Toffolon, M., & Piccolroaz, S. (2015a). A hybrid model for river water temperature as a function of air temperature and discharge. *Environmental Research Letters, 10*, 14011.
Toffolon, M., & Piccolroaz, S. (2015b). A hybrid model for river water temperature as a function of air temperature and discharge. *Environmental Research Letters, 10*, 114011.
van Vliet, M. T. H., Ludwig, F., Zwolsman, J. J. G., Weedon, G. P., & Kabat, P. (2011). Global river temperatures and sensitivity to atmospheric warming and changes in river flow. *Water Resources Research, 47*, W02544. https://doi.org/10.1029/2010WR009198.
van Vliet, M. T. H., Yearsley, J. R., Franssen, W. H. P., Ludwig, F., Haddeland, I., Lettenmaier, D. P., & Kabat, P. (2012). Coupled daily streamflow and water temperature modelling in large river basins. *Hydrology and Earth System Sciences, 16*, 4303–4321.
Webb, B. W. (1996). Trends in stream and river temperature. *Hydrological Processes, 10*(2), 205–226.
Webb, B. W., & Nobilis, F. (2007). Long-term changes in river temperature and the influence of climatic and hydrological factors. *Hydrological Sciences Journal, 52*(1), 74–85.
Webb, B. W., Clack, P. D., & Walling, D. E. (2003). Water-air temperature relationships in a Devon river system and the role of flow. *Hydrological Processes, 17*(15), 3069–3084.
Yearsley, J. R. (2009). A semi-Lagrangian water temperature model for advection-dominated river systems. *Water Resources Research, 45*, W12405. https://doi.org/10.1029/2008WR007629.

Part IV
Other Aspects

Chapter 13
Thermophiles vs. Psychrophiles: Cues from Microbes for Sustainable Industries

Monica Sharma

Contents

Abstract Though cellular architecture and functions show vast array of adaptive features to combat extreme temperature regime, enzymes are the key determinants of thermal adaption in both extremes of life, i.e., psychrophily or thermophily, as they drive the metabolism and cell cycle. Psychrophilic enzymes exhibit high specific activity at lower temperature range by disappearance of non-covalent stabilizing interactions (H bonding, hydrophobic interactions, salt bridges, etc.) and proline and arginine residues which cause improved flexibility (local/global) in conformation of enzymes. These enzymes have devised diverse ways to achieve the feat to live in extremity. Thermophilic enzymes work totally opposite to psychrophiles, i.e., by increasing proline number that causes proline isomerization which renders them to be more rigid and have higher arginine content which leads to increased salt bridge formation and extensive H bonding, etc. Oligomerization and heat shock proteins further give microbes stability against temperature.

M. Sharma (✉)
Department of Biotechnology, School of Life Sciences, Babasaheb Bhimrao Ambedkar University (A Central University), Lucknow, Uttar Pradesh, India
e-mail: dr.monikas@bbau.ac.in

R. C. Sobti et al. (eds.), *Environmental Biotechnology: For Sustainable Future*,
https://doi.org/10.1007/978-981-10-7284-0_13

Keywords Psychrophiles · Thermophiles · Oligomerization · Proline isomerization

Introduction

Nature is endowed with many exotic environments which were characteristic features of primitive Earth, and microbes living in extreme conditions have evolved themselves to grow in such harsh conditions like extreme temperatures (less than 0 °C to >100 °C), high salinity (high NaCl), extreme pH (<2 to >10), high pressure (>50 MPa), etc. (Gerday et al. 1997; Andrade et al. 1999). Microbes are broadly classified into four categories according to their thermal adaptations: psychrophiles, mesophiles, moderate thermophiles (60–80 °C), and hyperthermophiles (80 °C or above). More than 70% of the Earth's surface is made up of extremely cold ecosystems like the Arctic and the Antarctic regions and marine habitats. Some habitats of the Earth are extremely hot like volcanic areas, solfatara fields, geothermal power plants, hydrothermal vents, deep hot springs, and fumeroles (Deming 2002; Junge et al. 2004; Rodrigues and Tiedje 2008). Such environments exert a highly selective pressure on the organisms colonizing the ecosystem, and these organisms develop adaptation strategies to either survive or breed like true psychrophiles/ thermophiles. Much progress has been made in studies elucidating the molecular adaptation mechanisms of enzymes of thermophiles and hyperthermophiles; however, the molecular adaptation to cold is still inadequately understood. In the past two decades, few groups have tried to accumulate structural data on psychrophilic enzymes, and now we are in the position to shed some light on their functional and structural characteristics (Feller and Gerday 1997). These microbes have evolved a variety of mechanisms to adapt at extreme temperatures: physiological changes and structural changes in enzymes which enable metabolic fluxes. Psychrophiles produce antifreeze proteins and cold shock proteins and altered membrane composition (flexible), while thermophiles possess heat shock proteins and rigid membrane structure (Trent et al. 1994; Koga 2012). In this chapter we will discuss about molecular adaptation strategies of psychrophiles or thermophiles at the enzyme level.

The stability/folding or unfolding of the enzymes of psychrophiles and thermophiles depends on intrinsic stabilizing forces (e.g., salt bridges, hydrogen bonds, hydrophobic interactions) but in opposite directions and extrinsic stabilizing factors to address either activity at low temperatures or stability in hot environments (Feller 2010). In the following paragraphs, these factors will be discussed.

Flexibility and Rigidity: Activity and Stability Relationship

Enzymes of cold-adapted microbes have lower number of proline and arginine residues, hence increasing backbone flexibility, and have fewer H-bonding interactions (Paredes et al. 2011). Feller et al. (1994) studied triosephosphate isomerases, β-lactamase, and α-amylase of psychrophile *Alteromonas haloplanktis* and observed that these enzymes have less number of proline residues as compared to porcine. In these enzymes proline residues are either deleted or substituted by small amino acids like alanine essentially within loops and turn regions of protein and hence introduce flexibility of connecting region of adjacent secondary structure. In psychrophilic *Psychrobacter immobilis*, class C β-lactamases lack two proline residues in the connecting loops of alpha-5 and alpha-6 helices which are present in the vicinity of the active site. The other three conserved proline residues present in mesophilic lactamases are replaced by other amino acids. β-lactamases of *P. immobilis* possess only 10 arginine-mediated hydrogen bonding as compared with 16 in homologous mesophilic *Enterococcus cloacae* β-lactamase (Schiffer and Dötsch 1996).

Glycine has highly rotatable dihedral angles around N-C bond and therefore increases the degree of freedom of an unfolded polypeptide backbone. Feller et al. (1991) observed that unusual stacking of glycine is frequently seen in closer functional domains like active site of lipases, α-amylase, as well as in subtilisin calcium binding site (Feller et al. 1991). Menendez-Arias and Argos (1989) pointed out that glycine in mesophile is frequently substituted for alanine residue in thermophiles. All these result in reduced packing density at core and make enzymes increasingly flexible conformationally (Paredes et al. 2011; Gianese et al. 2002; Siddiqui and Cavicchioli 2006; Struvay and Feller 2012). It was proposed that enhanced flexibility is the major molecular mechanism for the evolution of cold-adapted enzymes. Though, this is not uniform in all psychrophilic enzymes. In many psychrophilic enzymes like trypsin of Atlantic salmon, carbonic anhydrase, isocitrate dehydrogenase, alcohol dehydrogenase, and iron superoxide dismutases are highly flexible near the active site (Tsigos et al. 1998;, Fedoy et al. 2007; Papaleo et al. 2008; Chiuri et al. 2009; Merlino et al. 2010), while psychrophilic enzyme uracil DNA glycosylases possess strong ion pairs near the C-terminal and exhibit flexible regions at distant places other than the active site (Olufsen et al. 2005). Psychrophilic alpha-amylases have greater overall flexibility as compared to theromphilic counterparts (Aghajari et al. 1998). It is inferred from the above studies that psychrophilic enzymes use different strategies to tackle colder environments and local and global protein flexibility influences the thermal stability divergently.

Thermophiles on the other hand are relatively rigid and exhibit compact packing density (Wrba et al. 1990; Fontana 1991; Jaenicke 1991). Statistical analyses of many thermophiles have shown that thermophilic enzymes have higher ratio of arginine, proline, and lysine (Low et al. 1973; Menendez-Arias and Argos 1989; Matthews 1991). Due to the charge resonance of guanidinium group present in arginine, it can form additional salt bridges (Mrabet et al. 1992) and can form multiple hydrogen bonds with backbone oxygen of carbonyl group (Borders Jr et al. 1994). Proline is usually the least abundant amino acid residue in protein due to pyrrolidine ring restricting it to form few conformations which impairs N-C bond rotations. Proline residues are present only in the turn region of proteins and decrease the entropy energies of unfolded states. Oligo-1,6-glucosidase of thermophilic *Bacillus thermoglucosidasius* showed extra thermostability due to presence of 14 additional residues of proline (Watanabe et al. 1991, 1994).

Wang et al. (2014a, b) protein engineered xylanase mutants of mesophile *Streptomyces* sp. strain S9 through substitution of valine, glycine, serine, and aspartic acid with proline or glutamic acid which led to improved thermal properties. The resultant enzyme also exhibited evident temperature optima shifts of about 17 °C upward, and their half-lives were markedly increased by more than nine times for thermal inactivation at 70 °C. The substitution of V81P and G82E introduced rigidity at the end of the β-barrel which was earlier prone to unfold and thus improved thermostability of engineered xylanase. Razvi and Scholtz (2006) also proposed that loop denaturation can be constrainted by proline introduction which cause reduction in entropic contribution (ΔS) of folding, hence resulting in higher thermal stability (T_S). Barzegar et al. (2009) did genomic X-ray analysis of three related alcohol dehydrogenases of yeast, horse liver, and *Thermoanaerobacter brockii*. It was found out that alcohol dehydrogenase (ADH) of horse liver and *T. brockii* has higher turn and loop percentage as compared to yeast ADH. On pairwise alignment, it was observed that TBADH/HLADH exhibit 2.5-fold greater similarity in comparison to TBADH/YADH pair. Multiple alignments revealed the presence of higher number of conserved proline in TBADH/HLADH pair. These proline residues occurred in surface loops, and as it is known, wherever proline is present, it introduces kink/turn in the main chain by forming H bonding. Proline residues more likely introduce more coils (turns/loops) in structure and hence proposed to be responsible for the increased thermostability by loop rigidification. Igarashi et al. (1998) generated the mutant of α-amylase with improved thermostability by proline substitution for the first time. Later many enzymes like alkaline protease, chitinase, and luciferase were protein engineered by site-directed mutagenesis by proline substitution (Igarashi et al. 1999; Gaseidnes et al. 2003; Yu et al. 2015). Figure 13.1

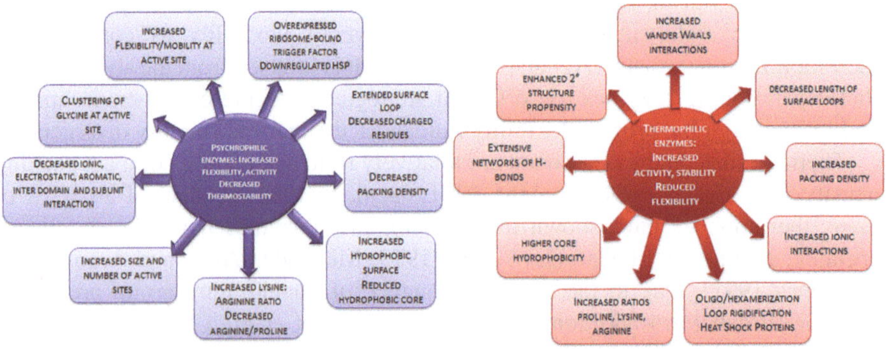

Fig. 13.1 Common structural modifications of enzymes of psychrophiles vs thermophiles

illustrates that there is no particular interaction which causes stabilization; rather, there are common structural modifications and a sheer change in the number of interactions which are responsible for thermostability/thermolability.

Hydrophobic Interactions

Proteins undergo spontaneous folding due to the presence of hydrophobic side chain clustering at the core, which is a major driving force to achieve stable folded conformation (Matthews 1991). Psychrophilic enzymes have a unique adaptation; they possess higher number of non-polar amino acid residues on the outer surface of the protein and destabilize the structured water layer found around it. Paredes et al. (2011) using bioinformatic tools presented evidences that cavity size of active sites of many psychrophilic enzymes is much larger as compared to thermophilic counterparts and active site contains more surrounding hydrophilic groups and hence has more water molecules, consequently increasing conformational dynamics of the enzyme and hence greater activity under the rigidifying cold climate. In thermophiles strongly hydrophobic amino acids form the core of the enzyme; hence more energy is needed to unfold it, making them thermostable. Glutamate dehydrogenase of *Pyrococcus furiosus* had higher content of hydrophobic amino acids and reduced content of polar and charged amino acids. It was also observed that the enzyme had lesser number of glycine residues as compared to the mesophilic counterpart (Maras et al. 1994). Goldstein (2007) reported the presence of lesser number of uncharged polar residues and charged and aromatic residues and higher number of hydrophobic residues in psychrophiles as compared to mesophiles. Various groups have also suggested that the cold-adapted proteins augment their elasticity by reducing strength and number of non-covalent interactions like hydrogen bonding, salt bridges, etc. (Gianese et al. 2001; Violot et al. 2005). Psychrophiles possess a higher number of glycine residue which enhance their flexibility and stability by raising the conformation entropy of the unfolded state.

Ion Binding

Most of the thermophilic enzymes are stabilized by tight ion binding or multiple ion binding sites (Teplyakov et al. 1990; Sen and Nilsson 2016), while psychrophilic enzymes possess loosely bound ions. Subtilisin from thermophiles possess asparagine a strong calcium binder at position 77 which is substituted by threonine in psychrophilic subtilisin (Davail et al. 1994). The alpha-amylase of *A. haloplanktis* a psychrophile has an unidentate lysine residue at position 337 instead of a bidentate arginine residue of porcine which binds to Cl ion (Feller et al. 1994). Pyrolysin of hyperthermophilic archaeon *P. furiosus* possesses two calcium binding sites, and mutation of calcium binding site causes destabilization of enzymes. This suggests that metal ions play a pivotal role in thermostability and pyrolysin activity (Zeng et al. 2014).

Electrostatic and Solvent Interactions

In many cold-active enzymes, a marked difference in electrostatic potential near active site was observed in comparison to thermophiles and mesophiles. Electrostatic potential difference is generally introduced by discrete substitution in non-conserved charged residues. Such difference facilitated enzyme-ligand interaction occur in citrate synthase (Russell et al. 1998), malate dehydrogenase (Kim et al. 1999), uracil-DNA glycosylase (Leiros et al. 2003), and trypsin (Smalås et al. 2000; Gorfe et al. 2000; Brandsdal et al. 2001). In *Aquaspirillium arcticum*, malate dehydrogenase enzyme possesses two sites, one for oxaloacetate binding having increased positive potential and NADH binding site characterized by significantly decreased negative surface potential (Kim et al. 1999). In psychrophilic enzymes, solvent-accessible area calculations have revealed that they have higher fraction of non-polar residues exposed to the surrounding medium. This causes entropy-driven destabilization due to reorganization of water molecules around uncovered hydrophobic patches. In some instances electrostatic potential of psychrophiles is negative at surface, and mostly isoelectric point (pI) is more acidic than thermophilic or mesophilic homologues. This phenomenon of charge generation causes increased solvent protein interactions which result in acquisition of flexible molecular edifice near zero/sub-zero temperature range (Feller et al. 1999).

Hydrogen Bonding, Salt Bridges, and Aromatic Interactions

Intramolecular and intermolecular hydrogen bonding is a prime factor in stabilizing native structure of the functional proteins (Creighton 1991). Commonly most of the thermophilic and mesophilic proteins exhibit increased hydrogen bonding and salt bridge pattern (Vogt and Argos 1997; Vogt et al. 1997; Vetriani et al. 1998; Berezovsky and Shakhnovich 2005); however psychrophilic enzymes lack extensive hydrogen bonding, e.g., trypsin enzyme of an Antarctic fish exhibited 20% lower intramolecular hydrogen bonding potential as compared to bovine protease. An alpha-amylase of the psychrophile *A. haloplanktis* lacked 12 surface bridges when compared with mesophilic counterpart (Walker et al. 1980). In psychrophiles, basic residues which form salt bridges were replaced by glutamine or asparagine amino acid. Subtilisin enzyme of thermophilic *Bacillus* sp. had higher salt bridges which gave enzyme better thermal stability than the parent strain.

Two aromatic side chains of the proteins show weakly polar interaction of enthalpic importance. Burley and Petskn (1988) reported that in extremophilic subtilisins aromatic-aromatic interactions can be correlated with optimal growth temperature of the parent *Bacillus* strains. In psychrophilic amylase also, aromatic-aromatic interactions were found to be strongly conserved to maintain confirmation of enzyme, but such interactions were rare in psychrophiles as compared to thermophilic homologues (Feller et al. 1994).

Protein Packaging

Hyperthermophiles and thermophiles exhibit decreased protein surface area in comparison to protein size; hence they are denser. Enzymes of *P. furiosus* exhibit multiple adaptations against thermal denaturation by having tightly packed hydrophobic cores with many ion pair interaction and buried atom to minimize unfavorable solvent interactions (Rice et al. 1996; Kumar and Nussinov 2001). However, these enzymes show highly conserved structure (active site), and structural variability (size of helices, beta sheets, ionic modification at terminal ends) is seen in the protein surfaces which react with the solvent (Ladenstein and Antranikian 1998; Kumar and Nussinov 2001). Again, tight packaging is observed, but this does not alter the final confirmation. Johns and Somero (2004) advocated that psychrophilic enzymes are more flexible in their structure to pay off for the "sub-zero effect" of freezing habitats. Either the whole enzyme can be flexible or it may be restricted to the site of catalysis (Collins et al. 2003; D'Amico et al. 2003). Crystallographic analysis of psychrophilic proteins indicated that enhanced flexibility is generally due to attenuation of strength of entropic and enthalpic contribution. Psychrophilic proteins possess increased number of amino acids with smaller side chains which points out to be weakly hydrophobic in nature (Armstrong et al. 2011). They also possess increased number of exposed hydrophobic residues at the surface and reduced number at the core; hence they have internal loose packing at the core which is related to their intrinsic flexibility at cold temperature (Violot et al. 2005).

Extrinsic Stabilizing Factors

Psychrophiles exhibit extreme temperature resistance due to metabolic fluxes, peculiarly by producing antifreeze proteins other than intrinsic stabilizing factors. Georlette et al. (2003) studied thermophilic DNA ligases of psychrophiles, mesophiles, and thermophiles and concluded that psychrophiles work opposite to mesophiles and thermophiles and there is a complex relationship between activity, flexibility, and stability. Psychrophiles have developed varied mechanisms of temperature compensation, but majority of them deal with cold environment by decreasing enthalpy and enhancing turnover number (K_{cat}) and catalytic efficiency (K_{cat},/K_M) (Low et al. 1973; Arpigny et al. 1994). Oligomerization is a well-documented strategy to stabilize enzymes which contribute significantly to thermostability (Jaenicke et al. 1996; Sterner et al. 1996; Villeret et al. 1998; Jaenicke and Bohm 1998; Dams and Jaenicke 1999). Though coil-coil interaction of alpha-helix region causes oligomerization in thermophilic enzyme, Tanaka et al. (2004) reported that in L-isoaspartyl O-methyltransferase from *Sulfolobus tokodaii* (StoPIMT), three interfacial contact regions known as major, minor, and coiled-coil contact regions affect the oligomerization and protein thermostability via hexamerization.

Geobacillus thermodenitrificans NG80-2 is a thermophilic facultative anaerobe that grows at 45–73 °C and has three putative superoxide dismutases (SODs) genes, Fe/Mn-SOD, Mn-SOD, and Cu/Zn-SOD. Of these three, Fe/Mn-SOD encodes additional 244-amino acid domain at N-terminal known as NTD. Wang et al. (2014a, b) reported that NTD-fused proteins of SOD contribute to thermophilicity in host. *Sulfolobus solfataricus* SOD thermostability was further strengthened by fusion of N-terminal domain (NTD) isolated from *G. thermodenitrificans* NG80-2. The NTD conferred heat-improved thermophilicity/thermostability, higher optimum activity temperature, broader pH permanence, and superior tolerance to organic media and inhibitors without altering SOD's oligomerization state (Li et al. 2016). Psychrophiles are reported to possess cold-adapted chaperones such as DnaK (Rina et al. 2000) and GroEL (Ferrer et al. 2003, 2004; Strocchi et al. 2006). When these chaperones were expressed in mesophilic *Escherichia coli*, they were able to grow at low temperature; hence it was inferred that these chaperones are crucial for thermal adaptation of microorganisms. Beside well-documented heat shock proteins (HSPs) involved in protein folding, a few studies have reported the presence of another primary cold shock protein, known as ribosome-bound trigger factor (TF) which provides thermal adjustment in cold habitat when expressed in *E. coli* (Kandror and Goldberg 1997; Hartl and Hayer-Hartl 2009). In several cold-adapted microbes, TF gene was found to be overexpressed (Qiu et al. 2006; Kawamoto et al. 2007; Piette et al. 2010; Mykytczuk et al. 2011), while HSP chaperones were downregulated (Goodchild et al. 2005; Rodrigues et al. 2008; Piette et al. 2011).

Extremophiles Relevance in Industries

Enzymes/biocatalysts are being used by mankind since ancient times unknowingly for the production of food and beverages. However in recent times, they have been studied more extensively, and significant understanding has been generated about their kinetics and physicochemical behavior. Diastase/amylase was the first discovered enzyme in the year 1833 (Payen and Persoz 1833). In the twentieth century, enzymes were used commercially in detergent industries. Many efforts were made to modify these proteins by protein engineering. Since then industrial enzymes have evolved into multibillion dollar global market. Dewan (2014) predicted that the global market of industrial enzymes is expected to reach 1.8 billion dollar by 2018 with 11.3% compounded annual growth rate.

Enzymes from thermophiles and psychrophiles have carved a niche in the market due to their substantial variability in adaptations and which expanded the range of their applications in enzymatically unfavorable conditions of industrial processes. They have opened a new vista in catalysis which is highly attractive, sustainable, cost-effective, and environment-friendly as compared to chemical catalysis. Thermophilic enzymes are of utmost economic importance in many biotechnological industries owing to their ability to produce thermostable biocatalyst as compared to their mesophilic counterparts. Thermostable enzymes are preferred because of their ability to work at elevated temperatures which reduced the cost of cooling infrastructure in industries, reduced risk of contamination, reduced viscosity of the reaction medium, and hence increased solubility and bioavailability of many organic compounds resulting in low power consumption in mixing and aeration in bioreactors. At higher temperatures, substrates' diffusion rate increases which causes higher rate of reaction. Many industries use thermostable enzymes like cellulases, xylanases, and pectinases as catalysts in wastewater treatment, bio-bleaching of paper pulp, animal feed production, biofuel production from fermentable sugars of cellulosic wastes, fruit juice clarification, natural fiber degumming, curing of raw coffee, cocoa, and tobacco, etc. (Table 13.1). Most of the industrial enzymes are produced by heterologous expression in a recombinant host either prokaryotes or eukaryotes (Gomes et al. 2016).

Psychrophiles adopt different mechanisms to combat the temperature extremes. Recently substantial efforts have been put forward to develop cold enzymes for many industries like molecular biology reagents, detergent industries, and food and beverage industries. Muñoz et al. (2017) studied the role of anti-freeze protein interactions with ice in Antarctic microorganisms and anticipated that they could revolutionize frozen food industry. Psychrophilic enzymes give edge over their meso-/thermophilic counterparts due to their high activity. In some processes like meat tenderization, their activity can be controlled by rising temperature after certain periods to stop its action. They can also be used as catalyst in processes to avoid formation of by-products due to rise in temperature. Psychrophilic enzymes work at low and slight moderate temperature and do not require much input of heat. These characteristics provide clear advantages for use of these enzymes in various industrial processes in a cost-effective manner and reduced environmental impact. Major market players of detergent industry like Procter & Gamble and Unilever are aiming for cold wash detergent by 2020, which would lead to reduced energy consumption and low carbon dioxide emissions as well as improved fabric texture after every wash. Psychrophilic enzymes are also becoming a part of food and beverage industry where high-temperature processes are replaced by energy-efficient low-temperature processing. Low-temperature processing is not only energy efficient but also prevents food contamination and spoilage, minimizes undesirable chemical reactions, and retains volatile and thermolabile flavor compounds (Horikoshi 1999; Pulicherla et al. 2011). A few selected industrially important psychrophilic enzymes are enlisted in Table 13.2.

Future Prospects

Enzymes encoded by thermophiles, hyperthermophiles (showing optimal growth at >80 °C), and psychrophiles have the same catalytic mechanisms as compared to their mesophilic counterparts. They usually retain their physicochemical properties even when cloned and expressed in mesophiles. Structural and sequence studies (sequence alignments, amino acid content comparison) have indicated that these

Table 13.1 Applications of important enzymes produced by thermophiles

Enzymes	Organism	Application
DNA polymerases		
DNA polymerase	*Thermus thermophilus*	Reverse transcription PCR
Taq DNA polymerase	*T. aquaticus*	PCR in situ hybridization
		Reverse transcription-PCR
Deep vent polymerase	*P. furiosus*	PCR technologies
Vent polymerase	*T. litoralis*	PCR technologies
C. therm DNA polymerase	*Carboxydothermus hydrogenoformans*	Reverse transcription PCR
DNA ligases		
Pfu DNA ligase	*P. furiosus*	Ligase chain reaction
Tcs DNA ligase	*T. scodoductus*	Ligase chain reaction
Amylases		
α-Amylases	*Desulfurococcus mucosus*	Enzymatic starch hydrolysis to form syrups.
	Pyrococcus furiosus	Starch liquefaction
	P. woesei	
	Pyrodictium abyssi	
	Staphylothermus marinus	
	Thermococcus profundus	
	Dictyoglomus thermophilum	
	Thermotoga maritima	
β-Amylase	*T. maritima*	Saccharification process in maltose syrup production
	Thermoanaerobacterium thermosulfurigenes	
	Clostridium thermocellum SS8	
Glucoamylase	*C. thermosaccharolyticum*	Starch saccharification
	Methanococcus jannaschii	
	T. thermosaccharolyticum	
α-Glucosidase	*Thermoanaerobacter ethanolicus*	Starch saccharification
Pullulanase	*T. maritima*	Debranching enzyme
	T. caldophilus	
	Fervidobacterium pennavorans	
Amylopullulanase	*Desulfurococcus mucosus*	Debranching enzyme
	ES4	
	Pyrococcus furiosus	
	T. celer	
	T. litoralis	
	T. hydrothermalis	
	T. ethanolicus	

(continued)

Table 13.1 (continued)

Enzymes	Organism	Application
Xylose isomerase	*T. thermosulfurigenes*	High fructose syrup production
	Thermoanaerobacterium strain JW/SL-YS 489	
	T. aquaticus	
	T. maritima	
	T. neapolitana	
Glucose isomerase	*F. gondwanense*	P reparation of fructose-based syrups
	Bacillus sp.	
Alkaline phosphatase	*T. neapolitana*	Enzyme-labeling applications
Carboxypeptidase	*S. solfataricus*	C-terminal sequencing
Serine protease	*Thermus* Rt41A strain	DNA and RNA purifications; cellular structures degradation prior to PCR
Protease S	*P. furiosus*	Protein fragmentation for sequencing
Methionine aminopeptidase	*P. furiosus*	Cleavage of the N-terminal Met in proteins
Pyroglutamate aminopeptidase	*P. furiosus*	Cleavage of the N-terminal L-pyroglutamate in proteins
Aldo-keto reductase	*T. maritima*	Production of primary alcohols
Amidase	*Geobacillus pallidus* MTCC-9225	Production of hydroxamic acid and organic acids
Nitrile hydratase	*Pseudonocardia thermophila*	Acryalmide production
Acetylxylan esterase	*Thermobifida fusca* NTU22	Paper and pulp industry
Dextranase	*Streptomyes sp.* NK458	Sugar mills
Xylanase	*Thermomonospora fusca*	Paper and pulp
	Kocuria sp. RM1	Baking
		Animal feed
Keratinase	*Actinomadura keratinilytica* strain	Leather industry
	Cpt29	Pharmaceutical
	T. curvata	
Protease	*Nocardiopsis prasina* HA-4	Leather industry and brewing
	Saccharomonospora viridis SJ-21	Pharmaceutical industry
		Detergent

enzymes share much similarity to their mesophilic homologues. The diverse array of mechanisms adopted by enzymes of these organisms should be studied in detail and could be used as guiding principle to engineer the already discovered catalysts. Else these enzymes can be further cloned and expressed in mesophilic counterparts and can be used as potent catalysts in many industrial processes.

A few studies have investigated the potential of some thermophilic facultative bacteria in bioremediation of petroleum-contaminated sites for both in situ and ex situ. Hence the potential of such microbes can be harnessed for bioremediation of contaminated sites as well as for microbial-aided oil recovery.

Table 13.2 Applications of important enzymes produced by psychrophiles

Enzymes	Organism	Application
Molecular biology reagents		
Alkaline phosphatases	Antarctic strain TAB5.	5′ end dephosphorylation of a linearized DNA
	Metagenomic library	
Uracil-DNA N-glycosylase	*Psychrobacter* sp.HJ147	Release of free uracil from uracil-containing DNA
Nucleases	*Shewanella* sp.	All types of DNA and RNA digestion
Recombinases		Recombinase polymerase amplification
DNA ligases (not commercially available)	*Pseudoalteromonas haloplanktis*	Catalyze formation of phosphodiester bonds in nicked double-stranded DNA molecules
Textile		
Amylases	–	Desizing of woven fabrics
Cellulases	–	Bio-finishing and dyeing of cellulosic fabrics
Food and beverages		
Pectinases		Beer and wine fermentation, bread making, fruit juice extraction and clarification
β-galactosidasesor lactases	Marine psychrophilic bacterium	Production of lactose-free foodstuffs
	P. haloplanktis	Production of tagatose
Xylanases	*P. haloplanktis* TAH3A	In bread making
	Flavobacterium sp.MSY-2	
	Sorangium cellulosum So9733–1	
	Flavobacterium johnsoniae	
Catalase	*Serratia* sp.	Textile, research, and cosmetic applications
Detergent		
Lipases	*Pseudomonas stutzeri* PS59	Remove oily stains
Proteases	*Glaciozyma antarctica*	Removing protein stains
Excellase		Dish washing
Amylase	*Bacillus cereus* GA6	Starch stains removal
	Alteromonas haloplanktis	
	Microbacterium foliorum GA2	
	Zunongwangia profunda	
Cellulases	Antarctic and Antarctic LTC	Lint removal
	Humicola insolens	
Pectinases	–	Pectin-stain removal
Mannanases	–	Mannan and gum removal

Conclusion

The thermal adaption evolved by microbes is quite astonishing, and if we learn exactly how they achieve this feat to combat extreme temperature adaptation, we can gain insight to protein engineering using directed evolution. Though the understanding of mechanisms of extremophiles is still in very primitive stage, with better know-how and use of latest techniques, the future of utilizing microbes with capacities to flourish in extreme temperatures is going to be very exciting and useful. We have only touched the tip of the iceberg as yet, and a whole lot is yet to be explored in this area.

References

Aghajari, N., Feller, G., Gerday, C., & Haser, R. (1998). Structures of the psychrophilic *Alteromonas haloplanctis* [alpha]-amylase give insights into cold adaptation at a molecular level. *Structure, 6*, 1503–1516.

Andrade, C. M. M. C., Pereira, N., Jr., & Antranikian, G. (1999). Extremely thermophilic microorganisms and their polymer-hydrolytic enzymes. *Revista de Microbiologia, 30*(4), 287–298.

Armstrong, B. D., Choi, J., López, C., Wesener, D. A., Hubbell, W., Cavagnero, S., & Han, S. (2011). Site-specific hydration dynamics in the nonpolar Core of a molten globule by dynamic nuclear polarization of water. *Journal of the American Chemical Society, 133*, 5987–5995.

Arpigny, J. L., Feller, G., Davail, S., Genicot, S., Narmx, E., Zekhnini, Z., & Gerday, C. (1994). Molecular adaptations of enzymes from thermophilic and psychrophilic organisms. In R. Gilles (Ed.), *Advances in comparative and environmental physiology* (Vol. 20, pp. 269–295). Berlin/Heidelberg: Springer.

Barzegar, A., Moosavi-Movahedi, A. A., Pedersen, J. Z., & Miroliaei, M. (2009). Comparative thermostability of mesophilic and thermophilic alcohol dehydrogenases: Stability-determining roles of proline residues and loop conformations. *Enzyme and Microbial Technology, 45*(2), 73–79.

Berezovsky, I. N., & Shakhnovich, E. I. (2005). Physics and evolution of thermophilic adaptation. *Proceedings of the National Academy of Sciences of the United States of America, 102*, 12742–12747.

Borders, C. L., Jr., Broadwater, J. A., Bekeny, P. A., Salmon, J. E., Lee, A. S., Eldridge, A. M., & Pett, V. B. (1994). A structural role for arginine in proteins: Multiple hydrogen bonds to backbone carbonyl oxygens. *Protein Science, 4*(3), 541–548.

Brandsdal, B. O., Smalås, A. O., & Åqvist, J. (2001). Electrostatic effects play a central role in cold adaptation of trypsin. *FEBS Letters, 499*, 171–175.

Burley, S. K., & Petskn, G. A. (1988). Weakly polar interactions in proteins. *Advances in Protein Chemistry, 39*, 125–189.

Chiuri, R., Maiorano, G., Rizzello, A., del Mercato, L. L., Cingolani, R., Rinaldi, R., Maffia, M., & Pompa, P. P. (2009). Exploring local flexibility/rigidity in psychrophilic and mesophilic carbonic anhydrases. *Biophysical Journal, 96*, 1586–1596.

Collins, T., Meuwis, M. A., Gerday, C., & Feller, G. (2003). Activity, stability and flexibility in glycosidases adapted to extreme thermal environments. *Journal of Molecular Biology, 328*, 419–428.

Creighton, T. E. (1991). Stability of folded conformations. *Current Opinion in Structural Biology, 1*, 5–16.

D'Amico, S., Marx, J. C., Gerday, C., & Feller, G. (2003). Activity–stability relationships in extremophilic enzymes. *The Journal of Biological Chemistry, 278*, 7891–7896.

Dams, T., & Jaenicke, R. (1999). Stability and folding of dihydrofolate reductase from the hyperthermophilic bacterium *Thermotoga maritima. Biochemistry, 38*, 9169–9178.

Davail, S., Feller, G., Narinx, E., & Gerday, C. (1994). Cold adaptation of proteins. Purification, characterization and sequence of the heat-labile subtilisin from the Antarctic psychrophilic *Bacillus* TA41. *The Journal of Biological Chemistry, 269*, 17448–17453.

Deming, J. W. (2002). Coping with our cold planet. *Applied and Environmental Microbiology, 74*(6), 1677–1686.

Dewan, S. S. (2014). *Global Markets for Enzymes in industrial applications*. Wellesley: BCC Research.

Fedoy, A. E., Yang, N., Martinez, A., Leiros, H. K., & Steen, I. H. (2007). Structural and functional properties of isocitrate dehydrogenase from the psychrophilic bacterium *Desulfotalea psychrophila* reveal a cold-active enzyme with an unusual high thermal stability. *Journal of Molecular Biology, 372*, 130–149.

Feller, G. (2010). Protein stability and enzyme activity at extreme biological temperatures. *Journal of Physics. Condensed Matter, 22*(32), 323101.

Feller, G., & Gerday, C. (1997). Psychrophilic enzymes: Molecular basis of cold adaptation. *Cellular and Molecular Life Sciences, 53*, 830–841.

Feller, G., Thiry, M., & Gerday, C. (1991). Nucleotide sequence of the lipase gene lip2 from the. antarctic psychrotroph *Moraxella* TA144 and site-specific. Mutagenesis of the conserved serine and histidine residues. *DNA and Cell Biology, 10*, 381–388.

Feller, G., Payan, F., Theys, F., Qian, M., Haser, R., & Gerday, C. (1994). Stability and structural analysis of α-amylase from the Antarctic psychrophile *Alteromonas haloplanctis* A23. *European Journal of Biochemistry, 222*, 441–447.

Feller, G., D'Amico, D., & Gerday, C. (1999). Thermodynamic stability of a cold-active α-amylase from the Antarctic bacterium *Alteromonas haloplanctis. Biochemistry, 38*, 4613–4619.

Ferrer, M., Chernikova, T. N., Yakimov, M. M., Golyshin, P. N., & Timmis, K. N. (2003). Chaperonins govern growth of Escherichia coli at low temperatures. *Nature Biotechnology, 21*(11), 1266–1267.

Ferrer, M., Lünsdorf, H., Chernikova, T. N., Yakimov, M., Timmis, K. N., & Golyshin, P. N. (2004). Functional consequences of single: Double ring transitions in chaperonins: Life in the cold. *Molecular Microbiology, 53*(1), 167–182.

Fontana, A. (1991). How nature engineers protein (thermo) stability. In G. di Prisco (Ed.), *Life under extreme conditions: Biochemical adaptations* (pp. 89–113). Berlin/Heidelberg: Springer.

Gaseidnes, S., Synstad, B., Nielsen, J. E., & Eijsink, V. G. H. (2003). Rational engineering of the stability and the catalytic performance of enzymes. *Journal of Molecular Catalysis B: Enzymatic, 21*, 3–8.

Georlette, D., Damien, B., Blaise, V., Depiereux, E., Uversky, V. N., Gerday, C., & Feller, G. (2003). Structural and functional adaptations to extreme temperatures in psychrophilic, mesophilic, and thermophilic DNA ligases. *The Journal of Biological Chemistry, 278*, 37015–37023.

Gerday, C., Aittaleb, M., Arpigny, J. L., Baise, E., Chessa, J. P., Garsoux, G., Petrescu, I., & Feller, G. (1997). Psychrophilic enzymes: A thermodynamic challenge. *Biochimica et Biophysica Acta, 1342*, 119–131.

Gianese, G., Argos, P., & Pascarella, S. (2001). Structural adaptation of enzymes to low temperatures. *Protein Engineering, 14*, 141–148.

Gianese, G., Bossa, F., & Pascarella, S. (2002). Comparative structural analysis of psychrophilic and meso- and thermophilic enzymes. *Proteins, 47*, 236–249.

Goldstein, R. A. (2007). Amino-acid interactions in psychrophiles, mesophiles, thermophiles, and hyperthermophiles: Insights from the quasi-chemical approximation. *Protein Science : A Publication of the Protein Society, 16*(9), 1887–1895. https://doi.org/10.1110/ps.072947007.

Gomes, E., de Souza, A. R., Orjuela, G. L., Da Silva, R., de Oliveira, T. B., & Rodrigues, A. (2016). Applications and benefits of thermophilic microorganisms and their enzymes for

industrial biotechnology. In M. Schmoll & C. Dattenböck (Eds.), *Gene expression systems in fungi: Advancements and applications. Fungal biology* (pp. 459–492). Cham: Springer.
Goodchild, A., Raftery, M., Saunders, N. F. W., Guilliaus, M., & Cavicchioli, R. (2005). Cold adaptation of the Antarctic archaeon, *Methanococcoides burtonii* assessed by proteomics using ICAT. *Journal of Proteome Research, 4*(2), 473–480.
Gorfe, A. A., Brandsdal, B. O., Leiros, H. K., Helland, R., & Smalas, A. O. (2000). Electrostatics of mesophilic and psychrophilic trypsin isoenzymes: Qualitative evaluation of electrostatic differences at the substrate binding site. *Proteins, 40*, 207–217.
Hartl, F. U., & Hayer-Hartl, M. (2009). Converging concepts of protein folding in vitro and in vivo. *Nature Structural & Molecular Biology, 16*(6), 574–581.
Horikoshi, K. (1999). Alkaliphiles: Some applications of their products for biotechnology. *Microbiology and Molecular Biology Reviews, 63*, 735–750.
Igarashi, K., Hatada, Y., Hagihara, H., Saeki, K., Takaiwa, M., Eumura, T., Ara, K., Ozaki, K., Kawai, S., Kobayashi, T., & Ito, S. (1998). Enzymatic properties amylase from an alkaliphilic *Bacillus* isolate of a novel liquefying and entire nucleotide and amino acid. *Journal of Microbiology, 33*, 57–61.
Igarashi, K., Ozawa, T., Ikawa-Kitayama, K., Hayashi, Y., Araki, H., Endo, K., Hagihara, H., Ozaki, K., Kawai, S., & Ito, S. (1999). Thermostabilization by proline substitution in an alkaline, liquefying α-amylase from *Bacillus* sp. strain KSM-1378. *Bioscience, Biotechnology, and Biochemistry, 63*, 1535–1540.
Jaenicke, R. (1991). Protein stability and molecular adaptations to extreme conditions. *European Journal of Biochemistry, 202*, 715–728.
Jaenicke, R., & Bohm, G. (1998). Stability of proteins in extreme environments. *Current Opinion in Structural Biology, 8*, 738–748.
Jaenicke, R., Schurig, H., Beaucamp, N., & Ostendorp, R. (1996). Structure and stability of hyperstable proteins: Glycolytic enzymes from hyperthermophilic bacterium *Thermotoga maritima*. *Advances in Protein Chemistry, 48*, 181–269.
Johns, G. C., & Somero, G. N. (2004). Evolutionary convergence in adaptation of proteins to temperature: A4-lactate dehydrogenases of Pacific damselfishes (*Chromis* spp.). *Molecular Biology and Evolution, 21*, 314–320.
Junge, K., Eicken, H., & Deming, J. W. (2004). Bacterial activity at −2 to −20°C in arctic winter time sea ice. *Applied and Environmental Microbiology, 70*(1), 550–557.
Kandror, O., & Goldberg, A. L. (1997). Trigger factor is induced upon cold shock and enhances viability of *Escherichia coli* at low temperatures. *Proceedings of the National Academy of Sciences of the United States of America, 94*(10), 4978–4981.
Kawamoto, J., Kurihara, T., Kitagawa, M., Kato, I., & Esaki, N. (2007). Proteomic studies of an Antarctic cold-adapted bacterium, *Shewanella livingstonensis* Ac10, for global identification of cold-inducible proteins. *Extremophiles, 11*(6), 819–826.
Kim, S. Y., Hwang, K. Y., Kim, S. H., Sung, H. C., Han, Y. S., & Cho, Y. (1999). Structural basis for coldadaptation. Sequence, biochemical properties, and crystal structure of malate dehydrogenasefrom a psychrophile *Aquaspirillium arcticum*. *Journal of Biological Chemistry, 274*, 11761–11767.
Koga, Y. (2012). Thermal adaptation of the archaeal and bacterial lipid membrane. *Archaea, 2012*, 789652.
Kumar, S., & Nussinov, R. (2001). How do thermophilic proteins deal with heat? *Cellular and Molecular Life Sciences, 58*, 1216–1233.
Ladenstein, R., & Antranikian, G. (1998). Proteins from hyperthermophiles: Stability and enzymatic catalysis close to the boiling point of water. *Advances in Biochemical Engineering/Biotechnology, 61*, 37–85.
Leiros, I., Moe, E., Lanes, O., Smalas, A. O., & Willassen, N. P. (2003). The structure of uracil-DNA glycosylase from Atlantic cod (*Gadus morhua*) reveals cold-adaptation features. *Acta Crystallographica. Section D, Biological Crystallography, 59*, 1357–1365.

Li, M., Zhu, L., & Wang, W. (2016). Improving the thermostability and stress tolerance of an archaeon hyperthermophilic superoxide dismutase by fusion with a unique N-terminal domain. *Springer plus, 5*, 241.

Low, P. S., Bada, J. L., & Somero, G. N. (1973). Temperature adaptations of enzymes: Roles of the free energy, the enthalpy and the entropy of activation. *Proceedings of the National Academy of Sciences of the United States of America, 70*, 430–432.

Maras, B., Valiante, S., Chiaraluce, R., Consalvi, V., Politi, L., de Rosa, M., Bossa, F., Scandurra, R., & Barra, D. (1994). The amino acid sequence of glutamate dehydrogenase from Pyrococcus furiosus, a hyperthermophilic archaebacterium. *Journal of Protein Chemistry, 13*, 253–259.

Matthews, C. R. (1991). The mechanism of protein folding. *Current Opinion in Structural Biology, 1*, 28–35.

Menendez-Arias, L., & Argos, P. (1989). Engineering protein thermal stability. Sequence statistics point to residue substitutions in α-helices. *Journal of Molecular Biology, 206*, 397–406.

Merlino, A., Russo Krauss, I., Castellano, I., de Vendittis, E., Rossi, B., Conte, M., Vergara, A., & Sica, F. (2010). Structure and flexibility in cold-adapted iron superoxide dismutases: The case of the enzyme isolated from *Pseudoalteromonas haloplanktis*. *Journal of Structural Biology, 172*, 343–352.

Mrabet, N. T., Van den Broeck, A., Van den Brande, I., Stanssen, P., Laroche, Y., Lambeir, A., Matthijssens, G., Jenkins, J., Chiadmi, M., van Tilbeurgh, H., Rey, F., Janin, J., Quax, W. J., Lasters, I., De Maeyer, M., & Wodak, S. J. (1992). Arginine residues as stabilizing elements in proteins. *Biochemistry, 31*, 2239–2253.

Muñoz, P. A., Márquez Sebastián, L., Gonzalez-Nilo Fernando, D., Marquez-Miranda, V., & Blamey Jenny, M. (2017). Structure and application of antifreeze proteins from Antarctic bacteria. *Microbial Cell Factories, 16*, 138. https://doi.org/10.1186/s12934-017-0737-2.

Mykytczuk, N. C. S., Trevors, J. T., Foote, S. J., Leduc, L. G., Ferroni, G. D., & Twine, S. M. (2011). Proteomic insights into cold adaptation of psychrotrophic and mesophilic *Acidithiobacillus ferrooxidans* strains. *Antonie van Leeuwenhoek, 100*(2), 259–277.

Olufsen, M., Smals, A., Moe, E., & Brandsdal, B. (2005). Increased flexibility as a strategy for cold adaptation a comparative molecular dynamics study of cold and warm-active uracil DNA glycosylase. *Journal of Biological Chemistry, 280*, 18042–18048.

Papaleo, E., Pasi, M., Riccardi, L., Sambi, I., Fantucci, P., & Gioia, L. (2008). Protein flexibility in psychrophilic and mesophilic trypsins. Evidence of evolutionary conservation of protein dynamics in trypsin-like serine-proteases. *FEBS Letters, 582*, 1008–1010.

Paredes, D. I., Watters, K., Pitman, D. J., Bystroff, C., & Dordick, J. S. (2011). Comparative void-volume analysis of psychrophilic and mesophilic enzymes: Structural bioinformatics of psychrophilic enzymes reveals sources of core flexibility. *BMC Structural Biology, 11*, 42.

Payen, A., & Persoz, J. F. (1833). Memoir on diastase, the principal products of its reactions, and their applications to the industrial arts. *Annales de Chimie Physique, 53*, 73–92.

Piette, F., D'Amico, S., Struvay, C., Mazzucchelli, G., Renaut, J., Tutino, M. L., Danchin, A., Leprince, P., & Feller, G. (2010). Proteomics of life at low temperatures: Trigger factor is the primary chaperone in the Antarctic bacterium *Pseudoalteromonas haloplanktis* TAC125. *Molecular Microbiology, 76*(1), 120–132.

Piette, F., D'Amico, S., Mazzucchelli, G., Danchin, A., Leprince, P., & Feller, G. (2011). Life in the cold: A proteomic study of cold-repressed proteins in the Antarctic bacterium *Pseudoalteromonas haloplanktis* TAC125. *Applied and Environmental Microbiology, 77*(11), 3881–3883.

Pulicherla, K. K., Mrinmoy, G., Kumar, S., & Sambasiva Rao, K. R. (2011). Psychrozymes–the next generation industrial enzymes. *Journal of Marine Science: Research and Development, 1*, 102. https://doi.org/10.4172/2155-9910.1000102.

Qiu, Y., Kathariou, S., & Lubman, D. M. (2006). Proteomic analysis of cold adaptation in a Siberian permafrost bacterium—*Exiguobacterium sibiricum* 255-15 by two-dimensional liquid separation coupled with mass spectrometry. *Proteomics, 6*(19), 5221–5233.

Razvi, A., & Scholtz, J. M. (2006). Lessons in stability from thermophilic proteins. *Protein Science: A Publication of the Protein Society, 15*(7), 1569–1578.

Rice, D. W., Yip, K. S., Stillman, T. J., Britton, K. L., Fuentes, A., Connerton, I., Pasquo, A., Scandurra, R., & Engel, P. C. (1996). Insights into the molecular basis of thermal stability from the structure determination of *Pyrococcus furiosus* glutamate dehydrogenase. *FEMS Microbiology Reviews, 18*, 105–117. https://doi.org/10.1111/j.1574-6976.1996.tb00230.x.

Rina, M., Pozidis, C., Mavromatis, K., Tzanodaskalaki, M., Kokkinidis, M., & Bouriotis, V. (2000). Alkaline phosphatase from the Antarctic strain TAB5. Properties and psychrophilic adaptations. *European Journal of Biochemistry, 267*, 1230–1238.

Rodrigues, D. F., & Tiedje, J. M. (2008). Coping with our cold planet. *Applied and Environmental Microbiology, 74*(6), 1677–1686. https://doi.org/10.1128/AEM.02000-07.

Rodrigues, D. F., Ivanova, N., He, Z., Huebner, M., Zhou, J., & Tiedje, J. M. (2008). Architecture of thermal adaptation in an *Exiguobacterium sibiricum* strain isolated from 3 million year old permafrost: A genome and transcriptome approach. *BMC Genomics, 9*, 547.

Russell, R. J., Gerike, U., Danson, M. J., Hough, D. W., & Taylor, G. L. (1998). Structural adaptations of the cold-active citrate synthase from an Antarctic bacterium. *Structure, 6*, 351–336.

Schiffer, C. A., & Dötsch, V. (1996). The role of protein-solvent interactions in protein unfolding. *Current Opinion in Biotechnology, 7*, 428–432.

Sen, S., & Nilsson, L. (2016). Thermostable subtilisin. In *Thermostable proteins: Structural stability and design*. CRC Press.

Siddiqui, K. S., & Cavicchioli, R. (2006). Cold-adapted enzymes. *Annual Review of Biochemistry, 75*, 403–433.

Smalås, A. O., Leiros, H. K., Os, V., & Willassen, N. P. (2000). Cold adapted enzymes. *Biotechnology Annual Review, 6*, 1–57.

Sterner, R., Kleemann, G. R., Szadkowski, H., Lustig, A., Hennig, M., & Kirchner, K. (1996). Phosphoribosyl anthranilate isomerase from *Thermotoga maritima* is an extremely stable and active homodimer. *Protein Science, 5*, 2000–2008.

Strocchi, M., Ferrer, M., Timmis, K. N., & Golyshin, P. N. (2006). Low temperature-induced systems failure in *Escherichia coli*: Insights from rescue by cold-adapted chaperones. *Proteomics, 6*(1), 193–206.

Struvay, C., & Feller, G. (2012). Optimization to low temperature activity in psychrophilic enzymes. *International Journal of Molecular Sciences, 13*(9), 11643–11665.

Tanaka, Y., Tsumoto, K., Yasutake, Y., Umetsu, M., Yao, M., Fukada, H., Tanaka, I., & Kumagai, I. (2004). How oligomerization contributes to the thermostability of an archaeon protein. Protein L-isoaspartyl-O-methyltransferase from *Sulfolobus tokodaii*. *Journal of Biological Chemistry, 279*(31), 32957–32967.

Teplyakov, A. V., Kuranova, I. P., Harutyunyan, E. H., Vainshtein, B. K., Friimmel, C., Hiihne, W. E., & Wilson, K. S. (1990). Crystal structure of thermitase at 1.4 A resolution. *Journal of Molecular Biology, 214*, 261–279.

Trent, J. D., Gabrielsen, M., Jensen, B., Neuhard, J., & Olsen, J. (1994). Acquired thermotolerance and heat shock proteins in thermophiles from the three phylogenetic domains. *Journal of Bacteriology, 176*(19), 6148–6152.

Tsigos, I., Velonia, K., Smonou, I., & Bouriotis, V. (1998). Purification and characterization of an alcohol dehydrogenase from the Antarctic psychrophile *Moraxella* sp. TAE123. *European Journal of Biochemistry, 254*, 356–362.

Vetriani, C., Maeder, D. L., Tolliday, N., Yip, K. S.-P., Stillman, T. J., Britton, K. L., Rice, D. W., Klump, H. H., & Robb, F. T. (1998). Protein thermostability above 100°C: A key role for ionic interactions. *Proceedings of the National Academy of Sciences of the United States of America, 95*, 12300–12305.

Villeret, V., Clantin, B., Tricot, C., Legrain, C., Roovers, M. S., Glansdorff, V. N., & Van Beeumen, J. (1998). The crystal structure of *Pyrococcus furiosus* ornithine carbamoyltransferase reveals a key role for oligomerization in enzyme stability at extremely high temperatures. *Proceedings of the National Academy of Sciences of the United States of America, 95*, 2801–2806.

Violot, S., Aghajari, N., Czjzek, M., Feller, G., Sonan, G. K., Gouet, P., Gerday, C., Haser, R., & Receveur-Brechot, V. (2005). Structure of a full length psychrophilic cellulase from

Pseudoalteromonas haloplanktis revealed by X-ray diffraction and small angle X-ray scattering. *Journal of Molecular Biology, 348*, 1211–1224.

Vogt, G., & Argos, P. (1997). Protein thermal stability: Hydrogen bonds or internal packing? *Folding & Design, 2*, S40–S46.

Vogt, G., Woell, S., & Argos, P. (1997). Protein thermal stability, hydrogen bonds, and ion pairs. *Journal of Molecular Biology, 269*, 631–643.

Walker, J. E., Wonacott, A. J., & Harris, J. I. (1980). Heat stability of a tetrameric enzyme, n-glyceraldehyde-3-phosphate dehydrogenase. *European Journal of Biochemistry, 108*, 581–586.

Wang, K., Luo, H., Tian, J., Turunen, O., Huang, H., Shi, P., Hua, H., Wang, C., Wang, S., & Yao, B. (2014a). Thermostability improvement of a *Streptomyces* xylanase by introducing proline and glutamic acid residues. *Applied and Environmental Microbiology, 80*, 2158–2165.

Wang, W., Ma, T., Zhang, B., Yao, N., Li, M., Cui, L., Li, G., Ma, Z., & Cheng, J. (2014b). A novel mechanism of protein thermostability: A unique N-terminal domain confers heat resistance to Fe/Mn-SODs. *Scientific Reports, 4*, 7284.

Watanabe, K., Chishiro, K., Kitamura, K., & Suzuki, Y. (1991). Proline residues responsible for thermostability occur with high frequency in the loop regions of an extremely thermostable oligo-1, 6-glucosidase from *Bacillus thermoglucosidasius* KP 1006. *The Journal of Biological Chemistry, 266*, 24287–24294.

Watanabe, K., Masuda, T., Ohashi, H., Mihara, H., & Suzuki, Y. (1994). Multiple proline substitutions cumulatively thermostabilize *Bacillus cereus* ATCC7064 oligo-1, 6-glucosidase refragable proof supporting the Proline Rule. *European Journal of Biochemistry, 226*, 277–283.

Wrba, A., Schweiger, A., Schultes, V., Jaenicke, R., & Závodszky, P. (1990). Extremely thermostable o-glyceraldehyde-3.phosphatedehydrogenase from the eubacterium *Thermotoga maritime*. *Biochemistry, 29*, 7584–7592.

Yu, H., Zhao, Y., Guo, C., Gan, Y., & Huang, H. (2015). The role of proline substitutions within flexible regions on thermostability of luciferase. *Biochimica et Biophysica Acta, 1854*, 65–72.

Zeng, J., Gao, X., Dai, Z., Tang, B., & Tang, X. F. (2014). Effects of metal ions on stability and activity of hyperthermophilic pyrolysin and further stabilization of this enzyme by modification of a Ca2+–binding site. *Applied and Environmental Microbiology, 80*(9), 2763–2772.

Chapter 14
Role of Solar Energy Applications for Environmental Sustainability

Atin K. Pathak, Kapil Chopra, Har Mohan Singh, V. V. Tyagi, Richa Kothari, Sanjeev Anand, and A. K. Pandey

Contents

A. K. Pathak · H. M. Singh · V. V. Tyagi (✉) · S. Anand
School of Energy Managment, Shri Mata Vaishno Devi University,
Katra, Jammu and Kashmir, India
e-mail: v.tyagi16@gmail.com

K. Chopra
School of Energy Managment, Shri Mata Vaishno Devi University,
Katra, Jammu and Kashmir, India

School of Mechanical Engineering, Shri Mata Vaishno Devi University,
Katra, Jammu and Kashmir, India

R. Kothari
Department of Environmental Sciences, Central University of Jammu,
Jammu and Kashmir, India

Babasaheb Bhimrao Ambedkar University, Lucknow, Uttar Pradesh, India

A. K. Pandey
Research Centre for Nano-Materials and Energy Technology (RCNMET), School of Science and Technology, Sunway University, Petaling Jaya, Selangor Darul Ehsan, Malaysia
e-mail: adarshp@sunway.edu.my

R. C. Sobti et al. (eds.), *Environmental Biotechnology: For Sustainable Future*,
https://doi.org/10.1007/978-981-10-7284-0_14

Abstract Energy and environment are the opposite sides of the same coin. Increasing energy production depends on the fossil fuel availability and is the main cause of the environmental degradation by emission of greenhouse gases. To overcome the environmental degradation problem, the whole world is moving towards the renewable energy technologies. The sun is the main direct source of all forms of energy present on the earth. The solar energy can prove to be the sustainable future for maintaining energy demand. Solar energy is the utmost auspicious technology because it can be used for heating as well as electricity production. This technology is the most mature technology and can be used at large or small scale as cleanest source of energy. This chapter deals with the different solar energy technologies mainly working towards the environmental sustainability and cleaning.

Keywords Energy · Environment · Solar energy · Environmental sustainability

Introduction

The world is moving very fast in the direction of industrialization and urbanization. The trend of lifestyle in the developing and underdeveloped countries is changing and moving towards the adaptation of modern techniques. There was a time when people were dependent upon the traditional cooking system, natural water resources and traditional transportation system; at that time issues associated with environmental pollution were very less. However, in present time demand of energy is increasing tremendously due to the adaptation of modern instruments in the industries (Panchal and Patel 2017). The main portion of energy production is coming from coal and fossil fuel-based power plants having serious environmental issues. The sustainability for the environment can be acquired by moving towards the adoption of renewable energy options for different applications, i.e. water heating, cooking, power generation, transportation, etc. Solar energy is the most important energy source among other renewable sources of energy (Tiwari and Tiwari 2017). It is clean and freely available abundantly on earth. Different conversion methods of solar energy have potential to compete for the projected energy demand worldwide in the coming future. In spite of great potential, the solar energy contribution globally is very less. There are many solar energy devices such as solar air/water heater, solar photovoltaic (PV), solar thermal devices, etc. which are commercialized widely and have a potential for energy demand reduction through zero emission. Therefore, the adoption of technologies based on solar energy would appreciably alleviate the matters related to climate change, energy security and unemployment. As per the ongoing research work for solar energy-based transport fuel, it can be seen that these technologies are very useful to reduce energy demand in this sector as well as carbon emission reduction. It is also anticipated that in the next few years the trend for solar energy-based energy generation will be followed by the developing countries due to the huge pressure for carbon emission reduction by the United Nations worldwide (Tyagi et al. 2016).

In this chapter, different solar energy-based technologies, i.e. solar collectors, solar photocatalysis and desalination systems (Fig. 14.1), are discussed with new achievements. These technologies are accepted widely for urban and industrial purposes like power generation, water heating and wastewater treatment.

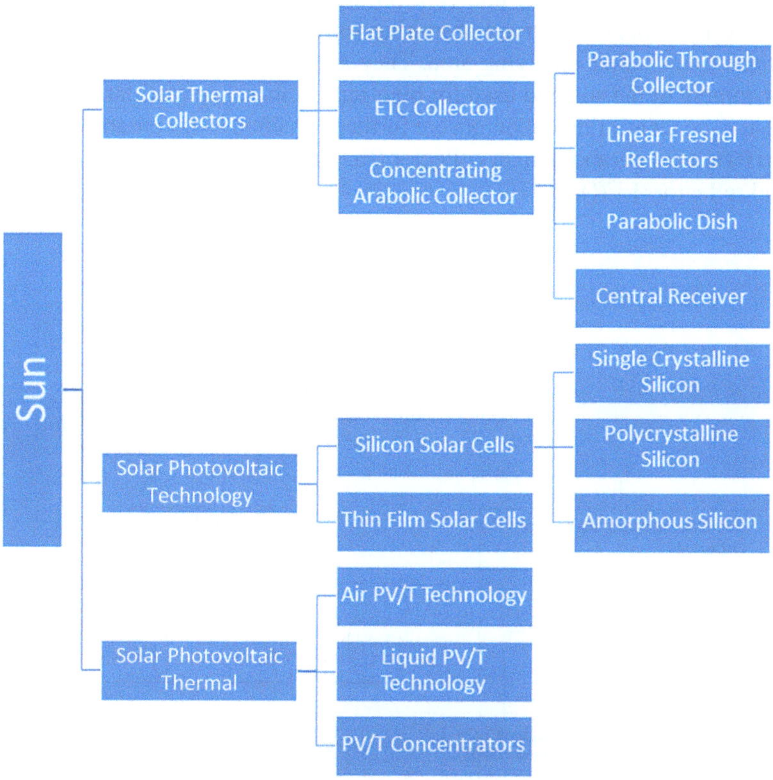

Fig. 14.1 Solar energy technological classification

Solar Energy Collectors

The demand of energy is growing with time, and the world's supply of conventional sources shrinking; a worthy option is harvesting of energy through solar collectors to meet growing energy demands. Solar collectors are the heat exchangers that convert solar energy either to thermal energy of working fluid or electrical energy for solar thermal and photovoltaic applications, respectively. In solar thermal applications, solar radiation absorbed by the solar collector as heat is transferred to circulating working fluid (water, air, oil, etc.), and heat is carried by the fluid and used for useful work. In case of photovoltaic (PV) applications, solar radiation are not only directly converted to electricity but waste thermal energy is also recovered by attaching PV board with recuperating tubes filled with carrier fluids.

As shown in Fig. 14.2, solar collectors are broadly classified into two categories: non-concentrating and concentrating collectors (De Winter 1990). Non-concentrating solar collectors exhibit the same area for interception and absorption of solar radiations. These collectors are also termed as stationary collectors used for low- and medium-temperature applications. These collectors are further classified into two subcategories: (a) flat plate solar collector (FPC) and (b) evacuated tube solar collector (ETC). Unlike non-concentrating type collectors, concentrating collector has a parabolic or curved absorbing surface to intercept and focus the solar radiations to a smaller receiving area. They are continuously moving according to the movement of the sun and used for high-temperature applications. Depending on the design, concentrating collectors are further classified as (a) parabolic trough collector (PTC), (b) parabolic dish collector (PDC), (c) solar tower (ST) and (d) linear Fresnel reflector (LFR).

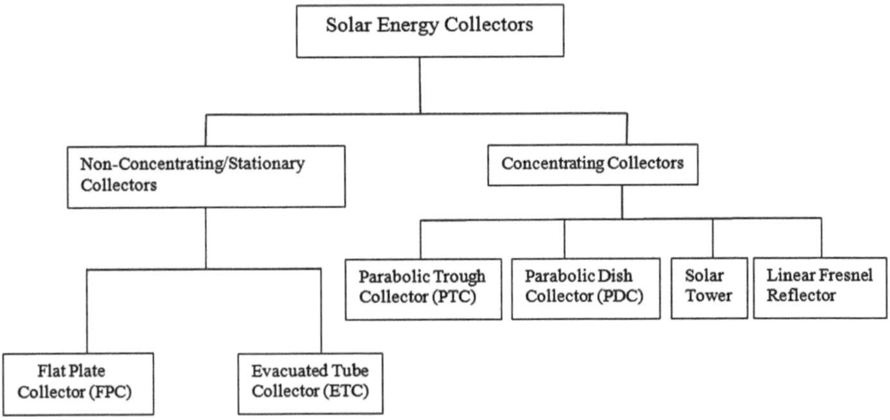

Fig. 14.2 Classification of solar collectors

Flat Plate Collector

The schematic diagram of conventional FPC is shown in Fig. 14.3. It is designed for low- and medium-temperature (60–100 °C) applications and can work on both diffuse and beam solar radiations. It does not require any sun-tracking device. Its maintenance and running cost are low. A conventional FPC is comprised of the following subcomponents:

- Glazing: It is a transparent envelope over the absorbing surface for reduction of radiation and convection losses to the atmosphere. It is opaque for long wavelength and transparent for short wavelength. More than one transparent cover can be utilized in FPC (Fig. 14.3a).
- Riser tubes and fins: The working fluid is to be heated through riser tubes and fins used for increasing the heat exchange between absorbing surface and riser tubes (Fig. 14.3b).
- Absorber plate: It is a black surface of a conductive material (copper, aluminium, etc.) used to absorb solar energy and transferring to fluid flowing in riser tubes (Fig. 14.3b).
- Manifold/header: Two manifolds, namely, upper and lower, used to allow discharging fluid after heating and fluid to be heated, respectively (Fig. 14.3b).
- Insulation: In order to diminish the conduction losses, an insulating or non-conducting material (polyurethane rock wool, glass wool, etc.) is also there at the bottom and sides of FPC (Fig. 14.3b).
- Casing or container: The numerous components of FPC placed in casing or cover to protect from moisture and dust (Fig. 14.3b).

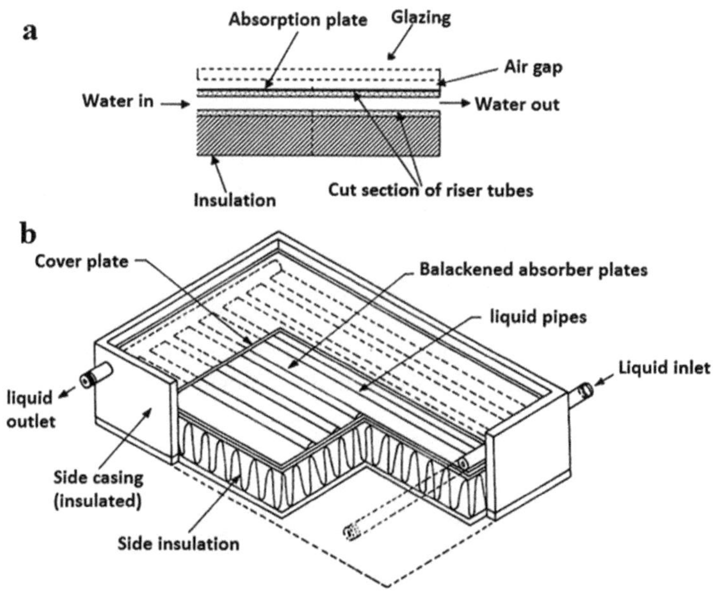

Fig. 14.3 (**a**) Side view and (**b**) cut section and front view of conventional FPC. (Tiwari and Tiwari 2017)

Velmurugan et al. (2016) presented the impact of different nanofluids on a parametric performance of FPC. It was found that exergy efficiencies of MWCNT/H_2O, CuO/H_2O, graphene/H_2O, TiO_2/H_2O, Al_2O_3/H_2O and SiO_2/H_2O increased by 29.32%, 16.67%, 21.46%, 6.97%, 10.86% and 5.74%, respectively, compared to water used as the working fluid. Besides these energy efficiencies, MWCNT/H_2O, CuO/H_2O, graphene/H_2O, TiO_2/H_2O, Al_2O_3/H_2O and SiO_2/H_2O also increased by 23.47%, 12.64%, 16.97%, 5.09% 8.28% and 4.08%, respectively, compared to water used as the working fluid. Experimental findings revealed that a drop in exergy generation rate was also recorded by 65.55%, 48.32%, 57.89%, 24.49%, 36.84% and 10.04% by using of MWCNT/H_2O, CuO/H_2O, graphene/H_2O, TiO_2/H_2O, Al_2O_3/H_2O and SiO_2/H_2O respectively, as compared to water used as working fluid. Gunjo et al. (2017) evaluated various performance parameters such as mass flow rate, absorber plate temperature, outlet temperature, solar insolation, thermal efficiency and ambient temperature of open- and closed-type flat plate collector. Parameters were measured through experiential investigation validated with computational fluid dynamics (CFD) under steady-state condition. It was found that thermal efficiency of proposed system enhanced with increase of solar insolation, mass flow rate and ambient temperature. However, drop in thermal efficiency is also observed by increasing of inlet temperature. It was also witnessed that developed model through CFD could predict the selected performance parameters of the proposed system with reasonable accuracy. Kabeel et al. (2016a, b, c) fabricated and designed a novel dual-function solar heating system (DFSHS) for heating of air or water individually or simultaneously. Maximum 74 °C outlet temperature and 73.68% collector efficiency were obtained when an experiment was conducted in the solar water heater (SWH) at a water mass flow rate of 0.0225 kg/s. Similarly, maximum 69.18 °C outlet temperature and 69.18% collector efficiency were obtained when an experiment was conducted in solar air heater (SAH) at an air mass flow rate of 0.0361 kg/s. The values evaluated for overall heat transfer coefficient for SAH and SWH were 19.91 and 1.06 W/m^2 k, respectively. Authors suggested using dual-function solar heater for air/water heating as per the requirement of the user. Kabeel et al. (2016a, b, c) studied the thermal performance of v-corrugated, finned and flat plate solar air heater at Tanta city, Egypt. It was witnessed that when flow rate of air is fixed to 0.062 kg/s, the highest value of outlet temperature v-corrugated plate solar heater air was 3.5 and 5 °C greater than that of finned and flat plate air heater, respectively. Also, it enhanced to be 5.5 and 8 °C when the mass flow rate of air is 0.009 kg/s. When flow rate of air was 0.062 kg/s, convective heat transfer coefficient of v-corrugated air heater was found to be 1.36 and 1.64 times that of finned and flat plate air heater, respectively.

Evacuated Tube Collector

It was observed by numerous researchers that ETC has much higher performance than FPC, especially at low solar radiations and temperature. Ayompe et al. (2011) compared the performance parameters of ETC equipped with heat pipe and FPC for a native (domestic) water heating system. It was found that for similar surrounding conditions, the ETC efficiencies were observed to be 50.3% and 37.1% and FPC efficiencies were found to be 60.7% and 46.1% for heat pipe ETC and FPC, respectively. An ETC consists of parallel evacuated glass tubes. An evacuated tube consists of two concentric tubes separated by vacuum in which the inner tube called as absorber tube is coated with the selective black coating and the outer tube is transparent. Solar radiations pass through the outer tube and are absorbed by absorber/inner tube. The vacuum between tubes reduces the convection and radiation losses from inner tube to ambient. Thus heat resides inside the inner tube and solar radiations collected efficiently (Fig. 14.4).

Nkwetta et al. (2013) have done improvement in output temperatures of solar collector by combining evacuated tubes (single walled), heat pipe and concentrator (internal or external). A heat pipe evacuated tube prototype for water heating was designed, and research has been carried out for theoretical and experimental outcomes by Zhao et al. (2010). Morthy (2010) did performance analysis of air conditioning system using heat pipe ETC (HP-ETC) system. Author concluded that energy produced by heat pipe ETC system is sufficient to power the air conditioning system. Author also found that efficiency of HP-ETC and system varied from 26%

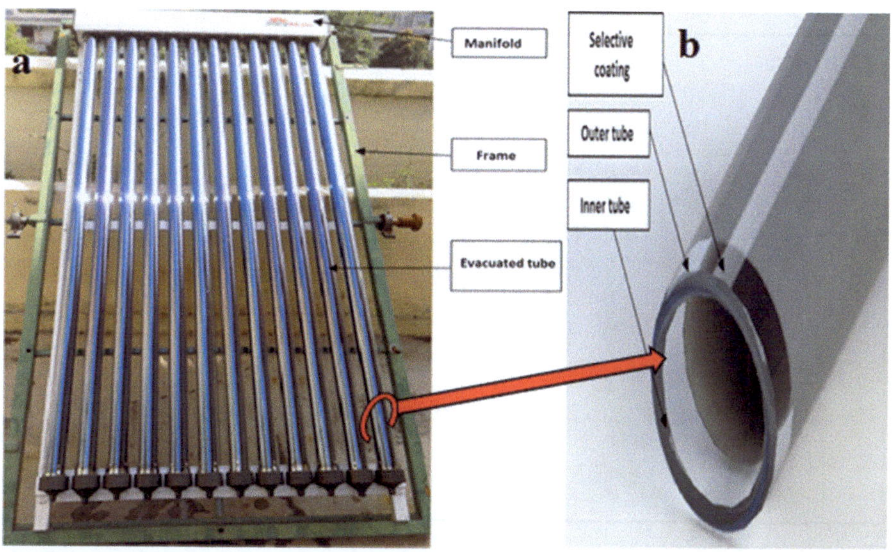

Fig. 14.4 (**a**) Evacuated. (Shah and Fubro 2004). (**b**) Evacuated tube solar collector. (Sabiha et al. 2015)

to 51% and 27% to 48%, respectively. Thus air condition system run by solar energy is a possible solution to overcome the pollution problems.

Concentrator Solar Collectors

There are different types of solar collectors designed and utilized by different researchers for varieties of applications. This section covers the various types of solar collectors.

Parabolic Dish Collector (PDC)

As shown in Fig. 14.5a, parabolic dish solar concentrator rotates about optical axis. The concentration ratio of PDC is very high. In PDC, due to highly reflective compound curvature, solar radiations are focused at a point, and heat losses take place because of reflective nature of radiations. Heat losses may be decreased by reducing the area of absorber. But, this result decreases the intercept factor which has optimal range of 0.95–0.98. Therefore selection of absorber was sized as such to obtain the optimum intercept factor. The high-grade thermal energy and efficiency are the features of PDC. These collectors are commonly used in high-temperature applications like water heating and cooking (Tiwari and Tiwari 2017).

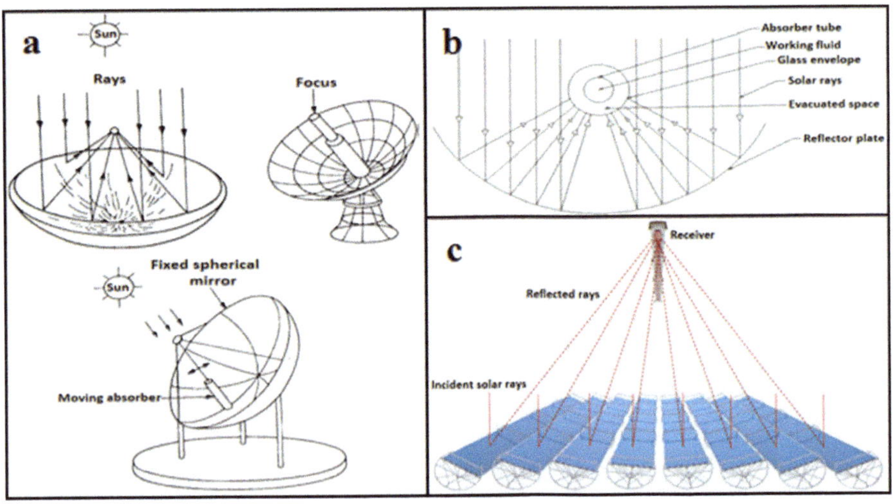

Fig. 14.5 Diagrammatic view of (**a**) parabolic solar dish collector (Tiwari and Tiwari 2017), (**b**) parabolic solar trough collector (Jebasingh and Herbert 2016) and (**c**) linear Fresnel reflector system

Parabolic Trough Collector (PTC)

The schematic diagram of parabolic solar trough collector is presented in Fig. 14.5b. PTC is comprised of reflector plate of parabolic shape and absorber tube with transparent cover. The working fluid is flowing through absorber tube which is permanently fixed at focus of PTC. The vacuum is created at the gap between absorber tube and transparent cover to minimize the heat losses. The whole assembly is positioned on a rigid structure, and tracking mechanism is attached with rigid structure for tracking of solar radiations (Jebasingh and Herbert 2016).

Garcia-Cortes et al. (2012) stressed on the reflected surface strongly linked with parabolic trough collector structure, and the main reason of deformation of solar collector is due to its own weight. Padilla et al. (2011) did heat transfer analysis of PTC absorber and found that at reduction of convection losses by 41.8%, increase in performance of system was observed.

Lupfert et al. (2001) studied the issue related to self-weight of PTC and done enhancement in performance by experimental testing of prototype. Rojas et al. (2008) stressed direct-steam generation based on capillary system on absorber tubes which can be used for various applications. Kalogirou (2004) saved approximately 24% coal consumption in thermal power plant by resolving the coupling of solar power plant and coal-based power plant.

Linear Fresnel Reflector (LFR)

The LFR system is comprised of parallel rows of linearly coupled reflector units, a tracking mechanism and a receiver fixed at the focus of reflected radiations from reflectors. The schematic diagram of LFR is shown in Fig. 14.5c. The LFR system mainly consists of the following subcomponents:

- Reflector units
- Linear cavity receiver
- Tracking mechanism

Reflector Units

A reflector unit consists of highly reflective low-iron mirror that reflects the radiations to a receiver; a corrugated sheet is a wavy structure generally made up of galvanized iron on which mirror is pasted. The corrugated sheet is provided with good support to the mirror.

Linear Cavity Receiver (LCR)

It consists of various numbers of absorbing tubes built up of stainless steel with a black selective coating. The flow velocity and operating pressure of working fluid decide the wall thickness and inner diameter of absorber tubes. These absorber tubes are placed in the cavity of an insulated casing. The inner side of insulated casing coated with anti-reflective coating and cavity is sealed with a glass cover to minimize the thermal losses. The receiver is mounted on a steel frame which is grouted to the concrete structure.

Tracking Mechanism

A tracking mechanism should be there to synchronize the reflector with the sun. The tracking system consists of sprocket and transmission system driven by a stepper motor. The tracking system is connected to the programmable logic controller (PLC) for tracking of a reflector. On the basis of date, time, latitude and longitude of location, an algorithm is implemented in PLC to evaluate the elevation and azimuth of the sun.

Solar Tower

In point focusing system known as central receiver system (CRS), high temperature of working fluid and high efficiencies can be obtained. The diagrammatic view of CRS system is presented Fig. 14.6; the working of such plant is based on the principle of concentration of incident solar radiations. The incident solar radiations on a group of sun-tracking reflectors reflect onto a receiver (Hennecke et al. 2008). The

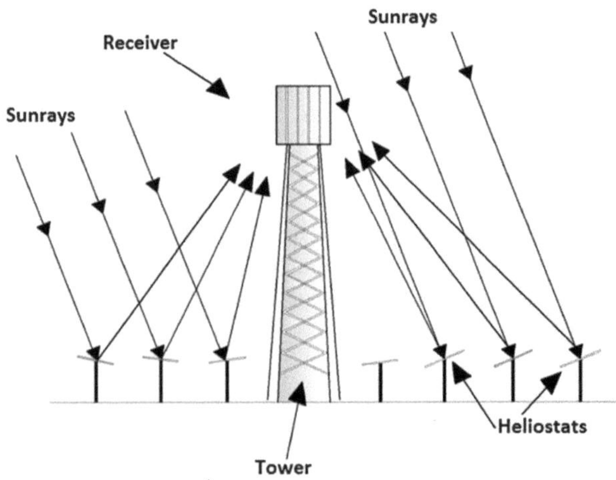

Fig. 14.6 Schematic diagram of central receiver system. (Islam et al. 2013)

temperature above 1200 °C can be achieved on the receiver. The receiver acts as a heat exchanger mounted on the top of the tower. The material selection for a receiver is such that it can withstand high-energy density and high-temperature changes (Alexopoulos and Hoffschmidt 2010). The concentrated solar radiations on receiver are converted into heat which is transferred for production of steam in the boiler.

Solar Energy Technologies for Sustainable Environment

Today demand for energy is increasing due to industrialization and urbanization. The wastewater production and need of hot water demand also increased in last few decades. Therefore, for treatment of wastewater, a different solar energy-based technology was developed which is discussed in this section with a different source of hot water production techniques.

Solar Photocatalysis-Based System

Wastewater mainly consists of inorganic and organic pollutants, and a number of technologies have been established and validated for organic pollutant removal from the wastewater. The developed technologies are equipped with advanced oxidation processes (AOPs). These technologies have their own advantages and disadvantages. The AOP-equipped technologies are used for remediation of organic contaminants, but the only drawback of this technology is it consumes high energy in comparison with other technologies. This electricity consumption increases the operating cost of treatment and also increases the general demand for electricity. Hence, to overcome their financial stress, use of solar energy by researchers attempts a fruitful choice for wastewater-based problems. The photocatalysis is one of the techniques, which comes under the category of AOP. This process helps in complete degradation of an organic pollutant into less harmful pollutant (CO_2 and H_2O) under mild climatic conditions (Parilti 2010).

Solar detoxification is one of the promising technologies that has potential to disinfect wastewater up to a certain limit. It requires sunlight as a primary source feed of energy that breaks down contaminants into lower molecules. The mixture of sunlight and catalyst has proven very effective for various wastewater treatments, and improves the quality of wastewater. The amalgam of contaminated water and catalyst is fed into the solar photocatalysis system. UV light activates the catalyst molecule and forms chemically reactive species which act as oxidizing agent into the system. These oxidizing agents come into contact with contaminants (pollutants); they break them into non-toxic by-products. A redox environment creates a photocatalysis system, and this is able to destroy industrial pollutants easily. Therefore, solar detoxification method is widely accepted and demonstrated in recent times. Figure 14.7 illustrates solar photocatalysis process. Chong et al. (2010)

worked with semiconductor photocatalysis technology for water and wastewater treatment and found that this technology aligns with "zero" waste. The main role in treatment of wastewater through solar photocatalytic process is played by free potent chemical oxidant, i.e. hydroxyl radical (HO*). Photocatalysis is using light as a catalyst to increase the rate of a photoreaction. Most known photoreactions begin when light creates a free radical in the reaction system. Semiconductor photocatalytic materials also have great potential, and they are showing "zero" waste scheme in the water with cost-effectiveness and being eco-friendly (Chong et al. 2010).

Heterogeneous photocatalytic oxidation, one of the AOPs, is a reliable technology which helps in remediation of organic pollutant present in wastewater at ambient climatic conditions. The excitation of a catalyst is induced in a classical photocatalytic process by ultraviolet radiations which separate charge from large band gap semiconductor and reaction is initiated.

Organic molecule + $O_2{}^{light} \rightarrow CO_2 + H_2O$ + Mineral acids

Sunlight is a form of electromagnetic radiation, and it consists of a wide spectrum of radiation (Fig. 14.7a). The total radiation of the sun is approximately equivalent to that of a blackbody at 5776 K. At this temperature, visible and infrared spectrum fit the blackbody emission. In terms of equivalent blackbody temperature of the sun, the UV region of solar radiation deviates greatly from the visible and infrared regions.

Light can have a lot of energy and is able to break chemical bonds. The photocatalytic detoxification process utilizes wavelength near the ultraviolet spectrum for the promotion of oxidative and reductive reactions (Inamdar and Singh 2008). Photocatalyst research is seeking attention these days in environmental and energy-related fields (Fujishima et al. 2000). The main property of catalyst utilized in solar photocatalytic process is for conversion of solar energy into chemical energy which reduces or oxidizes the materials by producing hydrogen (Maeda 2011) and hydrocarbons (Inoue et al. 1979) and removal of microbes and pollutants (Peller et al. 2007; Wolfrum et al. 2002) on wall surfaces and in air and water (Peral et al.

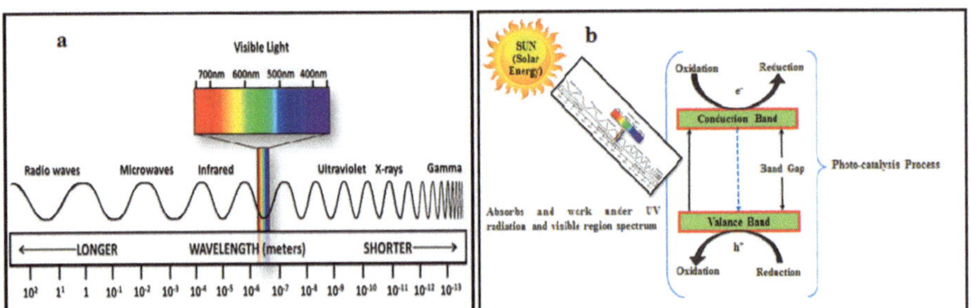

Fig. 14.7 (**a**) Electromagnetic spectrum. (**b**) Mechanism of photocatalysis. (Modified figure of Chong et al. 2010)

1997; Tryk et al. 2000). In different photocatalysts, utilized TiO_2 proves to be the most suitable for different applications because of its strong oxidizing nature; it is the most studied too by different researchers (Nosaka et al. 2003) for the decomposition of organic pollutants (Abe 2010; Tryk et al. 2000).

Solar photocatalysis is found to be one of the promising technologies for wastewater treatment and disinfection (COD, colour) for municipal wastewater (Inamdar and Singh 2008). The conventional method of treatment of wastewater is not able to remove contaminants (chemical or microbial) which affect the public health and environment. The increasing population and industrial activities need more and more water which is very tough due to decreasing water resources. Due to this, the new contaminants are increasing in water bodies (municipal, industrial wastewater) which are impossible for a conventional treatment facility to remove and treat (Kositizi et al. 2004). Zhou and Smith (2002) suggested that solar photocatalytic method (heterogeneous/homogeneous) is suitable for municipal wastewater, especially for small villages or communities.

Research is going on presently for developing new technologies using a different catalyst for wastewater treatment with the help of solar energy to make the process more technically, economically and environmentally feasible. Quinones et al. (2015) treated secondary municipal wastewater with the help of solar photocatalytic ozonation process using photo-Fenton and titanium dioxide (TiO_2) as a catalyst material in a pilot-scale compound parabolic collector. McLoughlin et al. (2004) have carried out an experiment for the treatment of wastewater with *Escherichia coli* K-12 and designed a reactor of 3 and 1 m^2 illuminated area comprised of non-tracking compound parabolic collectors. Tests were also carried out using suspension of TiO_2 at concentrations ranging from 0 to 9 mg/l. The final result comprises that *E. coli* K-12 removal was achieved up to the level of log 4 (approximately 95% removal) which is highest known in around 30 min with and without the use of TiO_2; the enhancement in the process was achieved using TiO_2 as catalyst only at the dose of 3 mg/l; the radiation used in the process was near UV range. Another study is being carried out by Bahnemann et al. (2013) on the mechanistic and kinetic aspects of wastewater treatment with the help of photocatalysts using solar photons and compared the various photocatalytic reactors available for the wastewater treatment process. The final results showed that thin-film fixed-bed reactor (TFFBR) was simple and low-cost reactor for the treatment of textile wastewater. Wang et al. (2013) worked on the solar photocatalytic fuel cell to enhance the process by using CdS quantum-dot-sensitized TiO_2 nanorod array as an effective photo-anode deposited onto glass. The use of TiO_2 nanorods of 1.2 μm array decorated with 10 nm CdS quantum dots helps to increase the trapping of solar radiation. The findings revealed that light irradiation, acid concentrations, pH of electrolyte and conductivity significantly influence the maximum power density output and short-circuit current of the reactor. The highest short-circuit current of 5.1 mA/cm^2 and maximum power density of 3980 mW/cm^2 were achieved with an electrolyte conductivity of 63.1 mS/cm. Bernabeu et al. (2011) found that TiO_2-based solar photocatalyst has the significant property to remove all the pollutants which emerged from the outlet of wastewater treatment plant in South East of Spain. The findings also revealed the removal of the nine emerging pollutants and faecal bacterial elimination very close to 100%.

Solar Desalination System

Desalination technology is one of the primitive technologies for freshwater production from saline water, but this system is being modified for wastewater treatment too. Desalination process can be segregated in many parts like a multistage flash, multi-effect distillation, vapour compression, ion exchange, solar desalination and many others. The major problem with the desalination process is its cost-ineffectiveness, and the conventional desalination process consumes a large amount of electricity/energy for working. Due to large consumption of energy by many desalination systems, solar desalination processes is in more use due to its low cost and its dependency on solar energy for working. Due to the dependency on solar energy, this technology is proving to be the sustainable and efficient technology for portable water generation without the involvement of any conventional energy source; this technology is in active use due to potable water crisis. The only need is to study the different configurations of solar still for better efficiency.

El-Agouz et al. (2015) with his co-workers tried to study the performance efficiency of the designed solar still with inclined surface with and without water closed loop. And, the result showed that efficiency of the closed-loop system is 57.2% higher than the open-loop solar inclined system. In the same context when observed that desalination system is affected due to unavailability of solar energy one third of the day. Kabeel et al. (2016a, b, c) tried to improve the overall performance of the system by enhancement in the productivity of freshwater by utilizing phase change material (PCM) as a thermal storage medium. They designed two solar stills of the same configuration but one with PCM and other without PCM; finally the result concluded that the freshwater productivity was about 7.54 L/m^2 with PCM as storage, while in simple solar still, it is only about 4.5 l/m^2. The overall productivity of PCM-utilized system is 67.18% higher than the conventional solar still. The exergy efficiency is calculated for solar still combined with thermal storage, and findings revealed that at night generation of brackish water and PCM is high with lowest exergy destruction (Asbik et al. 2016). The result also revealed that the instantaneous exergy efficiency of the system is less than 5% during the daytime, but in some cases, it also exceeds up to 80% at night. Thermo-physical property characterization of PCM was done for non-membrane-based solar desalination application, and the results show a great output for PCM to be utilized in desalination process and other purposes (Sarwar and Mansoor 2016). They tried to increase the productivity of the solar desalination process by utilization of thermal energy storage (PCM) integrated with the system, and results revealed that when PCM is integrated, productivity decreases with higher flow rate, but when flow rate decreases, the productivity of the system increases up to 49%. The solar irradiation also affects the production rate from about 0.75 L/day to 2.1 L/day. Also, suggested is that use of a higher range of PCM can be more effective and helpful for water production.

Kumar et al. (2016) reported system productivity increment by a hybrid plant which is made up of parabolic concentrator-concentric tubular solar still (CPC-CTSS). A set of 2-m-long concentric tubes with rectangular basins of the same length was fabricated with 2 m^2 area. The system is also incorporated with five types of PCM. The CPC-CTSS was equipped with cooling tower, and the heat rejected from cooling tower is used for rising temperature in the basin of the solar still. The PCM used in the system provides heat to the system after sunset to maintain basin temperature which leads to the increase in distillate output measured to be 2.7 L/m^2/day. Panchal et al. (2017) published a review on the utilization of different thermal energy storage materials for improvement of distillate output of solar still. They also reported that due to low productivity of the system, i.e. 3 l/day, this is unable to be used in industries and domestic applications. Hence, it is required to design a system with high yield and integrated with thermal energy storage to increase the yield by providing energy during off-sunshine hours. Sathyamurthy et al. (2015) and his group worked on the portable solar still with evaporation and condensation chambers. The system is tested in summer in Chennai and found that when it is used without storage material, it is minimal and when used with storage material, accumulated yield is 52% and productivity is found to be 34% more than normal. Gululothu et al. (2015) published and presented a paper in a conference on use of three different PCMs for solar distillation. The PCM used are sodium sulphate, sodium acetate and potassium dichromate and found that among all the three, sodium sulphate proves to be the best one. Kumar et al. (2016) studied experimentally on solar desalinating system with parabolic assisted and four modes of operations with different flow rate of cooling water to the top surface. Result revealed from the experiment that at high flow rate the efficiency of the system enhances and with PCM integration yield increases for few more hours of the day without sunshine too.

Types of Solar Desalination

Solar desalination process is classified as an active and passive solar still as per the mode of operation and modification used to run the system. Passive solar desalination is a technique in which no other external equipment is utilized for desalination process (Bloemer et al. 1965). While in case of active solar desalination, an extra-thermal energy is given to the basin for faster evaporation. In active solar desalination, collector/concentrator panel (Rai and Tiwari 1983; Kudish et al. 2003), conventional boilers and waste thermal energy (Tleimat and Howe 1966) are used as the external systems for desalination. The active and passive solar desalination processes are further classified as single-effect solar stills and multi-effect solar stills. Many types of research have been done on the solar desalination for its development for large-scale production of clean water (Kumar et al. 2015). Rajaseenivasan et al. (2013) reviewed some more of them such as solar stills of conventional type, single-slope integrated with passive condenser, solar still with double condensing chamber (Tiwari et al. 1997), vertical-type solar still (Coffey 1975; Kiatsiriroat 1989), conical solar still (Tleimat and Howe 1969), inverted absorber solar still (Suneja and Tiwari 1998) and multi-effect solar still (Barrera 1933; Tanaka et al. 2000). Figure 14.8 shows the classification of solar desalination.

Active solar stills are integrated with an additional heat source which helps in increasing the heat supply to the system and leads in increment of distillate output. This system works through an exchange of working fluid or preheated fluid from external installed heating system, i.e. evacuated tube collector, flat plate collector, etc. The only factor which should be taken into consideration while working with an active solar still is controlling the flow rate of heat transfer fluid as solar intensity and ambient temperature are continuously changing which also affect the input temperature of the fluid. Hamadou and Abdellatif (2014) tried to model an active solar still for seawater desalination. They suggest many factors to be taken into account, i.e. heat transfer, fluid rate, inlet temperature, relative humidity, ambient temperature, seawater rate, wind speed, solar radiation and basin depth which are important parameters that are also discussed by Panchal and Patel (2017), Tiwari and Madhuri (1987), Al-Hayeka and Badran (2004), Tamini (1987) and Sodha et al. (1980). The simplest form of solar desalination is single effect, very easy to construct; this system is economical as per maintenance, construction and operation are concerned. Many researchers used single effect solar desalination system for different paramateric studies. The only drawback of this system is low productivity of distillate output. Panchal and Shah (2014) tried to develop the double-basin solar still and El-Sebaii (2005) developed a triple-basin solar still for increased distillate output. In this section, we are discussing the active solar still.

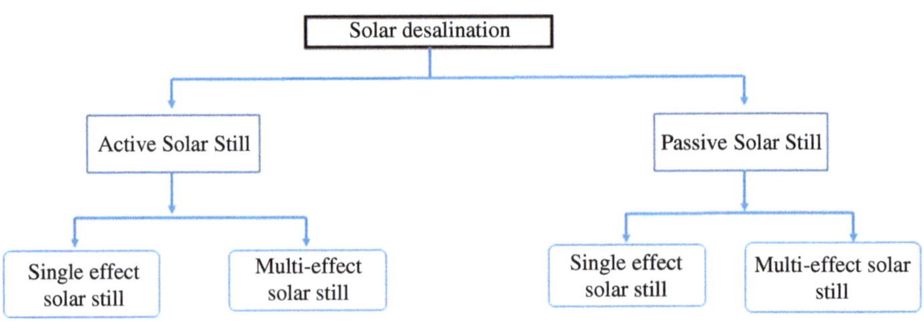

Fig. 14.8 Classification of solar desalination. (Panchal and Patel 2017)

Integrated with Solar Heater

Active solar still when coupled with different solar collectors gives better distillate output. The active solar still integrated with solar heaters can also be divided into many different types as discussed below (Table 14.1).

Flat Plate Collector Integrated with Desalination System

Lawrence and Tiwari (1990) stated that different solar thermal devices under natural circulation mode prove to be more effective than forced circulation mode in terms of economic viability and reliability. They developed a theoretical model for single-basin solar still joined with different solar collectors working on natural circulation mode and the effect of various parameters like temperature, depth of water mass, heat exchanger length and collector area. They conducted a study of single-basin still coupled with heat exchanger attached to flat plate collector. Their findings revealed that active solar still is more efficient than passive solar still in producing distillate with increasing water depth. The schematic representation of the designed system is shown in Fig. 14.9a. In the same direction was when Badran et al. (2005) integrated solar still with flat plate collector to study the effect of augmentation under local conditions on still. The designed system is equipped with conventional-type double-sloped solar still which is also equipped with flat plate collector, feeder tank and a constant head tank. They reported that when solar still system is coupled with FPC and operated for 24 h, still alone operated continuously for the same hours, but still connected to flat plate operated only during sunlight (8 am–5 pm). It has been found that the output of the system is different in all cases. The system is fed with two different types of water (saline and tap water). They found that when solar still is coupled with FPC, there is an extensive increase in temperature which leads to high productivity due to high evaporation rate. It was found that by utilizing tap and seawater, distillate output was increased up to 231% using tap water and 52% in case of seawater (Fig. 14.9b).

The study on the flat plate integration with single-basin solar still was conducted by Rai and Tiwari (1983). The experiment was done with forced circulation mode, and the result when compared with simple still was found to be 24% higher, but the output when compared with Badran et al. (2005) that forced active solar still is not efficient was found true as the system designed by Badran et al. (2005) runs on natural circulation mode which produces 231% higher efficiency than the normal one, whereas Rai and Tiwari (1983) found only 24% increment from forced mode of circulation (Fig. 14.10). But when Yadav (1991) used a forced mode circulation, the distillate output increased by around 5–10% when compared with thermosyphon mode still. This also, contradicts the Badran et al. (2005) statement as stated above.

Table 14.1 Description of flat plate collector integrated with desalination system

S. no.	Modification	Increase in output (%)	Productivity	Observation/findings/ advantages	References
1.	Heat exchanger and FPC	5		Active solar still is more efficient than passive solar still in producing distillate with increasing water depth	Lawrence and Tiwari (1990)
2.	Double-slope solar still with FPC	22.26	4.6 (L/m^2)	Increment of distilled water production is increased up to 231% by using tap water and 52% in case of feeding with saltwater	Badran et al. (2005)
3.	Single-basin solar collector coupled with FPC	24	~350 (Kg/day)	The daily distillate production of a coupled single basin still is 24% higher than that of an uncoupled one	Rai and Tiwari (1983)
4.	Improving the double-slope solar still performance by using flat plate collector and cooling glass cover	80.6	10.06 L/m^2	The daily output of the system can be obtained at a maximum by using 3 mm brine depth with cover cooling	Morad et al. (2015)
5.	The single-basin and FPCB is used with different materials: jute cloth, black gravel, and in combination	60	3.62 kg/m^{2}for conventional and 5.82 kg/m^2	The result of the experimental study revealed that FPCB proves to be more effective than the conventional one with evaporation rate	Rajaseenivasan et al. (2014)
6.	Flat plate solar collector is integrated with single-slope solar still	51–148		The system was evaluated in active and passive modes; water is sprayed from bottom or sprayed; hot air is circulated and heated from solar air collector and also equipped with spraying unit, perforated tube and external collector	Eltawil and Omara (2014)

(continued)

Table 14.1 (continued)

S. no.	Modification	Increase in output (%)	Productivity	Observation/findings/ advantages	References
7.	Novel 2 m^2 phase change flat plate solar collector	Better efficiency		It has been observed from different experimental data that maximum output is found at an inclination angle of 40° and a filament volume of 50% and also increased with increased mass flow rate	Martinopoulos et al. (2016)

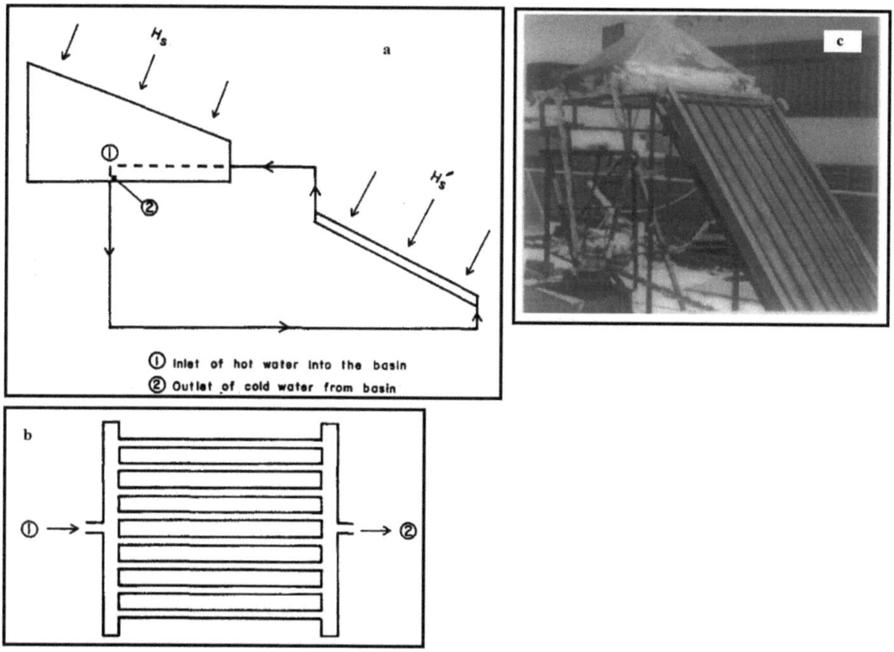

Fig. 14.9 (**a, b**) Representation of designed single-slope still coupled with FPC (Lawrence and Tiwari 1990). (**b**) Photographic view of the designed system of Badran et al. (2005), coupled with FPC

Another study by Morad et al. (2015) has tried to improve the double slope solar still performance by utilizing flat plate collector with and without glass cover cooling (Fig. 14.11). Other parameters were also taken into consideration like water temperature, glass cover temperature (outer and inner), depth of brine and thickness of glass cover. The active and passive still were compared simultaneously, and findings reveal that production of freshwater is 10.06 l/m^2 in case of active solar still

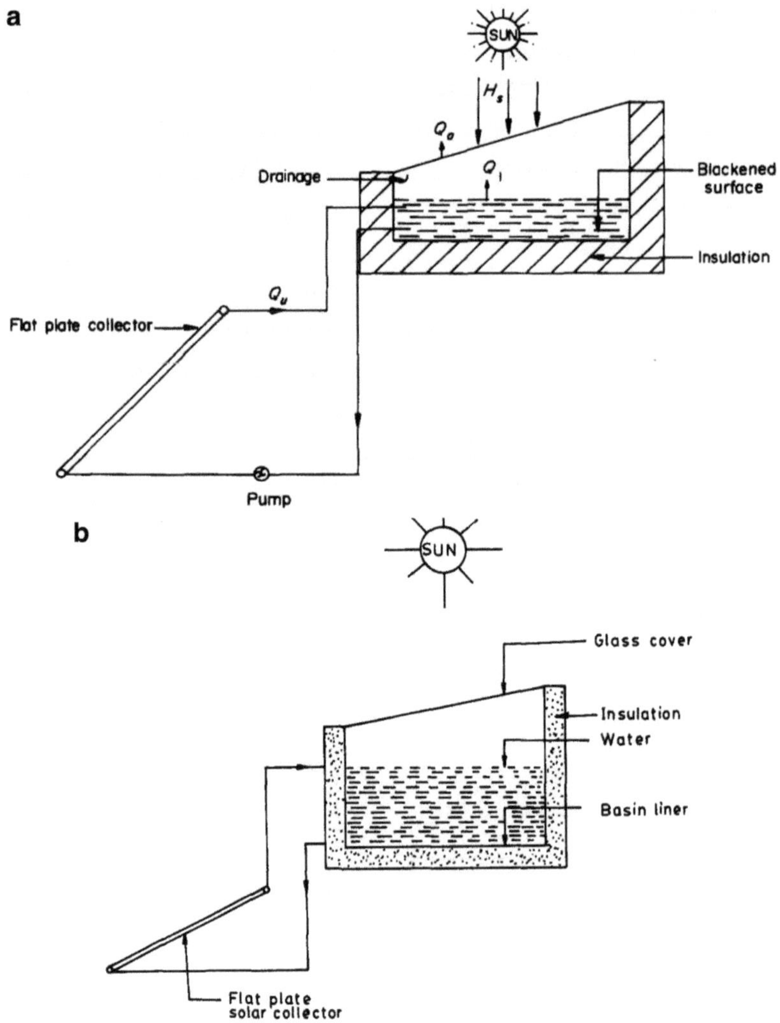

Fig. 14.10 (**a**) Sketch of single-basin solar still coupled with flat plate collector. (**b**) Schematic diagram of single-basin solar still coupled to flat plate solar collectors under thermosyphon mode (Rai and Tiwari 1983)

which was found more when compared with passive solar still which only produces 7.8 l/m^2 with an internal efficiency of 80.6% and 57.15% respectively. The whole experiment was conducted with set parameters of basin depth (1 cm) and thickness of glass cover (3 mm) and by providing flash tactic cover cooling for 5 min on and off concept. In another experiment solar FPC was integrated with solar still for water purification by Rajaseenivasan et al. (2014) (Fig. 14.12) in a climatic condition in Tamil Nadu, India. They fabricated the conventional still and single-slope flat plate collector basin with equal basin depth of 1 m^2. The experiment was mainly recorded and varied with the water depth of the basin, utilizing energy storage mate-

Fig. 14.11 Active solar still coupled with flat plate collector (FPC) (Morad et al. 2015)

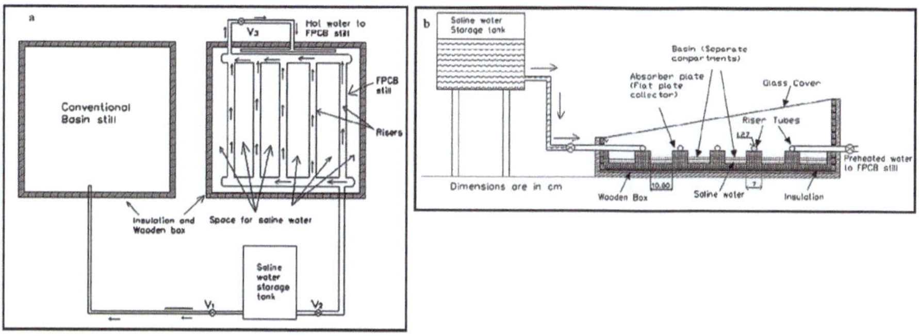

Fig. 14.12 (**a**) Schematic representation of top view of conventional and FPCB (**b**) Fabricated flat plate collector basin still (Rajaseenivasan et al. 2014)

rial and using wick and comparing both the fabricated stills. The comparative result revealed that FPCB is 60% more efficient than conventional type solar still. While use of jute cloth and black gravel gives higher distillate output of 1275 kg/m^2, the lowest was found with conventional still with only jute cloth as storage material. Viewing the present scenario of low availability of energy/electricity and pure water basically in remote areas, Eltawil and Omara (2014) (Figs. 14.13 and 14.14) tried to develop an enhanced single-slope still equipped with flat plate solar collector, external condenser, solar collector, spraying unit and perforated tubes. The system was run in active and passive modes and compared with the conventional still. The water was circulated in the designed system from the bottom in pumped mode forming fountains or sprayed in between the unit; also hot air was forced from bottom to burst as air bubble. The different motors and electric instruments used in system were run by solar photovoltaic panels. The productivity of the designed system when compared from conventional type solar still was found to be 51–148%, whereas 51% increment was found in productivity when system was attached with external condenser. The hot water is circulated in the system as active and passive

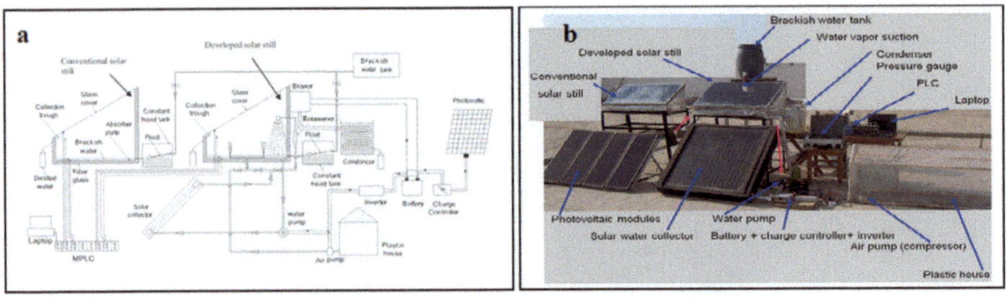

Fig. 14.13 (a) A schematic diagram of a hybrid solar still conventional and developed still (b) The photographic view of conventional solar still and developed solar still designed by Eltawil and Omara (2014)

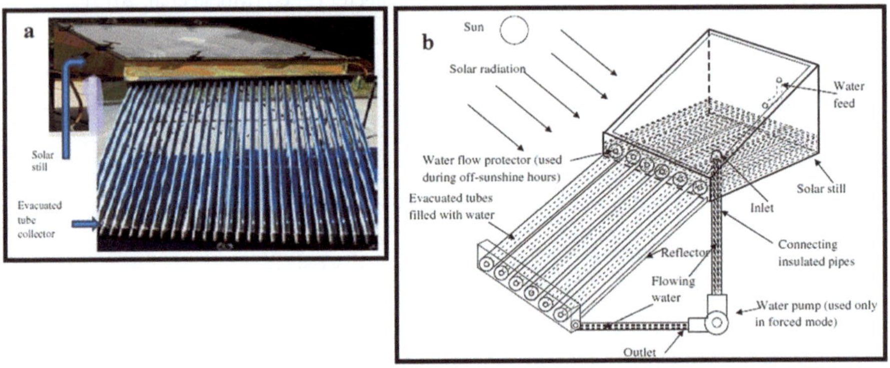

Fig. 14.14 (a) Experimental setup designed by Kumar et al. 2014 (b) Schematic diagram of integrated ETC desalinations (Kumar et al. 2014)

mode without condenser, and the productivity is found maximum in active mode with 82% in comparison with passive mode which is only about 56%. Martinopouls et al. (2016) evaluated a novel type of PCM-based solar collector for desalination, and they utilized solar thermal collector to overcome the current issues of desalination, i.e. cost-effectiveness and environmental. The designed system is approximately 2 m^2; the different parameters taken into consideration during experimental evaluation are ratio of collector volume to heat carrier volume and inclination angle, and the most effective parameter is mass flow rate of fluid. The final result showed better efficiency with 50% volume filament at a 40 °C inclination; the system efficiency also increases with increasing mass flow rates.

Evacuated Tube Collector Integrated with Desalination System

ETC-equipped single-slope solar still was designed to operate in a climatic condition in New Delhi. The maximum output was found to be 3.9 kg with 0.03 m basin depth and 0.06 kg/s mass flow rate, whereas 3.7 kg output was found with 0.01 m

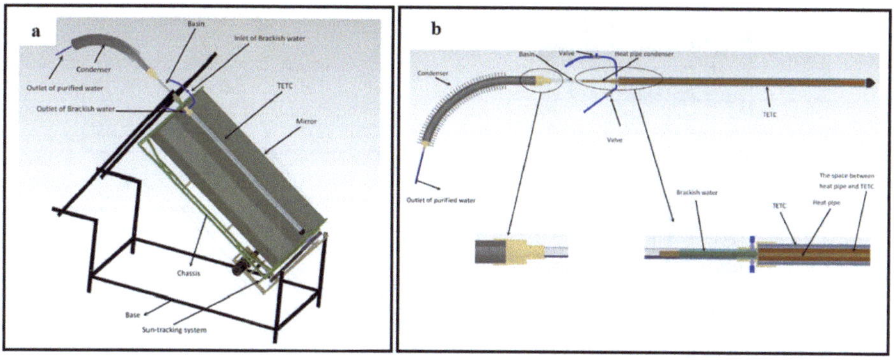

Fig. 14.15 (**a**) Schematic representation of the designed system for experimental setup by Mosleh et al. (2015) (**b**) Representation of different parts of designed system by Mosleh et al. (2015)

brine depth with same flow rate. The energy and exergy efficiency of the system for summer session was 33.8% and 2.6%, respectively (Kumar et al. 2014). The final result also revealed that the maximum average annual yield efficiency of the system is highest with a forced mode in comparison with a natural mode.

Mosleh et al. (2015) combined the heat pipe, evacuated tube collector and parabolic trough collector to improve the efficiency of the solar desalination (Fig. 14.15). They suggested that linear parabolic trough collector when integrated with the solar still increases the distillation process. For better distillate output, they utilized a twin-glass evacuated tube collector in combination with a heat pipe. They utilized aluminium foil and oil as a conducting material in space provided between twin-glass evacuated tube and heat pipe; the result revealed that better output was found with oil as conducting material which was around 65.2% and 0.933 kg/(m^2 h) distillate output, whereas 0.27 kg/(m^2 h) was the rate of production with 22.1% system efficiency with aluminium as a conducting material.

For better distillate output thermosyphon heat pipe and vacuum glass has been studied by Mamouri et al. (2014) and found that higher distillate output is observed with thermosyphon heat pipes which act as a fast and high-performance thermal conducting device. The system is also equipped with evacuated tube collector with flexible nature, which shows high performance in adverse conditions. The better output of the designed system is found at 2 cm depth with a maximum production rate and efficiency of 1.02 kg/(m^2 h) and 22.95, respectively.

Abad et al. (2013) in utilized a novel type of pulsating pipe with simple solar still and found that this device is having high-performance efficiency, fast responding and flexible thermal conducting which makes it a novel type (Fig. 14.16). The integration and use of these pipes have shown a noteworthy increment in the distillate output with a maximum production of 875 mL/(m^2 h), with an optimum depth of 1 cm with 40% filling ratio.

Behnam and Shafii (2016) have tried to work on the new design of solar desalination system equipped with air bubble column humidifier (ABCH) also fortified with ETC and thermosyphon heat pipes. The experimental result revealed from the paper

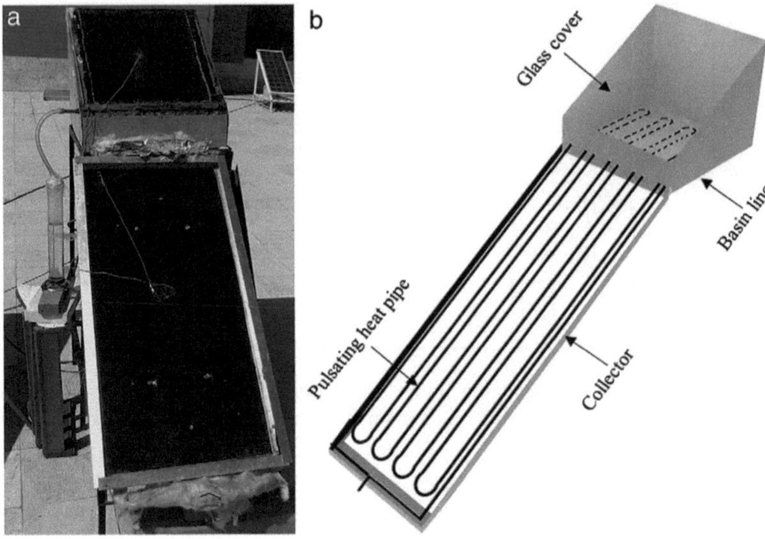

Fig. 14.16 (**a**) Photographic view of the experimental design of pulsating pipe with simple solar still (**b**) Solar still equipped with pulsating pipe (Abad et al. 2013)

shows that the novel humidification-dehumidification (HDH) uses the advantage of ETC-HP. The ETC-HP helps in humidifying the air and heating the water by creating effective mixing by air bubble column humidifier and also works as highly efficient thermal absorption and conductor device. The various parameters taken into consideration to investigate better system efficiency were initial depth of water, air flow rate in the humidifier and adding oil in between the ETCs and heat pipes. Better result was obtained with oil in between ETC and heat pipe with daily freshwater productivity of around 6.275 kg/day m^2 with overall system productivity of about 65%, whereas the water depth in the humidifier was found optimum as equal to heat pipe length of a condenser. The authors also suggested that daily productivity increases with air flow rate. The cost of production of freshwater from this was estimated to be 0.028 $/L. The designed system and schematic view are shown in Fig. 14.17. The integration of evacuated tube with desalination system is presented through Table 14.2.

Solar Concentrators Integrated with Desalination System

The active solar still is integrated with different solar collectors for enhancement of the distillate output. There are many researchers who have done work in the direction of distillate output enhancement as discussed in this section (Table 14.3).

Murtuza et al. (2017) found that parabolic trough collectors (PTSCs) are ahead in acceptance as a better technology. Technology has a great potential for utilization

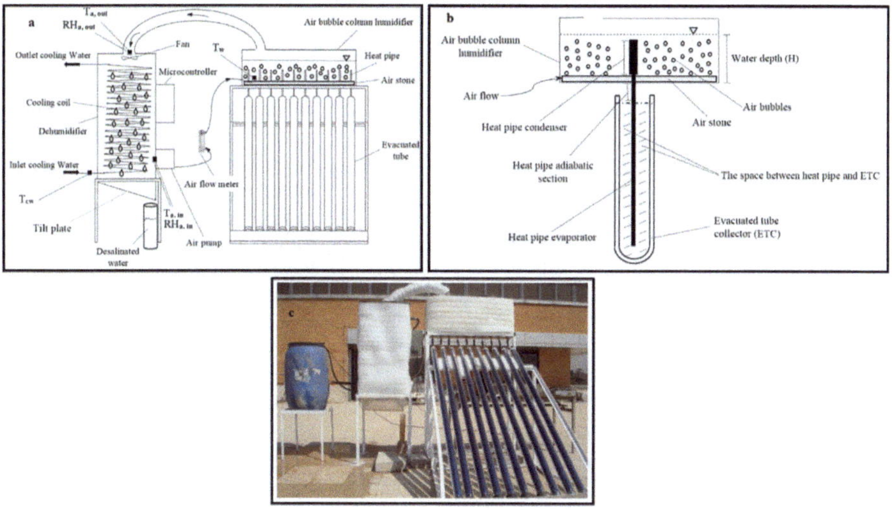

Fig. 14.17 (**a**) Schematic view of solar desalination system equipped with ABCH also fortified with ETC and thermosyphon heat pipes (**b**) Schematic view of ABCH (**c**) Photographic view of designed system by Behnam and Shafii (2016)

in different renewable energy options. These PTSCs are designed for 5 m length with stainless steel as a collecting substance and water as a working fluid. The result reveals that the better output from the system is obtained in months of February to May with different flow rates; the PTSC system when compared with photovoltaic was found with enhanced efficiency. Thus, they have shown different angles for utilization of PTSCs. This shows that PTSC system will prove a great opportunity for desalination purposes.

Solar Water Heating Systems

Solar water heating technology is most widely known for various solar thermal applications (Islam et al. 2013). The solar water heating system converts solar radiation into heat energy and transferred into working fluid (water, water antifreeze, or air). This system is possible replacement of water heating system run by fossil fuels and electricity (Sivakumar et al. 2012). Generally, this system is quite simple in structure because it requires only sun radiation to heat the water. It is based on the principle that heat absorbed by absorptive surface exposed to sunrays transfers it to working fluid which causes the rise of fluid temperature (Al-Badi and Albadi 2012). These systems can be divided into two groups, passive and active systems (Patel et al. 2012), as shown in Fig. 14.18.

Table 14.2 Description of evacuated tube collector integrated with desalination system

S. no.	Modification	Increase in output (%)	Productivity	Observation/findings/ advantages	References
1.	Solar still integrated with evacuated tube collector in a forced mode		3.9 kg	The average annual yield has been found maximum with forced mode in comparison with natural mode. The exergy and energy is found to be 33.8% and 2.6%, respectively	Kumar et al. (2014)
2.	The system is integrated with heat pipe, evacuated tube equipped with parabolic trough collector	22.1% with aluminium as conducting material and 65.2% with oil as conducting material	0.27 kg/(m^2 h) with aluminium and 0.97 kg/(m^2 h) with oil as conducting material	The system shows the better output with oil as a conducting material	Mosleh et al. (2015)
3.	Solar still is integrated with thermosyphon heat pipe and vacuum glass		1.02 kg/(m^2 h)	The better output in the designed system is found at 2 cm depth with a maximum production rate and efficiency of 1.02 kg/(m^2 h) and 22.95	Mamouri et al. (2014)
4.	Solar still integrated with pulsating pipe		875 mL/(m^2 h)	The maximum distillate output was found with 1 cm depth and 40% filling ratio	Abad et al. (2013)
5.	Solar desalination system integrated with humidifier for production of air bubble, heat pipe and evacuated tube	65%	6.275 kg/day m^2	The final result revealed that better result was obtain with oil in between ETC and heat pipe	Behnam and Shafii (2016)

Passive Solar Collector System

In passive solar water heating system, mechanical energy is not used for circulation of working fluid through collector, but working fluid is moved due to temperature gradient by absorption of solar radiations (Al-Abidi et al. 2012). In other words, in the passive solar water heating system without using mechanical device transfer, the working fluid movement between collector and water storage tank takes place as shown in Fig. 14.19a. This method is based on the natural convection heat transfer method. In this system, the fluid heated up by absorption of heat is collected by the

Table 14.3 Description of parabolic trough integrated with desalination system

S. no.	Modification	Increase in output (%)	Productivity	Observation/findings/advantages	References
1.	Parabolic trough solar collectors	–	93 °C–103 °C	The system can be utilized for different renewable energy sources one of which is desalination; the system efficiency was found maximum for the month of February to May	Murtuza et al. (2017)

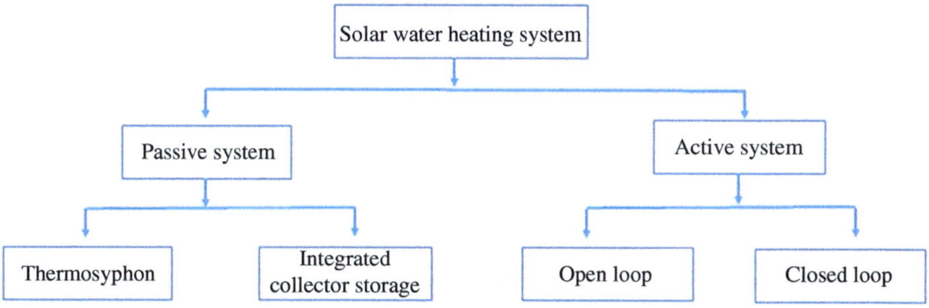

Fig. 14.18 Types of solar water heating systems (Patel et al. 2012)

solar collector, resulting in decrease of fluid density resulting in fluid moving to the top of the collector and gushing into water storage tank. Afterwards, hot water is supplied for further use and cold water is supplied from the bottom of a storage tank to the collector, and the cycle is continued.

Thermosyphon

The schematic arrangement of thermosyphon solar water heating system is shown in Fig. 14.19b. It is the most popular type of solar water heating system and commercially available in the market (Drosou et al. 2014).

The construction of thermosyphon solar water heating system is simple and requires less maintenance owing to an absence of moving parts. This system also heats drinking water or heat transfer fluid and transports it by natural convection method from collector to storage tank (Hossain et al. 2011). The thermosyphon starts when working fluid (water) in collector expands, causing heated water rising to the header of collector and finally into the storage tank. The cold water is continuously supplied to water storage tank to maintain the tank water level. Due to density difference, hot water is accumulated near the top of the storage tank as it is heated during daytime and cold water at the bottom of storage tank. The storage tank and sanitary fittings should be well insulated to minimize the heat losses. The size of

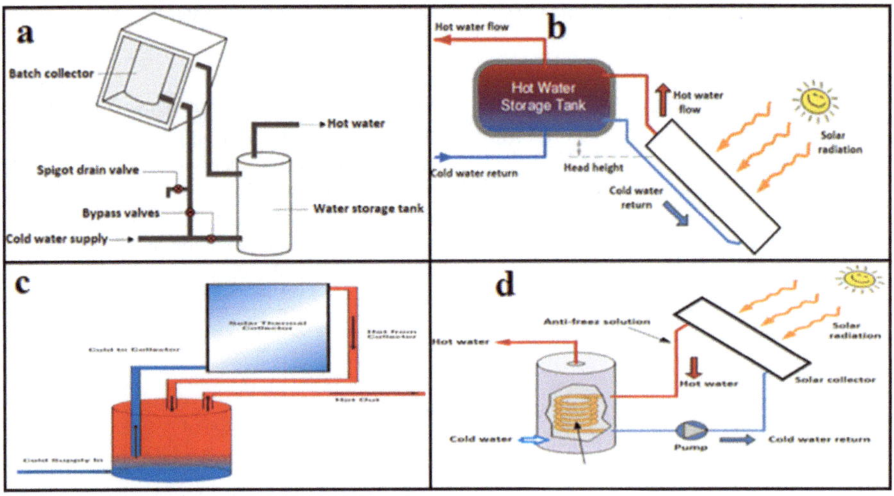

Fig. 14.19 Solar water heating systems (a) Passive solar water heating system (b) Thermosyphon solar hot water system (c) Open-loop solar water heating system (d) Closed-loop solar water heating system (Patel et al. 2012)

storage tank is such that it can store hot water for about 2 days to compensate for the poor solar radiation period.

Integrated Collector Storage

Another setup of the passive system is the integrated collector storage (ICS) system. In this system, solar collector and thermal storage unit combine together to form a single unit. It is different from the simple solar water heater but became very popular due to its low cost as a solar collector and thermal storage unit are integrated into the same construction. Its construction varies comprising of numerous tanks, where exposed surfaces absorb the energy and are enclosed in a heat insulated box with glass envelope on the top to allow solar radiation. The transparent cover and heat insulation are used for reduction of heat loss. These systems are simple and less expensive; however the loss of more heat and absence of freezing protection during night time are its weaknesses.

On the other hand, the distribution of hot water from solar water system through distribution equipment in a controlled manner is termed as an active solar water heating system (Chan et al. 2010). In other words, the active solar system uses pumps, controllers, and valves to circulate heat transfer fluid through solar collectors (Chen et al. 2009). Active solar water heating system is further classified as open-loop or a closed-loop system. As shown in Fig. 14.19c, open-loop or direct solar water heating system heats the household water through the solar collector and heated water is pumped into water storage tank.

Pipes are connected with stored water for the supply of hot water stored in a storage tank. Unlike open-loop, in a closed-loop or indirect solar active system shown in Fig. 14.19d, the water-antifreeze mixture is heated through the solar collector and pumped into water storage tank where a heat exchanger transfers its heat to the household water.

Conclusion

It is impossible to discuss all the techniques which help in maintaining a sustainable environment. The above chapter is a short overview for the development on solar technologies which produces zero carbon footprint and helps to maintain sustainable future. As we know, technologies based on solar energy have become most popular throughout the world. To attain this, huge investment is needed to overcome the problems associated with the solar industry. Presently, a large number of solar power plants are installed or to be installed in the near future in developing as well as developed countries. Here in this book chapter, the potential of solar technologies and their future prospects are reported and conclude that in spite of few limitations, among other renewable and conventional energy resources, solar energy is the most promising energy source to meet the increasing energy demand. Although renewable and non-renewable energy resources contributed a majority of global energy production, the solar energy sector has been making great progress.

- The PV technology is the most popular technology for a continuous supply of electricity. However, the overall cost of PV plant is high, and a novel approach is still necessary for the crucial development of PV technologies.
- However, concentrating solar power plants are more expensive than PV plant, but these plants are more appropriate for the regions of less frequent haze or clouds.
- The solar desalination and photocatalytic process help in maintaining low pollution levels by treating wastewater. This technology is proving to be the future of wastewater treatment due to their rapid growth and removal efficiency.
- Solar photocatalytic process and desalination require different type of solar collectors. These collectors help in improving their efficiency and could be integrated in the future for better output.
- Among other renewable and conventional resources, the researchers should also work on improving the competitiveness of solar technologies. The researchers effort should be dedicated to increase efficiency, availability, stability, manufacturability and reduction in the cost.

Solar energy creates a vicinity of a sustainable world which can be able to bear a load of energy at present as well as in future. Although, solar energy-harnessing technologies are still in nascent phase still, the above description of available technologies is reflecting importance in future for solar collectors, solar photocatalysis and integrated solar desalination systems regarding power generation, wastewater treatment and several other applications.

Future Prospects

Future recommendations are summarized based on the articles reviewed and analysed:

1. The solar technologies can be used in different industries for high-temperature supply.
2. The new systems can be designed for sterilization of instruments (thermal or direct).
3. Solar energy-based systems with some modifications could be used at large scale for wastewater treatment options.
4. The solar desalination technology and photocatalysis are a future of wastewater treatment at small and large scale as well.

Furthermore, integrated systems (with solar) should be promoted for wastewater treatment, sterilization and high-temperature supply for clean and green development at large scale.

Acknowledgements Financial assistance in the form of Junior Research Fellowship (JRF) to Atin Kumar Pathak under Inspire Fellowship scheme by the Department of Science and Technology, Ministry of Science and Technology (Govt. of India), New Delhi, which is gratefully acknowledged. One of the author Mr. Har Mohan Singh is thankful to Ministry of New and Renewable Energy for National Renewable Energy Fellowship. Dr. V. V. Tyagi is also highly thankful to University Grant Commission (Govt. of India) for providing startup research grant at Shri Mata Vaishno Devi University, Katra (J&K).

References

Abad, H. K. S., Ghiasi, M., Mamouri, S. J., & Shafii, M. B. (2013). A novel integrated solar desalination system with a pulsating heat pipe. *Desalination, 311*, 206–210.

Abe, R. (2010). Recent progress on photocatalytic and photoelectrochemical water splitting under visible light irradiation. *Journal of Photochemistry and Photobiology C: Photochemistry Reviews, 11*(4), 179–209.

Al-Abidi, A. A., Mat, S. B., Sopian, K., Sulaiman, M. Y., Lim, C. H., & Th, A. (2012). Review of thermal energy storage for air conditioning systems. *Renewable and Sustainable Energy Reviews, 16*(8), 5802–5819.

Al-Badi, A. H., & Albadi, M. H. (2012). Domestic solar water heating system in Oman: Current status and future prospects. *Renewable and Sustainable Energy Reviews, 16*(8), 5727–5731.

Alexopoulos, S., & Hoffschmidt, B. (2010). Solar tower power plant in Germany and future perspectives of the development of the technology in Greece and Cyprus. *Renewable Energy, 35*(7), 1352–1356.

Al-Hayeka, I., & Badran, O. O. (2004). The effect of using different designs of solar stills on water distillation. *Desalination, 169*(2), 121–127.
Asbik, M., Ansari, O., Bah, A., Zari, N., Mimet, A., & El-Ghetany, H. (2016). Exergy analysis of solar desalination still combined with heat storage system using phase change material (PCM). *Desalination, 381*, 26–37.
Ayompe, L. M., Duffy, A., Mc Keever, M., Conlon, M., & McCormack, S. J. (2011). Comparative field performance study of flat plate and heat pipe evacuated tube collectors (ETCs) for domestic water heating systems in a temperate climate. *Energy, 36*(5), 3370–3378.
Badran, A. A., Al-Hallaq, I. A., Salman, I. A. E., & Odat, M. Z. (2005). A solar still augmented with a flat-plate collector. *Desalination, 172*(3), 227–234.
Bahnemann, D. W., Lawton, L. A., & Robertson Peter, K. J. (2013). Chapter 16: The application of semiconductor photocatalysis for the removal of cyanotoxins from water and design concepts for solar photocatalytic reactors for large scale water treatment. In *New and future developments in catalysis* (pp. 395–415). Amsterdam/Boston: Elsevier.
Barrera, E. C. (1933). Double effect spherical solar still. *Sun World, 17*(1), 12–14.
Behnam, P., & Shafii, M. B. (2016). Examination of a solar desalination system equipped with an air bubble column humidifier, evacuated tube collectors and thermosyphon heat pipes. *Desalination, 397*, 30–37.
Bernabeu, A., Vercher, R. F., Santos-Juanes, L., Simon, P. J., Lardin, C., Martinez, M. A., Vicente, J. A., Gonzalez, R., Llosac, C., Arques, A., & Amat, A. M. (2011). Solar photocatalysis as a tertiary treatment to remove emerging pollutants from wastewater treatment plant effluents. *Catalysis Today, 161*(1), 235–240.
Bloemer, J. W., Eibling, J. A., Irwin, J. R., & Lof, G. O. (1965). A practical basin-type solar still. *Solar Energy, 9*(4), 197–200.
Chan, H. Y., Riffat, S. B., & Zhu, J. (2010). Review of passive solar heating and cooling technologies. *Renewable and Sustainable Energy Reviews, 14*(2), 781–789.
Chen, B. R., Chang, Y. W., Lee, W. S., & Chen, S. L. (2009). Long-term thermal performance of a two-phase thermosyphon solar water heater. *Solar Energy, 83*(7), 1048–1055.
Chong, M. N., Jin, B., Chow, C. W., & Saint, C. (2010). Recent developments in photocatalytic water treatment technology: A review. *Water Research, 44*(10), 2997–3027.
Coffey, J. P. (1975). Vertical solar distillation. *Solar Energy, 17*(6), 375–378.
DeWinter, F. (Ed.). (1990). *Solar collectors, energy storage, and materials* (Vol. 5). Cambridge, MA: MIT press.
Drosou, V. N., Tsekouras, P. D., Oikonomou, T. I., Kosmopoulos, P. I., & Karytsas, C. S. (2014). The HIGH-COMBI project: High solar fraction heating and cooling systems with combination of innovative components and methods. *Renewable and Sustainable Energy Reviews, 29*, 463–472.
El-Agouz, S. A., El-Samadony, Y. A. F., & Kabeel, A. E. (2015). Performance evaluation of a continuous flow inclined solar still desalination system. *Energy Conversion and Management, 101*, 606–615.
El-Sebaii, A. A. (2005). Thermal performance of a triple-basin solar still. *Desalination, 174*(1), 23–37.
Eltawil, M. A., & Omara, Z. M. (2014). Enhancing the solar still performance using solar photovoltaic, flat plate collector and hot air. *Desalination, 349*, 1–9.
Fujishima, A. K. I. R. A., Rao, T. N., & Tryk, D. A. (2000). TiO_2 photocatalysts and diamond electrodes. *Electrochimica Acta, 45*(28), 4683–4690.
Garcia-Cortes, S., Bello-Garcia, A., & Ordonez, C. (2012). Estimating intercept factor of a parabolic solar trough collector with new supporting structure using off-the-shelf photogrammetric equipment. *Applied Energy, 92*, 815–821.
Gugulothu, R., Somanchi, N. S., Reddy, K. V. K., & Gantha, D. (2015). A review on solar water distillation using sensible and latent heat. *Procedia Earth and Planetary Science, 11*, 354–360.
Gunjo, D. G., Mahanta, P., & Robi, P. S. (2017). Exergy and energy analysis of a novel type solar collector under steady state condition: Experimental and CFD analysis. *Renewable Energy, 114*, 655–669.

Hamadou, O. A., & Abdellatif, K. (2014). Modeling an active solar still for sea water desalination process optimization. *Desalination, 354*, 1–8.
Hennecke, K., Schwarzbozl, P., Alexopoulos, S., Gottsche, J., Hoffschmidt, B., Beuter, M., Koll, G., & Hartz, T. (2008). *Solar power tower Julich – The first test and demonstration plant for open volumetric receiver technology in Germany*. In Proceedings of the 14th biennial CSP solar PACES symposium, Las Vegas, Nevada.
Hossain, M. S., Saidur, R., Fayaz, H., Rahim, N. A., Islam, M. R., Ahamed, J. U., & Rahman, M. M. (2011). Review on solar water heater collector and thermal energy performance of circulating pipe. *Renewable and Sustainable Energy Reviews, 15*(8), 3801–3812.
Inamdar, J., & Singh, S. K. (2008). Techno- economic analysis of zero effluent discharge by use of solar detoxification at household level. *International Journal of Natural and Engineering Sciences, 1*, 208–211.
Inoue, T., Fujishima, A., Konishi, S., & Honda, K. (1979). Photoelectrocatalytic reduction of carbon dioxide in aqueous suspensions of semiconductor powders. *Nature, 277*(5698), 637–638.
Islam, M. R., Sumathy, K., & Khan, S. U. (2013). Solar water heating systems and their market trends. *Renewable and Sustainable Energy Reviews, 17*, 1–25.
Jebasingh, V. K., & Herbert, G. J. (2016). A review of solar parabolic trough collector. *Renewable and Sustainable Energy Reviews, 54*, 1085–1091.
Kabeel, A. E., Abdelgaied, M., & Mahgoub, M. (2016a). The performance of a modified solar still using hot air injection and PCM. *Desalination, 379*, 102–107.
Kabeel, A. E., Khalil, A., Shalaby, S. M., & Zayed, M. E. (2016b). Experimental investigation of thermal performance of flat and v-corrugated plate solar air heaters with and without PCM as thermal energy storage. *Energy Conversion and Management, 113*, 264–272.
Kabeel, A. E., Khalil, A., Shalaby, S. M., & Zayed, M. E. (2016c). Investigation of the thermal performances of flat, finned, and v-corrugated plate solar air heaters. *Journal of Solar Energy Engineering, 138*(5), 051004.
Kalogirou, S. A. (2004). Solar thermal collectors and applications. *Progress in Energy and Combustion Science, 30*(3), 231–295.
Kiatsiriroat, T. (1989). Review of research and development on vertical solar stills. *ASEAN Journal on Science and Technology for Development, 6*(1), 15.
Kositzi, M., Poulios, I., Malato, S., Caceres, J., & Campos, A. (2004). Solar photocatalytic treatment of synthetic municipal wastewater. *Water Research, 38*(5), 1147–1154.
Kudish, A. I., Evseev, E. G., Walter, G., & Priebe, T. (2003). Simulation study on a solar desalination system utilizing an evaporator/condenser chamber. *Energy Conversion and Management, 44*(10), 1653–1670.
Kumar, S., Dubey, A., & Tiwari, G. N. (2014). A solar still augmented with an evacuated tube collector in forced mode. *Desalination, 347*, 15–24.
Kumar, P. V., Kumar, A., Prakash, O., & Kaviti, A. K. (2015). Solar stills system design: A review. *Renewable and Sustainable Energy Reviews, 51*, 153–181.
Kumar, R. A., Esakkimuthu, G., & Murugavel, K. K. (2016). Performance enhancement of a single basin single slope solar still using agitation effect and external condenser. *Desalination, 399*, 198–202.
Lawrence, S. A., & Tiwari, G. N. (1990). Theoretical evaluation of solar distillation under natural circulation with heat exchanger. *Energy Conversion and Management, 30*(3), 205–213.
Lupfert, E., Geyer, M., Schiel, W., Esteban, A., Osuna, R., Zarza, E., & Nava, P. (2001). Eurotrough design issues and prototype testing at PSA. *Solar Engineerings, 2001*, 387–392.
Maeda, K. (2011). Photocatalytic water splitting using semiconductor particles: History and recent developments. *Journal of Photochemistry and Photobiology C: Photochemistry Reviews, 12*(4), 237–268.
Martinopoulos, G., Ikonomopoulos, A., & Tsilingiridis, G. (2016). Initial evaluation of a phase change solar collector for desalination applications. *Desalination, 399*, 165–170.
McLoughlin, O. A., Ibanez, P. F., Gernjak, W., Rodrıguez, S. M., & Gill, L. W. (2004). Photocatalytic disinfection of water using low cost compound parabolic collectors. *Solar Energy, 77*(5), 625–633.

Moorthy, M. (2010). *Performance of solar air-conditioning system using heat pipe evacuated tube collector*. Doctoral dissertation, UMP, National conference in mechanical engineering research and postgraduate studies. Pahang UMP Pekan.

Morad, M. M., El-Maghawry, H. A., & Wasfy, K. I. (2015). Improving the double slope solar still performance by using flat-plate solar collector and cooling glass cover. *Desalination, 373*, 1–9.

Mosleh, H. J., Mamouri, S. J., Shafii, M. B., & Sima, A. H. (2015). A new desalination system using a combination of heat pipe, evacuated tube and parabolic trough collector. *Energy Conversion and Management, 99*, 141–150.

Murtuza, S. A., Byregowda, H. V., & Imran, M. (2017). Experimental and simulation studies of parabolic trough collector design for obtaining solar energy. *Resource-Efficient Technologies, 3*(4), 414–421.

Mamouri, S. J., Derami, H. G., Ghiasi, M., Shafii, M. B., & Shiee, Z. (2014). Experimental investigation of the effect of using thermosyphon heat pipes and vacuum glass on the performance of solar still. *Energy, 75*, 501–507.

Nkwetta, D. N., Smyth, M., Zacharopoulos, A., & Hyde, T. (2013). Experimental field evaluation of novel concentrator augmented solar collectors for medium temperature applications. *Applied Thermal Engineering, 51*(1), 1282–1289.

Nosaka, Y., Nakamura, M., & Hirakawa, T. (2002). Behavior of superoxide radicals formed on TiO_2 powder photocatalysts studied by a chemiluminescent probe method. *Physical Chemistry Chemical Physics, 4*(6), 1088–1092.

Nosaka, A. Y., Kojima, E., Fujiwara, T., Yagi, H., Akutsu, H., & Nosaka, Y. (2003). Photoinduced changes of adsorbed water on a TiO2 photocatalytic film as studied by 1H NMR spectroscopy. *The Journal of Physical Chemistry B, 107*(44), 12042–12044.

Padilla, R. V., Demirkaya, G., Goswami, D. Y., Stefanakos, E., & Rahman, M. M. (2011). Heat transfer analysis of parabolic trough solar receiver. *Applied Energy, 88*(12), 5097–5110.

Panchal, H. N., & Patel, S. (2017). An extensive review on different design and climatic parameters to increase distillate output of solar still. *Renewable and Sustainable Energy Reviews, 69*, 750–758.

Panchal, H. N., & Shah, P. K. (2014). Enhancement of distillate output of double basin solar still with vacuum tubes. *Frontier Energy, 8*(1), 101.

Panchal, H., Patel, P., Patel, N., & Thakkar, H. (2017). Performance analysis of solar still with different energy-absorbing materials. *International Journal of Ambient Energy, 38*(3), 224–228.

Parilti, N. B. (2010). Treatment of a petrochemical industry wastewater by a solar oxidation process using the Box-Wilson experimental design method. *Ekoloji, 19*(77), 9–15.

Patel, K., Patel, P., & Patel, J. (2012). Review of solar water heating systems. *International Journal of Advanced Engineering Technology, 3*(IV), 146–149.

Peller, J. R., Whitman, R. L., Griffith, S., Harris, P., Peller, C., & Scalzitti, J. (2007). TiO_2 as a photocatalyst for control of the aquatic invasive alga, Cladophora, under natural and artificial light. *Journal of Photochemistry and Photobiology, A: Chemistry, 186*(2), 212–217.

Peral, J., Domenech, X., & Ollis, D. F. (1997). Heterogeneous photocatalysis for purification, decontamination and deodorization of air. *Journal of Chemical Technology & Biotechnology, 70*(2), 117–140.

Quinones, D. H., Alvarez, P. M., Rey, A., Contreras, S., & Beltran, F. J. (2015). Application of solar photocatalytic ozonation for the degradation of emerging contaminants in water in a pilot plant. *Chemical Engineering Journal, 260*, 399–410.

Rai, S. N., & Tiwari, G. N. (1983). Single basin solar still coupled with flat plate collector. *Energy Conversion and Management, 23*(3), 145–149.

Rajaseenivasan, T., Murugavel, K. K., Elango, T., & Hansen, R. S. (2013). A review of different methods to enhance the productivity of the multi-effect solar still. *Renewable and Sustainable Energy Reviews, 17*, 248–259.

Rajaseenivasan, T., Raja, P. N., & Srithar, K. (2014). An experimental investigation on a solar still with an integrated flat plate collector. *Desalination, 347*(2014), 131–137.

Rojas, D., Beermann, J., Klein, S. A., & Reindl, D. T. (2008). Thermal performance testing of flat-plate collectors. *Solar Energy, 82*(8), 746–757.

Sabiha, M. A., Saidur, R., Mekhilef, S., & Mahian, O. (2015). Progress and latest developments of evacuated tube solar collectors. *Renewable and Sustainable Energy Reviews, 51*, 1038–1054.

Sarwar, J., & Mansoor, B. (2016). Characterization of thermophysical properties of phase change materials for non-membrane based indirect solar desalination application. *Energy Conversion and Management, 120*, 247–256.

Sathyamurthy, R., El-Agouz, S. A., & Dharmaraj, V. (2015). Experimental analysis of a portable solar still with evaporation and condensation chambers. *Desalination, 367*, 180–185.

Shah, L. J., & Furbo, S. (2004). Vertical evacuated tubular-collectors utilizing solar radiation from all directions. *Applied Energy, 78*(4), 371–395.

Sivakumar, P., Christraj, W., Sridharan, M., & Jayamalathi, N. (2012). Performance improvement study of solar water heating system. *ARPN Journal of Engineering and Applied Sciences, 7*(1), 45–49.

Sodha, M. S., Kumar, A., Tiwari, G. N., & Pandey, G. C. (1980). Effects of dye on the performance of a solar still. *Applied Energy, 7*(1), 147–162.

Suneja, S., & Tiwari, G. N. (1998). Optimization of number of effects for higher yield from an inverted absorber solar still using the Runge-Kutta method. *Desalination, 120*(3), 197–209.

Tamini, A. (1987). Performance of a solar still with reflectors and black dye. *Solar & Wind Technology, 4*(4), 443–446.

Tanaka, H., Nosoko, T., & Nagata, T. (2000). Parametric investigation of a basin-type-multiple-effect coupled solar still. *Desalination, 130*(3), 295–304.

Tiwari, G. N., & Madhuri, G. N. (1987). Effect of water depth on daily yield of the still. *Desalination, 61*(1), 67–75.

Tiwari, G. N., & Tiwari, A. (2017). *Handbook of solar energy*. Singapor: Sprinter.

Tiwari, G. N., Kupfermann, A., & Aggarwal, S. (1997). A new design for a double-condensing chamber solar still. *Desalination, 114*(2), 153–164.

Tleimat, B. W., & Howe, E. D. (1966). Nocturnal production of solar distillers. *Solar Energy, 10*(2), 61–66.

Tleimat, B. W., & Howe, E. D. (1969). Comparison of plastic and glass condensing covers for solar distillers. *Solar Energy, 12*(3), 293IN3297IN5303–296IN4302IN6304.

Tryk, D. A., Fujishima, A., & Honda, K. (2000). Recent topics in photoelectrochemistry: Achievements and future prospects. *Electrochimica Acta, 45*(15), 2363–2376.

Tyagi, V. V., Pathak Atin, K., Singh, H. M., Kothari, R., Selvaraj, J., & Pandey, A. K. (2016). *Renewable energy scenario in Indian context: Vision and achievements*. 4th IET clean energy and technology conference (Vol. 8, p. 85). https://doi.org/10.1049/cp.2016.1342. ISBN: 978-1-78561-238-1.

Velmurugan, K., Christraj, W., Kulasekharan, N., & Elango, T. (2016). Performance study of a dual-function Thermosyphon solar heating system. *Arabian Journal for Science and Engineering, 41*(5), 1835–1846.

Wang, C., Liu, H., & Qu, Y. (2013). TiO2-based photocatalytic process for purification of polluted water: Bridging fundamentals to applications. *Journal of Nanomater, 14*. Article ID 319637.

Wolfrum, E. J., Huang, J., Blake, D. M., Maness, P. C., Huang, Z., Fiest, J., & Jacoby, W. A. (2002). Photocatalytic oxidation of bacteria, bacterial and fungal spores, and model biofilm components to carbon dioxide on titanium dioxide-coated surfaces. *Environmental Science & Technology, 36*(15), 3412–3419.

Yadav, Y. P. (1991). Analytical performance of a solar still integrated with a flat plate solar collector: Thermosiphon mode. *Energy Conversion and Management, 31*(3), 255–263.

Zhao, X., Wang, Z., & Tang, Q. (2010). Theoretical investigation of the performance of a novel loop heat pipe solar water heating system for use in Beijing, China. *Applied Thermal Engineering, 30*(16), 2526–2536.

Zhou, H., & Smith, D. W. (2002). Advanced technologies in water and wastewater treatment. *Journal of Environmental Engineering and Science, 1*(4), 247–264.

Chapter 15
Natural Sensitizers and Their Applications in Dye-Sensitized Solar Cell

A. K. Pandey, Muhammad Shakeel Ahmad, Nasrudin Abd Rahim, V. V. Tyagi, and R. Saidur

Contents

A. K. Pandey (✉)
Research Centre for Nano-Materials and Energy Technology (RCNMET), School of Science and Technology, Sunway University, Petaling Jaya, Selangor Darul Ehsan, Malaysia
e-mail: adarshp@sunway.edu.my

M. S. Ahmad
UM Power Energy Dedicated Advanced Center (UMPEDAC), Level 4, University of Malaya, Kuala Lumpur, Malaysia

N. A. Rahim
UM Power Energy Dedicated Advanced Center (UMPEDAC), Level 4, University of Malaya, Kuala Lumpur, Malaysia

Renewable Energy Research Group, King Abdulaziz University, Jeddah, Saudi Arabia

V. V. Tyagi
School of Energy Managment, Shri Mata Vaishno Devi University, Katra, Jammu and Kashmir, India

R. Saidur
Research Centre for Nano-Materials and Energy Technology (RCNMET), School of Science and Technology, Sunway University, Subang Jaya, Selangor Darul Ehsan, Malaysia

Department of Engineering, Lancaster University, Lancaster, UK

R. C. Sobti et al. (eds.), *Environmental Biotechnology: For Sustainable Future*,
https://doi.org/10.1007/978-981-10-7284-0_15

Abstract Many organic, inorganic and natural dye sensitizers have been trailed in the past in an attempt to reduce the cost, improve the performance and make the dye-sensitized solar cell (DSSC) technology more environment-friendly. Ruthenium-based complexes are by far the most efficient dye sensitizers and have been commercially used in DSSC technology and achieved approximately 12–14% conversion efficiency. But the problems associated with ruthenium complexes are high cost and toxicity which drive the researchers to identify new metal-free and environment-friendly dye sensitizers such as organic and natural sensitizers. In this regard, natural dye sensitizers due to their low-cost extraction and environment-friendly nature are becoming a new area of research in the field of DSSC technology. These dye sensitizers are naturally occurring dye pigments, such as chlorophyll, betanins, carotenoids, anthocyanins and tannins extracted from flowers, leaves, stems and roots of plants using water, acetone and/or alcohols. At present, the efficiency of natural dye sensitizers is quite low compared to ruthenium-based dye due to selective light absorption. Recently, highest recorded efficiency of 2% has been reported using cocktail of natural dyes extracted from flowers. Attempts have been made to improve the performance of natural dye sensitizers by making cocktails and/or by using a variety of solvents for the extraction of dye molecules.

Keywords Charge transfer · DSSC · Interface contact · Natural sensitizers · Pigments

Introduction

The industrialized modernization and human growth is much reliant on the highly concentrated energy sources such as fossil fuels which are already on the brink of extension. Figure 15.1a is showing the forecasted life of fossil fuel reserves. Another associated problem is the inappropriate technology to harness useful energy from fossil fuels which causes high emissions of greenhouse gases and increase in CO_2 concentration in global environment (Urban 2015). Figure 15.1b is showing CO_2 concentration for the past 400,000 years, depicting several cycles of variation from about 180 to 280 ppm. The CO_2 concentration has increased dramatically since the past 200 years due to industrial revolution and reached to 400 ppm (Sigman and Boyle 2000; Urban 2015).

The projected global energy demand at this rate of industrialization would reach up to 25–30TW in 2050 as compared to 17TW demand at present. The depletion of fossil fuels and their harmful environmental effects force the scientific communities to explore other renewable ways to ensure sustainable global development. Among

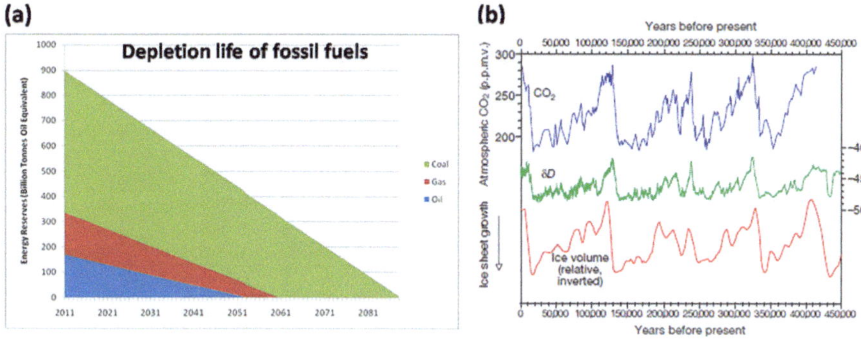

Fig. 15.1 **(a)** Depletion of fossil fuels and **(b)** global CO_2 emission history. (Sigman and Boyle 2000; Urban 2015)

other renewable resources such as biomass (potential harvestable energy, 5–7TW by all cultivatable land not used for food), wind (potential harvestable energy, 2–4TW extractable at 10 m and around 60TW at 1 km height), tidal (potential harvestable energy, 2TW gross currents), geothermal (potential harvestable energy, 12TW over land small fraction recoverable), hydroelectric (potential harvestable energy, 4–5TW gross, 1.6TW technically feasible) and nuclear (potential harvestable energy, 8TW builds one nuclear power plant every 1.5 days forever), it is very difficult or even impossible to cope with high-energy demand of the future. The most suitable alternative is solar energy. Earth receives 120,000TW of electromagnetic radiation per week. If solar photovoltaic panels possessing 10% efficiency are installed onto only 0.61% of the land on earth, it can provide 20TW of power which is twice the world's energy demand (Eisenberg and Nocera 2005).

Among solar energy-harvesting technologies, solar cells are the most suitable approach to convert energy directly from solar radiations to electricity. In general, solar cells are segmented into three classes based on their lifetime. Silicon-based solar cells are assumed to be the first-generation solar cells. These cells are based on the single crystal silicon which makes them expensive, and their manufacturing technologies require high temperatures which mostly come from burning of fossil fuels. Second-generation solar cells are based on polycrystalline silicon and thin films such as CdTe, copper indium gallium diselenide, etc. Second-generation solar cells are cost-effective, but the use of less abundant raw material and sophisticated vacuum technology along with their relatively less overall efficiency restricts their widespread use. The latest or third generation of solar cells totally differ from other two generations by optimizing the cost and better conversion efficiency. Most of the technology of third-generation solar cells is in research stage and includes organic solar cells (η~11%), quantum dots cells (η~10%), dye-sensitized solar cell (DSSC) (η~13%) and pervoskite-sensitized solar cells (η~22%). The efficiency chart of the solar cell technologies can be seen in Fig. 15.2.

The DSSC technology is probably the most attractive one in third-generation solar cell due to its operational simplicity, ease in manufacturing, utilization of earth-abundant raw materials and the capability to make flexible panels. The basic

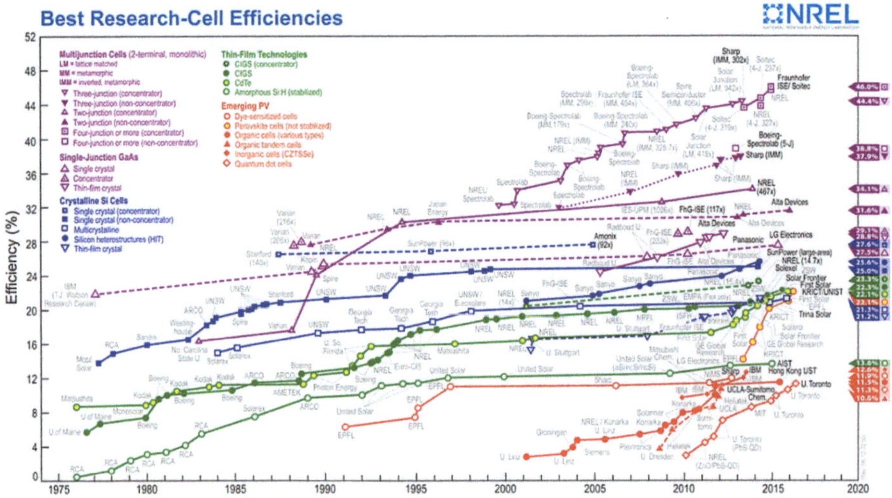

Fig. 15.2 Efficiency chart of various classes of solar energy conversion devices (Green et al. 2016; Ahmad et al. 2017a, b)

Fig. 15.3 Schematic diagram of basic DSSC assembly (Ahmad et al. 2017a, b; Ahmad et al. 2018)

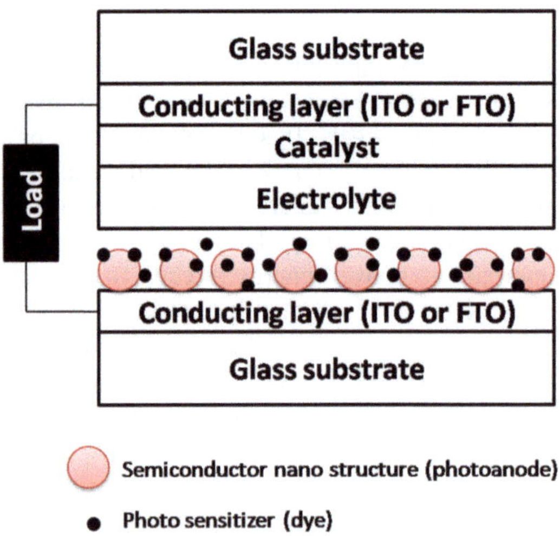

constituents of DSSC are (1) conducting substrate, (2) semi-conductor active layer, (3) dye sensitizer, (4) electrolyte and (5) catalyst. In general, dye sensitizer absorbs energy from incident light and oxidized, placing electron to the conduction band of semi-conductor material. Semi-conductor transfers electron to the external circuit through conducting substrate. Electrolyte gives its electron to the oxidized dye and accept electron from the external circuit with the help of catalyst (Pt) and gets reduced. This cycle keeps on repeating until the light source is available (Jafarzadeh et al. 2016). Figure 15.3 is showing the basics of DSSC.

Operation Principle

The DSSC device works on the principle of photosynthesis used by plants and many bacteria. Photosynthesis involves (1) capturing of sunlight, (2) using sunlight to make adenosine triphosphate (ATP) and reducing power in the form nicotinamide adenine dinucleotide phosphate (NADPH) and (3) use ATP and NADPH to power the production of organic molecules from carbon dioxide in the air (Force et al. 2003). On the other hand, DSSCs use sunlight to produce electrons and power the outer circuit. Dye sensitizer absorbs suitable radiations from the solar spectrum which causes excitation of electron. This excited electron is then transferred to the lower unoccupied molecular orbit (LUMO) of the semi-conductor. It is worth mentioning here that the LUMO of semi-conductor must be lower than the LUMO of dye sensitizer for successful transfer of electron from dye sensitizer to semi-conductor; otherwise due to thermodynamic unsuitability of electronic path, the electrons cannot be moved to LUMO of semi-conductor. The electron then transfers to transparent conducting oxide (TCO) layer and finally exits the device to the external circuit. The oxidized dye sensitizer due to electron deficiency picks electron from the redox mediator, i.e. electrolyte containing iodide/triiodide ions and regenerated. The electrolyte picks electron coming from outer circuit with the help of catalyst and gets regenerated. The whole process keeps on repeating until the sunlight is present. The process has been elaborated in detail (Eqs. 15.1, 15.2, 15.3, and 15.4) (Ahmad et al. 2017a, b; Lee et al. 2017). Figure 15.4 is depicting the complete process.

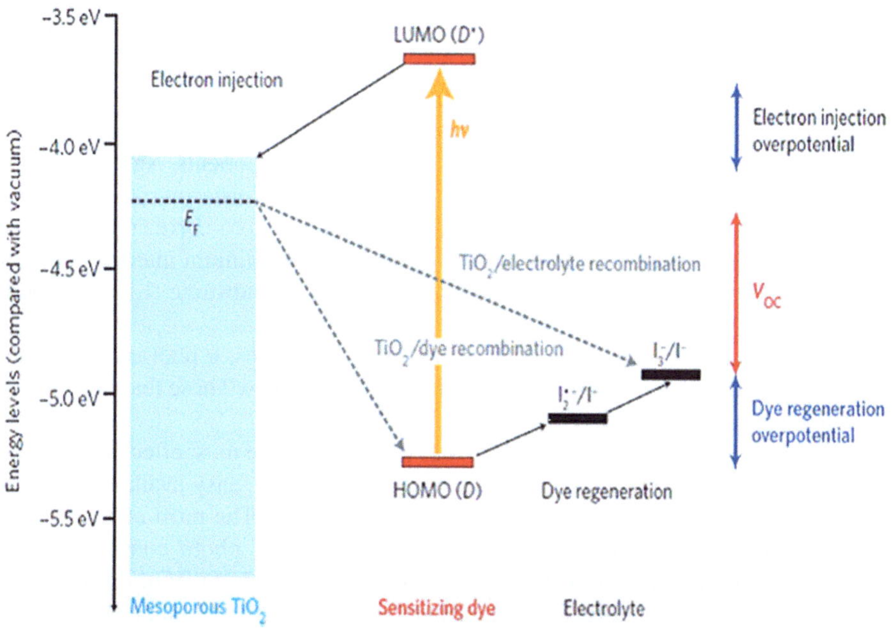

Fig. 15.4 DSSC working principle (Mozaffari et al. 2016; Ahmad et al. 2017a, b)

$$\text{DS (on TiO}_2\text{)} + h\nu \longrightarrow \text{DS* (on TiO}_2\text{)} \quad 15.1$$

$$\text{DS* (on TiO}_2\text{)} \longrightarrow \text{DS}^+ \text{ (on TiO}_2\text{)} + e^-\text{(injected)} \quad 15.2$$

$$\text{DS* (on TiO}_2\text{)} + 3/2\text{I}^- \longrightarrow \text{DS* (on TiO}_2\text{)} + ½\ \text{I}^{-3} \quad 15.3$$

$$\text{I}^{-3} + 2e^-\text{(cathode)} \longrightarrow 3^{\text{I-}}\text{(cathode)} \quad 15.4$$

However, during this electronic cycle, some undesirable reactions occur simultaneously, (1) recombination of injected electrons with the oxidized dye sensitizer or (2) with the electron deficient redox mediator. These recombination reactions are responsible for low photovoltaic efficiency of DSSCs. To minimize these unfavourable reactions, the electronic path must be thermodynamically suitable and kinetically favourable.

Components of DSSC and Their Role

Constituents of DSSC are shown in Fig. 15.3. A typical DSSC is comprised of four major components, i.e. (1) photoanode, (2) dye sensitizer, (3) electrolyte and (4) counter electrode. Understanding the function of these components is crucial for the researcher working in the field of DSSC.

Photoanode

Photoanode is the backbone of the DSSC that is why its material, shape and size have to be carefully selected and optimized with other components. An ideal photoanode should have high surface area; it should facilitate fast electron transfer ability, have high dye pickup ability, exhibit high resistance to photo-corrosion, and have the ability to absorb/scatter sunlight; and it should have optimum interface contact with the dye molecules and with conductive layer on the substrate (Li et al. 2006; Wang et al. 2006).

With these features and obviously some other properties, a photoanode would work ideally and give maximum photo conversion efficiency. These features though simplistic are still a challenge in the field of DSSC.

Wide bandgap annealed anatase TiO_2 has proved to be the most effective material for photoanode due to its cost effectiveness, good stability, easy availability, compatible optical and electronic properties and non-toxicity. The most efficient solar cell developed by TiO_2 delivered approximately 12–14% photo conversion efficiency. Various other metal oxides such as ZnO, SnO_2, Nb_2O_5, $SrTiO_3$, Zn_2SnO_4 and WO_3 have also been investigated (Kumavat et al. 2017; Shakeel Ahmad et al. 2017).

Dye Sensitizer

Among the basic components of DSSC, dye sensitizer is regarded as a very important component of the solar device. The dye anchors on the semi-conductor material and is responsible for absorption of incident light and gets oxidized. For proper functioning of solar device, the LUMO of the dye should be higher than the conduction band of semi-conductor material for proper injection of electron, and its higher occupied molecular orbit (HOMO) must be sufficiently low in energy (moving downward) than the redox potential of the iodide/triiodide electrolyte for ease in reduction (Ludin et al. 2014).

Many organic, inorganic and natural dye sensitizers have been tried in the past in an attempt to reduce the cost and improve the performance of DSSC devices. In general, the dye sensitizers have been classified into three categories, i.e. (1) metal complex dyes, (2) metal-free organic dyes and (3) natural dyes. Ruthenium-based complexes are by far the most efficient dyes and used commercially in DSSC technology. But the problems associated with ruthenium complexes are their high cost (ruthenium is a rare earth metal) and toxicity (Richhariya et al. 2017). Figure 15.5 depicts the chemical structure of the ruthenium-based dyes.

A novel Zn-prophyrin-based sensitizer has been recently reported as an alternative to ruthenium-based sensitizers, and 6.0% conversion efficiency has been claimed; the device displayed 13.6 $mAcm^{-2}$ current density, 0.70 V open circuit voltage and 0.63 fill factor (Lee et al. 2009). The metal-free organic dye sensitizers gained interest in the year 2000 due to their ease in structural modification and high extinction coefficient compared to Ru-based complexes. General designs for metal-free organic sensitizers are donor-π-acceptor, donor-acceptor-π-donor and donor-acceptor-π-acceptor, where π is conjugated spacer which acts as bridge between donor and acceptor. An alternative way to produce cost-effective and non-toxic dye sensitizer is through extraction of dyes from natural plant sources. The dyes are extracted from leaves, flowers, fruits and roots and termed as natural dye sensitizers.

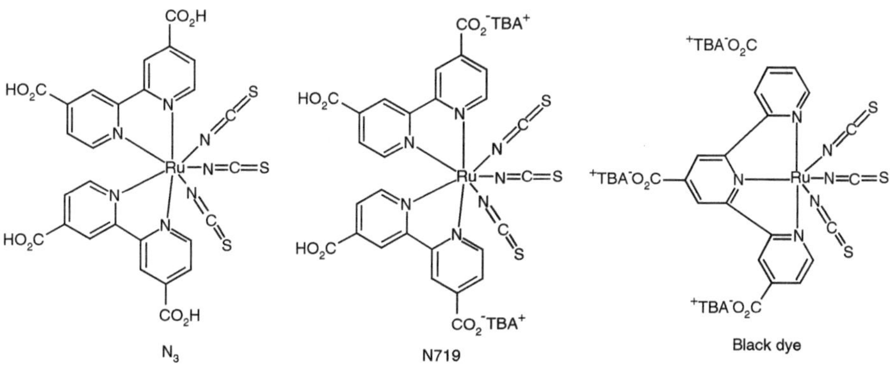

Fig. 15.5 Basic structure of common ruthenium-based dye sensitizers (Kumara et al. 2017)

The natural dyes have a number of advantages, i.e. cost-effective production, easy availability, non-toxic nature and 100% biodegradation. Natural dyes will be discussed in detail in subsequent sections.

Electrolyte

The electrolyte not only acts as electrically conducting medium but also plays one of the most important roles in regeneration of the oxidized dye sensitizer. An ideal electrolyte should have long-term stability, including chemical, optical, electrochemical, thermal and interfacial stability, must reduce the oxidized dye rapidly and guarantee the fast diffusion of charge carriers between porous nanocrystalline layer and the counter electrode in order to sustain the light energy conversion (Su'ait et al. 2015).

On the basis of their physical properties, electrolytes are categorized into three classes, i.e. (1) liquid, (2) quasi-solid and (3) solid electrolytes. Till date, the highest efficiency has been attained using iodide/triiodide liquid electrolyte and copper complex Cu(ll/l) (Freitag et al. 2017). The limitation of liquid electrolyte is its volatility and leakage. High-quality and expensive sealants need to be used to avoid electrolyte leakage issues. Sealants which also act as spacer to counter electrode and photo-anode and the sealing process account for 27% of total cost of DSSC (Mozaffari et al. 2016). To solve this issue, quasi-solid- and solid-state electrolytes have been developed, but their efficiency is still way too less compared to liquid electrolytes. Apart from iodide/triiodide electrolyte, Br^-/Br_3^-, $SCN^-/(SCN)_2$, $SeCN^-/(SeCN)_2$ and natural electrolytes have also shown promising results. In a recent study, agar, gelatine and DNA polymer-based natural electrolytes have been investigated along with natural dye sensitizer extracted from flowers of Hildegardia. Conversion efficiency of 3.38% has been claimed using gelatine-based electrolyte (Ahmad et al. 2017a, b).

Counter Electrode (CE)

The CE consisting of conducting layer and catalyst completes the device by connecting it to external circuit. An ideal counter electrode should reduce the electrolyte immediately after it oxidizes so as to improve the electron density in the device. Poor catalytic kinetics of catalyst towards electrolyte reduces the overall performance of DSSC due to less regeneration of dye molecules. Till date, Pt-based catalysts are by far the most efficient due to their high catalytic activity for redox reactions and high corrosion resistance. Other metals such as Zn and Co have also been investigated with promising results (Gao et al. 2017). Carbon-based catalysts, i.e. multiwalled carbon nanotubes (MWCNTs), graphite, graphene and activated carbon, have also been trialled with promising or even better results (Chen and Shao 2016). Recently eco-friendly catalysts extracted from natural sources have been prepared and investigated. In a study, carbonized mangosteen peels have been utilized to prepare catalyst for CE. A relative increase of 1.47% in conversion efficiency has been claimed compared to standard Pt-based CE (Maiaugree et al. 2015).

Transparent Conducting Oxide (TCO) Layer

TCO acts as electron transport layer and connect the device to external load. In general, the TCO layer should possess high electrical conductivity for easy electron transport. For single-facet DSSC, transparency of conducting layer is not important, but for bifacial DSSC (the bifacial DSSC created by Gratzel display high energy conversion per specific area), the transparency is of utmost importance which facilitates equal or near equal light transmittance from either front or rear sides.

Wide bandgap transparent oxides have been used to date for the fabrication of TCO layer which employ un-doped tin oxide (SnO_2) and/or SnO_2 with fluorine or indium. Fluorine-doped tin oxide (FTO) layer displayed highest performance due to its high transparency, optimum electrical conductivity, high adhesion to substrate and good mechanical and thermal properties. The associated problem with FTO-coated substrates is their high manufacturing cost. The TCO layer accounts for 9% of the total cost of DSSC. Another issue is the high deposition temperature (400 °C) which restricts its implication for low-temperature flexible DSSC. Other approaches to apply FTO coating on plastic substrates are low-temperature pyrolysis, sputter deposition from solid targets using techniques such as radio frequency (RF) magnetron sputtering and direct current (DC) reactive magnetron sputtering using various plasma atmospheres such as $Ar/O_2/CF4$ and Ar/O_2/Freon which causes additional cost to the device fabrication. Recently, graphene has been utilized as conducting layer to reduce cost and fabrication temperature (Lee et al. 2010; Mozaffari et al. 2016).

Natural Dye Sensitizers

Natural dye pigments have been classified into four broad categories, i.e. (1) chlorophyll, (2) carotenoids, (3) betalains and/or (4) flavonoids. Table 15.1 depicts the chemical structure of various classes and subclasses of natural dye pigments. These classes are discussed separately in subsequent sections.

Table 15.1 Chemical structure of natural dyes

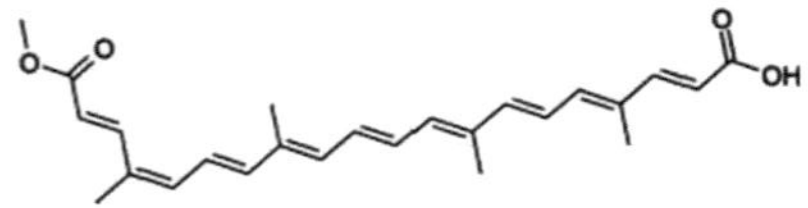	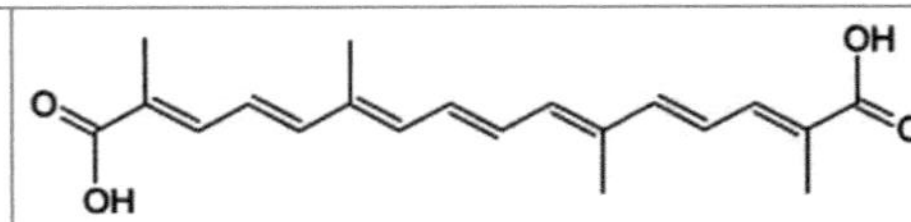
Bixin	Crocertin
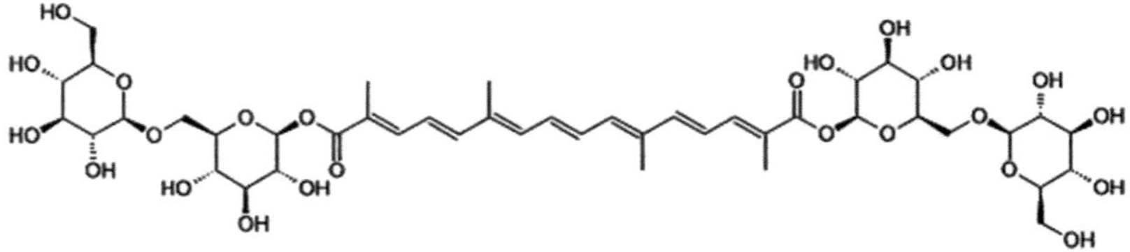	
Crocin	
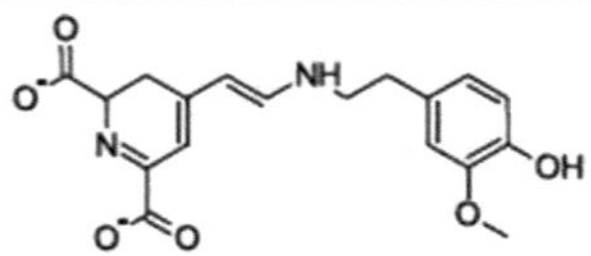	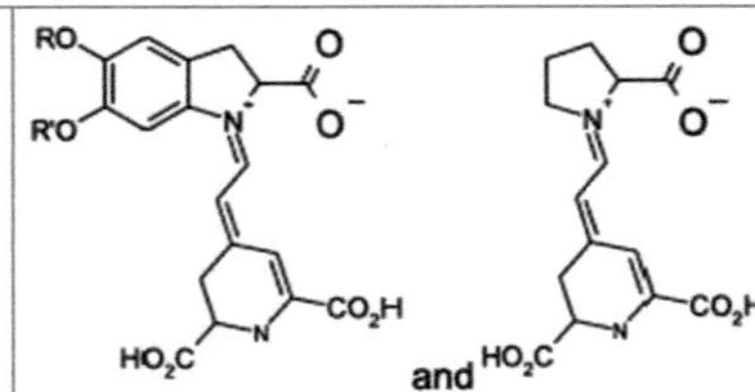
Betaxanthhin	Betalains
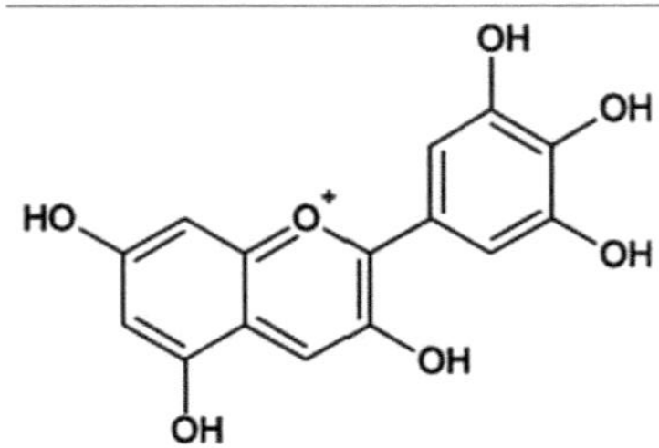	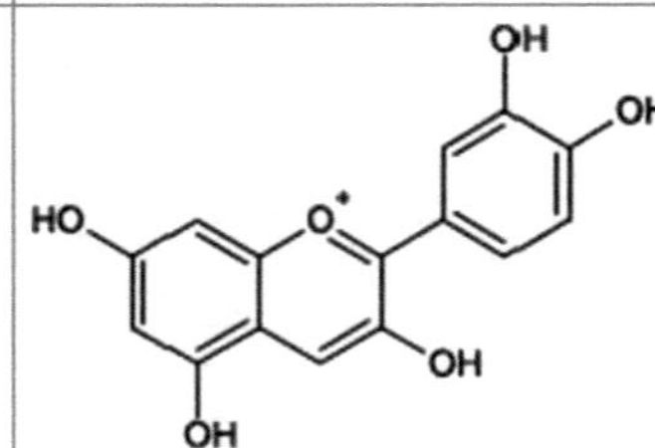
Delphtinidin (an anthocyanidin)	Cyanidin (an anthocyanidin)

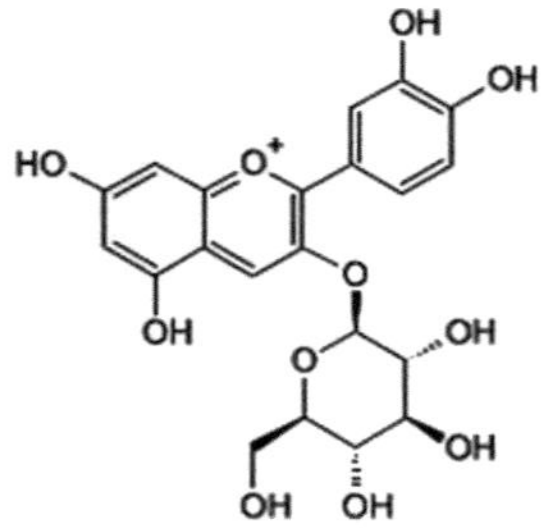	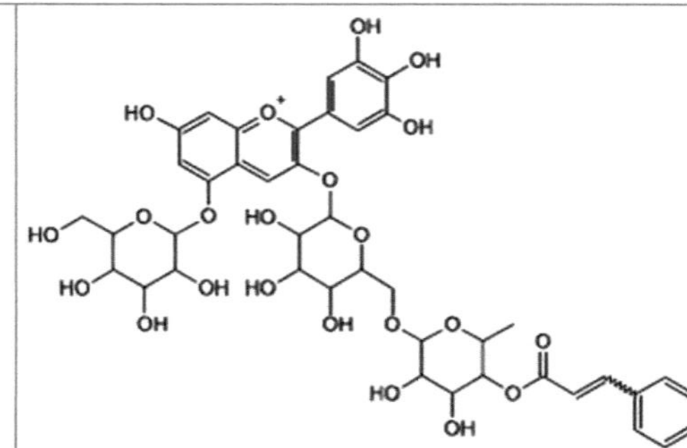
Cyanin (an anthocyanin)	Nasunin (an anthocyanin)
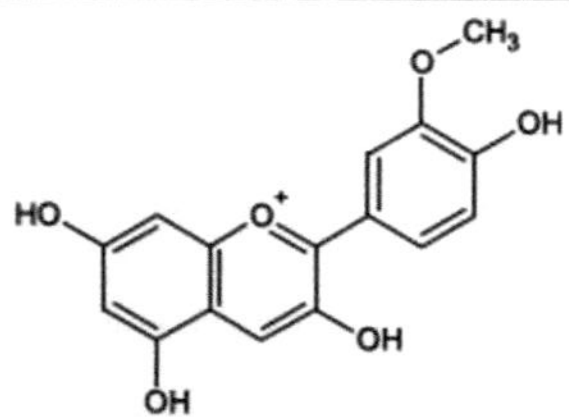	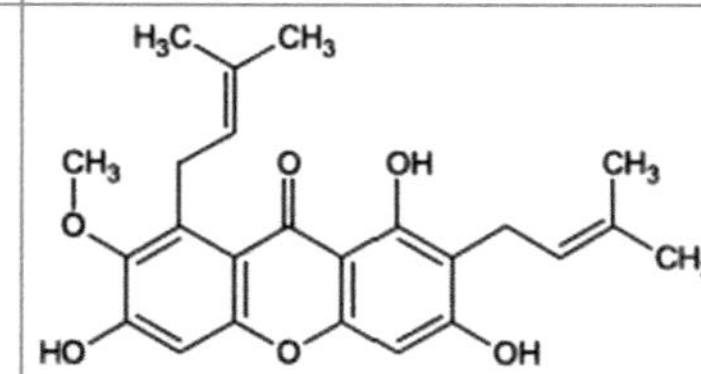
Peonidin (an anthocyanidin)	Aphla-mangosteen

(continued)

Table 15.1 (continued)

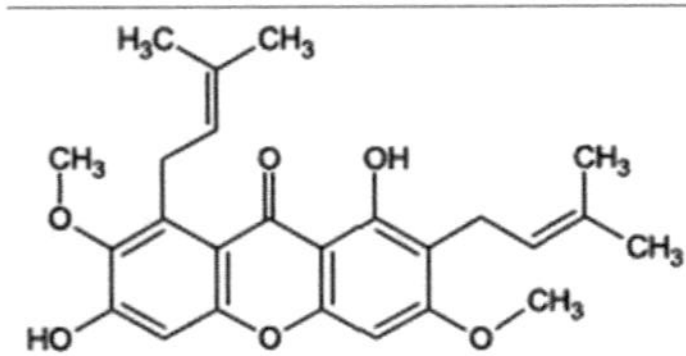	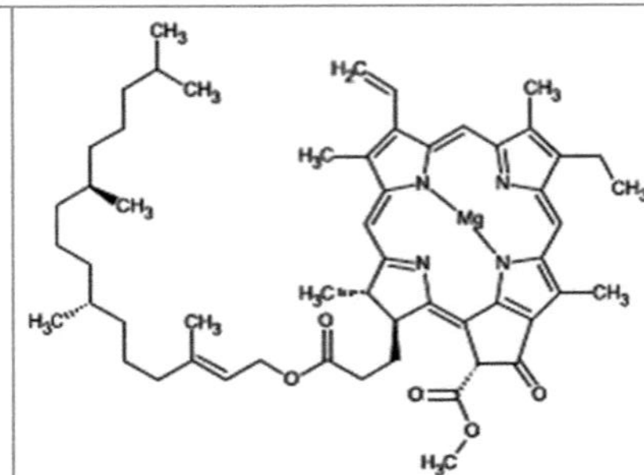
Beta-mangosteen	Chlorophyll a
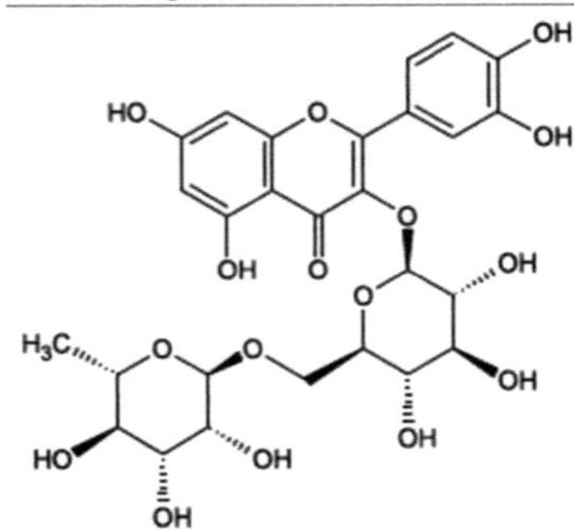	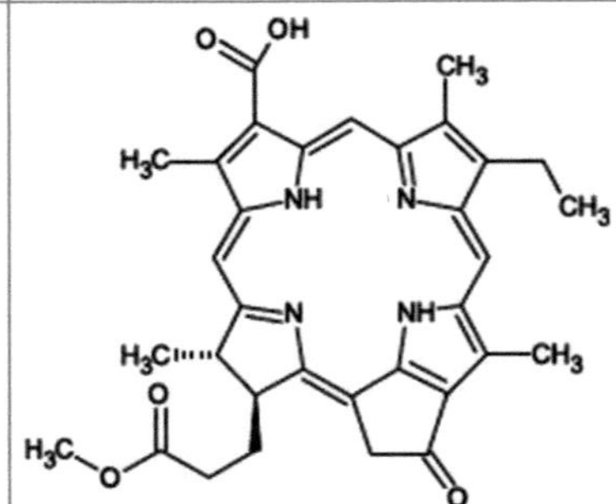
Rutin (a flavonol glycoside)	Modified cholorophyll

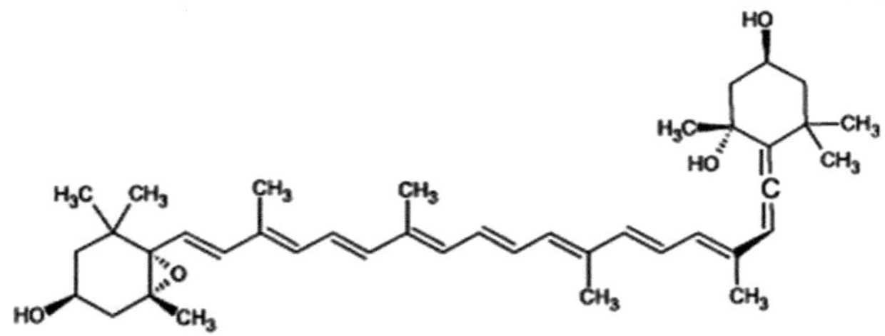

Neoxanthin

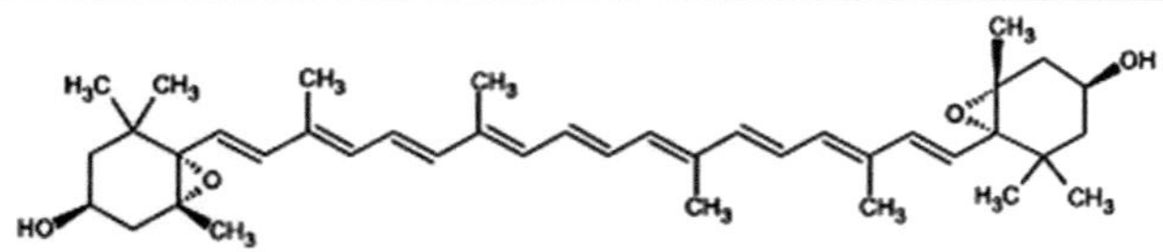

Violaxanthin

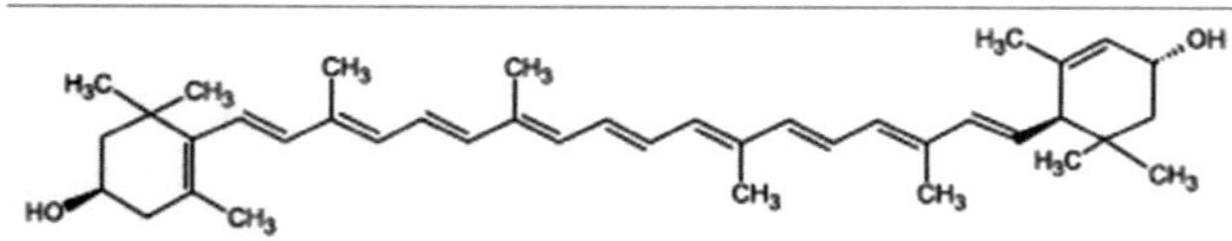

Lutein

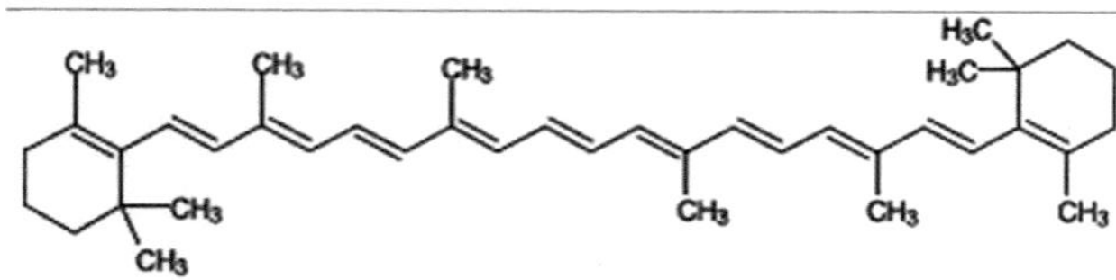

Beta-carotene

(continued)

Table 15.1 (continued)

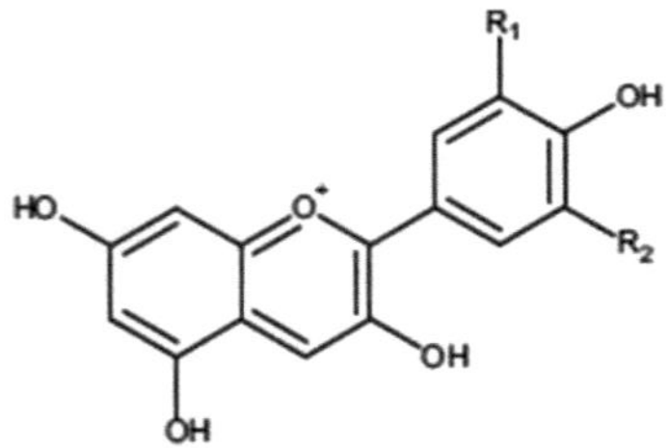	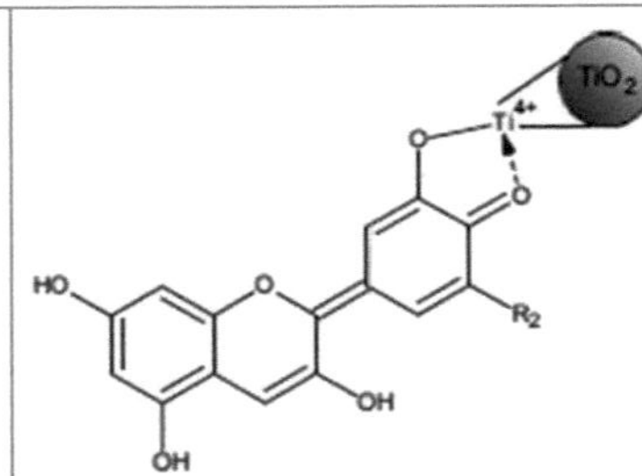
Most common anthocyanidins	Attachment (chelation) with semi-conductor

Hug et al. (2014) and Kumara et al. (2017)

Chlorophyll

Chlorophyll is a common green pigment found in plants and algae. It is essentially a chlorine pigment with magnesium ion in the centre of the chlorine ring. Moreover, chlorophyll a and chlorophyll b are the most used types. The basic difference is in the composition of side chain (in chlorophyll a, it is $-CH_3$, and in chlorophyll b it is CHO). Other less common types are chlorophyll *d*, *c1*, *c2* and *f*. Figure 15.6 depicts the UV-VIS spectroscopic analysis of chlorophyll a and b. Both the types absorb visible portion of solar spectrum from red, blue and violet wavelengths with a maximum absorption at 670 nm for chlorophyll a and 470 nm for chlorophyll b. Since the first evidence of its use as dye sensitizer in DSSC by Kay and Gratzel, many improvements and advantages have been reported by different research studies (Syafinar et al. 2015a, b; Shah et al. 2016).

The main disadvantage of using chlorophyll is its long chain length which causes steric hindrance and leads to low-electron transferability. Another issue is its low adsorption ability on TiO_2 semi-conductor. Some derivatives such as methyltrans-32-carboxypyropheophorbide alpha showed high binding ability with semiconductor surfaces via chelating and monodentate modes (Calogero et al. 2009).

Carotenoids

Carotenoids are organic pigments produced by plants and also found in some microorganisms. Carotenoids absorb red, orange and yellow wavelength from solar spectrum (Fig. 15.6). These are broadly classified into two categories, (1) xanthophylls which contain oxygen and (2) carotenes which are essentially pure hydrocarbons containing no oxygen. These pigments have been successfully employed in DSSCs as sensitizers and show excellent properties. The maximum claimed conversion

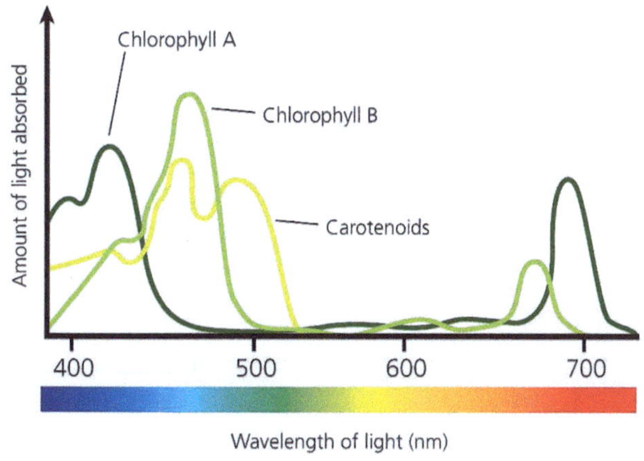

Fig. 15.6 UV-VIS spectra of chlorophyll a and b

efficiency of DSSCs using carotenoids is 2.6%, while the conversion efficiency of combination of carotenoids and chlorophyll is reported to be 4.2% (Wang et al. 2006; Zhu et al. 2010; Murillo et al. 2013).

Flavonoids

These are very important pigments present in almost all land plants. Flavonoids are essentially sugar compounds bonded with polyphenols and are responsible for colouration of many flowers and fruits. Other roles include attracting insects, photoprotection, antioxidation activity and enhancement of photosynthesis. The flavonoids can be classified into anthocyanins, aurones, flavones, chalones and flavonols. Among them anthocyanins are the most important pigments for DSSC applications due to their absorption in visible range and ability to transfer electrons efficiently. Anthocyanins are the basis for most orange, pink, red, magenta and blue colours. The light absorption properties of flavonoids are depicted in Fig. 15.7 (Stintzing and Carle 2004; Marais et al. 2006).

These pigments are also quick to make bonds with TiO_2 due to absorption of cyanine which leads to development of effective and strong complex via OH- displacement and formation of water molecules. The ability of fast charge transfer using anthocyanin extracted from blackberry has been successfully demonstrated using femtosecond analysis. Despite high electron transfer ability, the conversion efficiency was low due to possible agglomeration and high recombination reactions (Yamasaki et al. 1996; Cherepy et al. 1997).

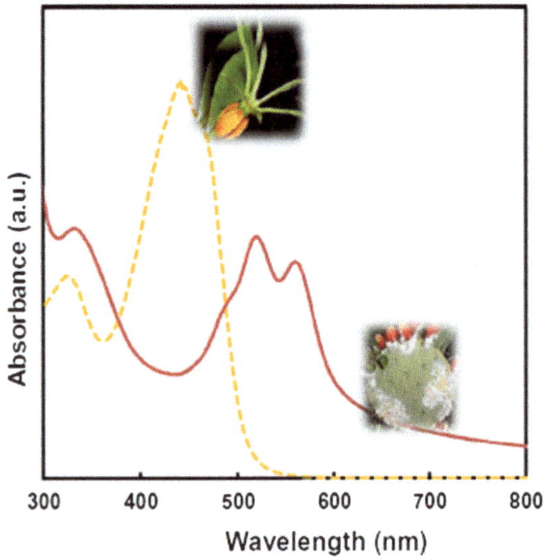

Fig. 15.7 UV-VIS spectroscopy of gardenia (yellow dot) and cochineal (red line) (Park et al. 2014)

Betalain

These pigments are an additional class of natural sensitizers and are present in flower petals, roots, stems and leaves of caryophyllales plants. They possess high extinction coefficient in visible range, and their redox properties are dependent on pH of the solution. They generally absorb red-purple wavelengths in solar spectrum (Calogero et al. 2009). UV-VIS spectroscopic analysis is presented in Fig. 15.8.

Various plants and their selected parts have been utilized for the extraction of natural dyes. Table 15.2 summarizes the plants used for the extraction of dyes along with detailed parameters of DSSCs. As can be observed from the data presented in Table 15.2, DSSCs using anthocyanin dyes show efficiencies less than 1%. On the other hand, modified chlorophyll compounds showed highest efficiencies.

Advancements in Natural DSSCs

Natural DSSCs due to their low efficiency are still unable to gain commercial importance. The low efficiency is due to limited light absorption capability and weak bonding of natural dyes with semi-conductor network. Various approaches have been tried to improve the incident photo conversion efficiency (IPCE) of natural DSSCs such as mixing of two or more natural dye pigments (dye cocktail) to increase light absorption capability, use of various solvents to extract the dyes from the natural sources in an attempt to improve dye extraction and bonding between dye molecule and semi-conductor and successive adsorption of natural dye to increase dye loading and/or pre-dye absorption (before coating semi-conductor material on conducting substrate) to enhance conversion efficiency.

In a study, a cocktail of natural anthocyanin dyes extracted from leaves and petals of *Ixora coccinea* have been investigated. The cocktail mixed in 1:4 ratio displayed highest conversion efficiency (0.80%) compared to the conversion efficiency of individual dye sensitizers. This improvement is believed to be due to increased

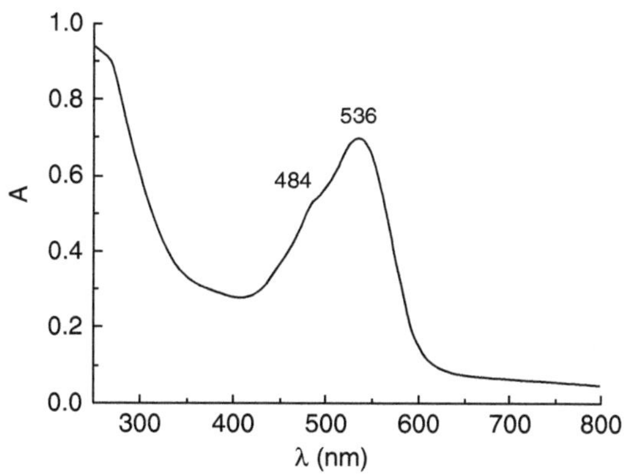

Fig. 15.8 UV-VIS spectra of red turnip extracts in 0.1 M HCL solution showing betaxanthin at 484 nm and betanin at 536 nm wavelengths (Calogero et al. 2009)

Table 15.2 Natural dyes and structural classes used in DSSC

Plant source	Structure/structural class	Jsc (mA/cm-2)	Voc (V)	η (%)	References
Red frangipani	Anthocyanins	0.94	0.49	*0.30*	Shanmugam et al. (2013)
Lxora sp.	Anthocyanins	6.26	0.35	*0.96*	Kumara et al. (2013a, b)
Begonia	Anthocyanins	0.63	0.53	*0.24*	Zhou et al. (2011)
Rhododendron	Anthocyanins	1.61	0.58	*0.24*	Zhou et al. (2011)
Marigold	Carotenoide	0.51	0.54	*0.23*	Zhou et al. (2011)
Perilla	Chlorophyll	1.36	0.52	*0.50*	Zhou et al. (2011)
China loropetal	Chlorophyll	0.84	0.51	*0.27*	Zhou et al. (2011)
Yellow rose	Carotenoide	0.74	0.60	*0.26*	Zhou et al. (2011)
Flowery knotweed	Rhein	0.60	0.55	*0.21*	Zhou et al. (2011)
Petunia	Chlorophyll	0.85	0.61	*0.32*	Zhou et al. (2011)
Violet	Anthocyanins	1.02	0.49	*0.33*	Zhou et al. (2011)
Rosella	Anthocyanins	1.63	0.40	*0.37*	Wongcharee et al. (2007)
Blue pea	Anthocyanins	0.37	0.37	*0.05*	Wongcharee et al. (2007)
Red bougainvillea glabra	Betacyanin	2.33	0.26	*0.49*	Hernandez-Martinez et al. (2011)
Violet bougainvillea	Betacyanin	1.86	0.23	*0.31*	Hernandez-Martinez et al. (2011)
Hibiscus rosa-sinensis	Anthocyanin	3.31	0.145	*1.08*	Yusoff et al. (2014)
Melastoma malabthricum L.	Anthocyanins	3.18	0.45	*0.83*	Kumara et al. (2014)
Blueberry	Anthocyanins	4.1	0.30	*0.69*	Teoli et al. (2016)
Tangerine peel	Flavones	0.74	0.59	*0.28*	Zhou et al. (2011)
Fructus lycii	Carotene	0.53	0.68	*0.17*	Zhou et al. (2011)
Mangosteen	α/β mangosteen	2.55	0.62	*0.92*	Zhou et al. (2011)
Mangosteen	Rutin	2.92	0.68	*1.17*	Zhou et al. (2011)
Capsicum	Carotenoid	0.23	0.41	*n.a.*	Hao et al. (2006)
Wild Sicilian prickly pear	Etalain	8.2	0.38	*1.19*	Calogero et al. (2010)
F. Sapindaceae	Anthocyanins	3.88	0.41	*0.56*	Kumara et al. (2013a, b)
Mulberry fruit	Anthocyanins	1.89	0.56	*0.55*	Chang and Lo (2010)
Ivy gourd fruit	Carotene	0.24	0.64	*0.08*	Shanmugam et al. (2013)
Canarium odontophyllum	Anthocyanins	2.45	0.39	*0.59*	Kumara et al. (2013a, b)
Eggplant skin	Nasunin	0.40	0.35	*0.48*	Calogero et al. (2012)
Codiaeum variegatum	Anthocyanins	4.03	0.44	*1.08*	Yusoff et al. (2014)
Spinach	Chlorophyll	0.47	0.55	*0.13*	Chang et al. (2010)

(continued)

Table 15.2 (continued)

Plant source	Structure/structural class	Jsc (mA/ cm-2)	Voc (V)	η (%)	References
Ipomoea	Chlorophyll	0.91	0.54	*1.18*	Chang et al. (2010)
Pomegranate	Chlorophyll	2.05	0.56	*0.59*	Chang and Lo (2010)
Ficus retusa	Chlorophyll, carotenoids	7.85	0.52	*1.18*	Lai et al. (2008)
Rhoeo spathacea	Chlorophyll, carotenoids	10.9	0.5	*1.49*	Lai et al. (2008)
Red cabbage	Anthocyanins	2.25	0.62	*0.97*	Chien and Hsu (2014)
Pandan leaves	Chlorophyll	1.91	0.48	*0.51*	Noor et al. (2014)
Black rice seeds	Anthocyanins	2.09	0.47	*0.56*	Noor et al. (2014)
Lawsonia inermis seed	Lawsone	2.99	0.50	*1.47*	Ananth et al. (2014)
Brown seaweed	Chlorophyll	0.80	0.36	*0.18*	Calogero et al. (2014)
Anthum graveolens leaves	Chlorophyll	0.96	0.57	*0.22*	Taya et al. (2013)
Arugula leaves	Chlorophyll	0.78	0.59	*0.20*	Taya et al. (2013)
Parsley leaves	Chlorophyll	0.53	0.44	*0.07*	Taya et al. (2013)
Amaranthus caudatus flower	Chlorophyll	1.82	0.55	*0.61*	Godibo et al. (2015)
Cordyline fruticosa leaves	Chlorophyll	1.30	0.61	*0.50*	Al-Alwani et al. (2016)
Pawpaw leaves	Chlorophyll	0.64	0.54	*0.20*	Kimpa et al. (2012)
Shiso leaves	Chlorophyll	3.52	0.43	0.59	Kumara et al. (2006)
Ocimum gratissimum leaves	Chlorophyll	0.04	0.46	0.02	Eli et al. (2016)
Spinach oleracea	Chlorophyll	0.33	0.59	0.08	Chang et al. (2010)
Red spinach leaves	Chlorophyll	1.00	0.50	0.58	Khan et al. (2012)
Papaya leaves	Chlorophyll	0.36	0.32	0.07	Suyitno et al. (2015)
Ipomoea leaves	Chlorophyll	0.85	0.49	0.23	Chang et al. (2010)
Azadirachta indica leaves	Chlorophyll	0.43	0.40	0.72	Swarnkar et al. (2015)
Basil leaves	Chlorophyll	1.39	0.58	0.40	Taya et al. (2015)
Ziziphus jujube leaves	Chlorophyll	3.18	0.65	1.07	Taya et al.(2015)
Mint flower	Chlorophyll	0.45	0.54	0.09	Monzir et al. (2017)
Lemon leaves	Chlorophyll	1.08	0.59	0.03	Maabong et al. (2015)
Morula leaves	Chlorophyll	0.05	0.47	0.01	Maabong et al. (2015)
Fig leaves	Chlorophyll	2.09	0.59	0.64	Taya et al. (2015)
Berry leaves	Chlorophyll	3.57	0.59	0.93	Taya et al. (2015)
Banana leaves	Chlorophyll	1.77	0.59	0.52	Taya et al. (2015)
Peach leaves	Chlorophyll	2.55	0.61	0.65	Taya et al. (2015)
Black tea leaves	Chlorophyll	0.39	0.55	0.08	Monzir et al. (2017)

(continued)

Table 15.2 (continued)

Plant source	Structure/structural class	Jsc (mA/cm-2)	Voc (V)	η (%)	References
Coccinia indica leaves	Chlorophyll	0.70	0.54	0.26	Priyadharsini et al. (2013)
Perilla	Chlorophyll	1.36	0.52	0.50	Zhou et al. (2011)
Petunia	Chlorophyll	0.85	0.61	0.32	Zhou et al. (2011)
Festuca ovina grass	Chlorophyll	1.18	0.54	0.46	Hernandez-Martinez et al. (2012)
Hierochloe odorata grass	Chlorophyll	2.19	0.59	0.46	Shanmugam et al. (2015)
Torulinium aegyptium grass	Chlorophyll	1.00	0.65	0.32	Shanmugam et al. (2015)
Moss bryophyte	Chlorophyll	5.78	0.60	1.97	Hassan et al. (2014)
Green algae	Chlorophyll	0.13	0.41	0.01	Taya et al. (2013)
Microalgae	Chlorophyll	2.53	0.55	0.90	Mohammadpour et al. (2014)
Bixa orellana L.	Bixin	1.10	5.70	0.59	Gómez-Ortíz et al. (2010)
Calafate fruit	Delphinidin	2.60	0.60	0.62	Polo and Iha (2006)
Jaboticaba skin	Peonidin	2.60	6.60	0.62	Polo and Iha (2006)
Tradescantia zebrina	Anthocyanins	0.63	3.50	0.55	Li et al. (2013)
Kopok	Anthocyanin/carotenoid	0.87	3.60	0.49	Li et al. (2013)
Canarium odontophyllum	Anthocyanin	2.45	3.85	0.62	Kumara et al. (2013a, b)
Gardenia fruit	Crocetine/crocin	0.56	1.29	0.35	Yamazaki et al. (2007)

light absorption capacity of mixture of two dyes (Zolkepli et al. 2015). In another study, a cocktail of purple cabbage and blueberries has been investigated with promising results (Syafinar et al. 2015a, b). Studies showed that there are few compatible combinations of natural pigments to be used as dye sensitizers for DSSCs. Other combinations though possess high light absorption ability leading to reduction in conversion efficiency. A recent study reported low conversion efficiency with the use of cocktail of *I.coccinea* and *Bougainvilla* compared to the performance of individual *I.coccinea* dye. This is due to unfavourable reactions between dye molecules and mismatch of adsorption capacity which leads to high recombination reactions (Lim et al. 2016). To avoid the drawbacks of mixtures, another approach, i.e. successive diffusion, has been investigated. In this study, natural dyes extracted from gardenia and cochineal have been adsorbed on the TiO_2 nanostructure successively (one after the other) and compared with individual dyes and cocktail of dyes. The successive adsorption of cochineal followed by gardenia showed highest conversion efficiency of 0.48% while successive adsorption of gardenia followed by cochineal showed 0.37% IPCE. On the other hand, the premixed cocktail of two dyes showed 0.31% IPCE which was even lower than the efficiency of individual dye extracted from gardenia, i.e. 0.35% (Park et al. 2014).

Solvent plays an important role in the extraction of pigments from the plant sources. They are thought to be responsible for efficient extraction of pigments and bonding of dye molecules with semi-conductor nanostructure. Solvents also improve the adsorption rate of dye molecules by forming suitable end group which can easily create strong bond with TiO_2- or ZnO- based nanostructure. Various solvents such as ethanol, methanol, water, isopropyl alcohol, have been investigated to improve dye loading, adsorption rate and contact between dye and semi-conductor network. In a recent study, methanol and ethanol with hot and normal water have been examined with turmeric as dye source. Ethanol displayed highest conversion efficiency, i.e. 0.33%, compared to methanol and water. The adsorption time was 2 h for dry turmeric-based dye extracted in ethanol (Hossain et al. 2017). In another detailed investigation, various solvents and their combinations have been examined with betalain, betaxanthins and chlorophyll dyes extracted from *Cordyline fruticosa* leaves, *Pandanus amaryllifolius* (pandan leaves) and *Hylocereus polyrhizus* (dragon fruit). The mixture of methanol and water with 3:1 ratio showed highest adsorption of *C.fruticosa*. For *P. amaryllifolius* the mixture of ethanol and water with 2:1 ratio displayed best results, and normal water exhibited best results for *H. polyrhizus* as solvent (Al-Alwani et al. 2015).

Another approach to improve the anchoring (binding) of dye molecules with semi-conductor network is to premix the dye with semi-conductor material before coating on to conducting substrate. In a study, *Lawsonia inermis* seeds were used as plant source with distilled water as extraction solvent. The extracted dye pigment was mixed in TiO_2 synthesis solution. The premix processing of dye showed improved performance in comparison to adsorption of dye on TiO_2 coating (Ananth et al. 2014). Figure 15.9 is showing the comparative efficiency of the premix and post-adsorbed dye on the TiO_2 nanostructure.

Future Prospects

The technology of dye-sensitized solar cell is becoming increasingly important in the scientific community due to its versatility, environment-friendly nature and ability to operate in indoor light conditions. Presently various natural dye sensitizers have been tried to improve the ability of the device to absorb more photons present in the solar spectrum. Natural dyes possess a decisive edge over other organic and inorganic dyes due to their cost-effective production, easy to maintain the supply and non-toxicity. At the same time, natural dyes are also at a disadvantage due to their very low conversion efficiency. Current focus of research is to improve the interfacial contact (between natural dye molecules and semi-conduction nanostructure), photo degradation and light harvesting ability using a combination of various dyes.

Another important aspect of natural dyes is their ability to harvest light while submerged under water. As underwater plants use chlorophyll and/or other compounds for absorbing light, there is a possibility to develop natural DSSCs which could produce useable energy while being submerged under water.

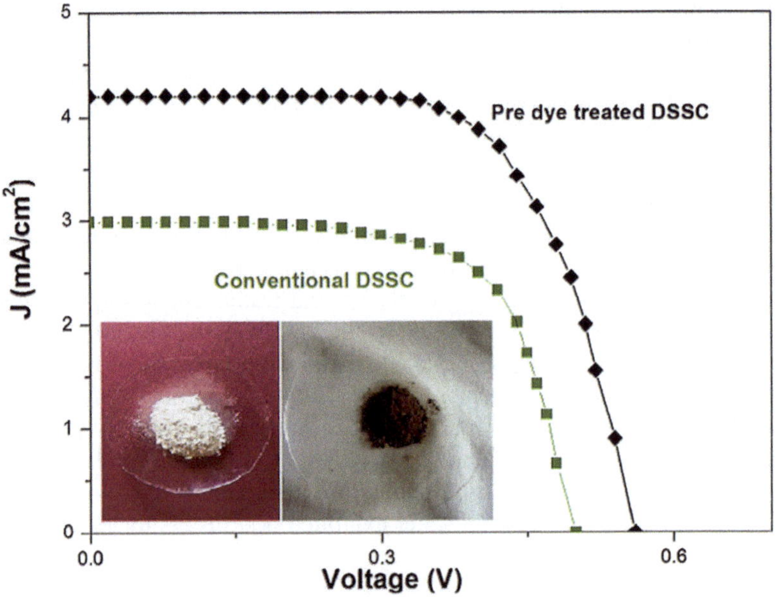

Fig. 15.9 Current-voltage (I-V) curves of premixed and post-adsorbed natural dye (Ananth et al. 2014)

Conclusion

Natural pigments due to their eco-friendly nature, 100% biodegradation and easy availability are promising sensitizers in the field of DSSC and have the potential to replace expensive metal complexes or organic dye sensitizers. Due to their non-toxic nature and low cost, these pigments are lucrative sensitizers for commercialization and public handling. The main drawback of natural dyes compared to other ruthenium (Ru)-based and organic dyes is low efficiency and stability. Therefore, further research is required to improve their performance. Various possible strategies have been tried and tested to improve the conversion efficiency such as preparation of various cocktail mixtures to improve electron density and use of various solvents to facilitate anchoring. Undoubtedly, current output is very low compared to other dye classes, but significant improvements have been made since the inception of concept of natural and environment-friendly dye sensitizers. The cost of natural dyes is negligible in comparison to Ru-based and organic dye sensitizers which makes them a potential candidate for the commercialization of DSSC technology.

Acknowledgment The authors thank the technical and financial assistance of UM Power Energy Dedicated Advanced Centre (UMPEDAC) and the Higher Institution Centre of Excellence (HICoE) Program Research Grant, UMPEDAC – 2016 (MOHE HICOE – UMPEDAC).

References

(2017). *Which regions of the electromagnetic spectrum do plants use to drive photosynthesis?* From https://support.heliospectra.com/portal/helpcenter/articles/which-regions-of-the-electromagnetic-spectrum-do-plants-use-to-drive-photosynthesis

Ahmad, M. S., Pandey, A., & Rahim, N. A. (2017a). Advancements in the development of TiO_2 photo anodes and its fabrication methods for dye sensitized solar cell (DSSC) applications. A review. *Renewable and Sustainable Energy Reviews, 77*, 89–108.

Ahmad, M. S., Pandey, A., & Rahim, N. A. (2017b). Effect of Nanodiamonds on the optoelectronic properties of TiO_2 Photoanode in dye-sensitized solar cell. *Arabian Journal for Science and Engineering*, 1–5.

Ahmad, M. S., Rahim, N. A., & Pandey, A. K. (2018). Improved electron transfer of TiO2 based dye sensitized solar cells using Ge as sintering aid. *Optik–International Journal for Light and Electron Optics, 157*(Supplement C), 134–140.

Al-Alwani, M. A. M., Mohamad, A. B., Kadhum, A. A. H., & Ludin, N. A. (2015). Effect of solvents on the extraction of natural pigments and adsorption onto TiO2 for dye-sensitized solar cell applications. *Spectrochimic Acta A Mol Biomol Spectrosc, 138*(Supplement C), 130–137.

Al-Alwani, M. A., Ludin, N. A., Mohamad, A. B., Kadhum, A. A. H., Baabbad, M. M., & Sopian, K. (2016). Optimization of dye extraction from Cordyline fruticosa via response surface methodology to produce a natural sensitizer for dye-sensitized solar cells. *Results in Physics, 6*, 520–529.

Ananth, S., Vivek, P., Arumanayagam, T., & Murugakoothan, P. (2014). Natural dye extract of lawsonia inermis seed as photo sensitizer for titanium dioxide based dye sensitized solar cells. *Spectrochimic Acta A Mol Biomol Spectrosc, 128*, 420–426.

Calogero, G., Di Marco, G., Caramori, S., Cazzanti, S., Argazzi, R., & Bignozzi, C. A. (2009). Natural dye sensitizers for photoelectrochemical cells. *Energy & Environmental Science, 2*(11), 1162–1172.

Calogero, G., Di Marco, G., Cazzanti, S., Caramori, S., Argazzi, R., Di Carlo, A., & Bignozzi, C. A. (2010). Efficient dye-sensitized solar cells using red turnip and purple wild Sicilian prickly pear fruits. *International Journal of Molecular Sciences, 11*(1), 254–267.

Calogero, G., Yum, J.-H., Sinopoli, A., Di Marco, G., Grätzel, M., & Nazeeruddin, M. K. (2012). Anthocyanins and betalains as light-harvesting pigments for dye-sensitized solar cells. *Solar Energy, 86*(5), 1563–1575.

Calogero, G., Citro, I., Di Marco, G., Minicante, S. A., Morabito, M., & Genovese, G. (2014). Brown seaweed pigment as a dye source for photoelectrochemical solar cells. *Spectrochimic Acta A Mol Biomol Spectrosc, 117*, 702–706.

Chang, H., & Lo, Y.-J. (2010). Pomegranate leaves and mulberry fruit as natural sensitizers for dye-sensitized solar cells. *Solar Energy, 84*(10), 1833–1837.

Chang, H., Wu, H., Chen, T., Huang, K., Jwo, C., & Lo, Y. (2010). Dye-sensitized solar cell using natural dyes extracted from spinach and ipomoea. *Journal of Alloys and Compounds, 495*(2), 606–610.

Chen, M., & Shao, L. L. (2016). Review on the recent progress of carbon counter electrodes for dye-sensitized solar cells. *Journal of Chemical Engineering, 304*(Supplement C), 629–645.

Cherepy, N. J., Smestad, G. P., Grätzel, M., & Zhang, J. Z. (1997). Ultrafast electron injection: Implications for a photoelectrochemical cell utilizing an anthocyanin dye-sensitized TiO_2 nanocrystalline electrode. *Journal of Physical Chemistry, B101*(45), 9342–9351.

Chien, C. Y., & Hsu, B. D. (2014). Performance enhancement of dye-sensitized solar cells based on anthocyanin by carbohydrates. *Solar Energy, 108*, 403–411.

Eisenberg, R., & Nocera, D. G. (2005). Preface: Overview of the forum on solar and renewable energy. *Inorganic Chemistry, 44*(20), 6799–6801.

Eli, D., Musa, G. P., & Ezra, D. (2016). Chlorophyll and betalain as light-harvesting pigments for nanostructured TiO_2 based dye-sensitized solar cells. *Journal of Energy & Natural Resources, 5*(5), 53–58.

Force, L., Critchley, C., & van Rensen, J. J. (2003). New fluorescence parameters for monitoring photosynthesis in plants. *Photosynthesis Research, 78*(1), 17–33.

Freitag, M., Teuscher, J., Saygili, Y., Zhang, X., Giordano, F., Liska, P., Hua, J., Zakeeruddin, S. M., Moser, J. E., Grätzel, M., & Hagfeldt, A. (2017). Dye-sensitized solar cells for efficient power generation under ambient lighting. *Nature Photonics, 11*(EPFL-ARTICLE-227875), 372–378.

Gao, C., Han, Q., & Wu, M. (2017). Review on transition metal compounds based counter electrode for dye-sensitized solar cells. *Journal of Energy Chemistry, 38*, 9828–9837.

Godibo, D. J., Anshebo, S. T., & Anshebo, T. Y. (2015). Dye sensitized solar cells using natural pigments from five plants and quasi-solid state electrolyte. *Journal of the Brazilian Chemical Society, 26*(1), 92–101.

Gómez-Ortíz, N., Vázquez-Maldonado, I., Pérez-Espadas, A., Mena-Rejón, G., Azamar-Barrios, J., & Oskam, G. (2010). Dye-sensitized solar cells with natural dyes extracted from achiote seeds. *Solar Energy Materials & Solar Cells, 94*(1), 40–44.

Green, M. A., Emery, K., Hishikawa, Y., Warta, W., & Dunlop, E. D. (2016). Solar cell efficiency tables (version 47). *Progress in Photovoltaics: Research and Applications, 24*, 3 (NREL/JA-5J00-65643).

Hao, S., Wu, J., Huang, Y., & Lin, J. (2006). Natural dyes as photosensitizers for dye-sensitized solar cell. *Solar Energy, 80*(2), 209–214.

Hassan, H., Abidin, Z., Careem, M., & Arof, A. (2014). Chlorophyll as sensitizer in I−/I3-based solar cells with quasi-solid-state electrolytes. *High Performance Polymers, 26*(6), 647–652.

Hernandez-Martinez, A. R., Estevez, M., Vargas, S., Quintanilla, F., & Rodriguez, R. (2011). New dye-sensitized solar cells obtained from extracted bracts of *Bougainvillea glabra* and spectabilis betalain pigments by different purification processes. *International Journal of Molecular Sciences, 12*(9), 5565–5576.

Hernandez-Martinez, A. R., Estevez, M., Vargas, S., Quintanilla, F., & Rodríguez, R. (2012). Natural pigment-based dye-sensitized solar cells. *International Journal of Applied Research and Technology, 10*(1), 38–47.

Hossain, M. K., Pervez, M. F., et al. (2017). Effect of dye extracting solvents and sensitization time on photovoltaic performance of natural dye sensitized solar cells. *Results in Physics, 7*(Supplement C), 1516–1523.

Hug, H., Bader, M., Mair, P., & Glatzel, T. (2014). Biophotovoltaics: Natural pigments in dye-sensitized solar cells. *Applied Energy, 115*, 216–225.

Jafarzadeh, M., Sipaut, C. S., Dayou, J., & Mansa, R. F. (2016). Recent progresses in solar cells: Insight into hollow micro/nano–structures. *Renewable and Sustainable Energy Reviews, 64*, 543–568.

Khan, M. A., Khan, S. M., Mohammed, M. A., Sultana, S., Islam, J. M., & Uddin, J. (2012). Sensitization of nanocrystalline titanium dioxide solar cells using natural dyes: Influence of acids medium on coating formulation. *American Academic & Scholarly Research Journal, 4*(5), 1.

Kimpa, M. I., Momoh, M., Isah, K. U., Yahya, H. N., & Ndamitso, M. M. (2012). Photoelectric characterization of dye sensitized solar cells using natural dye from pawpaw leaf and flame tree flower as sensitizers. *Materials Sciences and Applications, 3*(5), 281–286.

Kumara, G., Kaneko, S., Okuya, M., Onwona-Agyeman, B., Konno, A., & Tennakone, K. (2006). Shiso leaf pigments for dye-sensitized solid-state solar cell. *Solar Energy Materials and Solar Cells, 90*(9), 1220–1226.

Kumara, N., Ekanayake, P., Lim, A., Iskandar, M., & Ming, L. C. (2013a). Study of the enhancement of cell performance of dye sensitized solar cells sensitized with *Nephelium lappaceum* (*F Sapindaceae*). *Journal of Solar Energy Engineering, 135*(3), 031014.

Kumara, N., Ekanayake, P., Lim, A., Liew, L. Y. C., Iskandar, M., Ming, L. C., & Senadeera, G. (2013b). Layered co-sensitization for enhancement of conversion efficiency of natural dye sensitized solar cells. *Journal of Alloys and Compounds, 581*, 186–191.

Kumara, N., Hamdan, N., Petra, M. I., Tennakoon, K. U., & Ekanayake, P. (2014). Equilibrium isotherm studies of adsorption of pigments extracted from Kuduk-kuduk (*Melastoma malabathricum* L.) pulp onto TiO_2 nanoparticles. *Journal of Chemistry, 78*.

Kumara, N., Lim, A., Lim, C. M., Petra, M. I., & Ekanayake, P. (2017). Recent progress and utilization of natural pigments in dye sensitized solar cells: A review. *Renewable and Sustainable Energy Reviews, 78*, 301–317.

Kumavat, P. P., Sonar, P., & Dalal, D. S. (2017). An overview on basics of organic and dye sensitized solar cells, their mechanism and recent improvements. *Renewable and Sustainable Energy Reviews, 78*(Supplement C), 1262–1287.

Lai, W. H., Su, Y. H., Teoh, L. G., & Hon, M. H. (2008). Commercial and natural dyes as photosensitizers for a water-based dye-sensitized solar cell loaded with gold nanoparticles. *Journal of Photochemistry and Photobiology A: Chemistry, 195*(2), 307–313.

Lee, C. W., et al. (2009). Novel zinc porphyrin sensitizers for dye-sensitized solar cells: Synthesis and spectral, electrochemical, and photovoltaic properties. *Chemistry: A European Journal, 15*(6), 1403–1412.

Lee, K. S., Lee, H. K., Wang, D. H., Park, N.-G., Lee, J. Y., Park, O. O., & Park, J. H. (2010). Dye-sensitized solar cells with Pt-and TCO-free counter electrodes. *Chemical Communications, 46*(25), 4505–4507.

Lee, H., Kim, J., Kim, D. Y., & Seo, Y. (2017). Co-sensitization of metal free organic dyes in flexible dye sensitized solar cells. *Org Electron, 52*.

Li, B., Wang, L., Kang, B., Wang, P., & Qiu, Y. (2006). Review of recent progress in solid-state dye-sensitized solar cells. *Solar Energy Materials and Solar Cells, 90*(5), 549–573.

Li, N., Pan, N., Li, D., & Lin, S. (2013). Natural dye-sensitized solar cells based on highly ordered TiO2 nanotube arrays. *International Journal of Photoenergy, 2013*.

Lim, A., Ekanayake, P., Lim, L. B. L., & Bandara, J. M. R. S. (2016). Co-dominant effect of selected natural dye sensitizers in DSSC performance. *Spectrochimica Acta, Part A: Molecular and Biomolecular Spectroscopy, 167*(Supplement C), 26–31.

Ludin, N. A., Mahmoud, A. A.-A., Mohamad, A. B., Kadhum, A. A. H., Sopian, K., & Karim, N. S. A. (2014). Review on the development of natural dye photosensitizer for dye-sensitized solar cells. *Renewable and Sustainable Energy Reviews, 31*, 386–396.

Maabong, K., et al. (2015). Natural pigments as photosensitizers for dye-sensitized solar cells with TiO2 thin films. *International Journal of Renewable Energy Research, 5*(1), 54–60.

Maiaugree, W., Lowpa, S., et al. (2015). A dye sensitized solar cell using natural counter electrode and natural dye derived from mangosteen peel waste. *Scientific Reports, 5*, 15230.

Marais, J. P., Deavours, B., Dixon, R. A., & Ferreira, D. (2006). The stereochemistry of flavonoids. In *The science of flavonoids* (pp. 1–46). New York: Springer.

Mohammadpour, R., Janfaza, S., & Abbaspour-Aghdam, F. (2014). Light harvesting and photocurrent generation by nanostructured photoelectrodes sensitized with a photosynthetic pigment: A new application for microalgae. *Bioresource Technology, 163*, 1–5.

Monzir, S., Abuiriban, M., & Dahoudi, N. (2017). Dye-sensitized solar cells using fifteen natural dyes as sensitizers of nanocrystalline TiO/Sub 2. *Science, Technology and Development, 34*(3), 135–139.

Mozaffari, S., Nateghi., M. R., & Zarandi, M. B. (2016). An overview of the challenges in the commercialization of dye sensitized solar cells. *Renewable and Sustainable Energy Reviews, 71*, 675–686.

Murillo, E., Giuffrida, D., Menchaca, D., Dugo, P., Torre, G., Meléndez-Martinez, A. J., & Mondello, L. (2013). Native carotenoids composition of some tropical fruits. *Food Chemistry, 140*(4), 825–836.

Noor, M., Buraidah, M., Careem, M., Majid, S., & Arof, A. (2014). An optimized poly (vinylidene fluoride-hexafluoropropylene)–NaI gel polymer electrolyte and its application in natural dye sensitized solar cells. *Electrochimica Acta, 121*, 159–167.

Park, K. H., Kim, T. Y., Han, S., Ko, H. S., Lee, S. H., Song, Y. M., Kim, J. H., & Lee, J. W. (2014). Light harvesting over a wide range of wavelength using natural dyes of gardenia and cochineal for dye-sensitized solar cells. *Spectrochimica Acta Part A: Molecular and Biomolecular Spectroscopy, 128*, 868–873.

Polo, A. S., & Iha, N. Y. M. (2006). Blue sensitizers for solar cells: Natural dyes from Calafate and Jaboticaba. *Solar Energy Materials and Solar Cells, 90*(13), 1936–1944.

Priyadharsini, C. I., Prakasam, A., & Anbarasan, P. (2013). Enhanced photoelectrical performance of DSSC by co-doped SnO_2 nanoparticles. *International Letters of Chemistry, Physics and Astronomy, 12*, 82–93.

Richhariya, G., Kumar, A., Tekasakul, P., & Gupta, B. (2017). Natural dyes for dye sensitized solar cell: A review. *Renewable and Sustainable Energy Reviews, 69*(Supplement C), 705–718.

Shah, S., Buraidah, M., Teo, L., Careem, M., & Arof, A. (2016). Dye-sensitized solar cells with sequentially deposited anthocyanin and chlorophyll dye as sensitizers. *Optical and Quantum Electronics, 48*(3), 219.

Shakeel Ahmad, M., Pandey, A. K., & Rahim, N. A. (2017). Towards the plasmonic effect of Zn nanoparticles on TiO2 monolayer photoanode for dye sensitized solar cell applications. *Materials Letters, 195*, 62–65.

Shanmugam, V., Manoharan, S., Anandan, S., & Murugan, R. (2013). Performance of dye-sensitized solar cells fabricated with extracts from fruits of ivy gourd and flowers of red frangipani as sensitizers. *Spectrochimica Acta, Part A: Molecular and Biomolecular Spectroscopy, 104*, 35–40.

Shanmugam, V., Manoharan, S., Sharafali, A., Anandan, S., & Murugan, R. (2015). Green grasses as light harvesters in dye sensitized solar cells. *Spectrochimica Acta, Part A: Molecular and Biomolecular Spectroscopy, 135*, 947–952.

Sigman, D. M., & Boyle, E. A. (2000). Glacial/interglacial variations in atmospheric carbon dioxide. *Nature, 407*(6806), 859.

Stintzing, F. C., & Carle, R. (2004). Functional properties of anthocyanins and betalains in plants, food, and in human nutrition. *Trends in Food Science and Technology, 15*(1), 19–38.

Su'ait, M. S., Rahman, M. Y. A., & Ahmad, A. (2015). Review on polymer electrolyte in dye-sensitized solar cells (DSSCs). *Solar Energy, 115*, 452–470.

Suyitno, S., Saputra, T. J., Supriyanto, A., & Arifin, Z. (2015). Stability and efficiency of dye-sensitized solar cells based on papaya-leaf dye. *Spectrochimica Acta, Part A: Molecular and Biomolecular Spectroscopy, 148*, 99–104.

Swarnkar, A., Sahare, S., Chander, N., Gangwar, R. K., Bhoraskar, S., & Bhave, T. M. (2015). Nanocrystalline titanium dioxide sensitised with natural dyes for eco-friendly solar cell application. *Journal of Experimental Nanoscience, 10*(13), 1001–1011.

Syafinar, R., Gomesh, N., Irwanto, M., Fareq, M., & Irwan, Y. (2015a). Chlorophyll pigments as nature based dye for dye-sensitized solar cell (DSSC). *Energy Procedia, 79*, 896–902.

Syafinar, R., Gomesh, N., Irwanto, M., Fareq, M., & Irwan, Y. M. (2015b). Potential of purple cabbage, coffee, blueberry and turmeric as nature based dyes for Dye Sensitized Solar Cell (DSSC). *Energy Procedia, 79*(Supplement C), 799–807.

Taya, S. A., El-Agez, T. M., El-Ghamri, H. S., & Abdel-Latif, M. S. (2013). Dye-sensitized solar cells using fresh and dried natural dyes. *International Journal of Materials Science and Applications, 2*(2), 37–42.

Taya, S. A., El-Agez, T. M., Elrefi, K. S., & Abdel-Latif, M. S. (2015). Dye-sensitized solar cells based on dyes extracted from dried plant leaves. *Turkish Journal of Physics, 39*(1), 24–30.

Teoli, F., Lucioli, S., et al. (2016). Role of pH and pigment concentration for natural dye-sensitized solar cells treated with anthocyanin extracts of common fruits. *Journal of Photochemistry and Photobiology A: Chemistry, 316*, 24–30.

Urban, T. (2015). *How tesla will change the world. Wait but why*.

Wang, X.-F., Matsuda, A., Koyama, Y., Nagae, H., Sasaki, S.-i., Tamiaki, H., & Wada, Y. (2006). Effects of plant carotenoid spacers on the performance of a dye-sensitized solar cell using a chlorophyll derivative: Enhancement of photocurrent determined by one electron-oxidation potential of each carotenoid. *Chemical Physics Letters, 423*(4), 470–475.

Wongcharee, K., Meeyoo, V., & Chavadej, S. (2007). Dye-sensitized solar cell using natural dyes extracted from rosella and blue pea flowers. *Solar Energy Materials & Solar Cells, 91*(7), 566–571.

Yamasaki, H., Uefuji, H., & Sakihama, Y. (1996). Bleaching of the red anthocyanin induced by superoxide radical. *Archives of Biochemistry and Biophysics, 332*(1), 183–186.

Yamazaki, E., Murayama, M., Nishikawa, N., Hashimoto, N., Shoyama, M., & Kurita, O. (2007). Utilization of natural carotenoids as photosensitizers for dye-sensitized solar cells. *Solar Energy, 81*(4), 512–516.

Yusoff, A., Kumara, N., Lim, A., Ekanayake, P., & Tennakoon, K. U. (2014). Impacts of temperature on the stability of tropical plant pigments as sensitizers for dye sensitized solar cells. *Journal of Biophysics, 2014*, 1–8.

Zhou, H., Wu, L., Gao, Y., & Ma, T. (2011). Dye-sensitized solar cells using 20 natural dyes as sensitizers. *Journal of Photochemistry and Photobiology A: Chemistry, 219*(2), 188–194.

Zhu, C., Bai, C., Sanahuja, G., Yuan, D., Farré, G., Naqvi, S., Shi, L., Capell, T., & Christou, P. (2010). The regulation of carotenoid pigmentation in flowers. *Archives of Biochemistry and Biophysics, 504*(1), 132–141.

Zolkepli, Z., Lim, A., Ekanayake, P., & Tennakoon, K. (2015). Efficiency enhancement of cocktail dye of *Ixora coccinea* and *Tradescantia spathacea* in DSSC. *Journal of Biophysics, 2015*, 1–8.

MIX
Papier aus verantwortungsvollen Quellen
Paper from responsible sources
FSC® C105338

If you have any concerns about our products,
you can contact us on
ProductSafety@springernature.com

In case Publisher is established outside the EU,
the EU authorized representative is:
Springer Nature Customer Service Center GmbH
Europaplatz 3, 69115 Heidelberg, Germany

Printed by Libri Plureos GmbH
in Hamburg, Germany